Xizhi Shi

Blind Signal Processing

Theory and Practice

Xizhi Shi

Blind Signal Processing

Theory and Practice

With 139 figures

上海交通大學出版社
SHANGHAI JIAO TONG UNIVERSITY PRESS

Springer

Author
Xizhi Shi
Institute of Vibration Shock & Noise, Shanghai Jiao Tong University
200030, Shanghai, China
Email: xzshi@sjtu.edu.cn

ISBN 978-7-313-05820-1
Shanghai Jiao Tong University Press, Shanghai

ISBN 978-3-642-11346-8 e-ISBN 978-3-642-11347-5
Springer Heidelberg Dordrecht London New York

Library of Congress Control Number: 2009944101

Cover design: Frido Steinen-Broo, EStudio Calamar, Spain

Printed on acid-free paper

Springer is part of Springer Science+Business Media (www.springer.com)

Preface

In this information age, digital signal processing technology is playing an increasingly important role in various fields of science and technology. Blind signal processing (BSP) has been an active research area in signal processing because the "blind" property doesn't require on *a priori* knowledge about either the signal or training data. As a powerful statistical signal processing method, BSP has been successfully used in many fields, including acoustics, sonar, speech separation, image processing, biomedical signal processing, remote sensing, communications, geophysics, and economics.

Since the late 1990, students at Shanghai Jiao Tong University have actively probed this research area with much success "using for reference-innovation-practice" guiding principle. This book contains the results of this research, with the goal of furthering our academic work and providing a bridge for further domestic and international academic exchanges.

The book consists of three parts.

The first part, Chapters 1 to 5, gives the mathematical foundation of BSP, and introduces basic concepts, definitions, and representative algorithms.

The second part, Chapters 6 to 8, applies BSP using deconvolution, probability density estimation and joint diagolinization. In Chapter 6, the convolution mixtures and the techniques of deconvolution are explained. Compared with the linear instantaneous mixtures problem, the convolution mixtures problem matches well with the real environment. Because the convolution in the time-domain corresponds to the multiplication operation in the frequency-domain, the mixing process can be considered as the product of the finite impulse response (FIR) filter corresponding to the transmission channel and the z-domain expression of the mixed signal vector. Therefore, the blind separation of the convolution mixtures in the time-domain is changed to the blind separation of the instantaneous mixtures in the frequency-domain. The dual problems faced by the solution in the frequency-domain are the requirement for higher computation efficiency and the alignment connection of the separated frequency segment due to the permutation indeterminacy. The alignment method of the adjacent frequency segment developed in this book takes the correlation parameters as the criteria to correctly connect the signal of each frequency segment together. In Chapter 7, the estimation of the probability density function of the source function forms the central topic. In the case where the distribution function of the signal is unknown with enough sample data, the non-parametric estimation method of the probability density function can guarantee that the estimate converges to any complex unknown density. This property is consistent with the condition that the source signals are unknown in blind separation. The probability density function-based separating algorithm we developed is more effective than the popular EASI algorithm. For any signals with a complex distribution, increasing the accuracy of the separation algorithms should cause the estimated probability density function to approach the true probability density function of the signal as much as possible. For this reason, we developed the Gaussian mixtures model-based probability density estimator, and verified the performance of the algorithm with speech signals and ship-radiated noise signals. Meanwhile, the estimation method of the probability density function is extended to the blind deconvolution problem where Gram-Charlier expansion is introduced to approximate the probability density function. This approach is suitable to the hybrid mixed signal containing the super-Gaussian signal as well as the sub-Gaussian signal along with improved computation and real-time performance. In Chapter 8, we extend the joint approximate diagonalization method to the non-stationary signal through the time-frequency distribution function matrix instead of through the power spectrum function matrix. After the convolution mixing model is transformed to a higher dimensional instantaneous mixing model, the joint approximate block diagonalization is adopted to implement the blind deconvolution of the time-frequency feature of the source signals. Combining the developed non-parametric estimation method with the time-frequency distribution fully utilizes the limited samples to obtain more information in the time and frequency domains; therefore improving the performance of the algorithm.

The third part, Chapters 9 and 10, introduces extensions to BSP theory and provide applications in varied fields. In Chapter 9, we discuss in detail some key problems, including how to speed up the convergence, the underdetermined model, the complex valued method, and constrained independent component analysis. The developed second-order statistics based algorithm is more suitable to the convolution mixtures of the multi-input and multi-output system. On the basis of the prior distribution expressed by the Gaussian mixtures model, the

MoG-uICA algorithm with a parameter learning capability has been developed; that algorithm can recover the source signals with different distributions from the mixed signals without undue limitation. Recent study shows that on the basis of the state-space model of the signal, the particle filtering method combining with MoG-uICA algorithm is very effective to deal with the blind separation of non-Gaussian noise and non-linear mixtures. And the study of a complex-valued domain extends the application areas while it increases the computation speed of existing algorithms. In Chapter 10, we present a wide range of prospective problems for future research. Our illustrative areas have included speech, underwater acoustics, medical imaging, data compression, image feature extraction, and bioinformatics.

In surveying this book, readers can realize that key problems faced by blind processing include nonlinear mixtures, the effect of noise on the performance of separation, non-stationary signal models and the underdetermined model. Though some progress, as reported in this book, has been achieved regarding the above topics, there is still plenty of space for development in theory and applications.

Since the word “blind” excludes prior assumption on system characteristic, reliance is made of the cognitive model, constructed by statistical reasoning and machine learning, to perform blind separation of real signals. The cognitive model indicates how to transfer the indeterminacy of the mixed signal in the mixing process. The classical statistical reasoning method is the Bayesian theorem; however, it is only optimal in the case of a known prior probability distribution. In order to overcome the obstacle of the prior probability distribution, we need to develop the data-driven based statistical reasoning learning method under the Bayesian framework to find an effective approach method. The time series-based Bayesian reasoning can deal with the transfer of indeterminacy very well. If the blind separation problem is described as a distribution model, in which all hidden variables are deduced from observation variables, then the blind separation problem is reduced to the Bayesian structure learning process. In a general sense, Bayesian learning is an information-based knowledge reasoning system. The key to successful separation is how to select a suitable probability distribution to approach the initial distribution of the mixed signals. The non-parametric Bayesian method uses the random process to depict the infinite dimensional space, namely the Dirichlet process. The joint application of Dirichlet process mixtures-based general purpose density estimator and nonlinear decoupling technology of the indeterminacy transfer for system state and distribution estimation, can widen the successful application of blind signal processing.

I have written this book to show the partial research results by twelve Ph.D. students in recent years achieved in a collaborative, harmonious atmosphere. I would like to express my thanks to Hong Pan, Ze Wang, Xizhong Shen, Fengyu Cong, Wei Kong, Shanbao Tong, Yingyu Zhang, Yanxue Liang and Haixiang Xu for their respective contributions to Section 9.2; Section 1.4, Section 9.9, Section 10.5 and Section 10.6; Section 7.7 and Section 9.3; Section 2.6, Section 4.1, Section 9.6/9.10, and Section 10.1; Section 5.6, Section 7.5, Section 9.5/9.6, and Section 10.10; Section 10.2; Section 9.4 and Section 10.3; Section 10.7; Section 8.7.

I want to express my appreciation to the following experts for their concerns and support in the proposal and writing process; they are: Prof. Z. Y. He (Southeast University, China), Prof. C. H. Chen (Massachusetts University, Dartmouth, US), Prof. B. Z. Yuan (Northern Jiao Tong University, China), Prof. G. Meng, Prof. T. G. Zhuang, Prof. Z. Z. Han, Prof. Z. Z. Wang, Prof. J. Yang, Prof. Y. S. Zhu, Prof. L. Q. Zhang, and Prof. J. Chen of our laboratory.

Most of the early research results are taken from the related projects sponsored by the Department of Information Science National Natural Science Foundation of China. Without the enormous support from NSFC, it would have been difficult to complete the long-term research.

In early 2006, I had to stop my writing due to a sudden eye disease. Special thanks go to Prof. X. H. Sun for his treatment so that I could continue my writing.

Finally, I would like to express my sincere gratitude to Prof. C. H. Chen and Prof. Susan Landgraf for their help in editing the English version of the book.

Xizhi Shi
Shanghai Jiao Tong University
05/10/2010

Contents

Symbols

$A=[a_{ij}]$	mixing matrix
a_{ij}	*ij*th element of matrix $\boldsymbol{A}$
$\boldsymbol{C}$	mixing-separating composite matrix
$\boldsymbol{C}_x$	covariance matrix of $\boldsymbol{x}$
$\boldsymbol{C}_{xy}$	cross-covariance matrix of $\boldsymbol{x}$ and $\boldsymbol{y}$
$\boldsymbol{C}(\omega)$	matrix in frequency-domain
$\boldsymbol{D}$	diagonal matrix with full rank
$\boldsymbol{D}(t, f)$	distribution function matrix in time-domain
$\det(\boldsymbol{A})$	determinant of matrix $\boldsymbol{A}$
$\mathrm{diag}\ (d_1, d_2, \cdots, d_n)$	diagonal matrix with main diagonal elements $(d_1, d_2, \cdots, d_n)$
$E\{\cdot\}$	expectation operator
$\boldsymbol{G}$	generalized partition matrix
$\boldsymbol{H}^{-1}$	inverse of nonsingular matrix $\boldsymbol{H}$
$H(z)$	transfer function of discrete time filter
$\boldsymbol{H}(z)$	matrix transfer function of discrete time filter
$\boldsymbol{H}(\omega)$	mixing matrix in frequency-domain
$H(x)$	entropy of random variable x
$\boldsymbol{I}$	identity matrix
$\boldsymbol{I}_{n \times n}$	identity matrix of dimension $n \times n$
$I(x,y)$	mutual information of random variable $\boldsymbol{x}$ and $\boldsymbol{y}$
$\mathrm{Im}(\cdot)$	imaginary part of
$J(\boldsymbol{w})$	evaluation function
$K(x)$	kernel function
$KL(\cdot)$	Kullback-Leibler divergence (distance)
k	discrete-time or number of iterations
n	number of inputs and outputs, sequence number of discrete-time
N	data length
m	number of sensors
$p(x)$	probability density function of x
$\boldsymbol{P}$	permutation matrix
$\boldsymbol{P}(f)$	power spectrum matrix
$R_{xy}(m)$	cross-correlation function of discrete- time sequence $x(n)$ and $y(n)$ with time delay m

$R_{xy}(\tau)$	cross-correlation function of continuous-time function $x(t)$ and $y(t)$ with time delay τ
R^n	real space of dimension x
$\boldsymbol{R}_x$	autocorrelation matrix of $\boldsymbol{x}$
$\boldsymbol{R}_{xy}$	cross-correlation matrix of $\boldsymbol{x}$ and $\boldsymbol{y}$
$\boldsymbol{R}_x(\omega)$	correlation matrix of mixed signals in frequency-domain
$\boldsymbol{R}_y(\omega)$	correlation matrix of output signals in frequency-domain
$\mathrm{Re}(\cdot)$	real part of
$\boldsymbol{s}$	source signal vector
$s(t)$	continuous-time signal
$\boldsymbol{s}(k)=[s_1(k),\cdots,s_n(k)]^{\mathrm{T}}$	source signal vector of kth sample
$\boldsymbol{s}(z)$	z-transform of source signal vector $\boldsymbol{s}(k)$
$\mathrm{sgn}(x)$	sign function
$\mathrm{tr}(\boldsymbol{A})$	trace of matrix $\boldsymbol{A}$
$\boldsymbol{U}$	orthogonal matrix
$\boldsymbol{w}$	system parameters of state-space model
$\boldsymbol{W}=[w_{ij}]$	separating matrix, demixing matrix
$\boldsymbol{W}^{\mathrm{H}}=(\boldsymbol{W}^*)^{\mathrm{T}}$	complex conjugate transpose of $\boldsymbol{W}$
$\boldsymbol{W}(z)$	matrix transfer function of deconvolution filter
$\boldsymbol{W}(\omega)$	separating matrix in frequency domain
x,y	random variable
$\boldsymbol{x}(k)$	observed signals vector, mixed signals vector
$\Vert \boldsymbol{x} \Vert$	norm of vector $\boldsymbol{x}$
$\boldsymbol{y}$	separated (output) signals vector
$Y(w)$	frequency spectrum of output signal
Z^{-1}	inverse z-transform
δ_{ij}	Kronecker symbol
η	learning rate of discrete-time algorithm
$\Gamma(\cdot)$	Gamma function
λ	eigenvalue of correlation matrix
$\boldsymbol{\Lambda}$	diagonal matrix
$\phi(\cdot)$	contrast function, nonlinear activating function of neuron
ω	normalized angle frequency
∇	gradient operator
$\hat{\boldsymbol{\theta}}$	estimate of parameters $\boldsymbol{\theta}$

Chapter 1 Introduction

1.1 Introduction

Since the digital information age, signal processing has performed an increasingly important function in the areas of science and technology, among which blind signal processing (BSP), as one of the focal points, has great potential. The word "Blind" in BSP means that there are no training data and no prior knowledge of system parameters, so BSP can be utilized very generally.

BSP contains the following three areas: separation and extraction, deconvolution, and equalization of blind signal. And Independent component analysis (ICA) as a parallel method can be used in the three areas, and will be explained in the next chapters. BSP mainly concerns statistical and adaptive signal processing, so the theoretical basis is rigorous. However, the validity of model and algorithms is closely dependent on their specific applications, which afford additional information to BSP. The author's work expands along these application-dependent lines.

The book is composed of three parts. Part 1(Chapter 1~5) contains introductions, basic concepts and definitions, and some general algorithms; Part 2 (Chapter 6~8) concerns the progress of several important and typically related themes and recent progress; Part 3 (Chapter 9~10) introduces the expansion and development of basic models, as well as some application examples resulting from the author's researches.

This book can be used as a reference for researchers or textbook for graduate students.

1.2 Blind Source Separation

Let us consider the situation where multiple sensors receive different signals sent by several physical sources. The typical example is that several persons talk simultaneously in a room, in which a set of sensors have been placed in different locations,

X. Shi, *Blind Signal Processing*

so signals received by sensors are mixed ones composed with differently weighted original speech signals. The problem is to separate and identify the original speech signal out of the mixed one, which is called the cocktail party problem. The problem can be taken as impulse responses of a linear system. For speech, signals are transmitted in medium of air. And considering the reverberation effect, the cocktail party problem actually is a blind deconvolution problem of multi-channels. There is a similar multi-channels abnormality problem with underwater signals. Another example in this area is separation of EEG signal. The electrode on the head records the mixed signals from different sources. It is an instantaneous mixing problem without delay considerations to which BSP can be successfully applied.

To explain the model and characteristics of blind source separation method, we first discuss linear instantaneous mixing problem.

1.2.1 Linear Instantaneous Mixing Problem

As mentioned above, blind source separation is to recover the unobserved original signal from multiple signal mixtures. Generally, the observed signals come from the output of a set of sensors, each sensor receiving different original mixed signals, as shown in Fig. 1.1.

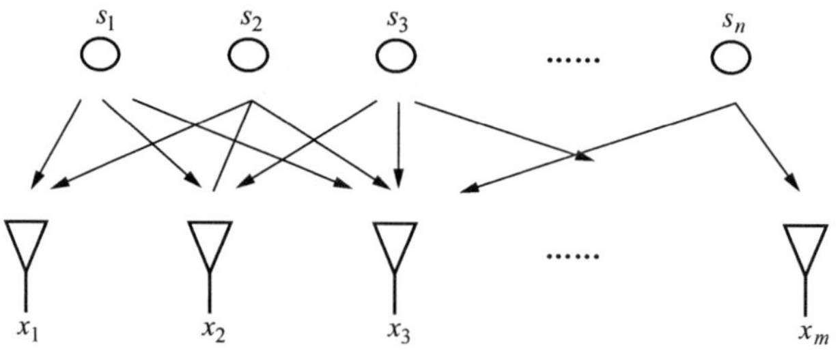

Fig. 1.1 Signal mixing process

In Fig. 1.1, signal source s_1, s_2, $\cdots$, s_n sends signals to sensors with the outputs x_1, x_2, $\cdots$, x_m. The transmission is assumed to be instantaneous, that is different signals arrive at each sensor simultaneously, and sensors receive linear mixtures from each signal sources. The output of i th sensor is

$$x_i(t) = \sum_{j=1}^{n} a_{ij} s_j(t) + n_i(t) \qquad i = 1,2,\cdots,m \tag{1.2.1}$$

where a_{ij} is mixing coefficient, $n_i(t)$ is observed noise of i th sensor. Another form with vector and matrix is

$$\boldsymbol{x}(t) = \mathbf{A}\boldsymbol{s}(t) + \boldsymbol{n}(t) \tag{1.2.2}$$

where $s(t) = [s_1(t), s_2(t), \cdots, s_n(t)]^T$ is an $n \times 1$ vector of source signal. Namely, $x(t)$ is an $m \times 1$ vector of mixed signals, $n(t)$ is an $m \times 1$ vector of noise, and matrix A is an $m \times n$ mixing matrix whose elements a_{ij} are mixing coefficients.

Without consideration of observation noise, in other words, assuming that noise can be ignored, Equation (1.2.2) can be rewritten as

$$x(t) = As(t) \tag{1.2.3}$$

Blind Signal source separation is to estimate source signal vector $s(t)$ or mixing matrix A without knowledge of both source signals and mixing coefficient a_{ij} by using only the mixed signal $x(t)$ received by the sensors. Signal source separation problem can also be represented as follows: in the case that the source signals vector $s(t)$ and mixing matrix A are unknown, find an $n \times m$ matrix W statisfying

$$y(t) = Wx(t) \tag{1.2.4}$$

that yields a reliable estimate for the source vector $s(t)$ or one of its components. generally, matrix W is called the separating matrix.

By combining Eqs. (1.2.3) and (1.2.4), we obtain

$$\begin{aligned} y(t) &= Wx(t) \\ &= WAs(t) \\ &= Cs(t) \end{aligned} \tag{1.2.5}$$

where $C = WA$ is an $n \times n$ matrix, this is called mixing-separating composite matrix.

Actually, the blind separation problem is a multiple solution problem, because for an observed signal $x(t)$, there may be an infinite group of mixing matrices A and the source signals $s(t)$ satisfying Eqs. (1.2.2) or (1.2.3). So to make the problem meaningful, some basic assumption must be made.

With regard to the different signal sources, $s_1, s_2, \cdots, s_n$ it is reasonable to assume that they are statistically independent. Let $p(s)$ represent the joint probability density function of source signal vector $s(t)$ and $p_1(s_1), p_2(s_2), \cdots, p_n(s_n)$ represent the marginal probability density function of then source signals respectively. Then the assumption of statistical independence among all components of the source signal vector can be written as

$$p(s) = p_1(s_1)p_2(s_2)\cdots p_n(s_n) = \prod_{i=1}^{n} p_i(s_i) \tag{1.2.6}$$

i.e. the joint probability density function of the source signal vector $s(t)$ equals to the product of the marginal probability density function of each component. This statistically independent assumption of source signals is the basis of most blind signal source separation algorithms.

Beside the statistically independent assumption regarding every component of source signal vector $\boldsymbol{s}(t)$, a reasonable assumption of mixing matrix $\boldsymbol{A}$ is needed. Obviously, the generalized inverse matrix $\boldsymbol{A}^{\#}$ could be obtained, then

$$\boldsymbol{s}(t) = \boldsymbol{A}^{\#}\boldsymbol{x}(t) - \boldsymbol{A}^{\#}\boldsymbol{n}(t) \tag{1.2.7}$$

without the consideration of noise, that is $\boldsymbol{n}(t) = \boldsymbol{0}$,

$$\boldsymbol{s}(t) = \boldsymbol{A}^{\#}\boldsymbol{x}(t) \tag{1.2.8}$$

To make sure that the blind separation problem is solvable, the generalized inverse matrix of $\boldsymbol{A}$ must exist, so it is assumed that mixing matrix $\boldsymbol{A}$ is of row full rank. In previous research, noise effects generally, were not taken into account. If noise is considered, it is assumed that the components of $\boldsymbol{n}(t)$ are independent of the signal source, as a white noise random process.

In general, there are several hypotheses for the blind source separation problem they are as follows:

Hypothesis 1 Mixing matrix $\boldsymbol{A} \in \boldsymbol{R}^{m \times n}$ is a column full rank matrix, i.e. $rank(\boldsymbol{A}) = n$.

Hypothesis 2 The source vector $\boldsymbol{s}(t)$ is a stationary random vector process with zero mean value, the components of vector are statistically independent and no more than one component of $\boldsymbol{s}(t)$ obeys Gaussian distribution.

Hypothesis 3 Noise vector $\boldsymbol{n}(t)$ is a random vector with zero mean value, statistically independent of the source vector $\boldsymbol{s}(t)$, or the noise vector can be ignored, i.e. $\boldsymbol{n}(t) = \boldsymbol{0}$.

The mixing model $\boldsymbol{x}(t) = \boldsymbol{A}\boldsymbol{s}(t)$ is actually a linear transformation of source vector $\boldsymbol{s}(t)$ when there is no noise. The invertibility of transformation $\boldsymbol{A}$ is insured by hypothesis 1. A random variable with zero mean value can always be obtained by subtracting the mean value in advance when the mean value of random variable is non-zero, so the suitable area of blind separation algorithm will not be affected by hypothesis 2. Because the linear mixtures of Gaussian random variables also have a Gaussian distribution which is symmetric without adding any information about the column vectors of mixing matrix $\boldsymbol{A}$, the blind separation is unsolvable when there is more than one component with Gaussian distribution. Thus, hypothesis 2 is needed to restrict that. Finally, the previous assumption of hypothesis 3 is used in most algorithms of statistical signal processing in accordance to the real situation.

In order to realize blind source separation, the resolvability of the mixing-separating composite matrix $\boldsymbol{C}$ must be determined.

The partition matrix can be defined as follows:

Let $N=\{1, 2, \cdots, n\}$ be the set of n natural numbers. N_i , $i=1, 2, \cdots, r$, is a subset of N and satisfies $N_i \cap N_j = \varnothing$, $i \neq j$, and $N_1 \cup N_2 \cup \cdots \cup N_r=N$. This decides, the set N into r disjoint subsets, where $r \leqslant n$. A $r \times n$ matrix $\boldsymbol{C}=(c_{ij})$ satisfying the following equation is called partition matrix.

$$c_{ij} = \begin{cases} 1, & j \in N_i \\ 0, & j \notin N_i \end{cases} \tag{1.2.9}$$

For given r and n , the total number of different partition matrices can be calculated as

$$\sum_{n_1+n_2+\cdots+n_r=n} \frac{n!}{n_1!n_2!\cdots n_r!} \tag{1.2.10}$$

The partition matrix is row full rank if and only if all subsets N_i , ($i=1, 2, \cdots, r$) of N are not empty sets. It is easy to see that the row full rank partition matrix $\boldsymbol{C}$ row consists of elements equal to 0 or 1, there is only one non-zero element in each column, and there is at least one non-zero element in each row. If subset N_k is empty, the k th row of partition matrix $\boldsymbol{C}$ consists of zero elements. Assume that the partition matrix $\boldsymbol{C}$ is row full rank in the following discussion.

Let $\boldsymbol{D}=\mathrm{diag}\{d_{11}, d_{22}, \cdots, d_{nn}\}$ be an $n \times n$ full rank diagonal matrix, i.e. the diagonal elements of $\boldsymbol{D}$ satisfy $d_{ii} \neq 0$, $i=1, 2, \cdots, n$, then the $r \times n$ matrix $\boldsymbol{G}=\boldsymbol{CD}$ is called generalized partition matrix. The generalized partition matrix $\boldsymbol{G}$ has the same zero elements as the corresponding partition matrix $\boldsymbol{C}$, while the non-zero elements can be any non-zero values (depending on the diagonal matrix $\boldsymbol{D}$).

If the mixing-separating composite matrix $\boldsymbol{C}$ in Eq. (1.2.5) is a partition matrix, Equation (1.2.5) can be written as

$$y_i = \sum_{j \in N_i} s_j \qquad i = 1,2,\cdots,r \tag{1.2.11}$$

here the source signals are divided into r groups, and the sum of the source signals in each group is an output of matrix $\boldsymbol{W}$.

If $r=n$, every subset N_i of set N will just have one element that element being different from the elements of the subsets, and the partition matrix becomes an elementary matrix of linear algebra, or permutation matrix, In such a matrix there is only one element equal to 1 in each row and column, and all the other elements are equal to zero. Equation (1.2.11) becomes

$$y_i = s_j \qquad i,j = 1,2,\cdots,n \tag{1.2.12}$$

Thus, all source signals can be separated by separating matrix $\boldsymbol{W}$ if $\boldsymbol{W}$ transforms the mixing-separating composite matrix into a permutation matrix, with the exception of the original permutation of signals in the source signals vector.

When the mixing-separating composite matrix $\boldsymbol{C}$ is a generalized partition matrix, Equation (1.2.11) can be written as

$$y_i = \sum_{j \in N_i} d_{jj} s_j \qquad i = 1,2,\cdots,r \tag{1.2.13}$$

i. e. the source signals are divided into r groups and the linear combination of source signals in each group is an output of separating matrix $\boldsymbol{W}$.

If r=n, the generalized partition matrix is equal to the product of a permutation matrix and full rank diagonal matrix. There is only one non-zero element for each row and column, and Eqs. (1.2.12) or (1.2.13) becomes

$$y_i = d_{jj} s_j \qquad i,j = 1,2,\cdots,n \tag{1.2.14}$$

i. e. when the mixing-separating composite matrix becomes a permutation matrix, all source signals can be separated by the separating matrix $\boldsymbol{W}$ except the original permutation of the source signal vector and real amplitude of the source signals.

Because the information in a signal is mainly contained in the waveform, how to recover the waveforms of each unknown source signal from mixed signal is the principal concern in blind source separation problems, and the original permutation of each signal and its real amplitude are of secondary concern. Indeed, both original permutation and real amplitude of the source signals are unsolvable for the case in which only the mixed signals are available and the mixing matrix and source signals are unknown.

From the above discussions, we can see that the task of blind source separation is to find a separating matrix $\boldsymbol{W}$, which will transform the mixing-separating composite matrix $\boldsymbol{C}$=$\boldsymbol{WA}$ into a generalized partition matrix or, more ideally, a generalized permutation matrix.

Obviously, the separating matrix $\boldsymbol{W}$ does not always exist to transform the mixing-separating composite matrix $\boldsymbol{C}$ to a generalized partition matrix. Theorem 1.1 shows that the existence of $\boldsymbol{W}$ completely depends on the mixing matrix $\boldsymbol{A}$.

Definition 1.1 Let the $m \times n$ matrix $\boldsymbol{A}$ is row full rank, i.e. rank($\boldsymbol{A}$)=m, if an $r \times m$ matrix $\boldsymbol{W}$ transform $\boldsymbol{C}$=$\boldsymbol{WA}$ into an $r \times n$ generalized partition matrix, then the matrix $\boldsymbol{A}$ is r row resolvable.

Obviously, if the matrix $\boldsymbol{A}$ is r row resolvable, then it is $r-1$ row resolvable too. For instance, define block diagonal matrix $\boldsymbol{B}$ by $\boldsymbol{B}$=diag$\{\boldsymbol{B}_1, \boldsymbol{I}\}$, where $\boldsymbol{B}_1$= $[1 \quad 1]$ and $\boldsymbol{I}$ is a $(r-2) \times (r-2)$ unit matrix, then $\boldsymbol{W}'$=$\boldsymbol{BW}$ exists to turn $\boldsymbol{C}'=\boldsymbol{W}'\boldsymbol{A}$ into a $(r-1)\times n$ generalized partition matrix.

If mixing matrix $\boldsymbol{A}$ is r row resolvable, then theoretically there exists a matrix $\boldsymbol{W}$ to divide n source signals into r groups, and the source signals can be completely separated when $r=n$.

Suppose the i th column of matrix $\boldsymbol{A}$ is $\boldsymbol{a}_i=[a_{1i}, a_{2i}, \cdots, a_{mi}]^{\mathrm{T}}$, $i=1, 2, \cdots, n$. Let set N_i $(i=1, 2, \cdots, r)$ containing n_i elements be a partition of $\{1, 2, \cdots, n\}$ $\sum_{i=1}^{r} n_i = n$. Define $G_i=\{\boldsymbol{a}_j \mid j \in N_i\}$, $i=1, 2, \cdots, r$. Suppose S_i is a linear space spanned by G_i, $i=1, 2, \cdots, r$, and S_{-k} is a direct sum of all the spaces S_i except S_k, i.e.

$$S_{-k} = S_1 \oplus \cdots \oplus S_{k-1} \oplus S_{k+1} \oplus \cdots \oplus S_r \tag{1.2.15}$$

the following theorem can be given:

Theorem 1.1 The necessary and sufficient conditions for an $m\times n$ matrix $\boldsymbol{A}$ to be a r row resolvable are that the n column vectors $\boldsymbol{a}_i$, $(i=1, 2, \cdots, n)$ of $\boldsymbol{A}$ can be divided into r disjoint groups G_i, $i=1, 2, \cdots, n$, and if $j \in N_k$, then $\boldsymbol{a}_j \in S_{-k}$.

This theorem states that if and only if the column vectors of matrix $\boldsymbol{A}$ can be divided into several disjoint groups, and the column vectors of each group are linearly uncorrelated with the column vectors of the other groups, then matrix $\boldsymbol{A}$ is r row resolvable.

The following corollaries can easily be obtained from Theorem 1.1:

Corollary 1.1 If $m\times n$ matrix $\boldsymbol{A}$ is column full rank, then the matrix $\boldsymbol{A}$ is r row resolvable for any $r \leqslant n$.

The condition $m \geqslant n$ is needed for Corollary 1.1 otherwise, the matrix $\boldsymbol{A}$ can't be column full rank. For $m<n$, the next corollary holds:

Corollary 1.2 If $m<n$, the necessary and sufficient conditions for an $m\times n$ matrix $\boldsymbol{A}$ to be r row resolvable are that $\boldsymbol{A}$ has r linearly independent column vectors and that every column vector besides the r independent column vectors is proportional to one of those r independent column vectors.

Corollary 1.1 is corresponds to the situation in which the number of mixed signals is not less than the number of sources, and Corollary 1.2 corresponds to the situation in which the number of mixed signals is less than the number of sources. Corollary 1.1 is an important conclusion for the related methods of blind source separation by separating matrix for the situation where the number of mixed signals is not less than the number of sources.

The next section discusses the separation problem with respect to convolutive mixtures.

1.2.2 Separation of Convolutive Mixtures

In the convolution mixing problem, m received signals compose an m-dimensional received signal vector $\boldsymbol{x}(k)=[x_1(k), \cdots, x_m(k)]^T$. Similarly, n source signals compose an n-dimensional source signal vector $\boldsymbol{s}(k)=[s_1(k), \cdots, s_n(k)]^T$, $n \leqslant m$. It is assumed that $\boldsymbol{x}(k)$ is obtained from $\boldsymbol{s}(k)$ by a linear convolution mixing model,which can be expressed as follows:

$$\boldsymbol{x}(k) = \sum_{p=0}^{\infty} \boldsymbol{H}_p \boldsymbol{s}(k-p) \tag{1.2.16}$$

where

$$\boldsymbol{H}_p = \begin{bmatrix} h_{11,p} & \cdots & h_{1n,p} \\ \vdots & \ddots & \vdots \\ h_{m1,p} & \cdots & h_{mn,p} \end{bmatrix} \tag{1.2.17}$$

$\boldsymbol{H}_p$ is the mixing filter matrix when at delay p, h_{ij} and represents impulse response from the jth source signal to the ith sensor. Since $s_i(k)$ and $s_j(k)$ $(i \neq j)$ are sent by two different signal sources, they are generally assumed to be mutually independent.

When the mixed signal vector $\boldsymbol{x}(k)$ passes through a group of linear separating filters $\boldsymbol{W}$, the estimated vector $\boldsymbol{y}(k)$ can be obtained from the source signal vector $\boldsymbol{s}(k)$. It can be written as follows:

$$\boldsymbol{y}^{(l)}(k) = \sum_{p=0}^{\infty} \boldsymbol{W}_p(l) \boldsymbol{x}(k-p) \tag{1.2.18}$$

where $\boldsymbol{y}^{(l)}(k)=[y_1^{(l)}(k),\cdots, y_n^{(l)}(k)]^T$ is the vector composed of n estimated signals calculated at the lth iteration, $\{\boldsymbol{W}_p(l), 0 \leqslant p \leqslant \infty\}$ is the matrix of the separating filter with delay p calculated at the lth iteration. When expressing the mixed signal and estimated signal are in the z domain, the following formulas can be obtained:

$$\boldsymbol{x}(k) = \boldsymbol{H}(z)[\boldsymbol{s}(k)] \tag{1.2.19}$$

$$\begin{aligned} \boldsymbol{y}^{(l)}(k) &= \boldsymbol{W}(z,l)[\boldsymbol{x}(k)] = \boldsymbol{W}(z,l)\boldsymbol{H}(z)[\boldsymbol{s}(k)] \\ &= \boldsymbol{C}(z,l)[\boldsymbol{s}(k)] \end{aligned} \tag{1.2.20}$$

where

$$\boldsymbol{H}(z) = \sum_{p=0}^{\infty} \boldsymbol{H}_p z^{-p}$$

$$\boldsymbol{W}(z,l) = \sum_{p=0}^{\infty} \boldsymbol{W}_p(l) z^{-p}$$

$$\boldsymbol{C}(z,l) = \boldsymbol{W}(z,l)\boldsymbol{H}(z)$$

are the z transforms of the impulse responses of the mixing filter, separating filter and the whole system filter channels respectively, z^{-1} is the delay operator, and $z^{-p}[s_i(k)]=s_i(k-p)$. Figure. 1.2 shows an example of a complete two-input two-output system.

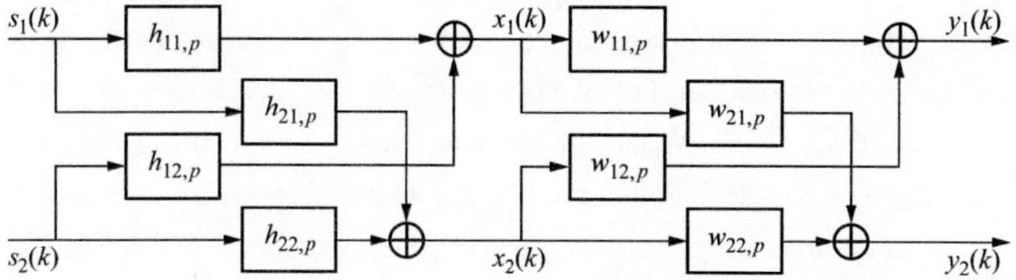

Fig. 1.2 Two-input two-output blind deconvolution system

The purpose of deconvolution is to adjust $\boldsymbol{W}(z, l)$ using an iterative algorithm, and let

$$\lim_{l\to\infty}\boldsymbol{C}(z,l) = \boldsymbol{PD}(z) \tag{1.2.21}$$

where $\boldsymbol{P}$ is an $n \times n$ dimensional permutation matrix, i.e. It has only one nonzero unit value in each line and each row; $\boldsymbol{D}(z)$ is a diagonal matrix, in which the (i, i) item is $d_{ij}z^{-\Delta_i}$ d_{ii} is a nonzero scalar; and the Δ_i is an integer. When Eq. (1.2.21) is true, the ith estimated signal $y_i(k)$ is the estimation of the jth source signal $s_j(k)$, i. e. $y_i(k) = d_{sj}(k-\Delta)$. It can be seen that blind deconvolution also has the intrinsic indeterminacy the same as in blind source separation.

The amplitude value of output signal is not indeterminate.

The delay of output signal is not indeterminate.

The permutation of output signals is not indeterminate.

As with instantaneous mixtures separation, the convolutive mixtures separation also depends on three initial hypotheses of statistical independence:

(1) The input signals are of mutually statistically independent, and each signal is i.i.d..

(2) The number of input signals is less than the number of output signals.

(3) The transfer function of hybrid channel does not have zero on unit circle.

1.3 Independent Component Analysis(ICA)

The original driver of ICA research was the Cocktail-Party problem mentioned at the beginning of the last section. As the rough approximation of the real mixing process, the ICA model is acceptable and it is the starting point of blind processing. Actually, in different situations of signal separation, the ICA model has different forms. Take an sound signal separation in the Cocktail-Party problem as an example, since the propagation velocity of the sound waves is slow, each source signal's time to reach the microphone is different, in addition there are echos from the walls. The model then to be established should be convolution mixing ICA model. When the signals are recorded by two microphones, different noises may increase the number of signal sources, in which case the problem becomes underdetermined. When the speaker walks or even shakes his head, it changes the mixing matrix; this adds nonstaionarity to the mixing process and requires a data sequence that is as short as possible to estimate mixing matrix. For convolution mixtures, however, the coefficients sequence of the filter must have sufficient length. In the basic model, noise is assumed to be zero or Gaussian, whereas noise in a real environment will be very high and may not be Gaussian, therefore a noisy ICA model must be established. Nevertheless, our discussion begins with the basic model.

The basic ICA model is a generative model, and it describes how the observed data is generated by a mixing process. Assume n mutually independent random variables $s_1, s_2, \cdots, s_n$, whose linear combination generates n random variables $x_1, x_2, \cdots, x_n$, i. e.

$$x_i = a_{i1}s_1 + a_{i2}s_2 + \cdots + a_{in}s_n \qquad i = 1,2,\cdots,n \tag{1.3.1}$$

where a_{ij}, i, j=1, 2, $\cdots$, n are real coefficients. For convenience, define $\boldsymbol{x}=[x_1, x_2, \cdots, x_n]^T$, $\boldsymbol{s}=[s_1, s_2, \cdots, s_n]^T$, and $\boldsymbol{A}$ as the matrix with elements a_{ij}. For the instantaneous mixing model, the time variable can be ignored and the composite model can be written as

$$\boldsymbol{x} = \boldsymbol{A}\boldsymbol{s} \tag{1.3.2}$$

To guarantee that the ICA model is solvable, several assumptions and limitations must be made.

The first assumption is statistical independence. It is the most important assumption for obtaining an estimable model and is also the main reason why the ICA method can be applied in many different fields. As described in the last section, independence TS defined in terms of probability density, i.e. It is assumed the joint probability density can be decomposed into the product of the component signal's marginal densities:

$$p(y_1, y_2, \cdots, y_n) = p_1(y_1)p_2(y_2)\cdots p_n(y_n) \tag{1.3.3}$$

The second assumption is non-Gaussian distribution. If the observation variable has Gaussian distribution, it is impossible to estimate the ICA model according to high-order information. In other words, because the mixing matrix $\boldsymbol{A}$ is orthogonal, $\boldsymbol{A}^{-1}=\boldsymbol{A}^{\mathrm{T}}$, it won't change the probability density function. The original distribution and the mixtures distribution are the same and the mixing matrix can't be deduced from the mixed signals. From the figures, the density of Gaussian variable has rotational symmetry, and does not contain any information in the column direction of $\boldsymbol{A}$, and so $\boldsymbol{A}$ can't be estimated.

The third assumption in order to simplify the estimation is that the unknown mixing matrix is square, i.e. the number of independent components equals that the number of observed mixed signals, When the number of independent components is less than the number of the observed mixed signals, the principal component analysis (PCA) can be utilized to reduce the dimension to make the mixing matrix square.

With the above three assumptions, the independent components $\boldsymbol{s}$ can be solved by computing the inverse matrix $\boldsymbol{B}$ of matrix $\boldsymbol{A}$, i. e

$$s = \boldsymbol{B}x \tag{1.3.4}$$

We can see from the above discussion and assumptions that ICA is equivalent to blind signal separation, and, ICA has becomes a commonly used blind signal separation method. The relevant algorithms and implementation of ICA will be further discussed in Chapter 3.

1.4 The Historical Development and Research Prospect of Blind Signal Processing

It is widely acknowledged that ICA originated from the H-J algorithm developed byHerault and Jutten in their seminar paper published in 1985, their method is called H-J algorithm[1]. It was promoted by the elicitation that the central nervous system can separate the information about an organism's movements. With a recurrent artificial neural network, the authors demonstrated the successful separation of two mixed signals. On the basis of H-J algorithm, Herault and Jutten presented a new notion-independent component analysis that was analogous to conventional principal component analysis. It is interesting to note that Herault and Jutten's understanding of blind source signal on separation was very superficial, and even some of the basic problems, including its solvability and possible solvable conditions, hadn't been resolved when they presented the H-J algorithm. They also didn't provide the convergence properties and stability analysis for this algorithm. Following

Herault and Jutten, many researchers have done systematic research on the H-J algorithm's convergence properties in the simplest case of two source signals and two mixed signals, and have presented analysis results. Sorouchyari[2] and Comon[3] performed independent analysis of H-J algorithm's convergence properties and stability using different methods and obtained different results. However, Deville was able to synthesize Sorouchyari's and Comon's results into a unified whole[4,5]. Through many researcher's efforts, the H-J algorithm's convergence properties under the simplest conditions problem, has been resolved completely.

The H-J algorithm first was researched using an artificial neural network to separate blind source, but this algorithm has some defects. First, It can't realize signal source blind separation successfully when some source signals are much weaker than others or the mixing matrix is similar to the singular matrix. In addition, the H-J algorithm doesn't have global stability, namely, the stability of the H-J artificial neural network depends on many factors such as the network's initial weights and mixing coefficient and the network's initial weights and mixing cofficient and the network's activation function etc. A solid theoretical framework for blind signal separation is required for further improvement of the basic H-J algorithm.

Tong *et al.* first elucidated the feasibility and uncertainty, namely the identifiability, of BSP in 1991, pointing out that BSP is an ill-conditioned problem with an infinite number of solutions. This unsolvability can be parameterized using an arbitrary full rank matrix[6,7]. When that matrix can be decomposed into the product of a full rank diagonal matrix and a permutation matrix(elementary matrix), the waveform of signal source can be restored. Tong *et al.* also proved that when each component of the source signal vector is mutually statistically independent, and if and only if each component of the vector being transformed is still mutually statistically independent, this transformation matrix can be decomposed to the product of a full rank diagonal matrix and a permutation matrix. In other words, the transformation changes the amplitude and permutation of each source signal's vector component, while its waveform isn't changed. In 1994, Comon proposed the same proposition and clearly defined the concept of ICA for the first time[8]. Both Tong's and Comon's work assumed that the number of mixed signals is equal to that of the source signals, but they have only addressed the conditions that source signals should meet for a solvable ICA problem, without touching the conditions that the mixing matrix should meet for a solvable blind separation. A comprehensive description about the separability and the principal of BSP was then given by Cao *et al.*[9], wherein the number of source signals could be less than, equal to, or more than that of the mixed signals, and there could be more than one Gaussian sources. Tong's and Comon's conclusion has become popular.

Comon's contribution lies that blind signal separation can be realized only if the statistical independence of mixed signal's components can be restored using a proper linear transformation. Comon also gave a strict mathematical definition of independent component analysis as originally presented by Herault and Jutten. He defined a contrast function in independent component analysis with reference to the concept of blind deconvolution. In short, the contrast function is a real valued scalar function that measures the statistical independence of several random variables. It is essentially a function of a random variable's probability density function. The function will have the maximum or minimum value if and only if each random variable measured by the contrast function is pairwise statistically independently. In this way, the problem of independent component analysis of random vectors is transformed into an optimization problem finding a maximum or minimum value for the contrast function.

The research of blind signal separation developed quickly after Comon's paper was published. In fact, many algorithms for blind signal separation are based on Tong's and Comon's result. Because of the complexity of the problem and for the sake of simplicity in the implementations, nearly all ICA algorithms assume that the number of sources equals the number of the mixed signals.

Some example of the many blind source separation algorithms based on the principle of independent component analysis that have been successively released are the following: the information maximization algorithm[10], the robust neural network algorithm[11,12], the nonlinear principal component analysis algorithm[13~17], the maximum likelihood method algorithm[18~24], the natural gradient method algorithm[25,26], the nonlinear Hebbian category learning algorithm[27,28], and the maximum negentropy algorithm[29~34], etc.

The "Jacobi" algorithm, one of the blind source separation algorithms explicitly using higher order statistics, was systematically studied by Cardoso and Comon. Both Cardoso and Comon proved that the problem of blind source separation can be translated into a problem of orthogonal calculation after prewhitening the processing to a mixed signal. Cardoso introduced the concept of the cumulant matrix and proved that the column vectors of the unknown mixing matrix can be solved by calculating the eigenvector of the cumulant matrix of the mixed signal. He further stated that eigenvectors of the cumulant matrix can be calculated by using the joint diagonalization method on one group of a cumulant matrix, thus finally estimating an unknown mixing matrix[35~37]. The matrix diagonalization based on blind source separation algorithms was summarized by Cardoso in Ref. [36].

The contrast functions are functions of probability density function of the mixed signal. The real contrast functions can't be calculated because the source signal is unknown. Comon estimated the real contrast functions using the Edgeworth expansions[38] of probability density function and higher order cumulant approximation.

Since Comon's work, the first famous independent component analysis algorithm based on neural network method was the information maximization method advanced by Bell and Sejnowski[10]. Bell and Sejnowski presented one kind of gradient algorithm for self-organizing learning, which can maximize the information transmitted through a nonlinear network. They explained that for the super-Gaussian source signal, maximizing the information transmitted through a nonlinear network is equivalent to minimizing mutual information, therefore, the statistical independence between each components of mixed signals can be restored by the information maximization algorithm, and blind source separation can be realized.

The information maximization principle was further successfully applied to blind deconvolution by Bell and Sejnowski, and a new blind deconvolution algorithm was developed. Thus, the blind source separation and blind deconvolution issue can be united by the information maximization principle. However, for sub-Gaussian source signal, the problem can only be solved when an expansion algorithm is developed. From the view of neural processing, this adaptive method is more plausible than the cumulant-based cost functions proposed by Comon.

To improve the stability of BSP especially when the mixing matrix is close to a singular matrix, Cichocki and Unvehauen developed a robust neural network-based BSP algorighm, including a feed-forward neural network and a recursive feed-backward network. Both types are capable of adjusting the magnitude of the output signal and the network weights automatically, and thus, the convergence of the separation is independent of the initial weights and the mixing matrix. Similar approaches are also explored by Macchi and Moreau[39] and Deville[4,5] *et al.*

Cardoso and Laheld studied one kind of blind source separation algorithm based on the equivariance concept that does not relying on initial condition or a specific mixing matrix. In brief, the separation ability of algorithm with an equivariance Property is independent of those two factors. They also presented the concept of relative gradient in Ref. [41], and developed an equivariance algorithm using relative gradient to minimize one specific contrast function.

A mari proposed a similar learning procedure based on the concept of the "natural gradient". In Euclidean space, each function's steepest descent direction is generally represented by the negative gradient, but this is not true when the function space is non-Euclidean. To handle this problem, Amari proposed the idea of the natural gradient. Representing the real steepest descent direction of a function in Riemannian space, the natural gradient is equal to the regular gradient in Euclidean space[42~45]. It is actually the same as the relative gradient proposed by Cardoso *et al.* and was incorporated into the latest Infomax algorithm to improve its stability.

According to the central limit theorem, the sum of many statistically independent random variables tends to become distributed normally. An intuitive way to achieve mutual independence is to maximize nongaussianity. A widely used measure for nongaussianity is the negentropy, which is equal to the difference between the differential entropy of a random vector and a Gaussian random vector that has the same covariance matrix as the former. Since it is quite hard to directly estimate negentropy, an approximation is usually required. Hyvärinen *et al.* gave several negentropy approximations, and using those approximations, they developed an ICA algorithm based on fixed-point iteration or Newton iteration called the FastICA[29~33,46].

When comparing other ICA algorithms, the findings show there is a large advantage in computing speed[47].

The fixed-point ICA algorithm can't do the blind separation online because it needs too many signal samples to estimate the mathematical expectation of certain random variables. To overcome this obstacle, Hyvärinen and Oja used a random gradient algorithm to maximize the contrast function and obtained a Hebbian class nonlinear ICA algorithm[27,28].

Most neural network-based ICA algorithms, like Infomax, choose their transfer functions according to the possible probability distribution. Consequently, an ICA network with a transfer function suitable for separating sub-Gassian signals would not work for super-Gaussian signals. Moreover, the separation could fail to separate the mixed signals when both sub-Gaussian and super-Gaussian signals are present in the source. To handle this issue, Lee *et al.* created the extended information maximization ICA algorithm[48]. This algorithm can estimate the cumulant of each output component on-line and adaptively selects the transfer function for either sub-Gaussian or super-Gaussian signals based on the cumulant computing results. Using similar idea, Douglas[49] and Choi[50] developed an improved ICA method that employs the natural gradient algorithm to separate sub-Gaussian and super-Gaussian signals simultaneously.

All the ICA algorithms described above are designed to recover all the sources in a given time. In some cases, however we may be interested in just a subset of the sources, and the rest of them could be treated as interference signals. In another words, it is not necessary to recover all the source signals. Delfosse and Loubaton found a way to extract just a subsection of unknown signals from the mixed signals by minimizing two contrast functions[51]. In this blind signal extraction algorithm, a single computing unit is used to extract one source signal from the mixtures at a time. The extracted component is subsequently removed from the mixtures using a compression algorithm, and the same computing unit is used to extract the next unknown signal from the refreshed mixtures. The same extraction and refreshing process is repeated until all the desired components have been extracted. Several other groups have developed various extensions with different extractions and compression algorithms[52~54]. Compared to regular BSP methods, the blind extraction algorithm has two advantages[25,27]. First, it is free of the problem of separating a mixtures of both sub-Gaussian and super-Gaussian signals because it only extracts one component at a time. Second, it can save computation time when the number of desired components is much smaller than the number of sources. The main disadvantage of blind extraction algorithms is that they need additional signal compression to clean up the extracted sources from the mixtures, which increases the total computation burden. Moreover, the order of the extracted components is still random, so there is no way to determine when the separation algorithm should stop. Cichocki put forward a blind extraction algorithm which can extract the sources according to the descending order of the absolute cumulant of the normalized sources. However, this method is still problematic when the signal features can't be described by the normalized cumulant.

As the blind source separation algorithm are proposed, researchers make made corresponding progress on those algorithms' stability. Sorouchyari[9], Comon[10] and Deville[11,12] *et al.* have done research on H-J algorithn's stability, but their results focused only on two source signals. Cardoso and Laheld researched the stability of the equivariant adaptive separation via independence, and presented stability analysis results for the case of any number of source signals and gave a qualitative explanation for the selection of the network activation function[50]. Amari and Chen *et al.* did special research on the stability of the blind source separation algorithm, the object of the research being to use the nature gradient to optimize the contract function of the maximum likelihood algorithm. Amari and Chen *et al.* set up the condition whereby the random natural gradient of the blind source separation gradually reached a stable point. They pointed out that the stability of the given signal blind separation depends on the statistical characteristic of unknown signals, and gave a method that changes the unstable separation algorithm into stable and valid blind source separation algorithm[55]. Cardoso did further research on the stability of blind source separation based on relative gradient, and derived convergent intervals of the algorithm, then they used the stability conditions of the algorithm to explain why there can be no more than one component that obeys the Gaussian distribution in the source signal vector[56~58].

BSP/ICA does not grow alone. It has been tightly linked to several other techniques, including factor analysis[59], principal component analysis[60~63], and projection pursuit[64,65]. Researchers have incorporated both ideas and methodologies others have used, and these tight connections have resulted in many new algorithms. A detailed description can be found in Ref. [34]. As BSP and blind deconvolution can be considered as two special cases of blind identification in a multi-input and multi-output linear invariant system, BSP also has intrinsic connections with blind deconvolution[8,66,67]. Comon pointed out that when the input of an unknown linear system is white noise, blind deconvolution can be considered as a special case of blind source separation where the mixed signal is the input of the unknown linear system and the mixing matrix is a Toeplitz triangle matrix composed of an impulse response function of the system. In this case, blind deconvolution becomes a restricted ICA process[8]. With a detailed exploration of the connection between BSP and blind deconvolution, Douglas and Haykin developed a new blind deconvolution algorithm by adopting the nature gradient method from BSP[67,68]. Amari and his colleagues further extended the idea of BSP into the multi-input multi-output linear system blind deconvolution problem. Since convolution in time domain is equal to the multiplication in the z-transform domain, blind deconvolution in a multi-input multi-output linear system turns into blind separation in z-transform domain. Using a contrast function in the z-transform domain with a functionality similar to that in regular BSP algorithms, Amari *et al.* designed a blind deconvolution approach that minimized the contrast function in the time domain. When the unknown system becomes a multi-input multi-output zero-order linear system, this blind deconvolution algorithm is the same as the natural gradient algorithm of blind signal source separation.

Amari and his colleagues generalized the thinking of the signal source blind separation algorithm to a multi-input multi-output linear system's blind deconvolution. Because convolution operation in the time domain is a product operation in the z-transform domain, multi-input multi-output linear system's blind deconvolution is trsnsformed completely into blind separation in the z-transform domain. Amari *et al.* first presented the contrast function in z-transform domain similar to blind signal separation algorithm using this Property and realized the system's blind deconvolution through minimizing the contrast function in the time domain. This blind deconvolution algorithm simplifies into the natural gradient algorithm of blind signal source separation when the unknown system is multi-input multi-output zero-order linear system. This not only points out the algorithm's relationship between blind signal source separation and the multi-input multi-output linear system's blind deconvolution but also leads to a conclusion that the corresponding blind deconvolution algorithm has uniformly varying properties similar to the natural gradient algorithm of blind signal source separation. The study of Douglas and Amari *et al.* reveals explicitly the relationship between blind separation and blind deconvolution, as well as blind separation, the single-input single-output linear system's blind deconvolution and the multi-input multi-output linear system' s blind deconvolution. This study makes it possible to apply the research result of signal source blind separation to blind deconvolution, and to some extent, deepens understanding to these two problems.

Since the middle of the 1990's, domestic researchers have also done active research in blind signal/image processing, and Z. Y. He, X. D. Zhang *et al.* published creative papers and presented new criteria and efficient algorithms[69~72].

While the basic theoretical framework has nearly been established after a decade research, several questions still exist regarding BSP/ICA, mainly including the nonlinear mixing and underdetermined problem, the noisy BSP/ICA, and the nonstationary problem. [73~77]

The difficulty of a nonlinear BSP/ICA is the high uncertainty (non-uniqueness) of the possible solutions as well as the indeterminacies of order and scale. Current research shows that finding appropriate solutions to nonlinear BSP needs an appropriate signal normalization process. Hyvarinen *et al.* found solutions to the nonlinear ICA problem using the generalized Gram-Schmit normalization. However, the number of possible solutions is still infinite. The main reason is that nonlinear BSP is still an ill-posed problem even with the assumption that the unknown sources are mutually independent. Extra constraints are basically required to seek a physically meaningful solution. A special case is post-nonlinear (PNL) mixing where the mixtures are generated by linear instantaneous mixtures followed by an unknown and invertible nonlinear distortion. Besides being realistic for many practical situations, this model reduces the indeterminacies of nonlinear BSP to be the same as ICA. As a result, with certain extensions general ICA could be directly employed to solve the nonlinear BSP. One major PNL method focuses on adapting and optimizing existing linear mixtures ICA algorithms for nonlinear BSP, as well as choosing an appropriate objective function. Another PNL approach is to separate the mixed signal based on some nonlinear features. An example of this is to use the self-organization feature map (SOM). A comprehensive view of nonlinear BSP can be found in Ref. [78] .

ICA becomes underdetermined when the number of signal sources is more than the number of sensors which makes it difficult when using standard ICA approaches. Overcomplete representations were used to solve the underdetermined problem by a proper chosen prior probability represented by basic coefficients. Here, the prior knowledge of source density and time structure is critical for solving the problem, unfortunately it is difficult to obtain in real practice. Sparse representation is another promising approach for the underdetermined problem[79]. It is usually a two-stage model. The first step is to estimate the mixing matrix; and the second step is to estimate the sources. If the sources are sparse enough, blind separation can be done directly. Preprocessing, including Wigner filtering or wavelet transform can be used to increase the sparsity followed by blind separation in time-frequency domain[80].

Dealing with noise is another difficult job for standard ICA. A common approach is to assume that signals and noise are mutually independent so that regular ICA applies directly. Unfortunately, it is nearly impossible to extract the signal in practice without any noise interference. How to improve the robustness of ICA extractions with respect to noise and to reduce the noise level if any exists is an important topic. One useful technique is self-adaptive denoising, which uses a referenced noise measured with additional sensor from the environment. Another alternatively pre-denoising the observed signal with a self-adaptive denoising method as Jafari *et al.* did for foetus electrocardiogram signal separation[81]. For auditory signal separation, it may be more difficult to generate the reference noise for ICA decomposition.

When the mixing process is time-variant, we encounter a non-stationarity problem. For example, the speakers move during sound recording. Because the mixing matrix varies with time, we should estimate a corresponding time-variant separation matrix. Everson and Roberts described the non-stationary mixing with a hidden state model and used the particle filtering technique to build a state density model that solved the problem[74]. Space-time self-adaptive processing and phase-only adaptive FIR filtering have also shown promise for solving the non-stationary ICA.

In summary, BSP/ICA addresses a very important signal processing strategy and is very useful for real signal separations. After nearly 20 years of research BSP/ICA has been successfully adopted in many applications, and the assumption is that future research will bring about more successful applications. Future development will still depend on the balance between blind and less blind in a mathematical way.

References

[1] Jutten C, Herault J (1991) Blind separation of sources, Part I: An adaptive algorithm based on neuromimetic architecture. Signal Processing 28(1): 1–10

[2] Sorouchyari E (1991) Blind separation of sources, Part Ⅲ: Stability analysis. Signal Processing 24(1): 21–29

[3] Comon P, Jutten C, Herault J (1991) Blind separation of sources, Part Ⅱ:

Problems statement. Signal Processing 24(1): 11–20

[4] Deville Y (1996) A unified stability analysis of the Herault-Jutten source separation neural network. Signal Processing 51(3): 229– 233

[5] Deville Y (1999) Analysis of the convergence properties of self-normalized source separation neural networks. IEEE Transactions on Signal Processing 47(5): 1272–1287

[6] Tong L, Liu R, Soon V C et al (1991) Indeterminacy and identifiability of bind identification. IEEE Transaction on Circuits and Systems 38(5): 499–509

[7] Tong L, Inouye Y, Liu R (1993) Waveform preserving blind estimation of multiple independent sources. IEEE Transactions on Signal Processing 41(7): 2461–2470

[8] Comon P (1994) Independent component analysis, a new concept? Signal Processing 36(3): 287–314

[9] Cao X R, Liu R W (1996) General approach to blind source separation. IEEE Transactions on Signal Processing 44(3): 562–571

[10] Bell A J, Sejnowski T J (1995) An information maximisation approach to blind separation and blind deconvolution. Neural Computation 7(6): 1004–1034

[11] Cichocki A, Unbehauen R, Rummert E (1994) Robust learning algorithm for blind separation of signals. Electronic Letters 30(17): 1386–1387

[12] Cichocki A, Unbehauen R (1996) Robust neural networks with on-line learning for blind identification and blind separation of sources. IEEE Transactions on Circuits and Systems 43(11): 894–906

[13] Karhunen J, Joutsensalo J (1994) Representation and separation of signals using nonlinear PCA type learning. Neural Networks 7(1): 113–127

[14] Karhunen J, Oja E, Wang L et al (1997) A class of neural networks for independent component analysis. IEEE Transactions on Neural Networks 8(3): 486–504

[15] Karhunen J, Pajunen P, Oja E (1998) The nonlinear PCA criterion in blind source separation: Relations with other approaches. Neurocomputing 22(1): 5–20

[16] Karhunen J, Wang L, Vigario R (1995) Nonlinear PCA type approaches for source separation and independent component analysis. In: Proceedings of ICNN'95, Paris, 1995, pp 995–1000

[17] Oja E (1997) The nonlinear PCA learning rule in independent component analysis. Neurocomputing 17: 25–45

[18] Pham D T, Garrat P, Jutten C (1992) Separation of a mixtures of independent sources through a maximum likelihood approach. In: Proceedings of EUSIPCO, Brussels, pp 771–774

[19] Pham D T (1996) Blind separation of instantaneous mixtures of sources via an independent component analysis. IEEE Transactions on Signal Processing 44(11): 2768–2779

[20] Pham D T, Garat P (1997) Blind separation of mixtures of independent sources through a quasi-maximum likelihood approach. IEEE Transactions on Signal Processing 45(7): 1712–1725

[21] Pearlmutter B A, Parra L C (1996) A context sensitive generalization of ICA. In: Proceedings of ICNIP, Hong Kong, 1996, pp 151–156

[22] Cardoso J F, Amari S I (1997) Maximum likelihood source separation: equivariance and adaptivity. In: Proceedings of SYSID'97, 11th IFAC Symposium on System Identification, Fukuoka, 1997, pp 1063–1068

[23] Cardoso J F (1997) Infomax and maximum likelihood for source separation. IEEE Letters on Signal Processing 4(4): 112–114

[24] Amari S I, Cichocki A, Yang H H (1996) A new learning algorithm for blind signal separation. In: Advances in Neural Information Processing Systems, Denver, Colorado, August 1996, pp 757–763

[25] Amari S I, Cichocki A (1998) Adaptive blind signal processing—neural network approaches. Proceedings of the IEEE 86(10): 2026–2048

[26] Yang H H, Amari S I, Cichocki A (1997) Adaptive on-line learning algorithms for blind separation—maximum entropy and minimum mutal information. Neural Computation 7(9): 1457–1482

[27] Oja E, Karhunen (1995) Signal separation by nonlinear Hebbian learning. In: Computational Intelligence. New York Press, New York, pp 83–97

[28] Hyvärinen A, Oja E (1998) Independent component analysis by general nonlinear Hebbian like learning rules. Signal Processing 64(3): 301–313

[29] Hyvärinen A (1997) One unit contrast functions for independent component analysis: A statistical analysis. In: Neural Networks for Signal Processing VII, Amelia Island, Florida, 1997, pp 388–

397

[30] Hyvärinen A (1997) A family of fixed point algorithms for independent component analysis. In: Proceedings of ICASSP'97, Munich, 1997, pp 3917–3920

[31] Hyvärinen A (1999) Fast and robust fixed point algorithms for independent component analysis. IEEE Transactions on Neural Networks 10(3): 626–634

[32] Hyvärinen A (1999) The fixed point algorithm and maximum likelihood estimation for independent component analysis. Neural Processing Letters

[33] Hyvärinen A Oja E (1997) A fast fixed point algorithm for independent component analysis. Neural Computation 9(7): 1483–1492

[34] Hyvärinen A (1999) Survey on independent component analysis. Neural Computing Surveys 2: 94–128

[35] Cardoso J F (1989) Source separation using higher order moments. In: Proceedings of ICASSP'89, Glasgow, 1989, pp 2109–2112

[36] Cardoso J F (1999) High-order contrast for independent component analysis. Neural Computation 11(1): 157–192

[37] Cardoso J F (1996) Jacobi angles for simultaneous diagonalization. SIAM Journal of Mathematical Analysis and Applications 17(1): 161–164

[38] Kendall M, Stuart A (1958) The Advanced Theory of Statistics. Charles Griffin & Company, London

[39] Macchi O,Moreau E (1997) Self adaptive source separation, Part Ⅰ: Convergence analysis of a direct linear network controlled by the Herault Jutten algorithm. IEEE Transactions on Signal Processing 45(4): 918–926

[40] Cardoso J F (1995) The equivariant approach to source separation. In: Proceedings of NOLTA, Las Vegas, Nevada, 1995, pp 55–60

[41] Cardoso J F, Laheld B (1996) Equivariant adaptive source separation. IEEE Transactions on Signal Processing 44(12): 3017–3030

[42] Amari S I (1998) Natural gradient works efficiently in learning. Neural Computation 10: 251–276

[43] Amari S I (1997) Neural learning in structured parameter spaces—natural riemannian gradient. In: Advances in Neural Information Processing. Denver, Colorado, 1997, pp 127–133

[44] Amari S I, Douglas S C (1998) Why natural gradient? In: Proceedings of ICASSP'98, Seattle, Washington, 1998, pp 1213–1216

[45] Douglas S C, Amari S I. Natural gradient adaptation. In: Haykin S (ed)

Unsupervised adaptive filtering. Wiley, New York

[46] Hyvärinen A (1998) New approximations of differential entropy for independent component analysis and projection pursuit. In: Advances in Neural Information Processing Systems. The MIT Press, Denver, pp 273–279

[47] Giannakopoulos X, Karhunen J, Oja E (1998) Experimental comparison of neural ICA algorithms. In: Proceedings of ICANN'98, Skövde, 1998, pp 651–656

[48] Lee T W, Girolami M, Sejnowski T J (1998) Independent component analysis using an extended infomax algorithm for mixed sub-gaussian and super-gaussian sources. Neural Computation 10(3): 609–633

[49] Douglas S C, Cichocki A, Amari S I (1997) Multichannel blind separation and deconvolution of sources with arbitrary distributions. In: Neural Networks for Signal Processing, Proceedings of 1997 IEEE Workshop (NNSP-97), Amelia Island, Florida, 1997, pp 436–445

[50] Choi S, Cichocki A, Amari S I (2000) Flexible independent component analysis. Journal of VLSI Signal Processing-Systems for Signal, Image, and Video Technology 26(1-2): 25–38

[51] Delfosse N, Loubaton P (1995) Adaptive blind separation of independent sources: A deflation approach. Signal Processing 45(1): 59–83

[52] Cichocki A, Thawonmas R, Amari S I (1997) Sequential blind signal extraction in order specified by stochastic properties. Electronics Letters 33(1): 64–65

[53] Kung S Y (1997) Independent component analysis in hybrid mixtures: KuicNet learning algorithm and numerical analysis. In: Proceedings of International Symposium of Multimedia Information Processing, Taipei, 1997, pp 368–381

[54] Kung S Y, Mejuto C. Extraction of independent components from hybrid mixtures: KuicNet learning algorithm and applications. In: Proceedings of ICASSP'98, Seattle, Washington, 1998, pp 1209–1212

[55] Amari S I, Chen T P, Cichocki A (1997) Stability analysis of adaptive blind source separation. Neural Networks 10(8): 1345–1351

[56] Cardoso J F (1998) On the stability of source separation algorithms. In: Proceedings of IEEE SSAP'98, Portland, Oregon, 1998

[57] Cardoso J F (1998) Blind signal separation: Statistical principles. Proceedings of the IEEE 9(10): 2009–2025

[58] Cardoso J F (2000) Entropic contrasts for source separation: Geometry and stability. In: Haykin S (ed) Unsupervised adaptive filtering. Wiley, New York

[59] Bernstein I H, Teng G, Garbin C P (1988) Applied multivariate analysis. Springer-Verlag, New York

[60] Jolliffe I T (1986) Principal component analysis. Springer, New York

[61] Oja E (1992) Principal components, minor components, and linear neural networks. Neural Networks 5(6): 927–935

[62] Diamantaras K I, Kung S Y (1996) Principal component neural networks: Theory and applications. Wiley, New York

[63] Chen T P, Amari S, Lin Q (1998) A unified algorithm for principal and minor components extraction. Neural Networks 11(3): 385–390

[64] Friedman J H (1987) Exploratory projection pursuit. Journal of the American Statistical Association 82(397): 249–266

[65] Jones M C, Sibson R (1987) What is projection pursuit? Journal of the Royal Statistical Society, Series A 150(1): 1–36

[66] Yang H H (1998) On-line blind equalization via on-line blind separation. Signal Processing 68(3): 271–281

[67] Douglas S C, Haykin S (1997) On the relationship between blind deconvolution and blind source separation. In: Proceedings of 31st Asilomar Conference on Signals Systems and Computers, Pacific Grove, CA, 1997, pp 1591–1595

[68] Haykin S (2000) Unsupervised adaptive filtering. Vol.2: Blind deconvolution. Wiley, New York

[69] Zhang X D, Bao Z (2001) Blind source separation. Acta Electronica Sinica 29 (12): 1766–1771

[70] Zhang X D, Zhu X L, Bao Z (2002) Phased learning based blind signal processing. Science in China, Series E 32(5): 693–703

[71] Liu J, He Z Y (2002) A survey of blind source separation and blind deconvolution. Acta Electronica Sinica 30(4): 570–576

[72] Zhang H Y, Shi X Z (2001) An effective algorithm for blind separation of arbitrary source signals. Acta Electronica Sinica 29(10): 1392–1396

[73] Lee T W (1998) ICA theory and applications. Kluwer Academic, Boston

[74] Girolami M (2000) Advances in independent component analysis. Springer, London

[75] Hyvärinen A, Karhunen J, Oja E (2001) Independent component analysis.

Wiley, New York

[76] Roberts S, Everson R (2001) Independent component analysis: Principles and practice. Cambridge University Press, New York

[77] Cichocki A, Amari S I (2002) Adaptive blind signal and image processing-learning algorithms and applications. Wiley, New York

[78] Jutten C, Karhunen J (2003) Advances in nonlinear blind source separation. In: Proceedings of ICA-2003, Nara, 2003

[79] Lewicki M S, Sejnowski T J (2002) Learning overcomplete representation. Neural Computation 12(2): 337–365

[80] Li Y, Cichocki A, Amari, S-I (2003) Sparse component analysis for blind source separation with less sensors than sources. In: Proceedings of ICA-2003, Nara, 2003, pp 89–94

[81] Jafari M G, Chambers J A (2003) Adaptive noise cancellation and blind source separation. In: Proceedings of ICA-2003, Nara, 2003, pp 627–632

Chapter 2 Mathematical Description of Blind Signal Processing

Mathematical definitions that are necessary to better understand blind signal processing are presented in this chapter. Due to limitation of the book's length, we will present theoretical results only, omitting the process of proof. Interested readers who want to know more about the process of proof can refer to Refs.[1~7].

2.1 Random Process and Probability Distribution

2.1.1 Random Process

For any physical phenomena, measured signal data can be classified into a deterministic signal or a random signal by whether it can be described in definite mathematical relations or not.

If the signal can be described in definite mathematical relations, it is defined as a deterministic signal and can be divided into a periodic signal or a non-periodic signal.

For those signals that cannot be described by definite mathematical relations, in other words, those signals with unexpectable results for each measurement, they are called random signals. The physical phenomena reflected in random signals are usually time-varying (or space-varying) processes called random process. For example, turbulent flow and jet noise are typical random processes.

Random process $X(t)$ is a collection of random variable $x(t)$ depending on parameter t. t belongs to a fixed real number set $\boldsymbol{I}$, written as $\{x(t), t \in \boldsymbol{I}\}$. At a specified time t_n, random variable is $x(t_n)$. It should be noticed that random variable is a function but not a variable usually defined in analysis. Take an observation of random process, a fixed random variable $x(t)$ is received, called a "reality" or "sample function" of random process, and the collection of sample function is defined as $x(t_1), x(t_2), \cdots, x(t_n)$ ensemble.

A random process $X(t)$, at fixed time t, has a determined one-dimensional distribution function

$$F(x,t) = P\{x(t) \leqslant x\} \tag{2.1.1}$$

For different time t_1, t_2, process $X(t)$ takes two random variable $x(t_1)$, $x(t_2)$ with two-dimensional distribution function

$$F(x_1,t_1;x_2,t_2) = P\{x(t_1) \leqslant x_1; x(t_2) \leqslant x_2\} \tag{2.1.2}$$

Generally, at any different time t_1, t_2, $\cdots$, t_n, $x(t_1)$, $x(t_2)$,$\cdots$,$x(t_n)$, build up an n-dimensional random vector, with an n-dimensional joint distribution function

$$\begin{aligned}&F(x_1,t_1;x_2,t_2;\cdots;x_n,t_n)\\&=P\{x(t_1) \leqslant x_1; x(t_2) \leqslant x_2;\cdots;x(t_n) \leqslant x_n\}\end{aligned} \tag{2.1.3}$$

due to the randomness of t_1, t_2, $\cdots$, t_n and n, the limited dimensional distribution function of random process $X(t)$ gives a complete probability description of random process $X(t)$.

According to whether the statistics of random variable changes along with time or not, random process can be divided to stationary process and non-stationary process. With the stationary process, its statistical Property do not vary as time passes. Its definition is that a random process $X(t)$, for any n, t_1, t_2, $\cdots$, t_n and τ, if

$$\begin{aligned}&F(x_1,t_1;x_2,t_2;\cdots;x_n,t_n)\\&=F(x_1,t_1+\tau,x_2,t_2+\tau,\cdots,x_n,t_n+\tau)\end{aligned} \tag{2.1.4}$$

exist, is called stationary process.

A stationary process, which is defined in Eq. (2.1.4), is called a strong stationary process if all (n-dimension) statistics are not time-varying. In practical applications, only mean, variance, the correlation function are used as basic statistics. If a stationary process follows two conditions: the ensemble mean of process remain constant overtime; the ensemble mean of product of variable value at any two times, i.e. correlation function $R_x(t, t+\tau)$, is only the function of time interval τ, then it is called a weak stationary process or generalized stationary process.

In order to compute the statistics of random process $X(t)$, all sample functions of $X(t)$ should be known. For one sample function of stationary process, when observed time T approximates infinity, its limit converges to the ensemble mean of infinitely repeated observation , that is

$$\overline{x(t)} = \lim_{T\to\infty}\frac{1}{T}\int_0^T x(t)\mathrm{d}t = m_x = E(X(t)) \tag{2.1.5}$$

similarly, for correlation function

$$\begin{aligned} R_x(\tau) &= \overline{x(t)x(t+\tau)} \\ &= \lim_{T\to\infty} \frac{1}{T}\int_0^T x(t)x(t+\tau)\mathrm{d}t \\ &= E(X(t)X(t+\tau)) \end{aligned} \tag{2.1.6}$$

the random process has ergodicity. The ergodic random process has an important practical meaning, because if observed time is long enough, one sample function is enough to describe the ensemble statistics of process. In engineering, most random process can be approximated to ergodic random process, and the following discussions focus on the ergodic random process.

If the statistics of random process are time-varying, it is called non-stationary process. The study of the non-stationary process is very complicated. A non-stationary process that can be described by stationary process after suitable transformation, is called quasi-stationary process, as it approximates the actual process.

2.1.2 Random Variable and Distribution

The cumulative distribution function F_x at $x=x_0$ of a continuous random variable x is defined as probability of $x \leqslant x_0$:

$$F_x(x_0) = P(x \leqslant x_0) \tag{2.1.7}$$

this function is non-negative, increases monotonically, and is continuous.

In most cases, density function is usually used to describe probability distribution. The probability density function (pdf) $p_x(x)$ of continuous random variable is derivative of cumulative distribution function:

$$p_x(x) = \left.\frac{\mathrm{d}F_x(x)}{\mathrm{d}x}\right|_{x=x_0} \tag{2.1.8}$$

A typical probability distribution is Gaussian distribution (the normal distribution), and its density function is as follows:

$$p_x(x) = \frac{1}{\sqrt{2\pi\sigma^2}}\exp\left(-\frac{(x-m_x)^2}{2\sigma^2}\right) \tag{2.1.9}$$

where parameter m_x is the mean confirming the peak position of symmetric density function; parameter σ is a standard deviation that affects the sharp peak level.

For n-dimensional vector

$$\boldsymbol{x} = [x_1, x_2, \cdots, x_n]^{\mathrm{T}} \tag{2.1.10}$$

where every component is a continuous random variable. The multiple probability density function $p_x(\boldsymbol{x})$ of $\boldsymbol{x}$ is defined as the derivative of the cumulative distribution function $F_x(\boldsymbol{x})$ to all components.

$$p_x(\boldsymbol{x}) = \frac{\partial}{\partial x_1}\frac{\partial}{\partial x_2}\cdots\frac{\partial}{\partial x_n}F_x(\boldsymbol{x}) \tag{2.1.11}$$

Obviously

$$\int_{-\infty}^{+\infty} p_x(\boldsymbol{x})\,\mathrm{d}\boldsymbol{x} = 1 \tag{2.1.12}$$

For two different random vector $\boldsymbol{x}$ and $\boldsymbol{y}$, a similar method can be used to compute joint distribution function:

$$F_{x,y}(\boldsymbol{x}_0, \boldsymbol{y}_0) = P(\boldsymbol{x} \leqslant \boldsymbol{x}_0, \boldsymbol{y} \leqslant \boldsymbol{y}_0) \tag{2.1.13}$$

where $\boldsymbol{x}_0$ and $\boldsymbol{y}_0$ are constant vectors with the same dimensions as $\boldsymbol{x}$ and $\boldsymbol{y}$ respectively. The differentiation of joint distribution function for all components of $\boldsymbol{x}$ and $\boldsymbol{y}$ can give a joint density function $p_{x,y}(\boldsymbol{x}, \boldsymbol{y})$, with this relation

$$F_{x,y}(\boldsymbol{x}_0, \boldsymbol{y}_0) = \int_{-\infty}^{\boldsymbol{x}_0}\int_{-\infty}^{\boldsymbol{y}_0} p_{x,y}(\xi, \eta)\,\mathrm{d}\eta\mathrm{d}\xi \tag{2.1.14}$$

Integrating joint density function for any random vector can get a marginal density $p_x(\boldsymbol{x})$ of $\boldsymbol{x}$:

$$p_x(\boldsymbol{x}) = \int_{-\infty}^{\infty} p_{x,y}(\boldsymbol{x}, \eta)\,\mathrm{d}\eta \tag{2.1.15}$$

and a marginal density $p_y(\boldsymbol{y})$ of $\boldsymbol{y}$:

$$p_y(\boldsymbol{y}) = \int_{-\infty}^{\infty} p_{x,y}(\boldsymbol{y}, \xi)\,\mathrm{d}\xi \tag{2.1.16}$$

Actually, the probability density function of random variables is not known; often the expectation value of random variable is used to replace it. Let $g(\boldsymbol{x})$ express any scalar, vector or matrix defined in random vector , and its expectation is defined as

$$E\{\boldsymbol{g}(\boldsymbol{x})\} = \int_{-\infty}^{\infty} \boldsymbol{g}(\boldsymbol{x})\,p_x(\boldsymbol{x})\,\mathrm{d}\boldsymbol{x} \tag{2.1.17}$$

the integral is taken with respect to each component or element of $\boldsymbol{x}$, so $E\{\boldsymbol{g}(\boldsymbol{x})\}$ and $g(\boldsymbol{x})$ are a vector or matrix with the same dimension.

The typical expectation value is mean vector and correlation matrix.

A mean vector is defined as

$$\boldsymbol{m}_x = E\{\boldsymbol{x}\} = \int_{-\infty}^{\infty} \boldsymbol{x}p_x(\boldsymbol{x})\,\mathrm{d}\boldsymbol{x} \tag{2.1.18}$$

In the case of a set of k samples $\boldsymbol{x}_1$, $\boldsymbol{x}_2$, $\cdots$, $\boldsymbol{x}_k$, the mean estimate is

$$\hat{\boldsymbol{m}}_x = \frac{1}{k}\sum_{j=1}^{k} \boldsymbol{x}_j \tag{2.1.19}$$

the correlation r_{ij} between ith component and jth component is second-order moment:

$$r_{ij} = E\{x_i x_j\} = \int_{-\infty}^{\infty} x_i x_j p_x(\boldsymbol{x})\mathrm{d}\boldsymbol{x} \tag{2.1.20}$$

it is the ith row and jth column element of the correlation matrix $\boldsymbol{R}_x$:

$$\boldsymbol{R}_x = E\{\boldsymbol{x}\boldsymbol{x}^{\mathrm{T}}\} \tag{2.1.21}$$

$\boldsymbol{R}_x$ is a symmetrical array and also a semi positive-definite matrix, that is for any n dimensional vector $\boldsymbol{a}$, $\boldsymbol{a}^{\mathrm{T}}\boldsymbol{R}_x\boldsymbol{a} \geqslant 0$, eigenvalue of $\boldsymbol{R}_x$ is real and non-negative.

If the mean vector is substracted, the covariance matrix corresponding to the correlation matrix is

$$\boldsymbol{C}_x = E\{(\boldsymbol{x} - \boldsymbol{m}_x)(\boldsymbol{x} - \boldsymbol{m}_x)^{\mathrm{T}}\} \tag{2.1.22}$$

with elements

$$c_{ij} = E\{(x_i - m_i)(x_j - m_j)\} \tag{2.1.23}$$

The covariance matrix $\boldsymbol{C}_x$ has the same Property as $\boldsymbol{R}_x$. Obviously, then, the following equation exists:

$$\boldsymbol{R}_x = \boldsymbol{C}_x + \boldsymbol{m}_x \boldsymbol{m}_x^{\mathrm{T}} \tag{2.1.24}$$

with the mean vector $\boldsymbol{m}_x$=0, the correlation matrix is identical to the covariance matrix.

Using joint density, expectation computation can be expanded into function $g(\boldsymbol{x}, \boldsymbol{y})$ of two different random vectors $\boldsymbol{x}$ and $\boldsymbol{y}$,

$$E\{g(\boldsymbol{x},\boldsymbol{y})\} = \int_{-\infty}^{\infty}\int_{-\infty}^{\infty} g(\boldsymbol{x},\boldsymbol{y}) p_{x,y}(\boldsymbol{x},\boldsymbol{y})\mathrm{d}\boldsymbol{x}\mathrm{d}\boldsymbol{y} \tag{2.1.25}$$

The most useful joint expectation computation is cross-correlation matrix:

$$\boldsymbol{R}_{xy} = E\{\boldsymbol{x}\boldsymbol{y}^{\mathrm{T}}\} \tag{2.1.26}$$

and cross-covariance matrix:

$$\boldsymbol{C}_{xy} = E\{(\boldsymbol{x} - \boldsymbol{m}_x)(\boldsymbol{y} - \boldsymbol{m}_y)^{\mathrm{T}}\} \tag{2.1.27}$$

Random vector $\boldsymbol{x}$ and $\boldsymbol{y}$ can have different dimensions, that is the cross-correlation matrix and cross-covariance matrix may not be square and symmetrical, but the following transposed relation exists:

$$\boldsymbol{R}_{xy} = \boldsymbol{R}_{yx}^{\mathrm{T}}, \quad \boldsymbol{C}_{xy} = \boldsymbol{C}_{yx}^{\mathrm{T}} \tag{2.1.28}$$

The density function of multiple Gaussian distribution of the n-dimension random vector $\boldsymbol{x}$ is

$$p_x(\boldsymbol{x}) = \frac{1}{(2\pi)^{n/2}(\det\boldsymbol{C}_x)^{1/2}}\exp\left(-\frac{1}{2}(\boldsymbol{x} - \boldsymbol{m}_x)^{\mathrm{T}}\boldsymbol{C}_x^{-1}(\boldsymbol{x} - \boldsymbol{m}_x)\right) \tag{2.1.29}$$

as long as mean vector $\boldsymbol{m}_x$ and covariance matrix $\boldsymbol{C}_x$ are known, a multivariate Gaussian distribution can be determined completely. Higher-order statistics don't provide any additional information for Gaussian distribution, so linear processing based on first-order and second-order statistics is optimal for

Gaussian data.

2.1.3 Non-Correlativity and Statistical Independence

If the cross-covariance matrix $\boldsymbol{C}_{xy}$ of random vector $\boldsymbol{x}$ and $\boldsymbol{y}$ is a zero matrix,

$$\boldsymbol{C}_{xy} = E\{(\boldsymbol{x}-\boldsymbol{m}_x)(\boldsymbol{y}-\boldsymbol{m}_y)^{\mathrm{T}}\} = \boldsymbol{0} \tag{2.1.30}$$

then the random vectors $\boldsymbol{x}$ and $\boldsymbol{y}$ are uncorrelated.

In the case that the random variable is scalar, if covariance C_{xy} is zero,

$$C_{xy} = E\{(x-m_x)(y-m_y)\} = 0 \tag{2.1.31}$$

or is equivalent to

$$R_{xy} = E\{xy\} = E\{x\}E\{y\} = m_x m_y \tag{2.1.32}$$

then random variables x and y are uncorrelated.

For a single n dimensional random vector $\boldsymbol{x}$, the correlation between components should be considered. One important condition is that different components of $\boldsymbol{x}$ are uncorrelated with each other, i.e. the covariance matrix

$$\boldsymbol{C}_x = E\{(\boldsymbol{x}-\boldsymbol{m}_x)(\boldsymbol{x}-\boldsymbol{m}_x)^{\mathrm{T}}\} = \boldsymbol{D} \tag{2.1.33}$$

where $\boldsymbol{D}$ is an $n \times n$ diagonal matrix:

$$\boldsymbol{D} = \mathrm{diag}\{c_{11}, c_{12}, \cdots, c_{nn}\} = \mathrm{diag}\{\sigma_{x_1}^2, \sigma_{x_2}^2, \cdots, \sigma_{x_n}^2\} \tag{2.1.34}$$

in which the n diagonal elements are a variance of component x_i, $\sigma_{xi}^2 = E\{(x_i - m_{xi})^2\} = c_{ii}$.

If a random vector has a zero-mean and unit covariance matrix, it is called white. Therefore, the white random vector satisfies the following conditions:

$$\boldsymbol{m}_x = \boldsymbol{0}, \quad \boldsymbol{R}_x = \boldsymbol{C}_x = \boldsymbol{I} \tag{2.1.35}$$

where $\boldsymbol{I}$ is an $n \times n$ unit matrix.

Transforming random vector $\boldsymbol{x}$ with the orthogonal matrix $\boldsymbol{T}$ results in

$$\boldsymbol{y} = \boldsymbol{T}\boldsymbol{x}, \quad \boldsymbol{T}^{\mathrm{T}}\boldsymbol{T} = \boldsymbol{T}\boldsymbol{T}^{\mathrm{T}} = \boldsymbol{I} \tag{2.1.36}$$

This transformation means a rotation in n dimension space; that is, only the coordinate changes while the norm and distance remain the same. Supposing $\boldsymbol{x}$ is white, we can get

$$\boldsymbol{m}_y = E\{\boldsymbol{T}\boldsymbol{x}\} = \boldsymbol{T}E\{\boldsymbol{x}\} = \boldsymbol{T}\boldsymbol{m}_x = 0 \tag{2.1.37}$$

and

$$\boldsymbol{C}_y = \boldsymbol{R}_y = E\{\boldsymbol{T}\boldsymbol{x}(\boldsymbol{T}\boldsymbol{x})^{\mathrm{T}}\} = \boldsymbol{T}E\{\boldsymbol{x}\boldsymbol{x}^{\mathrm{T}}\}\boldsymbol{T}^{\mathrm{T}} \tag{2.1.38}$$

it shows that $\boldsymbol{y}$ is also white, i.e. the orthogonal transformation doesn't change the white Property of the random vector.

The whitening process of data is a useful pre-processing step and will be discussed further in Section 2.5.

An important concept of blind signal process is statistic independence. In a word, in the case of two scalar random variables x and y, if values of y are obtained without any information of x, then random variable x is independent of y. For example, the random signals from two different physical processes don't have any relation between them.

Mathematically, statistical independence is defined in relation to probability density. If and only if

$$p_{xy}(x,y) = p_x(x)p_y(y) \tag{2.1.39}$$

random variable x and y is called independent. In other words, the joint probability density of x and y should be factorized into their own marginal density. Equivalently, the cumulative distribution function could be used to replace the probability density function to define independence.

Independent random variables satisfy the following basic properties:

$$E\{g(x)h(y)\} = E\{g(x)\}E\{h(y)\} \tag{2.1.40}$$

where $g(x)$ and $h(y)$ is absolutely integrable function of x and y. The existence of the above equation is due to

$$\begin{aligned} E\{g(x)h(y)\} &= \int_{-\infty}^{\infty}\int_{-\infty}^{\infty} g(x)h(y)p_{x,y}(x,y)\mathrm{d}x\mathrm{d}y \\ &= \int_{-\infty}^{\infty} g(x)p_x(x)\mathrm{d}x\int_{-\infty}^{\infty} h(y)p_y(y)\mathrm{d}y \\ &= E\{g(x)\}E\{h(y)\} \end{aligned} \tag{2.1.41}$$

Equation (2.1.40) shows that independence is a much stronger Property than non-correlativity.And non-correlativity defined in Eq.(2.1.32) is a special case in which $g(x)$ and $h(x)$ are linear functions and the second-order statistics (correlation or covariance) are considered only. If the random variable has Gaussian distribution, independence and non-correlativity is unitive.

Independence defined in Eq. (2.1.39) can be expanded to a higher dimension. The independence condition for x, y, z, ... is

$$p_{x,y,z,\cdots}(x,y,z,\cdots) = p_x(x)p_y(y)p_z(z)\cdots \tag{2.1.42}$$

and its basic Property is generalized to

$$E\{g_x(x)g_y(y)g_z(z)\cdots\} = E\{g_x(x)\}E\{g_y(y)\}E\{g_z(z)\}\cdots \tag{2.1.43}$$

where $g_x(x)$, $g_y(y)$ and $g_z(z)$ are arbitrary functions of random variables x, y and z, and their expectation value exists.

2.2 Estimation Theory

2.2.1 Basic Concept

Parameter estimation means to estimate parameters by using samples data. Parameter estimation can be classified into two kinds; One is the estimation of unknown constant parameters, called as the estimation of a deterministic parameter or classical estimation; the other is the estimation of a statistical parameter (random parameter), called as the estimation of the data parameter. Generally, vector $\boldsymbol{\theta}$ represents the parameter to be estimated, and $\hat{\boldsymbol{\theta}}$ is given for its estimate.

There are two estimations: One is point estimation and the other interval estimation. Point estimation gives out a single estimate of parameter and is called the point estimate. Interval estimation determines the interval where the parameter could be located and is called the confidence interval estimation.

Supposing $\boldsymbol{\theta}$ is a real parameter estimated by data set $\{x(n)\}$. When data number N is much larger, a good estimate $\hat{\boldsymbol{\theta}}$ should converge to true value $\boldsymbol{\theta}$ of parameter. For a consistent estimation, it satisfies

$$\lim_{N\to\infty} p_r(|\hat{\boldsymbol{\theta}} - \boldsymbol{\theta}| > \varepsilon) = 0 \tag{2.2.1}$$

or

$$\lim_{N\to\infty} p_r(|\hat{\boldsymbol{\theta}} - \boldsymbol{\theta}| < \varepsilon) = 0 \tag{2.2.2}$$

where p_r represents probability, and ε is a small positive number.

A method measuring the quality of parameter estimation adopts a Mean Squared Error (MSE):

$$\text{MSE} = E[(\hat{\boldsymbol{\theta}} - \boldsymbol{\theta})^2] \tag{2.2.3}$$

MSE can also be represented as the sum of variance and deviation of estimate:

$$\text{MSE} = E\{[\hat{\boldsymbol{\theta}} - E(\hat{\boldsymbol{\theta}})]^2\} + \{E(\hat{\boldsymbol{\theta}}) - \boldsymbol{\theta}\}^2 \tag{2.2.4}$$

the first element of the above equation is variance, while the second element is the square of deviation, and deviation is defined as

$$B(\boldsymbol{\theta}) = E(\hat{\boldsymbol{\theta}}) - \boldsymbol{\theta} \tag{2.2.5}$$

it shows the difference of mean value and true value of estimate. For all $\boldsymbol{\theta}$, if $B(\boldsymbol{\theta}) = 0$, the estimate is unbiased.

When we use parameter confidence interval with high probability of true value to describe the Property of estimate, two cases should be considered: one is that $\hat{\boldsymbol{\theta}}$ is subjected to Gaussian distribution, i.e. $\hat{\boldsymbol{\theta}} \sim N(\boldsymbol{\theta}, \alpha\boldsymbol{\theta}^2)$. The confidence interval of 100 $(1-\alpha)$ % is

$$\left(\frac{\hat{\boldsymbol{\theta}}}{1+\sqrt{\alpha}\xi(1-\alpha/2)},\frac{\hat{\boldsymbol{\theta}}}{1-\sqrt{\alpha}\xi(1-\alpha/2)}\right)$$

where $\xi(\alpha)$ is the value of the cumulative distribution function of $N(0,1)$ at α, or

$$P_r\{z \leqslant \xi(\alpha)\} = \alpha$$

where $z \sim N(0, 1)$.

The second case is that the scalar form of $\hat{\boldsymbol{\theta}}$ is subjected to χ^2 distribution, i. e. $\nu\hat{\boldsymbol{\theta}} / \boldsymbol{\theta} \sim \chi^2_\nu$. The confidence interval of $100(1-\alpha)\%$ is

$$\left(\frac{\nu\hat{\boldsymbol{\theta}}}{\chi^2_\nu(1-\alpha/2)},\frac{\nu\hat{\boldsymbol{\theta}}}{\chi^2_\nu(\alpha/2)}\right)$$

where $\chi^2_\nu(\alpha)$ is the value of the cumulative distribution function of χ^2_ν at α, or

$$p_r\{\chi^2_\nu \leqslant \chi^2_\nu(\alpha)\} = \alpha$$

Performance of the estimate is measured in variance and deviation. Because it is difficult to accurately represent variance, a lower bound of variance is expected. For an unbiased estimation, lower bound of variance is called a Cramer-Rao (CR) bound; that is, variance of any unbiased estimate cannot be lower than the CR bound that is computed from the probability density function of x.

If $p(x|\boldsymbol{\theta})$ is the conditional density of x, and $\partial p(x|\boldsymbol{\theta})/\partial\boldsymbol{\theta}$ exists, then

$$\mathrm{var}(\hat{\boldsymbol{\theta}}) = E(\hat{\boldsymbol{\theta}}-\boldsymbol{\theta})^2 \geqslant \frac{1}{E\left[\frac{\partial}{\partial\boldsymbol{\theta}}\ln p(x \mid \boldsymbol{\theta})\right]^2} \tag{2.2.6}$$

the denominator in Eq. (2.2.6) is computed at parameter $\boldsymbol{\theta}$ with true value using the following equation:

$$E\left[\frac{\partial}{\partial\boldsymbol{\theta}}\ln p(x \mid \boldsymbol{\theta})\right]^2 = -E\left[\frac{\partial^2}{\partial\boldsymbol{\theta}^2}\ln p(x \mid \boldsymbol{\theta})\right] \tag{2.2.7}$$

Generally, the denominator in the right side of Eq. (2.2.6) is written a $I(\boldsymbol{\theta})$, and is called Fisher information.

If $\hat{\boldsymbol{\theta}}_1$ and $\hat{\boldsymbol{\theta}}_2$ are unbiased estimates, a natural question arises how to make the choice between $\hat{\boldsymbol{\theta}}_1$ and $\hat{\boldsymbol{\theta}}_2$? Supposing $\hat{\boldsymbol{\theta}}_1$ has a smaller variance, then value of $\hat{\boldsymbol{\theta}}_1$ is closer to the true value $\hat{\boldsymbol{\theta}}$ than $\hat{\boldsymbol{\theta}}_2$. In other words, the probability of $\hat{\boldsymbol{\theta}}_1$ located in interval $(\boldsymbol{\theta}-\varepsilon, \boldsymbol{\theta}+\varepsilon)$ is higher than that of $\hat{\boldsymbol{\theta}}_2$

$$RE = \left[\frac{\mathrm{var}(\hat{\boldsymbol{\theta}}_1)}{\mathrm{var}(\hat{\boldsymbol{\theta}}_2)} \times 100\right]\% \tag{2.2.8}$$

we called it the "relative effectivity" of $\hat{\boldsymbol{\theta}}_2$ to $\hat{\boldsymbol{\theta}}_1$.

2.2.2 Maximum Likelihood Estimation

A general effective estimation is maximum likelihood estimation (MLE).

essentially, maximum likelihood estimation is to obtain a parameter, which maximizes the likelihood function, as an estimate. The log-likelihood function is defined as

$$L = \ln p(x_1, \cdots, x_N \mid \boldsymbol{\theta}) \tag{2.2.9}$$

the maximum likelihood estimation of $\boldsymbol{\theta}$ is written as $\hat{\boldsymbol{\theta}}_{\mathrm{ML}}$, and it can be obtained by making

$$\frac{\partial L}{\partial \boldsymbol{\theta}} = 0$$

Usually, $\boldsymbol{\theta}$ is a vector parameter, $\boldsymbol{\theta} = [\theta_1, \theta_2, \cdots, \theta_p]^{\mathrm{T}}$, and then $\hat{\boldsymbol{\theta}}_{i,\mathrm{ML}}$ is defined as

$$\left.\frac{\partial L}{\partial \theta_i}\right|_{\theta=\hat{\theta}_{i,\mathrm{ML}}} = 0 \quad (i = 1, 2, \cdots, p)$$

if samples $x_1, x_2, \cdots, x_N$ are independent, then

$$\begin{aligned} L &= \ln p(x_1, \cdots, x_N \mid \boldsymbol{\theta}) \\ &= \ln[p(x_1 \mid \boldsymbol{\theta}), \cdots, p(x_N \mid \boldsymbol{\theta})] \\ &= \sum_{i=1}^{N} \ln p(x_i \mid \boldsymbol{\theta}) \end{aligned} \tag{2.2.10}$$

Through solving

$$\frac{\partial L}{\partial \theta_1} = \frac{\partial}{\partial \theta_1} \sum_{i=1}^{N} \ln p(x_i \mid \boldsymbol{\theta})$$

$$\frac{\partial L}{\partial \theta_2} = \frac{\partial}{\partial \theta_2} \sum_{i=1}^{N} \ln p(x_i \mid \boldsymbol{\theta})$$

$$\frac{\partial L}{\partial \theta_p} = \frac{\partial}{\partial \theta_p} \sum_{i=1}^{N} \ln p(x_i \mid \boldsymbol{\theta})$$

we get $\hat{\theta}_{i,\mathrm{ML}}$ ($i=1, 2, \cdots, p$).

The maximum likelihood function has the following properties:

(1) The maximum likelihood estimation is usually not unbiased, and the deviation can be removed by multiplying some proper constant;

(2) The maximum likelihood estimation is consistent estimate;

(3) The maximum likelihood estimation provides effective estimate;

(4) For large N, $\hat{\boldsymbol{\theta}}_{\mathrm{ML}}$ is Gaussian distribution with mean $\boldsymbol{\theta}$ and variance

$$\sigma^2 = \frac{1}{N}\left[E\left\{\frac{\partial}{\partial \boldsymbol{\theta}}[\ln p(x_1, \cdots, x_N \mid \boldsymbol{\theta})]^2\right\}\right]^{-1}$$

For example, If N independently observed samples $x_1, x_2, \cdots, x_N$ of a Gaussian distribution of scalar random variable x, with mean μ and variance σ^2, the likelihood function is

$$p(x \mid \mu, \sigma^2) = (2\pi\sigma^2)^{-N/2} \exp\left[-\frac{1}{2\sigma^2} \sum_{i=1}^{N} (x_i - \mu)^2\right]$$

$$\ln p(x \mid \mu, \sigma^2) = -\frac{N}{2}\ln(2\pi\sigma^2) - \frac{1}{2\sigma^2}\sum_{i=1}^{N}(x_i - \mu)^2$$

and its log-likelihood function is

$$\frac{\partial}{\partial \mu}\ln p(x \mid \hat{\mu}_{\mathrm{ML}}, \hat{\sigma}^2_{\mathrm{ML}}) = \frac{1}{\hat{\sigma}^2_{\mathrm{ML}}}\sum_{i=1}^{N}(x_i - \hat{\mu}_{\mathrm{ML}}) = 0$$

Solving the likelihood equation

$$\hat{\mu}_{\mathrm{ML}} = \frac{1}{N}\sum_{i=1}^{N} x_i$$

will get the maximum likelihood estimate with mean μ

Similarly, solving the likelihood equation:

$$\frac{\partial}{\partial \sigma^2}\ln p(x \mid \hat{\mu}_{\mathrm{ML}}, \hat{\sigma}^2_{\mathrm{ML}}) = -\frac{N}{2\hat{\sigma}^2_{\mathrm{ML}}} + \frac{1}{2\hat{\sigma}^4_{\mathrm{ML}}}\sum_{i=1}^{N}(x_i - \hat{\mu}_{\mathrm{ML}})^2 = 0$$

get the maximum likelihood estimate with variance σ^2:

$$\hat{\sigma}^2_{\mathrm{ML}} = \frac{1}{N}\sum_{i=1}^{N}(x_i - \hat{\mu}_{\mathrm{ML}})^2$$

2.2.3 Linear Mean Square Estimation

In a maximum likelihood estimation, the conditional probability density of observed samples should be known, however, in practice, they are unknown. The linear mean square estimation and least-squares estimation in Section 2.2.4 provides an estimation without assuming probability distribution.

In linear mean square estimation (LMS), the estimate of the parameter to be determined is a linear sum of observed data, that is

$$\hat{\theta}_{\mathrm{LMS}} = \sum_{i=1}^{N} w_i x_i \tag{2.2.11}$$

where $w_1, \cdots, w_N$ are weight coefficients to be determined. In this section, the estimate is simplified as scalar. The principle of linear mean square estimation is to minimize the mean square error function $E(\hat{\theta}-\theta)^2$,

$$\begin{aligned}\min E(\hat{\theta} - \theta)^2 &= \min E\Big(\sum_{i=1}^{N} w_i x_i - \theta\Big)^2 \\ &= \min E(e)^2 \end{aligned}\tag{2.2.12}$$

where $e = \hat{\theta}-\theta$ is an estimate error. The condition for minimizing mean square error is

$$E(ex_i) = 0 \quad (i = 1, 2, \cdots, N) \tag{2.2.13}$$

Equation (2.2.13) is called orthogonal principle, i.e. if and only if the estimation errors are orthogonal to each given observed data, the mean square error is minimum.

To obtain weight coefficient, Eq. (2.2.13) is rewritten as

$$E\left[\left(\sum_{j=1}^{N} w_j x_j - \theta\right) x_i\right] = 0 \quad (i = 1,2,\cdots,N) \tag{2.2.14}$$

let

$$g_i = E\,(\theta x_i)$$

$$R_{ij} = E\,(x_i x_j)$$

Equation (2.2.14)can be simplified as

$$\sum_{j=1}^{N} R_{ij} w_j = g_i \tag{2.2.15}$$

Note

$$\boldsymbol{R} = \{R_{ij}\} = \{E(x_i x_j)\}$$

$$\boldsymbol{w} = [w_1, \cdots, w_N]^{\mathrm{T}}$$

$$\boldsymbol{g} = [g_1, \cdots, g_N]^{\mathrm{T}}$$

then Eq. (2.2.15) can be written as the following matrix equation:

$$\boldsymbol{R}\boldsymbol{w} = \boldsymbol{g} \tag{2.2.16}$$

where $\boldsymbol{R}$ is the autocorrelation matrix, and when $\boldsymbol{R}$ is nonsingular, weight vector can be solved by the following equation:

$$\boldsymbol{w} = \boldsymbol{R}^{-1}\boldsymbol{g} \tag{2.2.17}$$

2.2.4 Least-Squares Estimation

Consider the following matrix equation

$$\boldsymbol{x} = \boldsymbol{A}\boldsymbol{\theta} + \boldsymbol{\varepsilon} \tag{2.2.18}$$

where $\boldsymbol{x} = [x_1, \cdots, x_N]^{\mathrm{T}}$ is the observation vector, $\boldsymbol{A}$ is $N \times p\,(p \leqslant N)$ dimension coefficient matrix, and $\boldsymbol{x}$ and $\boldsymbol{A}$ are known; $\boldsymbol{\theta} = [\theta_1, \cdots, \theta_p]^{\mathrm{T}}$ is a p-dimension unknown parameter vector, $\boldsymbol{\varepsilon} = [\varepsilon_1, \cdots, \varepsilon_N]^{\mathrm{T}}$ is an N-dimension "Fitting Error" vector, which is also unknown. The question is how to use the known $\boldsymbol{x}$ and $\boldsymbol{A}$ to determine $\boldsymbol{\theta}$.

The criterion we used here is to minimize the square sum of error:

$$\sum_{i=1}^{N} \varepsilon_i^2 = \boldsymbol{\varepsilon}^{\mathrm{T}}\boldsymbol{\varepsilon} = (\boldsymbol{x} - \boldsymbol{A}\hat{\boldsymbol{\theta}})^2 \tag{2.2.19}$$

this estimation method is called least-squares estimation. Estimate $\hat{\boldsymbol{\theta}}_{\mathrm{LS}}$ *of* $\boldsymbol{\theta}$ is called lease square estimate of $\boldsymbol{\theta}$. Unwrapping equation $\boldsymbol{\theta}_N = \boldsymbol{\varepsilon}^{\mathrm{T}}\boldsymbol{\varepsilon}$, we have

$$\boldsymbol{\theta}_N = \boldsymbol{x}^{\mathrm{T}}\boldsymbol{x} - \boldsymbol{x}^{\mathrm{T}}\boldsymbol{A}\hat{\boldsymbol{\theta}} - \hat{\boldsymbol{\theta}}^{\mathrm{T}}\boldsymbol{A}^{\mathrm{T}}\boldsymbol{x} + \hat{\boldsymbol{\theta}}^{\mathrm{T}}\boldsymbol{A}^{\mathrm{T}}\boldsymbol{A}\hat{\boldsymbol{\theta}}$$

and the derivative of $\boldsymbol{\theta}_N$ with respect to $\hat{\boldsymbol{\theta}}$,

$$\frac{\mathrm{d}\boldsymbol{\theta}_N}{\mathrm{d}\hat{\boldsymbol{\theta}}} = -2\boldsymbol{A}^{\mathrm{T}}\boldsymbol{x} + 2\boldsymbol{A}^{\mathrm{T}}\boldsymbol{A}\hat{\boldsymbol{\theta}}$$

then $\hat{\boldsymbol{\theta}}$ must satisfy

$$\boldsymbol{A}^{\mathrm{T}}\boldsymbol{x} = \boldsymbol{A}^{\mathrm{T}}\boldsymbol{A}\hat{\boldsymbol{\theta}}$$

If $rank(\boldsymbol{A})=p$, $\boldsymbol{A}^{\mathrm{T}}\boldsymbol{A}$ is non-singular, and $\boldsymbol{\theta}$ is identifiable, $\hat{\boldsymbol{\theta}}_{\mathrm{LS}}$ can be determined uniquely by

$$\hat{\boldsymbol{\theta}}_{\mathrm{LS}} = (\boldsymbol{A}^{\mathrm{T}}\boldsymbol{A})^{-1}\boldsymbol{A}^{\mathrm{T}}\boldsymbol{x} \tag{2.2.20}$$

If $rank\ (\boldsymbol{A}) < p$, the same $\boldsymbol{A\theta}$ can be obtained from different value of $\boldsymbol{\theta}$, so $\boldsymbol{\theta}$ is called unidentifiable.

For more general cases, consider the following weighted error function:

$$Q(\boldsymbol{\theta}) = \boldsymbol{\varepsilon}^{\mathrm{T}}\boldsymbol{W}\boldsymbol{\varepsilon} \tag{2.2.21}$$

Similarly, minimize $Q(\boldsymbol{\theta})$to estimate $\boldsymbol{\theta}$. the weighted Least Square (WLS) estimate $\hat{\boldsymbol{\theta}}_{\mathrm{WLS}}$ must satisfy

$$\boldsymbol{A}^{\mathrm{T}}\boldsymbol{W}\boldsymbol{A}\hat{\boldsymbol{\theta}}_{\mathrm{WLS}} = \boldsymbol{A}^{\mathrm{T}}\boldsymbol{W}\boldsymbol{x}$$

Supposing $\boldsymbol{A}^{\mathrm{T}}\boldsymbol{W}\boldsymbol{A}$ is non-singular, then $\hat{\boldsymbol{\theta}}_{\mathrm{WLS}}$ can be determined by

$$\hat{\boldsymbol{\theta}}_{\mathrm{WLS}} = (\boldsymbol{A}^{\mathrm{T}}\boldsymbol{W}\boldsymbol{A})^{-1}\boldsymbol{A}^{\mathrm{T}}\boldsymbol{W}\boldsymbol{x} \tag{2.2.22}$$

The Gauss-Markov theorem indicates that the least square estimation is optimal in the sense of minimum variance when each component of the error vector has the same variance and is uncorrelated with each other. For weighted least square estimation, in order to get the optimal weighted least square estimate $\boldsymbol{\theta}$, the weighted matrix $\boldsymbol{W}$ should satisfy

$$\boldsymbol{W} = \boldsymbol{\Sigma}^{-1} \tag{2.2.23}$$

where $\boldsymbol{\Sigma}$ is a known positive definite matrix, and var($\boldsymbol{\varepsilon}$) has a general form of $\sigma^2\,\boldsymbol{\Sigma}$.

Besides the least square method and weighted least square method, there are two variants of least square: the generalized least square method and the ensemble least square method.

2.2.5 Bayesian Estimation

In the above discussion, the maximum likelihood estimate, linear mean square estimate, and least square estimate supposed that the parameter $\boldsymbol{\theta}$ is unknown deterministic constant, whereas the Bayesian estimation method supplies a probability estimation method. In this method, parameter $\boldsymbol{\theta}$ is random itself and is characterized by the priori probability density function of parameter $p(\boldsymbol{\theta})$. In practical cases, though, since we don't know the priori density accurately, some priori information, such as typical value, varying range and distribution form of parameter $\boldsymbol{\theta}_i$, can be utilized to make an assumption on $p(\boldsymbol{\theta})$ in the estimation process.

The bayesian theorem provides a probability method to compute parameter $\boldsymbol{\theta}$ according to priori probability and observed data, and it is defined asp

$$p(\boldsymbol{\theta} \mid x) = \frac{p(x \mid \boldsymbol{\theta}) p(\boldsymbol{\theta})}{p(x)} \tag{2.2.24}$$

where $p(x|\boldsymbol{\theta})$ is the conditional probability of observed data of given parameter $\boldsymbol{\theta}$, $p(\boldsymbol{\theta})$ is priori probability of $\boldsymbol{\theta}$, $p(\boldsymbol{\theta}\,|\boldsymbol{x})$ is the posteriori probability of $\boldsymbol{\theta}$, and it reflects confidence degree of existence of $\boldsymbol{\theta}$ after observing data $\boldsymbol{x}$. Compared with priori probability, posteriori probability reflects the effect of $\boldsymbol{x}$, whereas priori probability is not relevant to $\boldsymbol{x}$.

Consider a set of parameter $\boldsymbol{\theta}=[\theta_1, \theta_2, \cdots, \theta_p]^{\mathrm{T}}$ based on the observed data $\boldsymbol{x}$. The most probable parameter $\theta_t \in \boldsymbol{\theta}$ that makes posteriori probability $p(\boldsymbol{\theta}|\boldsymbol{x})$ maximum is called the Maximum A Posteriori (MAP) estimation.

In Eq. (2.2.24), the denominator is the probability of data $\boldsymbol{x}$, which is not relevant to parameter vector $\boldsymbol{\theta}$ and just normalizes the posteriori probability $p(\boldsymbol{\theta}|\boldsymbol{x})$. So, the MAP estimation is to maximize the numerator in Eq. (2.2.24).

$$p(\boldsymbol{\theta} \mid x) = p(x \mid \boldsymbol{\theta}) p(\boldsymbol{\theta}) \tag{2.2.25}$$

Similar to the maximum likelihood method, MAP estimation $\hat{\boldsymbol{\theta}}_{\mathrm{MAP}}$ can be obtained by solving the likelihood equation. Its log form is

$$\frac{\partial}{\partial \boldsymbol{\theta}} \ln p(\boldsymbol{\theta} \mid x) = \frac{\partial}{\partial \boldsymbol{\theta}} \ln p(x \mid \boldsymbol{\theta}) + \frac{\partial}{\partial \boldsymbol{\theta}} \ln p(\boldsymbol{\theta}) = 0 \tag{2.2.26}$$

compared with the maximum likelihood equation, Equation (2.2.26) has an additional item of $\frac{\partial}{\partial \boldsymbol{\theta}} \ln p(\boldsymbol{\theta})$. *If* $p(\boldsymbol{\theta})$ is a uniform distribution, MAP is equivalent to the maximum likelihood estimation. Therefore, the MAP estimation is a compromise between minimum mean the square error estimator and the maximum likelihood estimator.

2.3 Information Theory

In the last section, we have discussed the method of estimation theory for characterizing random variables. Another method is the information theory method with its emphasis on coding, i.e. coding observed data, then saving it in the computer and transferring through a communication channel. Proper coding depends on the statistical characteristics of data.

2.3.1 Entropy

Entropy is the basic concept in information theory. Shannon (1948) proposed entropy as the proper measure of a state's randomness. For discrete random variable x, the entropy is defined as

$$H(x) = -\sum_i p(x = a_i) \lg p(x = a_i) \tag{2.3.1}$$

where a_i is the probable value of x. For base-2 logarithm, the unit of entropy is *bits* while a unit is *nats* in natural logarithm.

A function f is defined as

$$f(p) = -p \lg p, \quad \text{for} \quad 0 \leqslant p \leqslant 1 \tag{2.3.2}$$

when $p=0$, the equation means $f(p)=\lim_{p \to 0}(-p \lg p)=0$.

This is a non-negative function, where $f(p)$ is zero for $p=0$ and $p=1$; and positive for other values. That is to say $H(x)$ is always greater than zero or equal to zero. $p=0$ and $p=1$ show a definite result with minimum randomness, so the entropy is smallest. For example, when using a coin experiment, the probability of face side and inverse side are equal, that is $p(x=a_1)=p(x=a_2)$, which has the largest entropy. The relation between entropy and randomness is reflected in the information coding length, where the value of entropy roughly gives out the averaging minimal coding length of a random variable.

2.3 .2 Differential Entropy

1. Differential entropy

Entropy defined for discrete random variables can be extended into continuous random variable and random vector, which is called differential entropy.

Differential entropy for a random variable with probability density $p(x)$ is defined as

$$h(x) = -\int_{-\infty}^{\infty} p(x) \lg p(x) \mathrm{d}x \tag{2.3.3}$$

Differential entropy has the same Property as entropy, but differential entropy might be negative, so it can't be described by coding length.

In the case that the increment δx converges to zero, the definition of entropy is

$$\begin{aligned} H(x) &= -\lim_{\delta x \to 0} \sum_{i=-\infty}^{\infty} p(x_i) \delta x \lg(p(x_i) \delta x) \\ &= -\lim_{\delta x \to 0} \left[\sum_{i=-\infty}^{\infty} p(x_i) \delta x \lg p(x_i) + \lg \delta x \sum_{i=-\infty}^{\infty} p(x_i) \delta x \right] \\ &= -\int_{-\infty}^{\infty} p(x) \lg p(x) \mathrm{d}x - \lim_{\delta x \to 0} \lg \delta x \int_{-\infty}^{\infty} p(x) \mathrm{d}x \end{aligned}$$

the first item of the right side of the equation is differential entropy and with the integral of $p(x)$ as 1, this equation can be rewritten as

$$h(x) = H(x) + \lim_{\delta x \to 0} \lg \delta x \tag{2.3.4}$$

the second item of the right side tends to negative infinity, so the differential entropy of the continuous random variable is negative infinite. For this problem, we can use the relative entropy which takes $\lg\delta x$ as reference to comparing these two differential entropies, thereby the relative entropy is kept as a positive value. The following section will discuss relative entropy.

2. Maximum entropy

Because entropy can be explained as the measure of randomness, in some constraints, the most random density in all probability density functions that satisfies constraints is maximum entropy density with the form

$$p(\xi) = A\exp\left(\sum_i a_i F^i(\xi)\right) \tag{2.3.5}$$

where A and a_i are determined by the following constraint conditions:

$$\int p(\xi) F^i(\xi) \mathrm{d}\xi = c_i, \quad \text{for } i = 1, 2, \cdots, m \tag{2.3.6}$$

For zero mean and unit variance random variables, there are two constraints and the maximum entropy distribution is Gaussian distribution:

$$p(\xi) = A\exp(a_1 \xi^2 + a_2 \xi) \tag{2.3.7}$$

Therefore we can conclude that for all random variables with unit variance, the Gaussian variable has the largest entropy. It means that entropy can be the measure of non-Gaussian. For example, with a random variable congregate in some value, its pdf has a spiky form and the entropy of distribution is small.

3. Negentropy

The measure of negentropy can be obtained from differential entropy, and for random vector $\boldsymbol{x}$, it is defined as

$$J(\boldsymbol{x}) = H(\boldsymbol{x}_{\text{gauss}}) - H(\boldsymbol{x}) \tag{2.3.8}$$

The entropy of Gaussian random vector $\boldsymbol{x}_{\text{gauss}}$, whose covariance matrix $\boldsymbol{\Sigma}$ is the same as that of $\boldsymbol{x}$, can be written as

$$H(\boldsymbol{x}_{\text{gauss}}) = \frac{1}{2}\lg|\det\boldsymbol{\Sigma}| + \frac{n}{2}[1 + \lg 2\pi] \tag{2.3.9}$$

where n is the dimension of $\boldsymbol{x}$.

For Gaussian variable, the maximum entropy distribution is unique and its negative entropy is zero, while for other variables, negative entropy is nonnegative. Negentropy has another important Property that is invariant for reversible linear transformation. For $\boldsymbol{y}=\boldsymbol{M}\boldsymbol{x}$, $E\{\boldsymbol{y}\boldsymbol{y}^{\mathrm{T}}\}=\boldsymbol{M}\boldsymbol{\Sigma}\boldsymbol{M}^{\mathrm{T}}$, negentropy can be computed as

$$
\begin{aligned}
J(\boldsymbol{Mx}) &= \frac{1}{2}\lg|\det(\boldsymbol{M\Sigma M}^{\mathrm{T}})| + \frac{n}{2}[1+\lg 2\pi] - (H(\boldsymbol{x}) + \lg|\det\boldsymbol{M}|) \\
&= \frac{1}{2}\lg|\det\boldsymbol{\Sigma}| + 2\frac{1}{2}\lg|\det\boldsymbol{M}| + \frac{n}{2}[1+\lg 2\pi] - H(\boldsymbol{x}) - \lg|\det\boldsymbol{M}| \\
&= \frac{1}{2}\lg|\det\boldsymbol{\Sigma}| + \frac{n}{2}[1+\lg 2\pi] - H(\boldsymbol{x}) \\
&= H(\boldsymbol{x}_{\mathrm{gauss}}) - H(\boldsymbol{x}) = J(\boldsymbol{x})
\end{aligned} \tag{2.3.10}
$$

2.3.3 Mutual Information

Mutual Information is a measure of information between two groups of random variables.

The joint entropy of two random variables x and y is defined as

$$H(x,y) = -\sum_i p(x=a_i, y=b_i)\lg p(x=a_i, y=b_i) \tag{2.3.11}$$

For an independent variable, if and only if $p(x, y)=p(x)p(y)$,

$$H(x,y) = H(x) + H(y) \tag{2.3.12}$$

Given $y=b_k$, the conditional entropy of x is the entropy of conditional probability $p(x|y=b_k)$

$$H(x \mid y=b_k) = -\sum_i p(x \mid y=b_k)\lg p(x \mid y=b_k) \tag{2.3.13}$$

Given y, the conditional entropy of x is an average of conditional entropy to y, and it is the measure for averaging uncertainty.

$$H(x \mid y) = -\sum_i p(x,y)\lg p(x \mid y=b_k) \tag{2.3.14}$$

Mutual information between x and y is that the sum of marginal entropies minus a joint entropy:

$$
\begin{aligned}
I(x,y) &\equiv H(x) + H(y) - H(x,y) \\
&= H(x) - H(x \mid y)
\end{aligned} \tag{2.3.15}
$$

Besides representing mutual information with entropy, a distance of relative entropy called Kullback-Leiber divergence can also be used to represent it and is defined as

$$D(p,q) = \sum_i p(x)\lg\frac{p(x)}{q(x)} \tag{2.3.16}$$

where $p(x)$ and $q(x)$ are two n dimensional probability density function respectively. It is the distance between two probability density functions, therefore, it is always nonnegative and is zero if and only if two functions have the same distribution.

According to the definition of mutual information, there is the following relationship between mutual information of random variables x and y and their

Kullback-Leiber divergence

$$I(x,y) = D(p(x,y) \mid p(x)p(y)) \tag{2.3.17}$$

Mutual information satisfies the following inequality:

$$I(x,y) \geqslant 0 \tag{2.3.18}$$

the sufficient and necessary condition of equality is statistically independent between x and y.

In blind signal separation problem, differential entropy, mutual information and Kullback-Leiber distance and their properties are usually used for measuring the independence of signals.

2.4 Higher-Order Statistics

Basic signal analysis is based on a linear model, so linear statistical tools can be used. In spectrum estimation, the content of the signal is considered as the superposition of statistically uncorrelated multi-harmonic components and is used to estimate the distribution of power in multiple frequency components, it omits the phase relationship between multiple frequencies. The information contained in the power spectrum exists mainly in autocorrelation sequence, and for the statistical description of Gaussian process with known mean , the estimation is complete. For the deviation from Gaussian and nonlinear condition, higher order statistics are used to obtain the information.

The third-order spectrum in the higher-order spectrum is called bispectrum, which is the Fourier transform of the third-order cumulative sequence in the stationary random process; whereas trispectrum is the Fourier transform of fourth-order cumulative sequence. A higher spectrum can be classified into moment spectrum and cumulant spectrum, where moment spectrum is suitable for periodic and non-periodic deterministic signals, the cumulant spectrum is suitable for stationary and nonstationary random signals.

2.4.1 Moment and Cumulant

For a given set of n real random variable $\{x_1, x_2, \cdots, x_n\}$, the joint moment of order $r=k_1+k_2+\cdots+k_n$ is

$$\begin{aligned}\mathrm{Mom}[x_1^{k_1}, x_2^{k_2}, \cdots, x_n^{k_n}] &\equiv E\{x_1^{k_1}, x_2^{k_2}, \cdots, x_n^{k_n}\} \\ &= (-j)^r \left.\frac{\partial^r \Phi(\omega_1, \omega_2, \cdots, \omega_n)}{\partial \omega_1^{k_1} \partial \omega_2^{k_2} \cdots \partial \omega_n^{k_n}}\right|_{\omega_1=\omega_2=\cdots=\omega_n}\end{aligned} \tag{2.4.1}$$

where

$$\Phi(\omega_1, \omega_2, \cdots, \omega_n) \equiv E\{\exp(j(\omega_1 x_1 + \omega_2 x_2 + \cdots + \omega_n x_n))\} \tag{2.4.2}$$

is joint characteristic function, and E{ • } is the expectation operation. For example, for two random variables $\{x_1, x_2\}$, the second-order moment is

$$\mathrm{Mom}[x_1, x_2]=E\{x_1 \cdot x_2\},\quad \mathrm{Mom}[x_1^2]=E\{x_1^2\},\quad \mathrm{Mom}[x_2^2]=E\{x_2^2\}$$

Joint cumulant (in other words, *Semi-invariants*) is defined as coefficients of Taylor expansion of second characteristic function at zero point, that is

$$\mathrm{Cum}[x_1^{k_1}, x_2^{k_2}, \cdots, x_n^{k_n}] \equiv (-j)^r \left. \frac{\partial^r \widetilde{\Psi}(\omega_1, \omega_2, \cdots, \omega_n)}{\partial \omega_1^{k_1} \partial \omega_2^{k_2} \cdots \partial \omega_n^{k_n}} \right|_{\omega_1=\omega_2=\cdots=\omega_n=0} \tag{2.4.3}$$

where $\widetilde{\Psi}(\omega_1, \omega_2, \cdots, \omega_n) \equiv \ln[\Phi(\omega_1, \omega_2, \cdots, \omega_n)]$.

The relation between moment and cumulant is shown in the following example. For random variable $x_1\}$, moment is

$$m_1=\mathrm{Mom}\ [x_1] = E\{x_1\}$$
$$m_2=\mathrm{Mom}\ [x_1, x_1] = E\{x_1^2\}$$
$$m_3=\mathrm{Mom}\ [x_1, x_1, x_1] = E\{x_1^3\}$$
$$m_4=\mathrm{Mom}\ [x_1, x_1, x_1, x_1] = E\{x_1^4\}$$

and cumulant is

$$c_1=\mathrm{Cum}\ [x_1] = m_1$$
$$c_2=\mathrm{Cum}\ [x_1, x_1] = m_2-m_1^2$$
$$c_3=\mathrm{Cum}\ [x_1, x_1, x_1] = m_3-3m_2m_1+2m_1^3$$
$$c_4=\mathrm{Cum}\ [x_1, x_1, x_1, x_1] = m_4-4m_3m_1-3m_2^2+12m_2m_1^2-6m_1^4 \tag{2.4.4}$$

above relation can be represented as the following equation:

$$\Phi(\omega_1) = 1 + \mathrm{j}\omega_1 m_1 - \frac{\omega_1^2}{2!} m_2 \cdots + \frac{(\mathrm{j}\omega)^k}{k!} m_k + \cdots \tag{2.4.5}$$

The general formula between $r=n$ order moment and joint cumulant Cum $[x_1, x_2, \cdots, x_n]$ of the random variable $\{x_1, x_2, \cdots, x_n\}$ is

$$\mathrm{Cum}[x_1, x_2, \cdots, x_n]$$
$$= \sum (-1)^{p-1}(p-1)! \cdot E\{\prod_{i \in s_1} x_i\} \cdot E\{\prod_{i \in s_2} x_i\} \cdot \cdots \cdot$$
$$E\{\prod_{i \in s_p} x_i\} \tag{2.4.6}$$

i.e. suming up all the partitions of $(s_1, s_2, \cdots, s_p)$, $p=1, 2, \cdots, n$. For example, integer set (1, 2, 3) can be partitioned into

$p=1$,	$s_1=\{1, 2, 3\}$		
$p=2$,	$s_1=\{1\}$,	$s_2=\{2,3\}$	
	$s_1=\{2\}$,	$s_2=\{1,3\}$	
	$s_1=\{3\}$,	$s_2=\{1,2\}$	
$p=3$,	$s_1=\{1\}$	$s_2=\{2\}$,	$s_3=\{3\}$

therefore, Equation (2.4.6) becomes

$$\text{Cum}\,[x_1, x_2, x_3] = E\{x_1x_2x_3\} - E\{x_1\}\cdot E\{x_2x_3\} - E\{x_2\}\cdot E\{x_1x_3\} - E\{x_3\}\cdot E\{x_1x_2\} - 2E\{x_1\}E\{x_2\}E\{x_3\} \quad (2.4.7)$$

the properties of moment and cumulant can be concluded as follows:

(1) $\text{Mom}\,[a_1x_1, a_2x_2, \cdots, a_nx_n] = a_1a_2\cdots a_n\text{Mom}\,[x_1, x_2, \cdots, x_n]$

and

$$\text{Cum}\,[a_1x_1, a_2x_2, \cdots, a_nx_n] = a_1a_2\cdots a_n\text{Cum}\,[x_1, x_2, \cdots, x_n]$$

where $(a_1, a_2, \cdots, a_n)$ is constant.

(2) Moment and cumulant is the symmetric function of independent variables; that is

$$\text{Mom}[x_1, x_2, x_3] = \text{Mom}[x_2, x_1, x_3] = \text{Mom}[x_3, x_2, x_1]\text{etc.}$$

(3) If random variable $\{x_1, x_2, \cdots, x_n\}$ can be divided into statistically independent groups,then n order cumulant is identical with zero.

$$\text{Cum}\,[x_1, x_2, \cdots, x_n] = 0$$

generally, $\text{Mom}\,[x_1, x_2, \cdots, x_n] \neq 0$.

For example, if two independent groups are $\{x_1, x_2, \cdots, x_\lambda\}$ and $\{x_{\lambda+1}, x_{\lambda+2}, \cdots, x_n\}$, their joint characteristic function is

$$\Phi(\omega_1, \omega_2, \cdots, \omega_n) = \Phi(\omega_1, \omega_2, \cdots, \omega_\lambda)\Phi(\omega_{\lambda+1}, \omega_{\lambda+2}, \cdots, \omega_n),$$

and the joint second characteristic function is

$$\widetilde{\Psi}(\omega_1, \omega_2, \cdots, \omega_n) = \widetilde{\Psi}_1(\omega_1, \omega_2, \cdots, \omega_\lambda) + \widetilde{\Psi}_2(\omega_{\lambda+1}, \omega_{\lambda+2}, \cdots, \omega_n)$$

(4) If the set of random variable $\{x_1, x_2, \cdots, x_n\}$ and $\{y_1, y_2, \cdots, y_n\}$ is independent, then

$$\text{Cum}\,[x_1+y_1, x_2+y_2, \cdots, x_n+y_n] = \text{Cum}\,[x_1, x_2, \cdots, x_n] + \text{Cum}\,[y_1, y_2, \cdots, y_n]$$

generally,

$$\text{Mom}\,[x_1+y_1, x_2+y_2, \cdots, x_n+y_n] = E\{(x_1+y_1)(x_2+y_2)\cdots(x_n+y_n)\} \neq \text{Mom}\,[x_1, x_2, \cdots, x_n] + \text{Mom}\,[y_1, y_2, \cdots, y_n]$$

however, for random variable $\{y_1, x_1, x_2, \cdots, x_n]$, there is

$$\text{Cum}\,[x_1+y_1, x_2, \cdots, x_n] = \text{Cum}\,[x_1, x_2, \cdots, x_n] + \text{Cum}\,[y_1, x_2, \cdots, x_n]$$

and

$$\text{Mom}\,[x_1+y_1, x_2, \cdots, x_n] = \text{Mom}\,[x_1, x_2, \cdots, x_n] + \text{Mom}\,[y_1, x_2, \cdots, x_n]$$

(5) If the set of random variable $\{x_1, x_2, \cdots, x_n\}$ is joint Gaussian, then all information about distribution is contained in the moment of order $n \leqslant 2$, and higher order moments don't provide new information. That is to say, for the Gaussian random variable, the joint cumulant of order $n > 2$ is identical with zero. Therefore, a cumulant in a higher-order than second can be used for the measurement of non-Gaussian or non-normality of time sequence.

2. 4. 2 Moment and Cumulant of Stationary Process

If $\{X(k)\}$, $k=\pm 1, \pm 2, \pm 3, \cdots$ is a real stationary random process with n-order moment,

$$\begin{aligned}&\text{Mom}\,[X(k), X(k+\tau_1), \cdots, X(k+\tau_{n-1})]\\&= E\,[X(k), X(k+\tau_1), \cdots, X(k+\tau_{n-1})]\end{aligned}$$

only depends on time difference $\tau_1, \tau_2, \cdots, \tau_{n-1}$, $\tau_i=0, \pm 1, \pm 2, \cdots$, for all i. Therefore, the moment of stationary random process is

$$m_n^x(\tau_1, \tau_2, \cdots, \tau_{n-1})=E[X(k), X(k+\tau_1), \cdots, X(k+\tau_{n-1})] \tag{2.4.8}$$

Similarly, the n order cumulant of $\{X(k)\}$ is

$$c_n^x(\tau_1, \tau_2, \cdots, \tau_{n-1}) \equiv \text{Cum}[X(k), X(k+\tau_1), \cdots, X(k+\tau_{n-1})] \tag{2.4.9}$$

Using Eqs. (2.4.6), (2.4.8) and (2.4.9), the relation between moment sequence and cumulant sequence can be obtained.

The first-order cumulant is

$$c_1^x = m_1^x = E\{X(k)\} \quad \text{(mean)} \tag{2.4.10}$$

The second-order cumulant is

$$\begin{aligned}c_2^x(\tau_1) &= m_2^x(\tau_1) - (m_1^x)^2 = m_2^x(-\tau_1) - (m_1^x)^2\\&= c_2^x(-\tau_1) \quad \text{(covariance sequence)}\end{aligned} \tag{2.4.11}$$

where $m_2^x(\tau_1)$ is the autocorrelation sequence.

The third-order cumulant is

$$c_3^x(\tau_1, \tau_2) = m_3^x(\tau_1, \tau_2) - m_1^x(m_2^x(\tau_1) + m_2^x(\tau_2) + m_2^x(\tau_2 - \tau_1)) + 2(m_1^x)^3 \tag{2.4.12}$$

where $m_3^x(\tau_1, \tau_2)$ is the third-order moment sequence.

The fourth-order cumulant is

$$\begin{aligned}c_4^x(\tau_1, \tau_2, \tau_3) = {}& m_4^x(\tau_1, \tau_2, \tau_3) - m_2^x(\tau_1)m_2^x(\tau_3 - \tau_2) +\\& m_2^x(\tau_2)m_2^x(\tau_3 - \tau_1) + m_2^x(\tau_3)m_2^x(\tau_2 - \tau_1) -\\& m_1^x[m_3^x(\tau_2 - \tau_1, \tau_3 - \tau_1) + m_3^x(\tau_2, \tau_3) +\\& m_3^x(\tau_2, \tau_4) + m_3^x(\tau_1, \tau_2)m_2^x(\tau_2 - \tau_1)] +\\& (m_1^x)^2[m_2^x(\tau_1) + m_2^x(\tau_2) + m_2^x(\tau_3) +\\& m_2^x(\tau_3 - \tau_1) + m_2^x(\tau_3 - \tau_2) + m_2^x(\tau_2 - \tau_1)] -\\& 6(m_1^x)^4\end{aligned} \tag{2.4.13}$$

If the $\{X[k]\}$ process is zero mean $m_1^x=0$, then the second-order cumulant and third-order cumulant are equal to the second-order moment and third-order moment respectively. However, to obtain fourth-order cumulant, the fourth-order moment and second-order moment are needed.

For non-Gaussian stationary random process $X(k)$, the nth order cumulant is written as (only for n=3, 4)

$$c_n^x(\tau_1, \tau_2, \cdots, \tau_{n-1})=m_n^x(\tau_1, \tau_2, \cdots, \tau_{n-1}-m_n^G(\tau_1, \tau_2, \cdots, \tau_{n-1})$$

where $m_n^x(\tau_1, \tau_2, \cdots, \tau_{n-1})$ is the n-order moment function of $X(k)$, $m_n^G(\tau_1, \tau_2, \cdots, \tau_{n-1})$ is the n-order moment function of the equivalent Gaussian process with the same mean and autocorrelation sequence. Obviously, if $X(k)$ is Gaussian, $m_n^x(\tau_1, \tau_2, \cdots, \tau_{n-1})=m_n^G(\tau_1, \tau_2, \cdots, \tau_{n-1})$, therefore, $c_n^x(\tau_1, \tau_2, \cdots, \tau_{n-1})=0$.

Let $\tau_1=\tau_2=\tau_3=0$ in the above order cumulant, and set $m_1^x=0$, then

$$r_2^x=E\{X^2(k)\}=c_2^x(0) \quad \text{(variance)}$$
$$r_3^x=E\{X^3(k)\}=c_3^x(0,0) \quad \text{(skew)}$$
$$r_4^x=E\{X^4(k)-3\,[r_2^x]\,\}=c_4^x(0,0,0) \quad \text{(kurtosis)} \qquad (2.4.14)$$

Normalized kurtosis is defined as $\frac{r_4^x}{[r_2^x]^2}$. In some references, kurtosis is also expressed as $k(x)$ for convenience.

In blind source separation problems, signals are always classified into three kinds by whether kurtosis of the random variable x is greater than zero, equal to zero or smaller than zero, that is, when $k(x) < 0$, then variable x is sub-Gaussian; when $k(x) > 0$, it is super-Gaussian; for Gaussian, then, it is $k(x)=0$. In the case of probability density, sub-Gaussian is more flat than Gaussian, and super-Gaussian shows a sharp peak and longer tail.

An important Property of kurtosis is its additivity. For two statistically independent random variables x and y,

$$k(x+y)=k(x)+k(y) \qquad (2.4.15)$$

but for a fourth-order moment, the additivity does not exist .

2.5 Preprocessing of Signal

Before blind separation of mixed signals, some preprocessing procedures are required. The most useful pre-processing steps are the elimination of mean and whitening (or sphering). For some special problems, band-limited filtering which contains low-pass filtering and band-pass filtering to eliminate a parts of noise or disturbance is also effective.

2.5.1 Zero Meaning of Signal

In most blind separation algorithms, the components of signal sources are supposed to be zero-mean random variable, so that the pre-elimination of signal mean before separation is needed to meet the presented mathematical model for the actual blind separation problem.

Suppose x is random variable with non-zero mean and is enough to replace x with $x_0=x-E(x)$. In actual computation, the arithmetic mean is used as a mathematical expectation to eliminating mean. Suppose $\boldsymbol{x}(t)=[x_1(t), x_2(t), \cdots, x_n(t)]^{\mathrm{T}}$, $t=1, 2, \cdots, N$ is N samples of random variable $\boldsymbol{x}$, then the following method is used for elimination of the mean:

$$x_{0i}(t) = x_i(t) - \frac{1}{N}\sum_{t=1}^{N} x_i(t), \quad i = 1,2,\cdots,N \tag{2.5.1}$$

2.5.2 Whitening of Signal

Whitening is an usual preprocessing method used in blind source separation. For some blind separation algorithms, whitening is also a necessary pre-processing step.

The so-called whitening of random vector $\boldsymbol{x}$ is to take some linear transform $\boldsymbol{T}$:

$$\tilde{\boldsymbol{x}} = \boldsymbol{T}\boldsymbol{x} \tag{2.5.2}$$

by which the correlation matrix of transformed random vector $\tilde{\boldsymbol{x}}$ satisfies $\boldsymbol{R}_{\tilde{x}} = E[\tilde{\boldsymbol{x}}\tilde{\boldsymbol{x}}^H] = \boldsymbol{I}$, that is, each components of $\tilde{\boldsymbol{x}}$ satisfies $E[\tilde{x}_i\tilde{x}_j] = \delta_{ij}$, where δ_{ij} is Kronecker delta function. Pre-whitening for the mixed signal is to eliminate correlation between each components of the signal, i.e. the second-order statistic between components of the whitened signal is independent. Therefore, matrix $\boldsymbol{T}$ in Eq. (2.5.2) is also called the whitening matrix.

In blind separation problems, real amplitudes of source signals are unsolvable, therefore, we can assume that all of source signals are random variables with unit variance and each component of vector $\boldsymbol{s}$ is statistically mutually independent, $E[\boldsymbol{s}\boldsymbol{s}^{\mathrm{T}}] = \boldsymbol{I}$. Let $\boldsymbol{U}=\boldsymbol{T}\boldsymbol{A}$, there is

$$E[\tilde{\boldsymbol{x}}\tilde{\boldsymbol{x}}^{\mathrm{T}}] = E[(\boldsymbol{U}\boldsymbol{s})(\boldsymbol{U}\boldsymbol{s})^{\mathrm{T}}] = \boldsymbol{U}E[\boldsymbol{s}\boldsymbol{s}^{\mathrm{T}}]\boldsymbol{U}^{\mathrm{T}} = \boldsymbol{U}\boldsymbol{U}^{\mathrm{T}} = \boldsymbol{I} \tag{2.5.3}$$

that is, matrix $\boldsymbol{U}$ is an orthogonal matrix. Actually, supposing that random vector $\boldsymbol{s}$ satisfies $E[\boldsymbol{s}\boldsymbol{s}^{\mathrm{T}}] = \boldsymbol{I}$, for any orthogonal matrix $\boldsymbol{U}$, there is $E[\tilde{\boldsymbol{x}}\tilde{\boldsymbol{x}}^{\mathrm{T}}] = E[(\boldsymbol{U}\boldsymbol{s})(\boldsymbol{U}\boldsymbol{s})^{\mathrm{T}}] = \boldsymbol{I}$. Usually whitening cannot recover statistical independency between each component of the source signal, and only recovering second-order non-correlation between signals is not enough for the blind separation of source signals.

Although whitening cannot ensure blind separation of source signals, it can simplify the blind separation algorithm or improve performance of blind separation algorithm. An example is given in Fig. 2.1.[6] In Fig. 2.1(a), five waveforms of source signals are illustrated, and mixing matrix $\boldsymbol{A}$ is a random matrix with uniform distribution in (0, 1). These five source signals mixed by the mixing matrix are shown in Fig. 2.1(b), and the waveforms of mixed signals

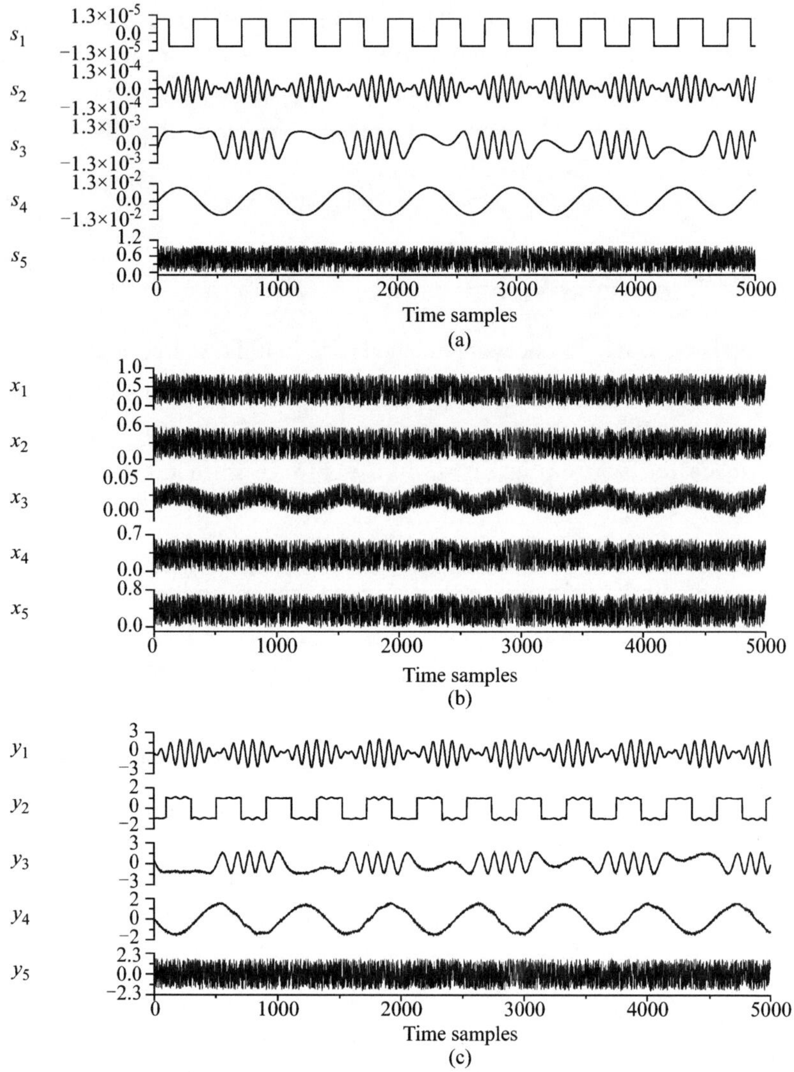

Fig. 2.1 Recovering waveforms of source signals through whitening
(a) source signals (b) mixed signals (c) mixed signals after whitening

after whitening are shown in Fig. 2.1(c). Comparing Fig. 2.1(a) with Fig. 2.1(c) shows that in this special example, whitening actually recovers waveforms of the source signals, and at the same time we can see that inderminacy in amplitude and permutation exist in recovered signal waveforms in Fig. 2.1(c).

Basically there are two kinds of signal whitening methods: one is realized through eigenvalue decomposition of the mixed signal correlation matrix; the other is realized through linear transformation by iterative algorithm. Statements are given as follows:

Suppose the correlation matrix of the mixed signal vector $\boldsymbol{x}$ is $\boldsymbol{R}_x$, then according to the properties of the correlation matrix, the eigenvalue decomposition (EVD) exists in $\boldsymbol{R}_x$:

$$\boldsymbol{R}_x = \boldsymbol{Q}\boldsymbol{\Sigma}^2\boldsymbol{Q}^{\mathrm{T}} \tag{2.5.4}$$

where matrix $\boldsymbol{\Sigma}^2$ is diagonal matrix, its diagonal element $\lambda_1^2, \lambda_2^2, \cdots, \lambda_n^2$ is the eigenvalue of matrix $\boldsymbol{R}_x$, and the column vector is the standard orthogonal eigenvector corresponding to these eigenvalues. So the whitening matrix is taken as

$$\boldsymbol{T} = \boldsymbol{\Sigma}^{-1}\boldsymbol{Q}^{\mathrm{T}} \tag{2.5.5}$$

set $\tilde{\boldsymbol{x}} = \boldsymbol{T}_x$, there is

$$\boldsymbol{R}_{\tilde{x}} = E[\tilde{\boldsymbol{x}}\tilde{\boldsymbol{x}}^{\mathrm{T}}] = \boldsymbol{T}E[\boldsymbol{x}\boldsymbol{x}^{\mathrm{T}}]\boldsymbol{T}^{\mathrm{T}} = \boldsymbol{T}\boldsymbol{R}_x\boldsymbol{T}^{\mathrm{T}} \tag{2.5.6}$$

substituting Eqs. (2.5.4) and (2.5.5) into Eq. (2.5.6), there is

$$\boldsymbol{R}_{\tilde{x}} = (\boldsymbol{\Sigma}^{-1}\boldsymbol{Q}^{\mathrm{T}})(\boldsymbol{Q}\boldsymbol{\Sigma}^2\boldsymbol{Q}^{\mathrm{T}})(\boldsymbol{\Sigma}^{-1}\boldsymbol{Q}^{\mathrm{T}})^{\mathrm{T}} = \boldsymbol{I} \tag{2.5.7}$$

therefore, through matrix $\boldsymbol{T}$ transformation, the correlation between components of the mixed signal does not exist again.

For the correlation matrix, its eigenvalue decomposition and singular value decomposition is equivalent. Therefore, the whitening matrix could be solved through the singular value decomposition (SVD) of correlation matrix. Usually, the numerical algorithm of singular value decomposition has better stability than eigenvalue decomposition, so SVD is recommended to obtain the whitening matrix.

In practice, the correlation matrix of the mixed signal can only be estimated from samples of a mixed signal vector. Supposing $\boldsymbol{x}(1), \boldsymbol{x}(2), \cdots, \boldsymbol{x}(N)$ is a group of samples of mixed signal random vector, the sample correlation matrix of the mixed signal can be estimated by the following equation:

$$\hat{\boldsymbol{R}}_x = \frac{1}{N-1}\sum_{i=1}^{N}\boldsymbol{x}(i)\boldsymbol{x}(i)^{\mathrm{T}} \tag{2.5.8}$$

it is not difficult to prove that $\hat{\boldsymbol{R}}_x$ is Hermitian and non-negative definite, so its eigenvalue is a non-negative real number and $\hat{\boldsymbol{R}}_x$ has eigenvalue decomposition similar to Eq. (2.5.4). In actual computation, eigenvalue decomposition (or singular decomposition) of $\hat{\boldsymbol{R}}_x$ is taken to get the whitening matrix. After the whitening transform, the correlation matrix of mixed signal vector samples becomes a unit matrix.

Another method for mixed signal whitening is realized through minimization of some cost function. Because the purpose of whitening is to find a whitening matrix $\boldsymbol{T}$ that makes the correlation matrix of the transformed new vector into a unit matrix, let $z(t)=T(t)x(t)$ and iteratively adjust the value of elements in matrix $T(t)$ to decrease the distance between the correlation matrix of the new vector $z(t)$ and unit matrix, finally realize mixed signals whitening. Let $z(t)=T(t)x(t)$, a feasible solution is the following iteration:

$$T(t+1) = T(t) - \lambda_t[z(t)z(t)^{\mathrm{T}} - I]T(t) \tag{2.5.9}$$

where λ_t is learning coefficient. It can be seen that when the iteration in Eq. (2.5.9) converged, there will be $E\,[z(t)z(t)^{\mathrm{H}}-I]=0$, i. e. , $R_z=I$, thus algorithm (2.5.9) can realize whitening of the mixed signal.

Besides eigenvalue decomposition and singular decomposition for whitening, sometimes an on-line method is adopted to realize whitening of the mixed signal, which is called equivariant adaptive whitening of the random vector.

Set T as the whitening matrix. Similarly, because the blind separation algorithm of source signals cannot recover the real amplitudes of source signals, suppose that all source signals are random variable with unit variance and zero mean, this setting is equivalent to the case where the correlation matrix of the source signal vector is a unit matrix,

$$R_s = E[ss^{\mathrm{T}}] = I \tag{2.5.10}$$

therefore

$$\begin{aligned} R_y &= E[yy^{\mathrm{T}}] \\ &= E[(Tx)(Tx)^{\mathrm{T}}] \\ &= E[(TAs)(TAs)^{\mathrm{T}}] \\ &= CE[ss^{\mathrm{T}}]C^{\mathrm{T}} \\ &= CC^{\mathrm{T}} \end{aligned} \tag{2.5.11}$$

where $C=TA$. After whitening, there will be $R_y=I$, so whitening of mixed signals can be realized through optimizing the following cost function:

$$\phi_w(C) = \frac{1}{4}\parallel R_y - I \parallel_{\mathrm{F}} \tag{2.5.12}$$

where $\parallel A \parallel_F = \left[\sum_{i=1}^{m}\sum_{j=1}^{n} | a_{ij} |^2\right]^{1/2}$ is the Frobeniius norm of the matrix. Obviously, if and only if $R_y=I$, there will be $\phi(C)=0$. Use $\parallel A \parallel_F^2=$ Trace $\{AA^{\mathrm{T}}\}$ and $R_y=CC^{\mathrm{T}}$, there is

$$\begin{aligned} \parallel R_y - I \parallel_F &= \mathrm{Trace}\{[R_y - I][R_y - I]^{\mathrm{T}}\} \\ &= \mathrm{Trace}\{R_yR_y^{\mathrm{T}} - R_y - R_y^{\mathrm{T}} + I\} \\ &= \mathrm{Trace}\{R_yR_y^{\mathrm{T}}\} - 2\mathrm{Trace}\{R_y\} + \mathrm{Trace}\{I\} \\ &= \mathrm{Trace}\{R_y^{\mathrm{T}}R_y\} - 2\mathrm{Trace}\{R_y\} + \mathrm{Trace}\{I\} \\ &= \mathrm{Trace}\{(CC^{\mathrm{T}})^{\mathrm{T}}(CC^{\mathrm{T}})\} - 2\mathrm{Trace}\{CC^{\mathrm{T}}\} + \mathrm{Trace}\{I\} \\ &= \mathrm{Trace}\{(CC^{\mathrm{T}})^2\} - 2\mathrm{Trace}\{CC^{\mathrm{T}}\} + \mathrm{Trace}\{I\} \end{aligned} \tag{2.5.13}$$

Through calculating the derivative of the matrix function, the negative gradient of cost function $\phi_w(\boldsymbol{C})$ is obtained:

$$\begin{aligned}-\nabla\phi_w(\boldsymbol{C}) &= -\frac{1}{4}\frac{\partial}{\partial \boldsymbol{C}}[\mathrm{Trace}\{(\boldsymbol{CC}^{\mathrm{T}})^2\} - 2\mathrm{Trace}\{\boldsymbol{CC}^{\mathrm{T}}\} + \mathrm{Trace}\{\boldsymbol{I}\}] \\ &= -\frac{1}{4}(4\boldsymbol{CC}^{\mathrm{T}}\boldsymbol{C} - 4\boldsymbol{C}) \\ &= (\boldsymbol{I} - \boldsymbol{CC}^{\mathrm{T}})\boldsymbol{C} \\ &= (\boldsymbol{I} - \boldsymbol{R}_y)\boldsymbol{C}\end{aligned} \tag{2.5.14}$$

Therefore, the following iteration algorithm realizes an on-line whitening of mixed signal:

$$\begin{aligned}\boldsymbol{C}(k+1) &= \boldsymbol{C}(k) + \mu_k(\boldsymbol{I} - \boldsymbol{R}_y)\boldsymbol{C}(k) \\ &= \boldsymbol{C}(k) + \mu_k(\boldsymbol{I} - E[\boldsymbol{yy}^{\mathrm{T}}])\boldsymbol{C}(k)\end{aligned} \tag{2.5.15}$$

In the above deduction process, the partial derivative of cost function $\phi_w(\boldsymbol{C})$ is processed in relation to the whitening-mixing matrix $\boldsymbol{C}$, and the mixing matrix $\boldsymbol{A}$ is just used to determine the initial value of algorithm and without any direct impact on the convergence of the algorithm; it means that this algorithm has equivariant Property.

2.5.3 Band-Limited Filtering

The effect of filtering is to improve the signal-to-noise ratio, consequently improving the quality of blind processing. In terms of information quantity, the design parameters of the filter, especially its bandwidth, are prior knowledge. The specific filtering operation will be discussed in Chapter 9, and readers can refer to Ref.[5] about the basic design method of digital filter.

2.6 Complex Nonlinear Function

In blind signal processing, the independent component and mixing matrix in a complex domain are usually involved; for example, in a convolutive mixtures separation, a normal method is to transform the convolution operation into a multiplication operation through Fourier transform to improve the efficiency of the algorithm. Real source signal becomes a complex signal after transformation. Moreover, the complex data, such as radar signals and magnetic resonance images, also requires the separation of complex sources. Therefore, it is necessary to give a simple explanation of a complex nonlinear function before going further.

Theoretically, in the nonlinear processing of complex function, the first problem involves the boundedness and analyticity in a complex domain. The Liouville Theorem gives a definite statement that if f is the integer function in a complex plane with a constant as bound, $f(z)$ is also constant in the whole plane. To solve this problem, there are mainly three methods. The first is defined as a complex split nonlinear function (CSNF)[8~10], which calculates a real part and an imaginary part respectively. Because both the real part and imaginary part of the complex function are real,the real nonlinear function in stead of a complex nonlinear function is used in the calculation. The boundedness can be ensured by this method, but analyticity is not satisfied. The second method is to calculate amplitude of variables, also using, of course, a real nonlinear function[11,12]. For some special problems, these two methods can provide satisfying results. However, in respect to extracting complex independent component to calculate higher-order statistics, these two methods are not so effective as directly using a complex nonlinear function. Therefore, the third method is to extract independent components from a complex domain by using a complex nonlinear function[8,13~15].

2.6.1 Expectation of Complex Random Variable and Complex Valued Function

A complex random variable can be represented as $z=u+\mathrm{j}_v$, where u and v are real random variables. In ICA, suppose that the real part and imaginary part of each independent component are irrelevant but have the same variance. The probability density of z is $f(z)=f(u, v) \in \Re$, with expectation $E\{z\}=E\{u\}+\mathrm{j}E\{v\}$. The covariance matrix of zero-mean complex random vector $z=(z_1, \cdots, z_n)$ is

$$E\{zz^{\mathrm{H}}\} = \begin{bmatrix} C_{11} & \cdots & C_{1n} \\ \vdots & \ddots & \vdots \\ C_{n1} & \cdots & C_{nn} \end{bmatrix} \tag{2.6.1}$$

where $C_{jk}=E\{z_j z_k^*\}$,and z^{H} is Hermite of z.

2.6.2 Decorrelation Lemma for Two Complex Random Vectors

Similar to the real random vector, the cross-covariance matrix of two complex random vectors $\boldsymbol{X}=\boldsymbol{X}_{\mathrm{R}}+\mathrm{j}\boldsymbol{X}_{\mathrm{I}}$ and $\boldsymbol{Y}=\boldsymbol{Y}_{\mathrm{R}}+\mathrm{j}\boldsymbol{Y}_{\mathrm{I}}$ is defined as

$$\boldsymbol{\Lambda}_{XY} = E[(\boldsymbol{X}-\boldsymbol{m}_X)(\boldsymbol{Y}-\boldsymbol{m}_Y)^{\mathrm{H}}] \tag{2.6.2}$$

where $\boldsymbol{m}_X=E(\boldsymbol{X})$,$\boldsymbol{m}_Y=E(\boldsymbol{Y})$. Simultaneously, the pseudo cross-covariance matrix of $\boldsymbol{X}$ and $\boldsymbol{Y}$ is defined as

$$\tilde{\boldsymbol{\Lambda}}_{XY} = E[(\boldsymbol{X}-\boldsymbol{m}_X)(\boldsymbol{Y}-\boldsymbol{m}_Y)^{\mathrm{T}}] \tag{2.6.3}$$

The cross-covariance between the real part and imaginary part of $\boldsymbol{X}$ and $\boldsymbol{Y}$ are

$$\mathrm{cov}[\boldsymbol{X}_{\mathrm{R}}, \boldsymbol{Y}_{\mathrm{R}}] = \frac{1}{2}\mathrm{Re}\{\boldsymbol{\Lambda}_{XY} + \widetilde{\boldsymbol{\Lambda}}_{XY}\};$$

$$\mathrm{cov}[\boldsymbol{X}_{\mathrm{I}}, \boldsymbol{Y}_{\mathrm{I}}] = \frac{1}{2}\mathrm{Re}\{\boldsymbol{\Lambda}_{XY} - \widetilde{\boldsymbol{\Lambda}}_{XY}\};$$

$$\mathrm{cov}[\boldsymbol{X}_{\mathrm{I}}, \boldsymbol{Y}_{\mathrm{R}}] = \frac{1}{2}\mathrm{Im}\{\boldsymbol{\Lambda}_{XY} + \widetilde{\boldsymbol{\Lambda}}_{XY}\};$$

$$\mathrm{cov}[\boldsymbol{X}_{\mathrm{R}}, \boldsymbol{Y}_{\mathrm{I}}] = -\frac{1}{2}\mathrm{Im}\{\boldsymbol{\Lambda}_{XY} - \widetilde{\boldsymbol{\Lambda}}_{XY}\} \tag{2.6.4}$$

If the above four cross-covariances in Eq. (2.6.4) are zero, then $\boldsymbol{X}$ and $\boldsymbol{Y}$ are uncorrelated.

Lemma 1 Two complex random vectors are uncorrelated, if and only if $\boldsymbol{\Lambda}_{XY}=0$, $\widetilde{\boldsymbol{\Lambda}}_{XY}=0$ both exist, that is, the cross-covariance and the pseudo cross-covariance matrix must be zero simultaneously .

2.6.3 Proper (inherent) Property of Complex Random Vector

Definition A complex random vector $\boldsymbol{X}=\boldsymbol{X}_{\mathrm{R}}+\mathrm{j}\boldsymbol{X}_{\mathrm{I}}$, when its pseudo cross-covariance matrix $\widetilde{\boldsymbol{\Lambda}}_{XX}$ is equal to zero, $\boldsymbol{X}$ is called proper. Obviously, the subvector of the proper complex random vector should be proper. A composite vector of two proper complex random vectors $\boldsymbol{X}_1$ and $\boldsymbol{X}_2$ is denoted as $\boldsymbol{X}$ with $\boldsymbol{X}_1$ and $\boldsymbol{X}_2$ as its sub-vector. If $\boldsymbol{X}$ is proper, then $\boldsymbol{X}_1$ and $\boldsymbol{X}_2$ are joint proper.

Now, deduct the existence condition of a proper complex random vector.

$\boldsymbol{\Lambda}_{\mathrm{RR}}=\mathrm{cov}\ [\boldsymbol{X}_{\mathrm{R}}, \boldsymbol{X}_{\mathrm{R}}]$, $\boldsymbol{\Lambda}_{\mathrm{RI}}=\mathrm{cov}\ [\boldsymbol{X}_{\mathrm{R}}, \boldsymbol{X}_{\mathrm{I}}]$, $\boldsymbol{\Lambda}_{\mathrm{IR}}=\mathrm{cov}\ [\boldsymbol{X}_{\mathrm{I}}, \boldsymbol{X}_{\mathrm{R}}]$, $\boldsymbol{\Lambda}_{\mathrm{II}}=\mathrm{cov}\ [\boldsymbol{X}_{\mathrm{I}}, \boldsymbol{X}_{\mathrm{I}}]$,and, $\boldsymbol{\Lambda}_{\mathrm{RI}}=\boldsymbol{\Lambda}_{\mathrm{IR}}^{\mathrm{T}}$, there is

$$\widetilde{\boldsymbol{\Lambda}}_{XX} = \boldsymbol{\Lambda}_{\mathrm{RR}} - \boldsymbol{\Lambda}_{\mathrm{II}} + \mathrm{j}(\boldsymbol{\Lambda}_{\mathrm{IR}} + \boldsymbol{\Lambda}_{\mathrm{IR}}^{\mathrm{T}}) \tag{2.6.5}$$

thus, when $\widetilde{\boldsymbol{\Lambda}}_{XX}$ is equal to zero, there is

$$\boldsymbol{\Lambda}_{\mathrm{RR}} = \boldsymbol{\Lambda}_{\mathrm{II}}, \quad \boldsymbol{\Lambda}_{\mathrm{IR}} = -\boldsymbol{\Lambda}_{\mathrm{IR}}^{\mathrm{T}} \tag{2.6.6}$$

It means that if and only if $\boldsymbol{X}_{\mathrm{R}}$ and $\boldsymbol{X}_{\mathrm{I}}$ have the same autocorrelation matrix and antisymmetric covariance matrix, then the complex random vector has proper characteristics. It is necessary to explain that the anti-symmetry of $\boldsymbol{\Lambda}_{\mathrm{IR}}$ means the elements in the main diagonal are equal to zero, that is, each component $X(k)$ of the proper complex random vector $\boldsymbol{X}$ has an uncorrelated real part and imaginary part, but it does not mean that the real part and imaginary part of $\boldsymbol{X}(k)$ and $\boldsymbol{X}(l)$ ($k \neq l$) are uncorrelated.

The proper complex random vector has good properties, and its decorrelation lemma is as follows.

Lemma 2 Two joint proper complex random vectors $\boldsymbol{X}=\boldsymbol{X}_{\mathrm{R}}+\mathrm{j}\boldsymbol{X}_{\mathrm{I}}$ and

$Y=Y_R+jY_I$ are uncorrelated if and only if their covariance matrix Λ_{XY} is equal to zero.

It can be seen that lemma 2 has less constraint conditions than lemma 1 due to the proper characteristics of the complex random vector, and it is useful for deducing algorithms.

2.6.4 Principle of Selection of Complex Nonlinear Function

The selection of a complex nonlinear function is very important and the following principles must be kept in mind.

Assuming for any complex nonlinear function *f*, the function mapping goes from complex area to complex area, i.e. f: $\mathbb{C}\to\mathbb{C}$, $\mathbb{C}$ representing a complex area. Through the use of a simple variable, the necessary and sufficient condition for the independence of a complex variable could be expanded. Suppose two independent complex random vectors, $X=X_R+jX_I$ and $Y=Y_R+jY_I$, similarly defining two complex nonlinear functions *g* and *h*, if

$$E\{g(X)h(Y)\}=E\{g(X)\}E\{h(Y)\} \tag{2.6.7}$$

i.e. if and only if $g(X)$ and $h(Y)$are uncorrelated for any complex nonlinear function g and h in the limited expectation, the two complex random vectors are mutually independent. Actually, $g(X)$ and $h(Y)$ are also two complex random vectors. If both are uncorrelated, lemma 1 or lemma 2 could be satisfied. This is the principle of selecting a complex nonlinear function.

For higher-order statistics in blind separation, the kurtosis of a complex random vector with zero-mean is defined as[8]

$$\text{kurt}(y)=E\{|y|^4\}-2(E\{|y|^2\})^2-|E\{y^2\}|^2=E\{|y|^4\}-2 \tag{2.6.8}$$

if *y* is Gaussian distribution, kurtosis is equal to zero.

In ICA separation algorithms, for example, the Infomax learning algorithm of a separating matrix has the following form:

$$\Delta W=\eta(I-\varphi(u)u^H)W \tag{2.6.9}$$

where nonlinear function $\varphi(\cdot)$ can be deducted from integer exponential function e^z; here z is a complex number, so this function is analytic. Compared with bounded but not analytic complex function (for example, a split nonlinear function), it is more effective in generating higher order statistics. Moreover, this function is saturated nonlinear, i.e. meaning that the farther it goes from its origin, lower the growth rate, so the oscillation characteristics will not be appeared in the response of real and imaginary parts.

The simulation shows [8] that the Trigonometric function tan (z) , reverse-trigonometric function arctan (z), arcsin (z), arccos (z) and hyperbolic function tanh (z), arctanh (z),arcsinh (z), and arccosh (z) are stable for broad inputs and mixturess; whereas sin (z), cos (z), arccot (z), sinh (z), cosh (z) and arccoth (z) are not stable. The approximations of four different complex nonlinear function are shown in Fig. 2.2.

The above complex nonlinear functions are based on Cartesian coordinates. Taking tanh (z) as an example, its general form is $\varphi(z)=\tanh(\eta z_{re})+j\tanh(\eta z_{im})$, where η is a gain parameter for nonlinearity control but will produce oscillations in convergence with some functions; another method is to use functions based on polar coordinates with the corresponding form of $\varphi(z)=\tanh(\eta|z|)e^{j\theta(z)}$, which is phase independent of the density of independent signal and makes the convergence process much smoother . If data are not too redundant, the difference between the performance of the two functions can be large[9]. For more information, please visit http: //mathworld.wolfram.com/.

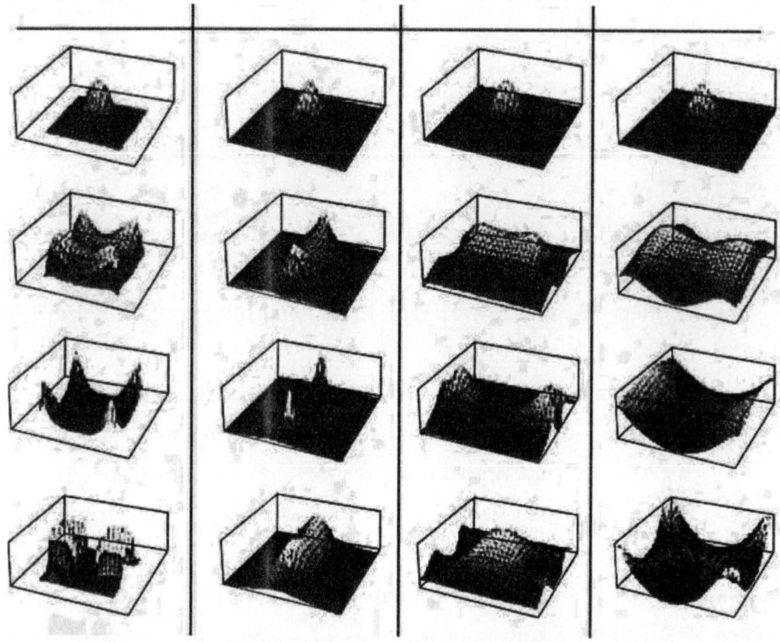

Fig. 2.2 Approximations of complex nonlinear functions

2.7 Evaluation Index

To evaluate the performance of an algorithm, two evaluation indexes are given: performance index (PI) and coding interference (ISI). Both correspond to instantaneous blind source separation algorithms and blind equalization algorithm respectively.

Performance index PI is defined as

$$\mathrm{PI}(\boldsymbol{C}) = \sum_{i=1}^{n}\left[\sum_{j=1}^{n}\frac{|c_{ij}|}{\max_k |c_{ik}|}-1\right]+\sum_{j=1}^{n}\left[\sum_{i=1}^{n}\frac{|c_{ij}|}{\max_k |c_{kj}|}-1\right] \quad (2.7.1)$$

where the mixing-separating matrix $\boldsymbol{C}=\boldsymbol{W}\boldsymbol{A}=(c_{ij})_{n\times n}$.

Coding interference is defined as

$$\mathrm{ISI}_i(t) = \frac{\sum_{j,l}|c_{ij}(l,t)|^2-\max_{j,l}|c_{ij}(l,t)|^2}{\max_{j,l}|c_{ij}(l,t)|^2} \quad (2.7.2)$$

where $c_{ij}(l, t)$ is the (i, j) element of the filter matrix of the equalizer system $\boldsymbol{C}(z,t)$ in l channel at t time. ISI shows the degree of demixing and deconvolution, and the lower the ISI is, the better the performance of the algorithm will be.

References

[1] Sorenson H (1980) Parameter estimation: Principles and problems. Marcel Dekker, New York

[2] Mendel J (1995) Lessons in estimation theory for signal processing, communications, and control. Prentice Hall, New Jersey

[3] Hyvarinen A, Karhunen J, Oja E (2001) Independent component analysis. Wiley, New York

[4] Cichocki A, Amari S I (2002) Adaptive blind signal and image processing—Learning algorithms and applications. Wiley, New York

[5] Shi X Z (2003) Signal processing and soft-computing. Higher Education Press, Beijing(Chinese)

[6] Zhang H Y (2001) Theoretical and experimental studies of blind source separation. Dissertation, Shanghai Jiao Tong University (Chinese)

[7] Shen X Z (2005) Study of algorithms of second and higher order blind signal processing. Dissertation, Shanghai Jiao Tong University (Chinese)

[8] Adali T, Taehwan K, Calhoun V (2004) Independent component analysis by complex nonlinearities. In: ICASSP '04. Montreal, Quebec, 2004,

5(17-21): 525–528

[9] Sawada H, Mukai R, Araki S et al (2003) A polar-coordinate based nonlinear function for frequency domain blind source separation. IEICE Transactions Fundamentals E86-A(3): 590–596

[10] Smaragdis P (1998) Blind separation of convolved mixturess in the frequency domain. Neurocomputing 22(1): 21–34

[11] Annemüller J, Sejnowski T J, Makeig S (2003) Complex spectral domain independent component analysis of electroencephalographic data. In: Proceedings of ICA-2003 , Nara, 2003

[12] Bingham E, Hyvärinen A (2000) A fast fixed-point algorithm for independent component analysis of complex-valued signals. Int. J. Neural Systems 10(1): 1–8

[13] Sawada H, Mukai R, Araki S et al (2002) A polar-coordinate based activation function for frequency domain blind source separation. In: Proceeding of ICASSP, Orlando, 2002: I-1001–1004

[14] Neeser F D, Massey J L (1993) Proper complex random processes with applications to information theory. IEEE Transactions on Information Theory 39(4): 1293–1302

[15] Eriksson J, Koivunen V(2004) Complex-valued ICA using second order statistics. In: Proceedings of the 2004 IEEE Signal Processing Society Workshop on Machine Learning for Signal Processing, Brazil, 2004, pp 183–191

Chapter 3 Independent Component Analysis

3.1 Problem Statement and Assumptions

Independent Component Analysis(ICA) is one of the methods for solving blind source separation in blind signal processing. This method seeks for a linear coordinate system to produce signals that are mutually statistically independent. Compared with Principal Component Analysis (PCA) based on correlation transform, ICA decorrelates signals and reduces the correlation in higher-order statistics. The existing blind processing algorithms are mainly based on ICA and will be unified to a certain extent through the research on ICA.

The basic ICA idea could be pursued back to more than 50 years ago. Darmois proposed in his paper that if source signals are mutually independent, estimations of source signals will be obtained through recovering the mutual independence of mixed signals, i.e. if y_1 and y_2 are two mutually independent random variables:

$$y_1 = \sum_{j=1}^{n} c_{1j} s_j ,$$

$$y_2 = \sum_{j=1}^{n} c_{2j} s_j \tag{3.1.1}$$

where s_j are mutually independent source signals. When both c_{1j} and c_{2j} are not zero and one signal at most in the source is Gaussian distribution, then the source signals s_j can be determined.

Comon (1994) described the concept of ICA in detail and proposed the contrast function related to mutual-information approximation minimization between sensors[1].

Definition A method is called ICA, which keeps the mutually statistical independency between each component of transformed new vector $\boldsymbol{y}$ after making the linear transformation $\boldsymbol{y}=\boldsymbol{W}\boldsymbol{x}$ for a random vector $\boldsymbol{x}$.

To clarify the Property of ICA, we discuss the following assumptions.

Current study on source signal separation generally assumes that the

X. Shi, *Blind Signal Processing*

number of outputs r of separating matrix $\boldsymbol{W}$, the number of mixed signals m and the number of source signals n are the same, i.e. $r=m=n$. At first, we will discuss the case in which the number of outputs of the separating matrix is equal to the number of source signals, $r=n$. For this case, we have the following theorem[2]:

Theorem 3.1 Suppose the mixing-separating compound matrix $\boldsymbol{C}$ is an $n \times n$ full-rank matrix, $\boldsymbol{y}=\boldsymbol{Cs}$, $\boldsymbol{s}=[s_1, s_2, \cdots, s_n]^{\mathrm{T}}$ is an n-dimensional random vector and each component is a mutually statistically independent random variable with zero mean value and unit variance. Each component of vector $\boldsymbol{y}=[y_1, y_2, \cdots, y_n]^{\mathrm{T}}$ is also a random variable with unit variance. If any two components of $\boldsymbol{y}$ are mutually statistically independent, i.e., for all $i, j=1, 2, \cdots, n$, $i \neq j$, the probability density functions of y_i and y_j satisfy the equation $f(y_i, y_j)=f(y_i)f(y_j)$, then we have

① Through the rearrangement of matrix components of $\boldsymbol{s}$ and $\boldsymbol{y}$, the matrix $\boldsymbol{C}$ can be changed into the form of a block diagonal matrix, $\boldsymbol{C}=\mathrm{diag}\{\boldsymbol{C}_1, \boldsymbol{C}_2, \cdots, \boldsymbol{C}_k, \boldsymbol{I}\}$, each block $\boldsymbol{C}_i$, $i=1, 2, \cdots, k$, is an orthogonal matrix with dimension larger than 1 and the dimension of unit matrix $\boldsymbol{I}$ may be zero.

② Components of random vector $\boldsymbol{s}$ corresponding to the unit block matrix can be Gaussian or non-Gaussion distribution.

③ Components of random vector $\boldsymbol{s}$ corresponding to blocks $\boldsymbol{C}_i$, $i=1, 2, \cdots, k$, must be Gaussian distribution.

From the above Theorem 3.1 we can obtain the following very useful following inference[1]:

Inference 3.1 Suppose all components of random vector $\boldsymbol{s}=[s_1, s_2, \cdots, s_n]^{\mathrm{T}}$ are mutually independent and the number of components subjected to Gaussian distribution is not more than one. $\boldsymbol{C}$ is an $n \times n$ full-rank matrix, $\boldsymbol{y}=\boldsymbol{Cs}$. Then the following three propositions are equivalent to each other[2]:

① Every two components of random vector $\boldsymbol{y}$, y_i, $i=1, 2, \cdots, n$ are statistically independent.

② All the components of random vector $\boldsymbol{y}$, y_i, $i=1, 2, \cdots, n$ are mutually statistically independent.

③ $\boldsymbol{C}$ is a generalized permutation matrix.

When the number of outputs of the separating matrix is less than the number of source signals, we have the following theorem[2]:

Theorem 3.2 Suppose $r<n$, and $\boldsymbol{C}$ is a $r \times n$ row full-rank matrix, $\boldsymbol{y}=\boldsymbol{Cs}$;

$s=[s_1, s_2, \cdots, s_n]^T$ is an n-dimensional random vector and each component is a mutually statistically independent random variable with zero mean value. Furthermore, when only one component of s subjected to Gaussian-distribution exists and every two components of random vector y are statistically independent, then we define such a matrix C as a generalized partition matrix.

In this case, if an $n \times n$ matrix A is column-full-rank, then matrix A is an n row decomposable matrix. It means that there is an $n \times n$ separating matrix that makes the matrix $C=WA$ a generalized permutation matrix. We know from Theorem 3.1 that though the mixing matrix A and source signals vector s are both unknown, the blind separation of the source can be realized if every two outputs of the separating matrix W are statistically independent.

From the discussions above we can obviously conclude that ICA when applied to mixed signals can realize the blind separation of source signals.

3.2 Contrast Functions

The discussion above has shown that ICA on mixed signals can realize the blind separation of source signals, especially when the number of mixtures and source signals are equal; as inference 3.1 stated, when the components of the output y of separating matrix W are pairwise statistically independent, they are mutually statistically independent. Meanwhile the mixing-separating compound matrix is equal to a generalized permutation matrix, i.e. a product of a permutation matrix and a full rank diagonal matrix.

In order to measure the independence among components of a random vector, we use the concept of contrast function and define it as follows[1]:

Denote $F=\{f_x, x \in \mathbf{R}^n\}$ as the set of the probability density function of all n-dimensional random vectors, then the contrast function of independent components analysis $\phi(\cdot)$ is a mapping from F to R, and satisfies the following conditions:

① The value of $\phi(f_x)$is independent of the permutation order of every components of random vector x, i.e., for any permutation matrix P, $\phi(f_{Px})=\phi(f_x)$.

② The value of $\phi(f_x)$ is independent of the absolute amplitudes of every components of random vector x, i.e. for any full-rank diagonal matrix D, $\phi(f_{Dx})=\phi(f_x)$.

③ If every component of random vector $\boldsymbol{x}$ is statistically independent, then for any reversible matrix $\boldsymbol{A}$, $\phi(f_{Ax}) \leqslant \phi(f_x)$.

④ If all components of random vector $\boldsymbol{x}$ are statistically independent, $\boldsymbol{P}$ is a permutation matrix and $\boldsymbol{D}$ is a nonsingular diagonal matrix, if and only if the matrix $\boldsymbol{A}$ has the decomposition $\boldsymbol{A} = \boldsymbol{PD}$, $\phi(f_{Ax}) = \phi(f_x)$.

From the definition of contrast functions, we know that if we find out a proper contrast function and minimize it, the ICA of random vectors as well as the blind separation of the signal source can be realized.

Because the probability density function of a random vector is unknown, a definite method must be used for estimation. Thus besides the four conditions defined above, practical contrast functions must be easy to estimate, approximate and optimize.

3.2.1 Common Contrast Functions

1. Maximum likelihood estimation contrast function

Maximum likelihood estimation is a very important parameter estimation method in statistics. Using one group realization $\{x_1, x_1, \cdots, x_T\}$ of random vector $\boldsymbol{x}$ estimates the parameter $\boldsymbol{\theta}$ in the parametric model $g(\boldsymbol{x}, \boldsymbol{\theta})$ of the probability density function of $\boldsymbol{x}$. The maximum likelihood method uses the maximum parameter value $\hat{\boldsymbol{\theta}}$ of the following logarithm likelihood function as the estimate $\boldsymbol{\theta}$:

$$L(\boldsymbol{\theta}) = \lg\left[\prod_{i=1}^{T} g(x_i, \boldsymbol{\theta})\right] = \sum_{i=1}^{T} \lg[g(x_i, \boldsymbol{\theta})] \tag{3.2.1}$$

i.e.

$$L(\hat{\boldsymbol{\theta}}) = \max_{\boldsymbol{\theta}} L(\boldsymbol{\theta}) \tag{3.2.2}$$

In blind separation problems, what is needed to estimate is the probability density function of mixed signal random vector $\boldsymbol{x}$ needs to be estimated, and the parameters of the probability density function is the probability density function $p(\boldsymbol{s})$ of mixing matrix $\boldsymbol{A}$ and source signal vector $\boldsymbol{s}$, i.e. $\boldsymbol{\theta}=(\boldsymbol{A}, p)$. According to the statistically independence between source signals, if we suppose the probability density function of source signal vector $\boldsymbol{s}= [s_1, s_2, \cdots, s_n]^{\mathrm{T}}$ is

$$p(\boldsymbol{s}) = p_1(s_1) \times p_2(s_2) \times \cdots \times p_n(s_n) = \prod_{i=1}^{n} p_i(s_i) \tag{3.2.3}$$

then the parametric model of the probability density function of mixed signal vector $\boldsymbol{x}=\boldsymbol{As}$ is

$$\hat{g}(\boldsymbol{x}; \boldsymbol{\theta}) = \hat{g}(\boldsymbol{x}; \boldsymbol{A}, p) \tag{3.2.4}$$

Suppose T samples of mixed signals $\boldsymbol{x}$, $\boldsymbol{X}= [\boldsymbol{x}(1), \boldsymbol{x}(2), \cdots, \boldsymbol{x}(T)]$, are statistically independent, that is

$$g(\boldsymbol{X}) = g(\boldsymbol{x}(1)) \times g(\boldsymbol{x}(2)) \times \cdots \times g(\boldsymbol{x}(T)) = \prod_{i=1}^{T} g(\boldsymbol{x}(i)) \quad (3.2.5)$$

where $g(\boldsymbol{x})$ is the real probability density function of mixed signals $\boldsymbol{x}$. Then we can obtain the normalized logarithmic likelihood function relative to sample set $\boldsymbol{X}$:

$$L_T(\boldsymbol{A}, p) = \frac{1}{T}\sum_{i=1}^{T} \lg[\hat{g}(\boldsymbol{x}; \boldsymbol{A}, p)] \quad (3.2.6)$$

Equation (3.2.6) is the sample average of function lg [$\hat{g}(\boldsymbol{x}; \boldsymbol{A}, f)$]. According to the theorem of great numbers, $L_T(\boldsymbol{A}, p)$ will converge in probability to the mathematical expectation of lg [$\hat{g}(\boldsymbol{x}; \boldsymbol{A}, p)$] , that is

$$\begin{aligned} l(\boldsymbol{A}, p) &= \lim_{T\to\infty} L_T(\boldsymbol{A}, p) \\ &= \lim_{T\to\infty} \frac{1}{T}\sum_{i=1}^{T} \lg[\hat{g}(\boldsymbol{x}; \boldsymbol{A}, p)] \\ &= E\{\lg[\hat{g}(\boldsymbol{x}; \boldsymbol{A}, p)]\} \\ &= \int g(\boldsymbol{x}) \lg[\hat{g}(\boldsymbol{x}; \boldsymbol{A}, p)] \mathrm{d}\boldsymbol{x} \end{aligned} \quad (3.2.7)$$

substituting $\hat{g}(\boldsymbol{x}; \boldsymbol{A}, p) = \frac{\hat{g}(\boldsymbol{x}; \boldsymbol{A}, p)}{g(\boldsymbol{x})} g(\boldsymbol{x})$ into the last equation of Eq. (3.2.7), we have

$$\begin{aligned} l(\boldsymbol{A}, p) &= \int g(\boldsymbol{x}) \lg[\hat{g}(\boldsymbol{x}; \boldsymbol{A}, p)] \mathrm{d}\boldsymbol{x} \\ &= -\int g(\boldsymbol{x}) \lg \frac{g(\boldsymbol{x})}{\hat{g}(\boldsymbol{x}; \boldsymbol{A}, p)} \mathrm{d}\boldsymbol{x} + \int g(\boldsymbol{x}) \lg[g(\boldsymbol{x})] \mathrm{d}\boldsymbol{x} \\ &= -D(\boldsymbol{x} \mid \boldsymbol{A}\boldsymbol{s}) - H(\boldsymbol{x}) \end{aligned} \quad (3.2.8)$$

where $D(\boldsymbol{x} \mid \boldsymbol{A}\boldsymbol{s})$ is the Kullback-Leibler divergence between the probability density functions of random vector $\boldsymbol{x}$ and $\boldsymbol{A}\boldsymbol{s}$, and $H(\boldsymbol{x})$is the differential entropy of the probability density function of $\boldsymbol{x}$. Suppose $\boldsymbol{y}=\boldsymbol{A}^{-1}\boldsymbol{x}$ and we use the invariance of Kullback-Leibler divergence in the invertible transform of sample space, we have

$$\begin{aligned} l(\boldsymbol{A}, p) &= -D(\boldsymbol{x} \mid \boldsymbol{A}\boldsymbol{s}) - H(\boldsymbol{x}) \\ &= -D(\boldsymbol{A}^{-1}\boldsymbol{x} \mid \boldsymbol{s}) - H(\boldsymbol{x}) \\ &= -D(\boldsymbol{y} \mid \boldsymbol{s}) - H(\boldsymbol{x}) \end{aligned} \quad (3.2.9)$$

Because $H(\boldsymbol{x})$is a constant parameter independent of $\boldsymbol{A}$,which is the estimation of the mixing matrix, we can obtain the following contrast function from Eq. (3.2.9):

$$\phi_{\mathrm{ML}}(\boldsymbol{y}) = D(\boldsymbol{y} \mid \boldsymbol{s}) \quad (3.2.10)$$

Therefore, when it is applied to blind signal separation problems, the maximum likelihood methods of parametric estimation of the probability density function can be explained as follows. In order to compute matrix $\boldsymbol{A}$, we must make the probability density function of random vector $\boldsymbol{A}^{-1}\boldsymbol{x}$ as close as possible to the assumed probability density function of source signal vector, i.e. $p(\boldsymbol{s})$ in Eq. (3.2.3) in relation to the Kullback-Leibler divergence.

2. Mutual information contrast function

The above maximum likelihood contrast function $\phi_{ML}(\boldsymbol{y})$ is obtained under the condition that we suppose the source signal vector is subjected to the probability distribution $p(\boldsymbol{s})$; therefore, we should optimize $\phi_{ML}(\boldsymbol{y})$through the simultaneous adjustment of mixing matrix $\boldsymbol{A}$ and $p(\boldsymbol{s})$.

Because $\phi_{ML}(\boldsymbol{y})=D(y|\boldsymbol{s})=D(y|\tilde{y})+D(\tilde{y}|\boldsymbol{s})$ and $D(y|\tilde{y})$ are independent of $\boldsymbol{s}$, the optimization of $\phi_{ML}(\boldsymbol{y})$ is equivalent to the optimization of $D(\tilde{y}|\boldsymbol{s})$. We can obtain the minimum, $\min_{s} D(\tilde{y}|\boldsymbol{s})=0$, through making $\boldsymbol{s}=\tilde{y}$. After minimization, we just need to minimize mixing matrix $\boldsymbol{A}$. We consider $D(y\,|\tilde{y})$ as a contrast function, that is

$$\phi_{MI}(y) = D(y \mid \tilde{y}) \tag{3.2.11}$$

Because $D(y|\tilde{y})$ is the mutual information between each components of $\boldsymbol{y}$, we call $\phi_{ML}(\boldsymbol{y})$ a mutual information contrast function. Mutual information is used to measure statistical independence between components of random vectors. In ICA, the contrast function $\phi_{ML}(\boldsymbol{y})$ is a natural choice.

3. Other contrast functions

Besides $\phi_{ML}(\boldsymbol{y})$ and $\phi_{ML}(\boldsymbol{y})$, ICA uses other contrast functions such as negentropy. As mentioned above, negentropy is defined as the Kullback-Leibler divergence between the probability density functions of a random vector and a corresponding Gaussian random vector with the same mean and covariance matrix. The basic idea of utilizing negentropy as an ICA contrast function is that, according to the central limit theorem, multiple random variable mixtures tend to Gaussian distribution, and blind separation can be realized through non-Gaussian processing by minimizing the negentropy of mixed signals.

Suppose the mixed signal has been pre-whitened, that is to say, all components of the random vector of the mixed signal are uncorrelated; then by using $\phi_{ML}(\boldsymbol{y})$, $\phi_{ML}(\boldsymbol{y})$ and negentropy etc., some new contrast functions are needed to be optimized to obtain new contrast functions under the condition of $E\,[yy^{\mathrm{T}}]=\boldsymbol{I}$. Because only orthogonal transform can ensure $E\,[yy^{\mathrm{T}}]=\boldsymbol{I}$, such a class of contrast functions is also called an orthogonal contrast function.

3.2.2 Computation of Contrast Function

Because the source signals and their probability density functions are unknown, the above discussed maximum likelihood contrast function, mutual information contrast function, negentropy, and orthogonal contrast function can't be computed in real applications. Two solutions exist; they are called the signal-processing method and the neural-network-based method respectively.

The signal processing method utilizes higher-order statistics such as each-order cumulant to approximate and estimate the probability density function of source[3]. Suppose $h_2(u)=u^2-1$ and $h_4(u)=u^4-6u^2+3$ are respectively the second and fourth order Hermite polynomials, and u is a random variable with a probability density function $p(u)$ close to standard Gaussian distribution $n(u)=(2\pi)^{-1/2}\exp(-u^2/2)$; then we can obtain the following approximation of $p(u)$ by using the Edgeworth expansion of probability density function around Gaussian function[3]:

$$p(u) \approx n(u)\left(1+\frac{\sigma^2(u)-1}{2}h_2(u)+\frac{k(u)}{4!}h_4(u)\right) \tag{3.2.12}$$

where $\sigma^2(u)$ and $k(u)$ are the variance and kurtosis of u respectively. If u and v are two quasi-Gaussian random variables in the real number domain, the Kullback-Leibler (KL) divergence can be approximated as

$$D(u \mid v) \approx \frac{1}{4}[\sigma^2(u)-\sigma^2(v)]^2+\frac{1}{48}[k(u)-k(v)]^2 \tag{3.2.13}$$

We know that Eq. (3.2.13) uses the difference of second-order and fourth-order cumulants to approximate the Kullback-Leibler divergence of the respective probability density functions. Equation (3.2.13) can be extended to the case of random vectors $\boldsymbol{U}$ and $\boldsymbol{V}$. Note $R_{ij}^U=\text{cum}\,[U_i, U_j]$ and $Q_{ijkl}^U=\text{cum}\,[U_i, U_j, U_k, U_l]$, results in

$$D(\boldsymbol{U} \mid \boldsymbol{V}) \approx \frac{1}{4}\sum_{ij}(R_{ij}^U-R_{ij}^V)^2+\frac{1}{48}\sum_{ijkl}(Q_{ijkl}^U-Q_{ijkl}^V)^2 \tag{3.2.14}$$

by using the approximation of Eq. (3.2.14), we can estimate $\phi_{\text{ML}}(\boldsymbol{y})$ and $\phi_{\text{MI}}(\boldsymbol{y})$, as well as the orthogonal contrast function, from $\phi_{\text{ML}}(\boldsymbol{y})$ and $\phi_{\text{MI}}(\boldsymbol{y})$. The negentropy contrast function can also be estimated using the Edgeworth expansion and high-order cumulants approximation methods. In addition, besides using the Edgeworth expansion, the Gram-Charlier expansion of probability density function can also be used to estimate and approximate contrast functions[4].

Neural-network-based blind separation methods use an iterative algorithm to adjust net weights, which are the elements w_{ij} of the separating (demixing) matrix $\boldsymbol{W}$, and deduce the iterative algorithm through optimizing contrast functions. In the deducing process, no detailed forms of the probability density function of the source signal are given. Instead, proper nonlinear functions are used to replace these measures related to the probability density function of the source signal. This function is generally an evaluation function which is defined as $\varphi(s)=-f'(s)/f(s)$, i.e. the ratio of the derivative of the probability density function to its probability density function. Different from a common neural network, a blind separation neural network works while being trained. The blind separation is realized after the algorithm converges, but the networks need renewed training for different mixed signals[5~7].

3.3 Information Maximization Method of ICA

An important ICA method is based on mutual information minimization, and the mutual information is used to construct a unified information theory framework to measure statistical correlations. Bell and Sejnowski (1995) proposed a single layer feedforward network learning algorithm that utilizes information maximization to obtain output $\boldsymbol{y}$ from the linear mixture $\boldsymbol{x}$ of independent source $\boldsymbol{s}$. They pointed out that the maximization of joint entropy $H(\boldsymbol{y})$ of network output components $y_i=g_i(u_i)$ can approximately minimize the mutual information between each component.

The joint entropy of network output is

$$H(y_1, \cdots, y_N) = H(y_1) + \cdots + H(y_N) - I(y_1, \cdots, y_N) \qquad (3.3.1)$$

where $H(y_i)$ is the marginal entropy of each output, $H(y_i)=-E(\lg p(y_i))$, and $I(y_1, \cdots, y_N)$ is their mutual information. The maximization of joint entropy means the maximization of marginal entropy and the minimization of mutual information. For bounded random variables $y_1, \cdots, y_N$, when the mutual information is zero, $H(y_1, \cdots, y_N)$ will reach its maximum and the marginal distribution will be uniform.

The parameters used to determine the maximum joint entropy have two groups, i.e. nonlinear $y_i=g_i(u_i)$ and the weight coefficient $\boldsymbol{W}$. Once the nonlinear functions are selected, the remaining parameter is $\boldsymbol{W}$. The derivation with respect to $\boldsymbol{W}$ is

$$\frac{\partial H(\boldsymbol{y})}{\partial \boldsymbol{W}} = \frac{\partial}{\partial \boldsymbol{W}}(-D(p_1(\boldsymbol{y}) \mid p(\boldsymbol{y})) \qquad (3.3.2)$$

this derivative is related to KL distance between uniform distribution $p_1(\boldsymbol{y})$

and uniform estimation $p(\boldsymbol{y})$. If $p(\boldsymbol{y})=p_1(\boldsymbol{y})$, then $I(\boldsymbol{y})=0$, $H(\boldsymbol{y})$ will reach the maximum. Because $g_i(u_i)$ is a reversible mapping from u_i to y_i, the KL distance in the equation equals that KL distance between estimation $p(\boldsymbol{u})$ and the source $p(\boldsymbol{s})$.

$$D(p_1(\boldsymbol{y}) \mid p(\boldsymbol{y})) = D(p(\boldsymbol{s}) \mid p(\boldsymbol{u})) \tag{3.3.3}$$

because no correlation could be introduced by nonlinearity, if we require the mutual information between output components to be zero, $I(y_1, \cdots, y_N)=0$, then the mutual information $I(u_1, \cdots, u_N)$ before nonlinear mapping must also be zero. The relationship between y_i, u_i and the nonlinear transfer function is

$$p(y_i) = \frac{p(u_i)}{\left|\frac{\partial g_i(u_i)}{\partial u_i}\right|} \tag{3.3.4}$$

for uniform distribution y_i, we have

$$p(u_i) = \left|\frac{\partial g_i(u_i)}{\partial u_i}\right| \tag{3.3.5}$$

rewriting Eq. (3.3.1) gives

$$H(\boldsymbol{y}) = -E\left\{\lg \frac{p(u_1)}{\left|\frac{\partial g_1(u_1)}{\partial u_1}\right|}\right\} + \cdots - E\left\{\lg \frac{p(u_N)}{\left|\frac{\partial g_N(u_N)}{\partial u_N}\right|}\right\} - I(\boldsymbol{y}) \tag{3.3.6}$$

or

$$H(\boldsymbol{y}) = -\sum_{i=1}^{N} E\left\{\lg \frac{p(u_i)}{\left|\frac{\partial g_i(u_i)}{\partial u_i}\right|}\right\} - I(\boldsymbol{y}) \tag{3.3.7}$$

the derivative with respect to joint entropy is

$$\frac{\partial H(\boldsymbol{y})}{\partial \boldsymbol{W}} = \frac{\partial}{\partial \boldsymbol{W}}(-I(\boldsymbol{y})) - \frac{\partial}{\partial \boldsymbol{W}}\sum_{i=1}^{N} E\left\{\lg \frac{p(u_i)}{\left|\frac{\partial g_i(u_i)}{\partial u_i}\right|}\right\} \tag{3.3.8}$$

this equation gives a clear relationship between the maximization of output joint entropy and the minimization of the mutual entropy of output components. We can learn that when the density function $p(u_i)$ of the estimated source u_i equals the derivative of nonlinear activation function g_i, the mutual information will be directly minimized. If the above condition can't be satisfied, $I(\boldsymbol{y}) \neq 0$, the second term of the right part of Eq. (3.3.8) will be the error term. Therefore, the minimization of $I(\boldsymbol{y})$ depends on how to estimate the density function by the nonlinear $g_i(u_i)$.

If the error term of Eq. (3.3.8) can be ignored, we calculate the gradient of $H(\boldsymbol{y})$ as

$$\frac{\partial H(\boldsymbol{y})}{\partial \boldsymbol{W}} = \frac{\partial}{\partial \boldsymbol{W}}(-E\{\lg|\boldsymbol{J}|\}) \tag{3.3.9}$$

to obtain the maximum of joint entropy $H(\boldsymbol{y})$. For the training set of data $\boldsymbol{x}$, the expectation term can be ignored when the learning rule is approximated. The nonlinear mapping between output density function $p(y)$ and input density function $p(\boldsymbol{x})$ can be expressed by the Jacobian determinant as follows

$$p(\boldsymbol{y}) = \frac{p(\boldsymbol{x})}{|\boldsymbol{J}(\boldsymbol{x})|} \tag{3.3.10}$$

and

$$\boldsymbol{J}(\boldsymbol{x}) = \det \begin{bmatrix} \frac{\partial y_1}{\partial x_1} & \cdots & \frac{\partial y_1}{\partial x_n} \\ \vdots & & \vdots \\ \frac{\partial y_n}{\partial x_1} & \cdots & \frac{\partial y_n}{\partial x_n} \end{bmatrix} \tag{3.3.11}$$

in the determinant, each partial derivative has the form of

$$\frac{\partial y_i}{\partial x_j} = w_{ij} \frac{\partial y_i}{\partial u_j} \tag{3.3.12}$$

the Jacobian determinant can be rewritten as

$$\boldsymbol{J}(\boldsymbol{x}) = \det(\boldsymbol{W}) \prod_{i=1}^{N} \left| \frac{\partial y_i}{\partial u_i} \right| \tag{3.3.13}$$

substituting into Eq. (3.3.9), we have

$$\begin{aligned} \frac{\partial H(\boldsymbol{y})}{\partial \boldsymbol{W}} &= \frac{\partial}{\partial \boldsymbol{W}} \lg\left(|\det(\boldsymbol{W})| \prod_{i=1}^{N} \left| \frac{\partial y_i}{\partial u_i} \right| \right) \\ &= \frac{\partial}{\partial \boldsymbol{W}} \lg|\det(\boldsymbol{W})| + \frac{\partial}{\partial \boldsymbol{W}} \sum_{i=1}^{N} \lg \left| \frac{\partial y_i}{\partial u_i} \right| \\ &= \frac{\partial}{\partial \boldsymbol{W}} \lg|\det(\boldsymbol{W})| + \sum_{i=1}^{N} \frac{\partial}{\partial \boldsymbol{W}} \lg \left| \frac{\partial y_i}{\partial u_i} \right| \end{aligned} \tag{3.3.14}$$

the first term in the above equation is

$$\frac{\partial}{\partial \boldsymbol{W}} \lg|\det(\boldsymbol{W})| = \frac{(\mathrm{adj}\boldsymbol{W})^{\mathrm{T}}}{\det(\boldsymbol{W})} = (\boldsymbol{W}^{\mathrm{T}})^{-1} \tag{3.3.15}$$

and the second term can be calculated:

$$\frac{\partial}{\partial w_{ij}} \sum_{i=1}^{N} \lg \left| \frac{\partial y_i}{\partial u_i} \right| = \frac{1}{\frac{\partial y_i}{\partial u_i}} \frac{\partial^2 y_i}{\partial u_i^{\,2}} x_j \tag{3.3.16}$$

define the derivative of nonlinear y_i with respect to u_i as the approximation of the source density function $p(u_i)$:

$$p(u_i) = \frac{\partial y_i}{\partial u_i} \tag{3.3.17}$$

Equation (3.3.16) can be rewritten as

$$\frac{\partial}{\partial \boldsymbol{W}} \sum_{i=1}^{N} \lg \left| \frac{\partial y_i}{\partial u_i} \right| = \frac{1}{p(\boldsymbol{u})} \cdot \frac{\partial p(\boldsymbol{u})}{\partial \boldsymbol{u}} \boldsymbol{x}^{\mathrm{T}} \tag{3.3.18}$$

Then we have the Infomax learning rule (Bell and Sejnowski, 1995)

$$\frac{\partial H(\boldsymbol{y})}{\partial \boldsymbol{W}} = (\boldsymbol{W}^{\mathrm{T}})^{-1} + \left(\frac{\frac{\partial p(\boldsymbol{u})}{\partial \boldsymbol{u}}}{p(\boldsymbol{u})} \right) \boldsymbol{x}^{\mathrm{T}} \tag{3.3.19}$$

A more effective way to compute the entropy gradient is the natural gradient (Amari 1996, Cardoso and laheld, 1996), a right multiplication with $\boldsymbol{W}^{\mathrm{T}}\boldsymbol{W}$:

$$\Delta \boldsymbol{W} \propto \frac{\partial H(\boldsymbol{y})}{\partial \boldsymbol{W}} \boldsymbol{W}^{\mathrm{T}}\boldsymbol{W} = \left[\boldsymbol{I} + \left(\frac{\frac{\partial p(\boldsymbol{u})}{\partial \boldsymbol{u}}}{p(\boldsymbol{u})} \right) \boldsymbol{u}^{\mathrm{T}} \right] \boldsymbol{W} \tag{3.3.20}$$

Further define the nonlinearity or the evaluation function as

$$\varphi(\boldsymbol{u}) = -\frac{\left(\frac{\partial p(\boldsymbol{u})}{\partial \boldsymbol{u}}\right)}{p(\boldsymbol{u})} \tag{3.3.21}$$

then

$$\Delta \boldsymbol{W} \propto [\boldsymbol{I} - \varphi(\boldsymbol{u})\boldsymbol{u}^{\mathrm{T}}]\boldsymbol{W} \tag{3.3.22}$$

3.4 Maximum Likelihood Method and Common Learning Rule

Gaeta and Lacoume (1990), Pham and Garrat (1992), Pearlmutter and Parra (1996), Mackay (1996), and Cardoso (1997) proposed common learning rules in the maximum likelihood method.

The probability density function of $\boldsymbol{x}$ can be expressed as (Cardoso, 1997)

$$p(\boldsymbol{x}) = |\det(\boldsymbol{W})| \, p(\boldsymbol{u}) \tag{3.4.1}$$

where $p(\boldsymbol{u}) = \prod_{i=1}^{N} p_i(u_i)$ is the supposed distribution of $p(\boldsymbol{s})$. Taking the logarithm likelihood

$$L(\boldsymbol{u}, \boldsymbol{W}) = \lg|\det(\boldsymbol{W})| + \sum_{i=1}^{N} \lg p_i(u_i) \tag{3.4.2}$$

Computing the maximization of logarithm of likelihood relative to $\boldsymbol{W}$, and gives the learning rule

$$\Delta \boldsymbol{W} \propto [(\boldsymbol{W}^{\mathrm{T}})^{-1} - \varphi(\boldsymbol{u})\boldsymbol{x}^{\mathrm{T}}] \tag{3.4.3}$$

where

$$\varphi(\boldsymbol{u}) = -\frac{\frac{\partial p(\boldsymbol{u})}{\partial \boldsymbol{u}}}{p(\boldsymbol{u})} = \left[-\frac{\frac{\partial p(u_1)}{\partial u_1}}{p(u_1)}, \cdots, -\frac{\frac{\partial p(u_N)}{\partial u_N}}{p(u_N)} \right] \tag{3.4.4}$$

This still computes the maximization of the logarithm of likelihood in natural gradients

$$\Delta W \propto \frac{\partial L(u,W)}{\partial W} W^{T}W = [I - \varphi(u)u^{T}]W \tag{3.4.5}$$

Compared with Eq. (3.4.3), W^TW rescales the gradient, which simplifies the learning rule and speeds up the convergence velocity. If $g_i(u_i)=\tanh(u_i)$, then $\varphi(u)=2\tanh(u)$,and learning the rule becomes

$$\Delta W \propto [I - 2\tanh(u)u^{T}]W \tag{3.4.6}$$

Both theory and experience show that this algorithm is limited only to the separation of super-Gaussian distribution sources rather than the mixtures of sub-Gaussian and super-Gaussian distribution sources. One way to solve this problem is the nonlinear modeling of parameters to match the distribution of sources.

Girolami (1997) used the density function estimation of parameters to handle sub-Gaussian and super-Gaussian, and gives two learning rules:

$$\Delta W \propto \begin{cases} [I - \tanh(u)u^{T} - uu^{T}]W & \textit{for super-Gaussian source} \\ [I + \tanh(u)u^{T} - uu^{T}]W & \textit{for sub-Gaussian source} \end{cases} \tag{3.4.7}$$

the difference between the two rules is the sign of the tanh functions, which can be selected by the sign of kurtosis.

Another parameter model defines the mixture density function as a function of b

$$p(u) \propto \mathrm{sech}^2(u+b) + \mathrm{sech}^2(u-b) \tag{3.4.8}$$

for b=0, this parameter density function is proportional to the hyperbolic Cauchy distribution and suitable for the separation of super-Gaussian sources. for b=2, this parameter density function has a bimodal distribution and is suitable for the separation of sub-Gaussian sources. Thus, the learning algorithm of the mixtures source distribution is

$$\Delta W \propto [I + 2\tanh(u)u^{T} - 2\tanh(u+b)u^{T} - 2\tanh(u-b)u^{T}]W \tag{3.4.9}$$

3.5 FastICA Algorithm

The main advantage of gradient computation based on the negentropy maximization method is that it uses neural networks to do learning and makes it adapt fast in a time-varying environment. But the convergence speed of such a method is lower and depends on the selection of the learning rate. If the learning rate is not well selected, the convergence can even be broken. Thus a faster and more reliable learning algorithm is needed. A fixed-point iterative algorithm called FastICA is suitable for the requirement.

As mentioned above, the central limit theorem tells us that the distribution of the sum of independent random variables is closer to Gaussian distribution than the distribution of any one random variable. Therefore, the non-Gaussian Property is the basis of ICA estimation.

For ICA estimation, the most classical non-Gaussian measure is kurtosis

$$\mathrm{Kurt}(x) = E(x^4) - 3(E(x^2))^2 \tag{3.5.1}$$

suppose x has an unity variance to further simplify the kurtosis expression:

$$\mathrm{Kurt}(x) = E(x^4) - 3 \tag{3.5.2}$$

For Gaussian distribution, the fourth-order moment equals $3(E(x^2))^2$, and the kurtosis of the Gaussian random variable is zero. Therefore, using the non-Gaussian Property of kurtosis results in kurtosis maximization through transformation and is the simplest ICA estimation method. The main disadvantage of directly using kurtosis as a non-Gaussian measure is that it is very sensitive to external factors, and any unsuitable value in samples will enlarge the kurtosis. Therefore, the kurtosis is not a robust measure.

Another non-Gaussian measure is negentropy. According to the information theory, among all the random variables with the same variance, Gaussian variables have the biggest entropy; that is to say, Gaussian distribution is the most random one. In Chapter 2, the negentropy of random vector $\boldsymbol{x}$ is defined as

$$J(\boldsymbol{x}) = H(\boldsymbol{x}_{\text{gauss}}) - H(\boldsymbol{x}) \tag{3.5.3}$$

The negentropy value is always non-negative, and if and only if $\boldsymbol{x}$ has Gaussian distribution, the negentropy is zero. For a reversible linear transform, negentropy is also invariant. The problem encountered by using negentropy is difficult to compute. So a certain approximation method is needed when we estimate the probability density function according to its definition.

A useful approximation is the generalization of a higher-order moment:

$$J(x) \approx k_1(E\{G^1(x)\})^2 + k_2(E\{G^2(x)\} - E\{G^2(\nu)\})^2 \tag{3.5.4}$$

where G^1 and G^2 are non-quadratic functions, k_1 and k_2 are both positive constants, and ν is a Gaussian variable with zero-mean and unit variance. Variable x can be assumed to have zero mean and unit variance.

If we only use one non-quadratic function G, Equation (3.5.4) is changed to

$$J(x) \propto [E\{G(x)\} - E\{G(\nu)\}]^2 \tag{3.5.5}$$

there are multiple choices for G and the following forms are proven to be useful:

$$G_1(x) = \frac{1}{a_1}\lg \cosh a_1 x \tag{3.5.6}$$

$$G_2(x) = -\exp(-x^2/2) \tag{3.5.7}$$

where $1 \leqslant a_1 \leqslant 2$ is a constant, which usually equals 1.

After defining the non-Gaussian measure, we come to discussion of practical methods of maximization.

A fixed-point iterative method proposed is

$$\boldsymbol{W} = E\{\boldsymbol{x}g(\boldsymbol{W}^{\mathrm{T}}\boldsymbol{x})\} \Leftrightarrow (1+\alpha)\boldsymbol{W} = E\{\boldsymbol{x}g(\boldsymbol{W}^{\mathrm{T}}\boldsymbol{x})\} + \alpha\boldsymbol{W} \tag{3.5.8}$$

Coefficient α can be searched by an approximate Newton method.

According to the Kuhn-Tucker condition, under the constraint of $E\{(\boldsymbol{W}^{\mathrm{T}}\boldsymbol{x})^2\} = \|\boldsymbol{W}\|^2 = 1$, the optimal value of $E\{\mathrm{G}(\boldsymbol{W}^{\mathrm{T}}\boldsymbol{x})$ is at the point

$$E\{\boldsymbol{x}g(\boldsymbol{W}^{\mathrm{T}}\boldsymbol{x})\} + \beta\boldsymbol{W} = 0 \tag{3.5.9}$$

where β is a constant. When the left part of the equation is labeled as $\boldsymbol{F}$, the Jacobian matrix becomes

$$J\boldsymbol{F}(\boldsymbol{W}) = E\{\boldsymbol{x}\boldsymbol{x}^{\mathrm{T}}g'(\boldsymbol{W}^{\mathrm{T}}\boldsymbol{x})\} + \beta\boldsymbol{I} \tag{3.5.10}$$

Becausc thc data has gone through the whitening process, the first term of the above equation can be approximated as

$$E\{\boldsymbol{x}\boldsymbol{x}^{\mathrm{T}}g'(\boldsymbol{W}^{\mathrm{T}}\boldsymbol{x})\} \approx E\{\boldsymbol{x}\boldsymbol{x}^{\mathrm{T}}\}E\{g'(\boldsymbol{W}^{\mathrm{T}}\boldsymbol{x})\} = E\{g'(\boldsymbol{W}^{\mathrm{T}}\boldsymbol{x})\}\boldsymbol{I} \tag{3.5.11}$$

Therefore, the Jacobian matrix is changed into a diagonal matrix, and the approximate Newton iteration becomes

$$\boldsymbol{W} \leftarrow \boldsymbol{W} - [E\{\boldsymbol{x}g(\boldsymbol{W}^{\mathrm{T}}\boldsymbol{x})\} + \beta\boldsymbol{W}]/[E\{g'(\boldsymbol{W}^{\mathrm{T}}\boldsymbol{x})\} + \beta] \tag{3.5.12}$$

Multiplying both sides by $\beta + E\{g'(\boldsymbol{W}^{\mathrm{T}}\boldsymbol{x})\}$ simplifies equation (3.5.12) to

$$\boldsymbol{W} \leftarrow E\{\boldsymbol{x}g(\boldsymbol{W}^{\mathrm{T}}\boldsymbol{x})\} - E(g'(\boldsymbol{W}^{\mathrm{T}}\boldsymbol{x}))\boldsymbol{W} \tag{3.5.13}$$

That's the basic fixed-point iterative algorithm of FastICA.

Thus we can summarize the entire FastICA algorithm as follows:

Setp 1 Whiten the data to give $\boldsymbol{x}$;

Setp 2 Select the initial value of random weight vector $\boldsymbol{W}$;

Setp 3 Select non-quadratic function G, such as the derivatives of Eqs (3.5.6) and (3.5.7),

$$g_1(u) = \tanh(a_1 u)$$
$$g_2(u) = u\exp(-u^2/2)$$

let

$$\boldsymbol{W} \leftarrow E\{\boldsymbol{x}g(\boldsymbol{W}^{\mathrm{T}}\boldsymbol{x})\} - E\{g'(\boldsymbol{W}^{\mathrm{T}}\boldsymbol{x})\}\boldsymbol{W}$$

Setp 4 Repeat the iteration until the algorithm converges.

For the estimation of multiple independent components, the method of maximizing non-Gaussian Property must be extended. Corresponding to different independent components, $\boldsymbol{W}_i$ should be orthogonal in the whitening space, i.e. $E\{(\boldsymbol{W}_i^{\mathrm{T}}\boldsymbol{x})(\boldsymbol{W}_j^{\mathrm{T}}\boldsymbol{x})\}=\boldsymbol{W}_i^{\mathrm{T}}\boldsymbol{W}_j$, and we can adopt the deflationary orthogonalization method and symmetric orthogonalization method.

The deflationary orthogonalization method uses the Gram-Schmidt method, which utilizes one-independent-component algorithm to estimate the p independent components one by one. After each iteration, the former estimated $p-1$ vector projections, $(\boldsymbol{W}_p^{\mathrm{T}}\boldsymbol{W}_j)\boldsymbol{W}_j, j=1, \cdots, p-1$ are subtracted from $\boldsymbol{W}_p$, that is

$$\boldsymbol{W}_p \leftarrow \boldsymbol{W}_p - \sum_{j=1}^{p-1}(\boldsymbol{W}_p^{\mathrm{T}}\boldsymbol{W}_j)\boldsymbol{W}_j \tag{3.5.14}$$

The shortcoming of such a method is that the estimation error of the first vector will be accumulated with estimations of subsequent vectors.

The symmetric orthogonalization method computes the square root of the matrix

$$\boldsymbol{W} \leftarrow (\boldsymbol{W}\boldsymbol{W}^{\mathrm{T}})^{-1/2}\boldsymbol{W} \tag{3.5.15}$$

where the reverse square root$(\boldsymbol{W}\boldsymbol{W}^{\mathrm{T}})^{-1/2}$ can be obtained from its eigenvalue decomposition:

$$(\boldsymbol{W}\boldsymbol{W}^{\mathrm{T}})^{-1/2} = \boldsymbol{E}\mathrm{diag}(d_1^{-1/2}, \cdots, d_m^{-1/2})\boldsymbol{E}^{\mathrm{T}} \tag{3.5.16}$$

3.6 Natural Gradient Method

In blind source separation and ICA, the mutual independence of outputs $\boldsymbol{y}(k)$ reflects the quality of separation. Usually we use the Kullback-Leibler divergence as the measure of independence and minimize it. We denote the probability density function of random variable $\boldsymbol{y}=\boldsymbol{W}\boldsymbol{x}$ as $p_y(\boldsymbol{y}, \boldsymbol{W})$ and create a probability density function $q(\boldsymbol{y})$, in which all components are statistically independent, as a reference. If and only if $p_y(\boldsymbol{y}, \boldsymbol{W})$ and $q(\boldsymbol{y})$ have the same distributions, the K-L divergence is zero and does not change with the nonlinear transform of variable y_i. The natural measure of independence is expressed as

$$K_{pq} = E\{\rho(\boldsymbol{y},\boldsymbol{W})\} = \int_{-\infty}^{\infty} p_y(\boldsymbol{y})\lg\frac{p_y(\boldsymbol{y})}{q(\boldsymbol{y})}\mathrm{d}\boldsymbol{y} \tag{3.6.1}$$

where $\rho(\boldsymbol{y}, \boldsymbol{W})$ is an objective function, $q(\boldsymbol{y}) = \prod_{i=1}^{n} q_i(y_i)$.

If it is expressed by the differential entropy of the output $\boldsymbol{y}$

$$H(\boldsymbol{y}) = -\int_{-\infty}^{\infty} p_y(\boldsymbol{y})\lg p_y(\boldsymbol{y})\mathrm{d}\boldsymbol{y}$$

then the expression of mutual information of K-L divergence is

$$D(p|q) = -H(\boldsymbol{y}) - \sum_{i=1}^{n} \int_{-\infty}^{\infty} p_y(\boldsymbol{y}) \lg q_i(y_i) \mathrm{d}\boldsymbol{y} \tag{3.6.2}$$

suppose $\boldsymbol{y}=\boldsymbol{W}\boldsymbol{x}$, the differential entropy can be expressed as

$$H(\boldsymbol{y}) = H(\boldsymbol{x}) + \lg|\det\boldsymbol{W}| \tag{3.6.3}$$

where $H(\boldsymbol{x}) = -\int_{-\infty}^{\infty} p_x(\boldsymbol{x}) \lg p_x(\boldsymbol{x}) \mathrm{d}\boldsymbol{x}$, then

$$\begin{aligned} D(p|q) &= E\{\rho(\boldsymbol{y},\boldsymbol{W})\} \\ &= -H(\boldsymbol{x}) - \lg|\det(\boldsymbol{W})| - \sum_{i=1}^{n} E\{\lg(q_i(y_i))\} \end{aligned} \tag{3.6.4}$$

Using the regular online learning algorithm of random gradient

$$\Delta\boldsymbol{W}(k) = \boldsymbol{W}(k+1) - \boldsymbol{W}(k) = -\eta(k) \frac{\partial\rho[\boldsymbol{y}(k),\boldsymbol{W}]}{\partial\boldsymbol{W}} \tag{3.6.5}$$

where $\eta(k)$ is the learning rate dependent of k, and $\partial\rho/\partial\boldsymbol{W}$ is an $n\times n$ gradient matrix with elements $\partial\rho/\partial w_{ij}$, through the differential matrix operation, we obtain

$$\Delta\boldsymbol{W}(k) = \eta(k)[\boldsymbol{W}^{-\mathrm{T}}(k) - f[\boldsymbol{y}(k)]\boldsymbol{x}^{\mathrm{T}}(k)] \tag{3.6.6}$$

where $f(\boldsymbol{y}) = [f_1(y_1), f_2(y_2), \cdots, f_n(y_n)]^{\mathrm{T}}$ is a column vector and its ith element is

$$f_i(y_i) = -\frac{\mathrm{d}\lg q_i(y_i)}{\mathrm{d}y_i} = -\frac{\mathrm{d}q_i(y_i)/\mathrm{d}y_i}{q_i(y_i)} = -\frac{q'_i(y_i)}{q_i(y_i)} \tag{3.6.7}$$

and $q_i(y_i)$ is the approximate model of the density function of source signal $\{s_i\}$. The gradient $-\partial\rho/\partial\boldsymbol{W}$ means the fastest descending direction of the objective function in Euclidean parameter space. Such a parameter space consists of all the nonsingular $n\times n$ matrices $\boldsymbol{W}$ and is a multiplicative group.

But in the curved space, the measurement of distance can't be done by the Euclidean norm. Instead, Riemannian geometry should be introduced; that is, a Riemannian metric tensor is used for measuring distance. $\boldsymbol{G}(\boldsymbol{W})$ is an $n\times n$ positive definite matrix with elements $g_{ij}(\boldsymbol{W})$, and such a metric tensor indicates the internal curvature of a specific manifold in N-dimensional space. For vectors $\boldsymbol{W}$ and $\boldsymbol{W}+\delta\boldsymbol{W}$, define the distance

$$\begin{aligned} d_{\mathrm{W}}(\boldsymbol{W},\boldsymbol{W}+\delta\boldsymbol{W}) &= \sqrt{\sum_{i=1}^{N}\sum_{j=1}^{N} \delta w_i \delta w_j g_{ij}(\boldsymbol{W})} \\ &= \sqrt{\delta\boldsymbol{W}^{\mathrm{T}}\boldsymbol{G}(\boldsymbol{W})\delta\boldsymbol{W}} \end{aligned} \tag{3.6.8}$$

in Euclidean coordinates $\boldsymbol{V}$, $\boldsymbol{G}(\boldsymbol{V})=\boldsymbol{I}$ is a unit matrix, and the distance is simplified as

$$d_E(\boldsymbol{V},\boldsymbol{V}+\delta\boldsymbol{V}) = \|\delta\boldsymbol{V}\|_2 = \sqrt{\sum_{i=1}^{N}\delta v_i^2} = \sqrt{\delta\boldsymbol{V}^{\mathrm{T}}\delta\boldsymbol{V}} \tag{3.6.9}$$

i.e. the Euclidean norm. When the manifold Property can be described by the transform from vector $\boldsymbol{V}$ in Euclidean orthogonal space to $\boldsymbol{W}$, the form of $\boldsymbol{G}(\boldsymbol{W})$ can be determined by the following formula:

$$d_E^2(\mathbf{V},\mathbf{V}+\delta\mathbf{V}) = d_W^2(\mathbf{W},\mathbf{W}+\delta\mathbf{W}) \tag{3.6.10}$$

For example, in polar coordinate space, $\boldsymbol{V}=[x\ y]^{\mathrm{T}}$, $x=r\cos\theta$, $y=r\sin\theta$, then the Riemannian metric tensor of $\boldsymbol{W}$ is

$$\mathbf{G}(\mathbf{W}) = \begin{bmatrix} 1 & 0 \\ 0 & r^2 \end{bmatrix}$$

In spherical coordinate space, $\boldsymbol{V}=[x\ y\ z]^{\mathrm{T}}$, $\boldsymbol{W}=[r\theta\varphi]^{\mathrm{T}}$ and

$$\mathbf{V} = \begin{bmatrix} r\cos\theta \\ r\sin\theta\cos\varphi \\ r\sin\theta\sin\varphi \end{bmatrix}$$

then

$$\mathbf{G}(\mathbf{W}) = \begin{bmatrix} 1 & 0 & 0 \\ 0 & r^2 & 0 \\ 0 & 0 & r^2\sin^2\theta \end{bmatrix}$$

Amari *et al* introduced a natural Riemannian metric for $\boldsymbol{W}$ space and proved that in the Riemannian space with parameter $\boldsymbol{W}$, the real fastest descending direction is not the standard negative gradient $-\partial\rho/\partial\boldsymbol{W}$ but

$$-\frac{\partial\rho(\mathbf{y},\mathbf{W})}{\partial\mathbf{W}}\mathbf{W}^{\mathrm{T}}\mathbf{W} = [\boldsymbol{I} - f(\mathbf{y})\mathbf{y}^{\mathrm{T}}]\mathbf{W} \tag{3.6.11}$$

This natural gradient modifies the search direction of the standard gradient, according to the Riemannian structure in parameter space, but does not move the local minimum point, thus ensuring the parameter estimation of the global optimum. The natural gradient has the isotropic convergence Property and no limitation of quadratic approximation on the objective function as the Newton method; therefore, it is suitable for more nonlinear models of objective functions and neural-networks.

The learning algorithm first proposed by Amari and Cichocki is

$$\begin{aligned} \Delta\mathbf{W}(k) &= -\eta(k)\frac{\partial\rho(\mathbf{y},\mathbf{W})}{\partial\mathbf{W}}\mathbf{W}^{\mathrm{T}}\mathbf{W} \\ &= \eta(k)[\boldsymbol{I} - f(\mathbf{y}(k))\mathbf{y}^{\mathrm{T}}(k)]\mathbf{W}(k) \end{aligned} \tag{3.6.12}$$

This learning rule is basically the same as the relative gradient algorithm proposed by Cardoso and Laheld, because the basic Riemannian measure is based on the Lie group structure, and the learning rule keeps an equivariant Property.

In order to solve the scale indeterminacy problem, some modifications are made to the basic natural gradient learning rule of Eq. (3.6.12). Learning rules impose some constraints on the recovered signal amplitudes, i.e. $E[f_i(y_i)y_i]=1$. However, when the sources are nonstationary signals, such as speech and biomedical signals, these constraints will introduce a fast variation of amplitudes of the separating matrix, which results in non-stability in some cases. To solve such a problem, nonholonomic constraints are introduced to the natural gradient learning algorithm. Such constraints are based on a geometric consideration, i.e. the variation direction of separating matrix $\boldsymbol{W}$ should be orthogonal to the separating matrix equivalence class resulting from the scale indeterminacy. Because these constraints are nonholonomic, the modified algorithm can adapt itself for the fast variation and/or intermittent variation of signal amplitude. This learning algorithm can be written as

$$\Delta \boldsymbol{W}(k) = \eta(k)[\boldsymbol{\Lambda}(k) - f(\boldsymbol{y}(k))\boldsymbol{y}^{\mathrm{T}}(k)]\boldsymbol{W}(k) \tag{3.6.13}$$

where $\boldsymbol{\Lambda}=\mathrm{diag}(\lambda_1, \lambda_2, \cdots, \lambda_n)$ is a diagonal positive definite matrix.

Another modification uses the modified gradient learning rule

$$\Delta \boldsymbol{W}(k) = \eta(k)[\boldsymbol{\Lambda}(k) - \boldsymbol{y}(k)\boldsymbol{g}^{\mathrm{T}}(\boldsymbol{y}(k))]\boldsymbol{W}(k) \tag{3.6.14}$$

The learning algorithm summarizing all of the above modifications is

$$\begin{aligned}\Delta \boldsymbol{W}(k) = \eta(k)[\boldsymbol{\Lambda}(k) - \alpha\{f(\boldsymbol{y}(k))\boldsymbol{y}^{\mathrm{T}}(k)\} + \\ \beta\{\boldsymbol{y}(k)\boldsymbol{g}^{\mathrm{T}}(\boldsymbol{y}(k))\}\boldsymbol{W}(k)\end{aligned} \tag{3.6.15}$$

where diagonal matrix $\boldsymbol{\Lambda}$ is used to eliminate the diagonal value inside the bracket and α and β are both proper parameters determined adaptively to ensure the best performance.

For the detailed discussions on necessary conditions of algorithm convergence and nonlinear parameter updating properties, please refer to Refs[8~11].

3.7 Hidden Markov Independent Component Analysis

For most signals in nature, no matter what the time sequence or image, they have their own nonstationary properties. Speech and EEG signals are both typical nonstationary signals. In Chapter 9, we will discuss how to use second-order statistics for separation of speech signals. This section introduces a nonstationary time sequence analysis method based on a probability generative framework, and this method combines the Hidden Markov Model (HMM) and ICA into a single generative model.

1. Hidden Markov Model

The Hidden Markov Model is a process that uses a statistical method to describe time-varying signals. In this model, the transition from one state to another state in the Markov chain depends on the statistical Property of that state (or state transition probability), and the observation produced by a state depends on the probability of the observation (state generating probability). In observation, the state transition is hidden; therefore, it is a double randomizing process[12~13].

HMM has N discrete hidden states $\boldsymbol{Q}=\{q_1, q_2, \cdots, q_N\}$, an initial state probability vector $\pi=[\pi_1, \pi_2, \cdots, \pi_N]$, a state transition probability matrix $\boldsymbol{A}=[a_{ij}]$, and a generative probability matrix $\boldsymbol{B}=[b_k(x_t)$ of a corresponding state k and a data point x_t. The observed parameters of the ICA model should be included in $\boldsymbol{B}$. All the parameters of HMM can be expressed as parameter set $\boldsymbol{\theta}=\{\boldsymbol{A}, \boldsymbol{B}, \pi\}$. At time t, giving the observation sequence $\boldsymbol{x}_1^T$, the probability at state k is $p_t(k)$. It should be pointed out that π only decides the initial state of observed HMM, and $\boldsymbol{B}$ is related directly to observations; thus, its influence is greatest.

HMM modeling uses original data to optimize and estimate the model parameters, which is also called the model training process. It includes the following three problems:

① Given the observation sequence $\boldsymbol{x}_1^T$ and $\boldsymbol{\theta}$, how can we compute probability distribution $p(\boldsymbol{x}_1^T\boldsymbol{\theta})$, which corresponds to determining the generative possibility of the observed sequence from model $\boldsymbol{\theta}$.

② Given the observation sequence $\boldsymbol{x}_1^T$, how can we determine a reasonable state sequence $\boldsymbol{I}=\{i_1, i_2, \cdots, i_T\}$to produce $\boldsymbol{x}_1^T$ optimally, such that $\boldsymbol{I}$ reveals the hidden state transferring rule.

③ For given $\boldsymbol{x}_1^T$, how can we adjust $\boldsymbol{\theta}$ elements to maximize $p\ (\boldsymbol{x}_1^T|\boldsymbol{\theta})$.

to solve the above problems, forward probability and backward probability algorithms are introduced.

the forward probability is defined as

$$\alpha_t(j) = p\{x_1, x_2, \cdots, x_t, i_t = q_j \mid \boldsymbol{\theta}\} \tag{3.7.1}$$

its recursive formula is

$$\begin{aligned} &\alpha_1(j) = \pi_j b_j(x_1) \\ &\alpha_{t+1}(j) = \Big[\sum_{i=1}^{N} \alpha_t(i) a_{ij}\Big] b_j(x_{t+1}) \qquad 1 \leqslant j \leqslant N, t = 1,2,\cdots,T-1 \end{aligned} \tag{3.7.2}$$

the backward probability is defined as

$$\beta_t(j) = p\{x_{t+1}, x_{t+2}, \cdots, x_T, i_t = q_j \mid \boldsymbol{\theta}\} \tag{3.7.3}$$

from the definition

$$\beta_t(j) = \sum_{j=1}^{N} a_{ij} b_j(x_{t+1}) \beta_{t+1}(j), \quad t = T-1, T-2, \cdots, 1; 1 \leqslant j \leqslant N \tag{3.7.4}$$

when $t=T$, suppose

$$\beta_T(j)=1$$

Solutions of above three problems by using forward and backward probability algorithms are respectively

①

$$p\{\boldsymbol{x}_1^{\mathrm{T}} \mid \boldsymbol{\theta}\} = \sum_{i=1}^{N} \sum_{j=1}^{N} \alpha_t(i) a_{ij} b_j(x_{t+1}) \beta_{t+1}(j) \qquad 1 \leqslant t \leqslant T-1 \tag{3.7.5}$$

or

$$p\{\boldsymbol{x}_1^T \mid \boldsymbol{\theta}\} = \sum_{i=1}^{N} \alpha_T(i) \tag{3.7.6}$$

② The problem is equivalent to $p\{\boldsymbol{x}_1^T, \boldsymbol{I}^*\} = \max_I \{\boldsymbol{x}_1^T, \boldsymbol{I}\}$, which can be solved by the Viterbi algorithm.

Step1:

$$\delta_1(i) = \pi_i b_i(x_1) \qquad 1 \leqslant i \leqslant N \tag{3.7.7}$$

Step2:

$$\delta_t(j) = \max_{1 \leqslant i \leqslant N} [\delta_{t-1}(i) a_{ij}] b_j(x_t) \qquad (2 \leqslant t \leqslant T; 1 \leqslant j \leqslant N) \tag{3.7.8}$$

Step3:

$$p\{\boldsymbol{x}^T, \boldsymbol{I}^*\} = \max_{1 \leqslant i \leqslant N} [\delta_t(i)] \tag{3.7.9}$$

$$i_T^* = \arg \max_{1 \leqslant i \leqslant N} [\delta_T(i)] \tag{3.7.10}$$

Step4: For $t=T-1, T-2, \cdots, 1$, solve the following equation:

$$i_T^* = \arg \max_{1 \leqslant i \leqslant N} [\delta_t(i) a_i i_{t+1}^*] \tag{3.7.11}$$

and finally determine the best state permutation $\boldsymbol{I}^* = (i_1^*, i_2^*, \cdots, i_T^*)$.

③ According to the definition, $\boldsymbol{\pi}$, $\boldsymbol{A}$, $\boldsymbol{B}$must satisfy the following three constraints.

$$\begin{gathered} \sum_{i=1}^{N} \pi_i = 1 \\ \sum_{j=1}^{N} a_{ij} = 1 \\ \sum_{k=1}^{N} b_i(k) = 1 \end{gathered} \tag{3.7.12}$$

Starting with the optimizing objective function, we solve

$$a_{ij} = \frac{\sum_{t=1}^{T-1} \alpha_t(i) a_{ij} b_j(x_{t+1}) \beta_{t+1}(j)}{\sum_{t=1}^{T-1} \alpha_t(i) \beta_t(i)} \qquad i, j = 1, 2, \cdots, N$$

$$b_i(k) = \frac{\sum \alpha_t(i)\beta_t(i)}{\sum_{t=1}^{T-1} \alpha_t(i)\beta_t(i)}$$

$$\pi_i = \frac{\alpha_1(i)\beta_1(i)}{\sum_{i=1}^{N} \alpha_1(i)\beta_1(i)} \qquad 1 \leqslant i \leqslant N \tag{3.7.13}$$

Given $\boldsymbol{\theta}$, the forward and backward probability algorithms[12] can be used to compute the probability of state k probability at time t.

2. Generative Model

In principle, the observation models can be any probability model such as Gaussian or non-Gaussian models in speech recognition. We will consider the ICA model now.

For M sources and N receivers, the observation variables can be expressed as

$$\boldsymbol{x}_t = \boldsymbol{A}\boldsymbol{s}_t \tag{3.7.14}$$

where $\boldsymbol{A}$ is mixing matrix, $\boldsymbol{s}_t$ is the source with components $s_i(t)$, expressed by the separating matrix

$$\boldsymbol{s}_t = \boldsymbol{W}\boldsymbol{x}_t \tag{3.7.15}$$

According to the maximum likelihood framework, the likelihood of observations can be computed from the likelihood of source density:

$$p(\boldsymbol{x}_t) = \frac{p(\boldsymbol{s}_t)}{|\boldsymbol{J}|} \tag{3.7.16}$$

where, the denominator is the absolute value of the Jacobian determinant. Because the transformation in ICA is linear, $|\boldsymbol{J}| = 1/|\det(\boldsymbol{W})|$ and $p(\boldsymbol{s}_t) = \prod_i p(s_i(t))$.

The above logarithm likelihood can be written as

$$\lg p(\boldsymbol{x}_t) = \lg|\det(\boldsymbol{W})| + \sum_{i=1}^{M} \lg p(s_i(t)) \tag{3.7.17}$$

For sub-Gaussian,Gaussian and super-Gaussian mixtures cases, a general density is exponential power distribution or generalized exponential (GE) functions:

$$p(s) = G(s; R, \beta) = \frac{R\beta^{1/R}}{2\Gamma(1/R)} \exp(-\beta|s|^R) \tag{3.7.18}$$

where $\Gamma(\cdot)$ is γ function, and the width of the density function is determined by $1/\beta$. When R=2, it is Gaussian distribution and when $R \to \infty$, it is a uniform distribution.

A hidden Markov ICA model has a demixing matrix $\boldsymbol{W}_k$ and density parameter vector $\boldsymbol{R}_k$, $\boldsymbol{\beta}_k$ corresponding to each state k, and an optimization process is required. Define an HMM auxiliary function

$$Q = \sum_k Q_k \tag{3.7.19}$$

where

$$Q_k = \lg|\det(\boldsymbol{W}_k)| + \frac{1}{p(k)}\sum_t p_t(k)\sum_t \lg p(s_i(t)) \tag{3.7.20}$$

$$p(k) = \sum_t p_t(k) \tag{3.7.21}$$

The demixing matrix learns according to the following gradient-based learning rule:

$$\frac{\partial Q_k}{\partial \boldsymbol{W}_k} = \boldsymbol{W}_k^{-\mathrm{T}} + \frac{1}{p(k)}\sum_t p_t(k)\boldsymbol{z}_t \boldsymbol{x}_t^{\mathrm{T}} \tag{3.7.22}$$

where $z_i(t) = \partial \lg p(s_i(t))/\partial s_i(t)$.

With right multiplication by $\boldsymbol{W}_k^{\mathrm{T}}\boldsymbol{W}_k$, the learning equation under the natural gradient is

$$\left[\frac{\partial Q_k}{\partial \boldsymbol{W}_k}\right] = \boldsymbol{W}_k + \frac{1}{p(k)}\sum_t p_t(k)\boldsymbol{z}_t \boldsymbol{y}_t^{\mathrm{T}} \tag{3.7.23}$$

where

$$\boldsymbol{y}_t = \boldsymbol{W}_k^{\mathrm{T}}\boldsymbol{s}_t$$

The above gradient is the same as that of standard ICA and each data point needs only to multiply the response $p_t(k)/p(k)$ corresponding to state k in the updated equation.

For a GE source, the logarithm likelihood is

$$\lg p(s) = \lg\left[\frac{R\beta^{1/R}}{2\Gamma(1/R)}\right] - \beta|\boldsymbol{s}|^R \tag{3.7.24}$$

Taking the partial derivative

$$z(\boldsymbol{s}) = \frac{\partial \lg p(\boldsymbol{s})}{\partial \boldsymbol{s}} = -\operatorname{sign}(\boldsymbol{s})R\beta|\boldsymbol{s}|^{R-1} \tag{3.7.25}$$

Substituting it into Eq. (3.7.23), we have

$$\left[\frac{\partial Q_k}{\partial \boldsymbol{W}_{ij}}\right] = \boldsymbol{W}_{ij} - \frac{1}{p(k)}\sum_t p_t(k)\operatorname{sign}(s_i(t))R_i\beta_i|s_i(t)|^{R-1}y_j(t) \tag{3.7.26}$$

where $y_j(t) = \sum_i \boldsymbol{W}_{ij}s_i(t)$.

Substituting Eq. (3.7.24) into Eq. (3.7.20), we obtain the updated expression of GE source parameters

$$Q_k = \lg|\det(\boldsymbol{W}_k)| + \frac{1}{p(k)}\sum_t p_t(k)\sum_i \lg\left[\frac{R_i\beta_i^{1/R_i}}{2\Gamma(1/R_i)}\right] - \beta_i|s_i(t)|^{R_i} \tag{3.7.27}$$

Estimate the partial derivatives $\partial Q_k/\partial \beta_i$ and $\partial Q_k/\partial R_i$ respectively, and we can then estimate the source density parameters at each state

$$\frac{1}{\beta_i} = \frac{R_i}{p(k)} \sum_t^T p_t(k) \mid s_i(t) \mid^{R_i} \tag{3.7.28}$$

$$\frac{1}{R_i} + \frac{1}{R_i^2} \lg \beta_i + \frac{1}{R_i^2} \varphi(1/R_i) - \frac{\beta_i}{p(k)} \sum_{t=1}^T p_t(k) \mid s_i(t) \mid^{R_i} \lg \mid s_i(t) \mid = 0 \tag{3.7.29}$$

where $\varphi(x) = \Gamma'(x)/\Gamma(x)$ is a double γ function.

References

[1] Comon P (1994) Independent component analysis, a new concept? Signal Processing 36: 287–314

[2] Cao X R, Liu R W (1996) General approach to blind source separation. IEEE Transactions on Signal Processing 44(3): 562–571

[3] Cardoso J F (1999) High-order contrast for independent component analysis. Neural Computation 11(1): 157–192

[4] Yang H H, Amari S I, Cichocki A (1997) Adaptive on-line learning algorithms for blind separation—Maximum entropy and minimum mutual information. Neural Computation 7(9): 1457–1482

[5] Lee T W (1998) ICA theory and applications. Kluwer Academic, Boston

[6] Robert S, Everson R (2001) Independent component analysis: Principles and practice. Cambridge University Press, New York

[7] Hyvarinen A, Oja E (1997) A fast fixed-point algorithm for independent component analysis. Neural Computation 9(7): 1483–1492

[8] Amari S I (1998) Natural gradient works efficiently in learning. Neural Computation 10(2): 251–276

[9] Amari S I (1997) Neural learning in structured parameter spaces—Natural Riemannian gradient. In: Advances in Neural Information Processing, Denver, Colorado, 1997, 9: 127–133.

[10] Amari S I, Douglas S C (1998) Why natural gradient? In: Proceedings of ICASSP'98, Seattle, Washington, 1998, 2: 1213–1216

[11] Douglas S C, Amari S I (2000) Natural gradient adaptation. In: Haykin S, (ed) Unsupervised adaptive filtering. Wiley, New York

[12] Rabiner L R (1989) A tutorial on hidden Markov models and selected

applications in speech recognition. Proceedings of the IEEE 77(2): 257–286

[13] Girolami M (2000) Advances in independent component analysis. Springer, New York

Chapter 4 Nonlinear PCA & Feature Extraction

4.1 Principal Component Analysis & Infinitesimal Analysis

The theorem of principal component analysis (PCA) was proposed by Pearson in 1901. As an important statistics method, it was applied by Hotelling in the research of correlation structures of random multi-variates in 1933. Oja proposed a fast algorithm for computing the eigenvalue and eigenvector of the covariance matrix and a Hebb learning rules-based neural networks model for adaptive extractions of maximum principal components in 1982[1,2]. Afterwards Sanger proposed a neural network model for adaptive extraction of multiple principal components. Kung proposed an APEX (Adaptive Principal Component Extraction) multiple principal component learning algorithm[3], which improved the efficiency of the PCA method. Related to linear transform, the basic idea of PCA is: Obtain a group of new features from the original features through linear transform, which keeps the same number of features, and the first few features contain the main information of the original features. Then by keeping the main information, the number of features decreases, and signal subspace is separated from noise subspace, so some of noise can be removed. Specifically, by the orthogonal eigenvector matrix of autocorrelation matrix of samples, we can reconstruct signals and remove correlations, then find the sample with the largest variance and also the highest energy.

Suppose n variables x_i with length N, which are not totally independent:

$$E(x_i) = 0, \quad i = 1, 2, \cdots, n; \quad \boldsymbol{X} = \{x_1, x_2, \cdots, x_n\}^{\mathrm{T}} \tag{4.1.1}$$

where $E(\cdot)$ is the expectations operation, x_i is the ith variable, $\boldsymbol{X}$ is $n \times N$ variable matrix, superscript T is the transpose operation. Let $\boldsymbol{R}_x$ be the autocorrelation matrix of n, not totally independent variables:

$$\boldsymbol{R}_x = E(\boldsymbol{X}\boldsymbol{X}^{\mathrm{T}}), \quad \boldsymbol{R}_x \boldsymbol{V} = \boldsymbol{V}\boldsymbol{\Lambda} \tag{4.1.2}$$

where $\boldsymbol{V}$ is $n \times n$ eigenvector matrix of $\boldsymbol{R}_x$, of which the column vectors are normalizing orthogonal; $\boldsymbol{\Lambda}$ is the diagonal matrix of the eigenvalues of $\boldsymbol{R}_x$, the

X. Shi, *Blind Signal Processing*

elements on the diagonal are λ_i, i=1, 2, ⋯, n. Then construct n uncorrelated new vectors:

$$\boldsymbol{Y} = \boldsymbol{V}^{\mathrm{T}}\boldsymbol{X}\ , \qquad \boldsymbol{Y} = \{y_1, y_2, \cdots, y_n\}^{\mathrm{T}} \tag{4.1.3}$$

where $\boldsymbol{Y}$ is $n \times N$ new matrix. To sum up, we have

$$\boldsymbol{R}_y = E(\boldsymbol{Y}\boldsymbol{Y}^{\mathrm{T}}) = E(\boldsymbol{V}^{\mathrm{T}}\boldsymbol{X}\boldsymbol{X}^{\mathrm{T}}\boldsymbol{V}) = \boldsymbol{V}^{\mathrm{T}}\boldsymbol{R}_x\boldsymbol{V} = \boldsymbol{V}^{\mathrm{T}}\boldsymbol{V}\boldsymbol{\Lambda} = \boldsymbol{\Lambda} \tag{4.1.4}$$

It can be concluded that the eigenvalues λ_i (i=1, 2, ⋯, n) are non-negative, and each subspace of each new variable is isolated and uncorrelated. PCA is implemented through eigenvector corresponding to large eigenvalues; however, infinitesimal analysis is implemented through those corresponding to smaller ones.

First, after arranging λ_i (i=1, 2, ⋯, n) in descending order, a definition of the amount of information η is given by

$$\eta = \frac{\lambda_p + \lambda_{p+1} + \cdots + \lambda_n}{\lambda_1 + \lambda_2 + \cdots + \lambda_n} \tag{4.1.5}$$

where the denominator is the sum of all eigenvalues, the numerator is the sum of cigenvalues from the pth to nth, of which in total the number is $m=n-p+1$. By Eq. (4.1.5), it can be seen clearly that the eigenvalues are the mean square of the samples of the reconstructed signal. Let

$$\tilde{\boldsymbol{Y}} = \boldsymbol{W}^{\mathrm{T}}\boldsymbol{X} = [y_p, y_{p+1}, \cdots, y_n]^{\mathrm{T}} \tag{4.1.6}$$

where $\boldsymbol{W}=\{v_p, v_{p+1}, \cdots, v_n\}$, v_i is the eigenvector corresponding to λ_i ($i=p$, $p+1$, ⋯, n), $\tilde{\boldsymbol{Y}}$ is the $m \times N$ matrix constructed by the eigenvectors corresponding to smaller eigenvalues, which have less information.

From Eq. (4.1.5), when selecting larger eigenvalue, the amount of information is larger, and the operation is PCA.

The infinitesimal analysis (or minor principal component analysis) is just opposite to PCA. While PCA is reconstructing the origin signal using the eigenvectors corresponding with large eigenvalues to extract the principal component, infinitesimal analysis is using those corresponding with small ones to extract the minor component. The calculations of infinitesimal analysis are similar with PCA; however, by selecting the amount of information, eigenvalues and corresponding eigenvectors are different.

As an example of a solution of neural networks—adaptive algorithm of neural networks extracting principal components, Kung posed an APEX (Adaptive Principal Component Extraction) multi-principal-component learning algorithm in 1994. Similarly, with lattice adaptive algorithm in principle, the weight of the following neurons is the recursion of the former ones. Thus, computational complexity is reduced greatly and computation can be realized more easily. The most significant feature of an adaptive algorithm of neural networks with principal component extraction is that the mth neuron can extract the largest principal component, which is orthogonal with the previous $(m-1)$ principal component denoted by $(m-1)$ trained neurons. The algorithm is shown in Fig. 4.1.

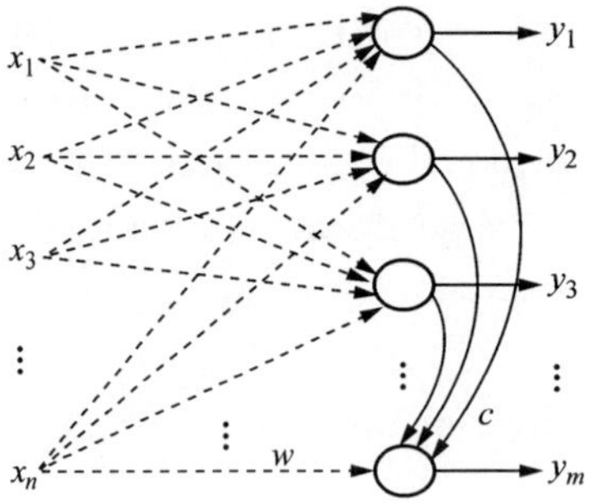

Fig. 4.1 APEX algorithm model

In the figure, $\boldsymbol{Y}(t)=[y_1(t), y_2(t), \cdots, y_{m-1}(t)]^{\mathrm{T}}$ is the output of corresponding neurons; $\boldsymbol{W}(t)=\{w_{ij}(t)\}(i=1, 2, \cdots, m-1; j=1, 2, \cdots, n)$ is corresponding with the connecting weight matrix of the previous $m-1$ neurons in the output layer; $\boldsymbol{W}_m(t)=[w_{m1}(t), w_{m2}(t), \cdots, w_{mn}(t)]$ is the connecting weight vector of the neurons in the output layer; and $\boldsymbol{C}(t)=[c_1(t), c_2(t), \cdots, c_{m-1}(t)]$ is the connecting weight vector between the previous $m-1$ neurons and the mth one.

The relation of the input/output of the networks is

$$\begin{aligned} \boldsymbol{Y}(t) &= \boldsymbol{W}(t)\boldsymbol{X}(t) \\ y_m(t) &= \boldsymbol{W}_m(t)\boldsymbol{X}(t) - \boldsymbol{C}(t)\boldsymbol{Y}(t) \end{aligned} \tag{4.1.7}$$

where

$$\boldsymbol{C} = \begin{bmatrix} 0 & \cdots & \cdots & \cdots & 0 \\ c_{21} & 0 & \cdots & \cdots & 0 \\ \vdots & & \ddots & & \vdots \\ c_{m-1,1} & \cdots & c_{m-1,m-2} & 0 & 0 \\ c_{m,1} & \cdots & \cdots & c_{m,m-1} & 0 \end{bmatrix}$$

The recursive algorithm of APEX is

$$\boldsymbol{W}_m(t+1) = \boldsymbol{W}_m + \beta[y_m(t)\boldsymbol{X}^{\mathrm{T}}(t) - y_m^2(t)\boldsymbol{W}_m(t)] \tag{4.1.8}$$

$$\boldsymbol{C}(t+1) = \boldsymbol{C}(t) + r[y_m(t)\boldsymbol{Y}^{\mathrm{T}}(t) - y_m^2\boldsymbol{C}(t)] \tag{4.1.9}$$

The APEX algorithm can be expressed as the following steps:

Step 1 Let $m=1$;

Step 2 Choose the initial value $\boldsymbol{W}_m(0)$ and $\boldsymbol{C}(0)$ randomly;

Step 3 Choose the proper step size for β and r;

Step 4 Calculate $\boldsymbol{W}_m(t)$ and $\boldsymbol{C}(t)$ by Eqs. (4.1.8) and (4.1.9);

Step 5 Discriminate $\|\boldsymbol{W}_m(t+1)-\boldsymbol{W}_m(t)\|$ and $\|\boldsymbol{C}(t+1)-\boldsymbol{C}(t)\|$;

Step 6 If the error is less than the preset value, let $m=m+1$, and go back to step 2;

Step 7 If the error is larger than the preset value, go back to step 4;

Step 8 When $m>p$ (p is the number of principal components required), stop searching.

4.2 Nonlinear PCA and Blind Source Separation

As discussed in the last section, the basic idea of PCA is to minimize the mean square error between m-dimension random vector $\boldsymbol{x}$ and its orthogonal projection on the lower dimension subspace. By denoting the orthogonal bases of the n-dimension subspace as $\boldsymbol{w}_1, \boldsymbol{w}_2, \cdots, \boldsymbol{w}_n$, the projection of $\boldsymbol{x}$ on subspace is $\sum_{i=1}^{n}(\boldsymbol{w}_i^{\mathrm{T}}\boldsymbol{x})\boldsymbol{w}_i$, then the minimal mean square error criterion of PCA is

$$\min E\left\{\left\|\boldsymbol{x}-\sum_{i=1}^{n}(\boldsymbol{w}_i^{\mathrm{T}}\boldsymbol{x})\boldsymbol{w}_i\right\|^2\right\} \tag{4.2.1}$$

The solution of the above optimization problem is the eigenvectors of the sample covariance matrix $\boldsymbol{C}_x=E(\boldsymbol{x}\boldsymbol{x}^{\mathrm{T}})$, which are denoted as $\boldsymbol{e}_1, \boldsymbol{e}_2, \cdots, \boldsymbol{e}_n$. The linear factor in the summation formula is the principal component $\boldsymbol{e}_i^{\mathrm{T}}\boldsymbol{x}$.

From past studies, the PCA method using neural learning criteria can be of different nonlinear expansion[4,5]. One specific form of non-linear PCA is suitable to blind source separation, and it is implemented through nonlinear minimal PCA criteria:

$$J(\boldsymbol{W}) = E\left\{\left\|\boldsymbol{x}-\sum_{i=1}^{n}g_i(\boldsymbol{w}_i^{\mathrm{T}}\boldsymbol{x})\boldsymbol{w}_i\right\|^2\right\} \tag{4.2.2}$$

which is proposed by Lei Xu and also called "minimal mean square reconstruction"[6]. Compared with Eq. (4.2.1), the nonlinear function $g_i(\cdot)$, which is usually some odd function such as $g(t)=\tanh(t)$ or $g(t)=t^3$, is applied on the linear factor.

For pre-whitening data, the nonlinear PCA criteria in Eq. (4.2.2) has a close relation with those criteria of kurtosis maximization/minimization, maximum likelihood, Bussgang, and so on. Suppose the dimension of the input vector $\boldsymbol{x}$ is decreased by whitening and vector $\boldsymbol{z}$ has a lower dimension, Equation (4.2.2) can be written as

$$J(\boldsymbol{W}) = E\{\|\boldsymbol{z}-\boldsymbol{W}^{\mathrm{T}}\boldsymbol{g}(\boldsymbol{W}\boldsymbol{z})\|^2\} \tag{4.2.3}$$

the function $\boldsymbol{g}(\boldsymbol{W}\boldsymbol{z})$is a column vector with elements $g_1(\boldsymbol{w}_1^{\mathrm{T}}\boldsymbol{z}), \cdots, g_n(\boldsymbol{w}_n^{\mathrm{T}}\boldsymbol{z})$. As $\boldsymbol{W}$ is $n\times n$ orthogonal matrix, where $\boldsymbol{W}\boldsymbol{W}^{\mathrm{T}}=\boldsymbol{W}^{\mathrm{T}}\boldsymbol{W}=\boldsymbol{I}$, the item in the parenthesis thus can be written as

$$\|\boldsymbol{z}-\boldsymbol{W}^{\mathrm{T}}\boldsymbol{g}(\boldsymbol{W}\boldsymbol{z})\|^2=[\boldsymbol{z}-\boldsymbol{W}^{\mathrm{T}}\boldsymbol{g}(\boldsymbol{W}\boldsymbol{z})]^{\mathrm{T}}[\boldsymbol{z}-\boldsymbol{W}^{\mathrm{T}}\boldsymbol{g}(\boldsymbol{W}\boldsymbol{z})]$$

$$= [z - W^{T}g(Wz)]^{T}W^{T}W[z - W^{T}g(Wz)]$$
$$= \| Wz - WW^{T}g(Wz) \|^{2}$$
$$= \| y - g(y) \|^{2}$$
$$= \sum_{i=1}^{n} [y_i - g_i(y_i)]^2$$

where $y = Wz$. Therefore, the criteria $J(W)$ turns to

$$J(W) = \sum_{i=1}^{n} E\{[y_i - g_i(y_i)]^2\} \tag{4.2.4}$$

Suppose $g_i(y)$ is the quadratic odd function

$$g_i(y) = \begin{cases} y^2 + y, & \text{if } y \geqslant 0 \\ -y^2 + y, & \text{if } y < 0 \end{cases}$$

The form of criteria (4.2.4) is

$$J(W) = \sum_{i=1}^{n} E\{(y_i - y_i \pm y_i^2)^2\} = \sum_{i=1}^{n} E(y_i^4) \tag{4.2.5}$$

which is just the kurtosis criteria. This means the nonlinear minimal PCA criteria is equivalent to minimization of the kurtosis sum of y_i for this function.

Meanwhile, for the maximum likelihood methods,

$$\lg L(W) = \sum_{k=1}^{K} \sum_{i=1}^{n} \lg p_i(w_i^{T} z(t)) \tag{4.2.6}$$

where K in the summation symbol represents the length of input samples. For large K, it tends to

$$J(W) = \sum_{i=1}^{n} E\{\lg p_i(y_i)\} \tag{4.2.7}$$

where $y_i = w_i^{T} z$. By criteria equation (4.2.4), those two criteria have an equivalent relation

$$\lg p_i(y) = \alpha - \beta[y - g_i(y)]^2 \tag{4.2.8}$$

where α and β are positive constants, which do not affect the expression of criteria term (4.2.4). Therefore, for any given density $p_i(y)$, the problem is to find a proper nonlinear function $g_i(y)$.

Similarly, the equivalent relation with the Bussgang method in blind equalization can be derived, which can be found in Ref. [5].

Once the nonlinear function $g_i(y)$ is determined, the next problem is the solution of the minimization of nonlinear PCA criteria. One simple random gradient type learning algorithm is the nonlinear subspace criteria. It is

$$\Delta W = \mu[F(Wx)Wrx^{T} + g(Wx)r^{T}] \tag{4.2.9}$$

where $r = x - W^{T}g(Wx)$ is the residual error, $F(Wx) = \mathrm{diag}\,[g'(w_1^{T}x), \cdots, g'(w_n^{T}x)]$, and $g'(y)$ is the derivative of $g(y)$. If the residual error is much smaller than the input vector, the first term in the right parentheses can be ignored and we have the learning rule:

$$\Delta W = \mu g(y)[x^{\mathrm{T}} - g(y^{\mathrm{T}})W] \tag{4.2.10}$$

The criteria equation (4.2.2), as well as this learning rule, can be applied to the separation of signals. It is better to separate the input by whitening preprocessing than without preprocessing, as it is closer to the objective function of ICA.

4.3 Kernel PCA

Kernel PCA, which was proposed by Scholkopf *et al* [8,9], is a nonlinear PCA method based on a support vector machine[10]. Its basic idea is that for the observation data with nonlinear changes, it may usually be mapped to a high dimensional linear space of F. Through the support vector machine method, a mapping from input space to F is constructed with simple kernel functions to find a computable solution for completing the nonlinear PCA in input space.

The goal of the decorrelation of data is achieved by diagonalization of the covariance matrix. Because the feature space is linear after mapping, the covariance matrix can be written as

$$\boldsymbol{C} = \frac{1}{M}\sum_{j=1}^{M}\Phi(\boldsymbol{x}_j)\Phi(\boldsymbol{x}_j)^{\mathrm{T}} \tag{4.3.1}$$

where $\Phi(\cdot)$ is a nonlinear function which maps the input vector from the input space onto F space. To diagonalize the covariance matrix, the eigenvalue problem should be solved in the F space, i.e. $\lambda \boldsymbol{V}=\boldsymbol{C}\boldsymbol{V}$. Let the eigenvalue $\lambda \geqslant 0$ and eigenvector $\boldsymbol{V} \in F$, with all the solutions of $\boldsymbol{V}$ in the spanning space by $\Phi(\boldsymbol{x}_1), \cdots, \Phi(\boldsymbol{x}_M)$. This problem is equivalent to

$$\lambda(\Phi(\boldsymbol{x}_k)\cdot \boldsymbol{V}) = (\Phi(\boldsymbol{x}_k \cdot \boldsymbol{C}\boldsymbol{V})) \qquad k=1,\cdots,M \tag{4.3.2}$$

and existing coefficients $\alpha_1, \cdots, \alpha_M$, that

$$\boldsymbol{V} = \sum_{i=1}^{M}\alpha_i\Phi(\boldsymbol{x}_i) \tag{4.3.3}$$

Substituting Eqs. (4.3.1) and (4.3.3) into Eq. (4.3.2):

$$\lambda\sum_{i=1}^{M}\alpha_i(\Phi(\boldsymbol{x}_k)\cdot\Phi(\boldsymbol{x}_i))$$
$$=\frac{1}{M}\sum_{i=1}^{M}\alpha_i\Big(\sum_{j=1}^{M}(\Phi(\boldsymbol{x}_k)\cdot\Phi(\boldsymbol{x}_j))(\Phi(\boldsymbol{x}_j)\cdot\Phi(\boldsymbol{x}_i))\Big) \tag{4.3.4}$$

It needs to be pointed out that this eigenvalue problem only includes the dot product of the mapping vector in the eigenspace. While the eigenspace is of a high dimension, the computing time for the dot product is very long. Thus the kernel function of the support vector machine is introduced, so that neither the completeness of mapping $\Phi(\cdot)$ nor the dot product in higher dimensional eigenspace F is necessary.

Define an $M \times M$ matrix $\boldsymbol{K}$, with elements

$$k_{ij} \triangleq (\Phi(\boldsymbol{x}_i) \cdot \Phi(\boldsymbol{x}_j)) \tag{4.3.5}$$

$\boldsymbol{K}$ is symmetric and Eq. (4.3.4) can be rewritten as

$$M\lambda\boldsymbol{\alpha} = \boldsymbol{K\alpha} \tag{4.3.6}$$

where column vector $\boldsymbol{\alpha}=[\alpha_1, \cdots, \alpha_M]^{\mathrm{T}}$. PCA in eigenspace F is equivalent to solving the eigenvalue problem of Eq. (4.3.6), which results in the eigenvector $\boldsymbol{\alpha}^1, \boldsymbol{\alpha}^2, \cdots, \boldsymbol{\alpha}^M$ corresponding to eigenvalues $\lambda^1 \geqslant \lambda^2 \geqslant \cdots \geqslant \lambda^M$. The dimension will be decreased by retaining only the first L eigenvectors.

To extract the principal component, the projection of one measuring point $\Phi(\boldsymbol{x})$ on the eigenvector in F should be calculated:

$$(\boldsymbol{V}^k \cdot \Phi(x)) = \sum_{i=1}^{M} \alpha_i^k (\Phi(\boldsymbol{x}_i) \cdot \Phi(\boldsymbol{x})) \tag{4.3.7}$$

and for calculating the projection of Eqs. (4.3.6) and (4.3.7), construct a kernel function

$$\kappa(\boldsymbol{x}, \boldsymbol{y}) = \Phi(\boldsymbol{x}) \cdot \Phi(\boldsymbol{y}) \tag{4.3.8}$$

where $\boldsymbol{x}$ and $\boldsymbol{y}$ are a pair of vectors of the input space, and the kernel $\kappa(\boldsymbol{x}, \boldsymbol{y})$ can be chosen by a functional analysis method[11]. The kernel used in a support vector machine includes the polynomial kernel $\kappa(\boldsymbol{x}, \boldsymbol{y})=(\boldsymbol{x} \cdot \boldsymbol{y})^d$, radial basis function $\kappa(\boldsymbol{x}, \boldsymbol{y})=\exp(-\|\boldsymbol{x}-\boldsymbol{y}\|^2/2\sigma^2)$, and Sigmoid kernel $\kappa(\boldsymbol{x}, \boldsymbol{y})=\tanh(k(\boldsymbol{x}, \boldsymbol{y})+\theta)$. The polynomial kernel of exponent d corresponds with mapping to the eigenspace, which is spanned by all products of the d table values of the input pattern.

Figure 4.2 illustrates the basic concepts of kernel PCA.

The entire process of kernel PCA is:

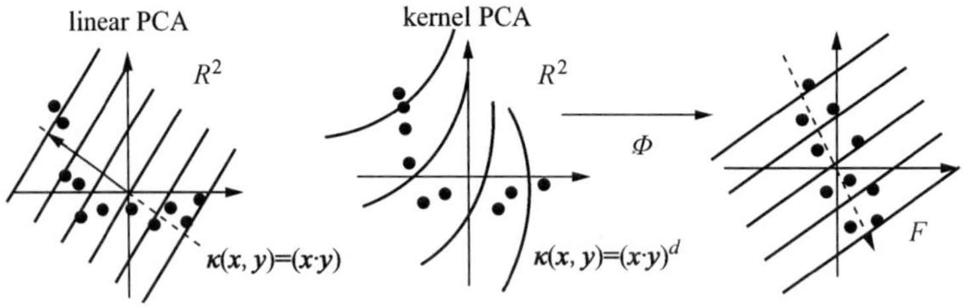

Fig. 4.2 Basic concept of kernel PCA

① Calculate the dot product matrix of Eq. (4.3.5);

② Diagonalize $\boldsymbol{K}$, solve Eq. (4.3.6);

③ Normalize the eigenvector, expand coefficient α^n;

④ Calculate the projection of a measuring point $\Phi(\boldsymbol{x})$ on the eigenvector in F by Eq. (4.3.7), and extract the principal component corresponding to kernel κ.

4.4 Neural Networks Method of Nonlinear PCA and Nonlinear Complex PCA

4.4.1 Basic Note

Nonlinear PCA is introduced in the literature of chemical engineering by Kramer in 1991. As a classical multivariate statistical analysis method, PCA is applied to cases, where the amount of data is large to decrease the dimension of the data set and extract features. For a linear method, it is impossible to explain the greatest global variance of data and to approach the local data clustering at the same time, but the nonlinear method can integrate them. As described in the previous section, Oja, Diamantaras and Kung proposed the neural networks method of PCA. Similarly, a classical PCA can be nonlinearly generalized by a neural networks model to process a nonlinear PCA. Because several local minimums exist in the evaluation function, the optimization starting from different initial parameters usually converges to different minimums and results in the instability of nonlinear PCA. One solution is to normalize the evaluation function with a weighting compensation[12], which will be explained in this section.

4.4.2 Nonlinear PCA

For data form $\boldsymbol{x}(t)= [x_1, x_2, \cdots, x_1]$, where each variable $x_i(i=1, 2, \cdots, l)$ is a time sequence including n measurments, PCA can be considered as the linear combination of x_i and vector $\boldsymbol{a}$

$$u(t)=\boldsymbol{a} \cdot \boldsymbol{x}(t) \tag{4.4.1}$$

which results in the minimum of

$$\langle \| \boldsymbol{x}(t) - \boldsymbol{a}u(t) \|^2 \rangle \tag{4.4.2}$$

in the formula, $\langle \cdot \rangle$ denotes sample or time average, u is the first principal component, a time sequence, and $\boldsymbol{a}$ is the first eigenvector of the data covariance matrix, describing a space pattern. Similarly, the second principal and so on can be extracted from the residue $\boldsymbol{x}-\boldsymbol{a}u$.

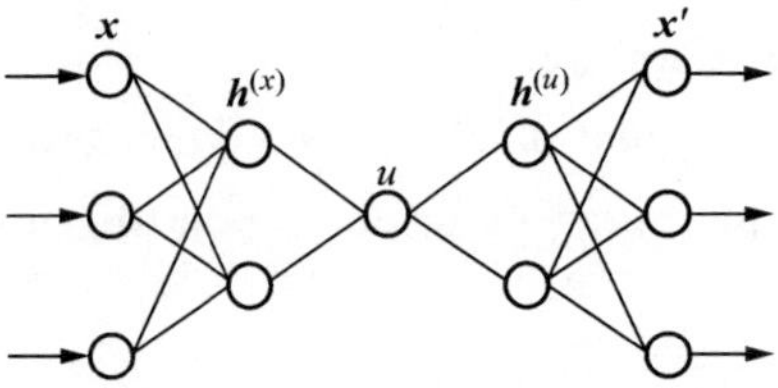

Fig. 4.3 Neural networks model for nonlinear PCA

The main difference between PCA and nonlinear PCA is that while nonlinear PCA allows the nonlinear mapping from $\boldsymbol{x}$ onto $\boldsymbol{u}$, PCA only allows linear mapping. In the following discussion, the neural networks nonlinear PCA has three hidden layers besides the input and output layer, as shown in Fig. 4.3.

In this model, the transfer function f_1 mapping from the input column vector $\boldsymbol{x}$ with long l to the first hidden layer is denoted by the column vector $\boldsymbol{h}^{(x)}$ with a long m:

$$h_k^{(x)} = f_1((\boldsymbol{W}^{(x)}\boldsymbol{x}+\boldsymbol{b}^{(x)})_k) \tag{4.4.3}$$

where $\boldsymbol{W}$ is $m \times l$ weight matrix and $\boldsymbol{b}$ is the bias parameter column vector. The transfer function f_2 mapping from the first hidden layer to the second hidden layer with a unique neuron generates the nonlinear principal component u:

$$u = f_2(\boldsymbol{w}^{(x)}\boldsymbol{h}^{(x)}+\overline{\boldsymbol{b}}^{(x)}) \tag{4.4.4}$$

where $\overline{\boldsymbol{b}}^{(x)}$ is the offset parameter of a single neuron.

The mapping from the second hidden layer u to the third hidden layer $\boldsymbol{h}^{(u)}$ generates

$$h_k^{(u)} = f_3((\boldsymbol{w}^{(u)}u+\boldsymbol{b}^{(u)})_k), \quad k=1,2,\cdots,m \tag{4.4.5}$$

so then the output column vector $\boldsymbol{x}'$ with long l is generated from $\boldsymbol{h}^{(u)}$ through f_4 mapping:

$$x'_i = f_4((\boldsymbol{W}^{(u)}\boldsymbol{h}^{(u)}+\overline{\boldsymbol{b}}^{(u)})_i) \tag{4.4.6}$$

where $\overline{\boldsymbol{b}}^{(u)}$ is the offset column vector of output layer.

By the mean square error criteria, the evaluation function $J=\langle \| \boldsymbol{x}-\boldsymbol{x}' \|^2\rangle$ should be minimized for the parameter set $\boldsymbol{W}^{(x)}$, $\boldsymbol{b}^{(x)}$, $\boldsymbol{w}^{(x)}$, $\overline{\boldsymbol{b}}^{(x)}$, $\boldsymbol{w}^{(u)}$, $\boldsymbol{b}^{(u)}$, $\boldsymbol{W}^{(u)}$ and $\overline{\boldsymbol{b}}^{(u)}$.

In the above mapping, f_1 and f_3 are nonlinear functions, which can be chosen as hyperbolic tangents, and f_2, f_4 usually are chosen as identity functions. Applying normalization condition $\langle u^2\rangle =1$, the evaluation function can be approximately

$$J = \langle \| \boldsymbol{x}-\boldsymbol{x}' \|^2\rangle + (\langle u^2\rangle - 1)^2 \tag{4.4.7}$$

the choice of the hidden unity number m in the model should adhere to the principle of simplicity.Though the increase of m can improve the ability of the model, it results in over-fitting the noise in the model. The generalization of Eq. (4.4.1) can be written as $\boldsymbol{u}=f(\boldsymbol{x})$, where f is a nonlinear mapping of feed-forward networks. If the nonlinear function of output layer is $\boldsymbol{g}$, the remaining amount $\boldsymbol{x}-\boldsymbol{g}(u)$ can be input to the same network for extracting the second nonlinear principal component, and so on to the high order principal component.

To solve the problem of local minima, the evaluation function is standardized by weighting compensation to restrict the nonlinearity of solution. The modified evaluation function is

$$J = \langle \| \boldsymbol{x} - \boldsymbol{x}' \|^2 \rangle + (\langle u^2 \rangle - 1)^2 + p \sum_{ki} (w_{ki}^x)^2 \tag{4.4.8}$$

where p is the weighting compensation parameter, and larger p increases the concavity of the evaluation function, by which the $\boldsymbol{W}^{(x)}$ is forced to decrease amplitude and leads to a more smooth and less nonlinear solution than smaller p. For large enough p, the possibility of over-fitting is greatly reduced, and the solution of the optimization process is guaranteed to converge to a global minimum. It is indicated by experiment that p of about 1 is suitable.

4.4.3 Complex Nonlinear PCA

The above nonlinear PCA can be generalized to complex nonlinear PCA, so that it can be applied to complex number data, a two-dimensional vector field and the real number data complexificated by its Hilbert transform, to extract the nonlinear feature and decrease the dimension. Complex nonlinear PCA uses the same neural networks structures as well; however, the weight parameter and bias parameter are all complex variables. Comparing the real number model with the same number of parameters, the first pattern of complex nonlinear PCA is a much smaller mean square error. Refering to the conventional method of singular value decomposition in PCA, the singular value decomposition of complex $\boldsymbol{Z}=\boldsymbol{X}+\mathrm{j}\boldsymbol{Y}$ is

$$\boldsymbol{Z} = \boldsymbol{U}\boldsymbol{\Lambda}_r\boldsymbol{V}^{\mathrm{H}} \tag{4.4.9}$$

where $\boldsymbol{Z}$ is $m \times n$ complex matrix, m is the number of variables and n is the number of measurement. Suppose $m \leqslant n$, the rank $r \leqslant m$. $\boldsymbol{U}$ is $m \times r$ matrix, of which the columns are the eigenvectors of $\boldsymbol{Z}\boldsymbol{Z}^{\mathrm{H}}$. $\boldsymbol{V}$ is $n \times r$ matrix, of which the columns are the eigenvectors of $\boldsymbol{Z}^{\mathrm{H}}\boldsymbol{Z}$. $\boldsymbol{\Lambda}_r$ is $r \times r$ matrix, of which the elements on the diagonal are singular values computed from the square roots of the r non-zero eigenvalues of $\boldsymbol{Z}\boldsymbol{Z}^{\mathrm{H}}$ and $\boldsymbol{Z}^{\mathrm{H}}\boldsymbol{Z}$. $\boldsymbol{V}^{\mathrm{H}}$ is the complex conjugation transpose of $\boldsymbol{V}$. As the linear correlation of m-r columns of $\boldsymbol{Z}$, the rank $r \leqslant m$, the jth row of

$\boldsymbol{\Lambda}_r\boldsymbol{V}^{\mathrm{H}}$ gives the score of the complex principal component or the jth pattern.

All features explained by $\boldsymbol{Z}$ can be described by a subspace spanned by the r linearly independent columns of $\boldsymbol{V}$. Therefore, there exists a transform described by the complex function $\boldsymbol{G}$, which projects the r coordinate of the row subspace of $\boldsymbol{Z}$ given by $\boldsymbol{\Lambda}_r\boldsymbol{V}^{\mathrm{H}}$ to the predicted value matrix $\boldsymbol{Z}_{\mathrm{pred}}\,(\boldsymbol{m}\times\boldsymbol{n})$

$$\boldsymbol{Z}_{\mathrm{pred}} = \boldsymbol{G}(\boldsymbol{\Lambda}, \boldsymbol{V}^{\mathrm{H}}) \tag{4.4.10}$$

for complex principal component analysis, the transform $\boldsymbol{G}(\boldsymbol{\Lambda}_r, \boldsymbol{V}^{\mathrm{H}})$ generates complex principal component $\lambda_j\boldsymbol{v}_j^{\mathrm{H}}$ corresponding to the $r\,(m\times n)$ rank 1 matrix and correlation for each eigenvector $\boldsymbol{u}_j$

$$\boldsymbol{Z}_{\mathrm{pred}} = \sum_{j=1}^{r} \boldsymbol{u}_j\lambda_j\boldsymbol{v}_j^{\mathrm{H}} \tag{4.4.11}$$

in which, the first pattern explains the variance of the maximum in Z, and then the second pattern until pattern r explains the variance of the minimum.

If the column space of error matrix $\boldsymbol{Y}=(\boldsymbol{Z}-\boldsymbol{G}\,(\boldsymbol{\Lambda}_r\boldsymbol{V})^{\mathrm{H}})$ is perpendicular to the subspace of $\boldsymbol{G}(\boldsymbol{\Lambda}_r\boldsymbol{V}^{\mathrm{H}})$, the minimum length solution is between predicted values $\boldsymbol{Z}_{\mathrm{pred}}$ and $\boldsymbol{Z}$. By the meaning of least square, it is equivalent to the minimization of the error quadratic sum of the object function or evaluation function J

$$J = \sum_{i=1}^{m}\sum_{j=1}^{n} |\,(\boldsymbol{Z})_{ij} - (\boldsymbol{G}(\boldsymbol{\Lambda}_r\boldsymbol{V}^{\mathrm{H}}))_{ij}\,|^2 \tag{4.4.12}$$

for linear function $\boldsymbol{G}$, this equation can be solved through Eq. (4.4.11); however, it can only be done by neural networks for nonlinear $\boldsymbol{G}$. The self-associated feedforward neural network structure, which is suitable for a complex domain, also can be applied in the nonlinear generalization of complex PCA, as shown in Fig. 4.4[13,14].

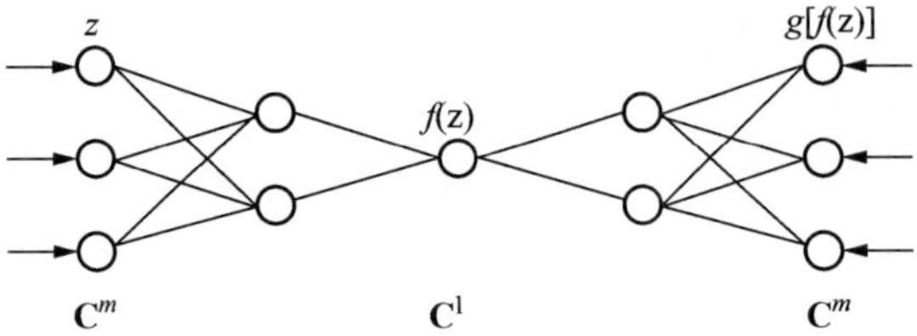

Fig. 4.4 The structure of neural networks for nonlinear complex PCA

This structure includes forward mapping f: $\mathscr{C}^m \to \mathscr{C}^1$, in which the bottleneck neuron gives out the nonlinear complex principal component, and the backward mapping g: $\mathscr{C}^1 \to \mathscr{C}^m$, which returns to the original data space from the nonlinear complex principal component. For nonlinear complex PCA, f and g are nonlinear continuous functions. Different from real domain, the unboundedness of the analysis function should be noticed in the complex region. For example, the widely used nonlinear transfer function is hyperbolic tangent; however, the singularity of tanh (z) in each $\left(\frac{1}{2}+l\right)\pi$, $l \in \mathcal{N}$ will lead to a divergent solution if a constraint is not applied[15]. To guarantee the boundedness and analysis of the transfer function almost everywhere, some constraints can be applied to $\boldsymbol{Z}$. If the magnitude of $\boldsymbol{Z}$ is in a circle with radius $\pi/2$, the singularity will not cause any problem and the boundedness is satisfied. In practice, the dot product of input and weight vector may be larger than $\pi/2$; thus, their magnitude should be constrained.

With implementation, the scale of input data is derived by dividing each element with the maximum amplitude of the rth row elements of $\boldsymbol{Z}$, and the weight of the hidden layer is randomly initialized with small amplitude. Restricting the dot product of the input vector and weight vector to about 0.1, and choosing the weighting compensation item p between 0.01 and 0.1 yields,

$$J=\sum_{j=1}^{n} \| z_j - z'_j \|^2 + p\Big(\sum_{l=1}^{q} \| \boldsymbol{W}_l^{(1)} \|^2 + \| \boldsymbol{W}_l^{(2)} \|^2 + \sum_{l=1}^{q} \| \boldsymbol{W}_l^{(3)} \|^2\Big) \tag{4.4.13}$$

where $\boldsymbol{W}^{(i)}$, $i=1, \cdots, 3$, is the weight vector of each layer, and q is the number of hidden layer neurons, appropriately between 3 and 6. As J is a real function with complex weight, the optimization of J is equivalent to finding the minimum gradient of J about the real and imaginary parts of weight. Therefore, the derivative should be applied on the real and imaginary parts respectively.

References

[1] Oja E (1982) A simplified neuron model as a principal component analyzer. Journal of Mathematical Biology 15: 267–273

[2] Oja E (1992) Principal components, minor components, and linear neural networks. Neural Networks 5: 927–935

[3] Diamantaras K I, Kung S Y (1996) Principal component neural networks: Theory and Applications. Wiley, New York

[4] Oja E (1997) The nonlinear PCA learning rule in independent components analysis. Neurocomputing 17(1): 25–46

[5] Karhunen J, Pajunen P, Oja E (1998) The nonlinear PCA criterion in blind source separation: Relations with other approaches. Neurocomputing 22(1): 5–20

[6] Xu L (1993) Least mean square error reconstruction principle for self-organizing neural nets. Neural Networks 6: 627–648

[7] Jollife I (1986) Principal components analysis. Springer, New York

[8] Scholkopf B, Smola A, Muller K (1998) Nonlinear component analysis as a kernel eigenvalue problem. Neural Computation 10(5): 1299–1319

[9] Scholkopf B, Smola A, Muller K (1999) Kernel principal component analysis. In: Advances in Kernel Methods. The MIT Press, Cambridge, MA

[10] Vipnik V (1995) The nature of statistical learning theory. Springer, New York

[11] Romdhani S, Gong S, Psarrou A (1999) A multi-view nonlinear active shape model using kernel PCA. In: BMVC.Nottingham, 1999, pp 483–492

[12] Hsieh W W (2001) Nonlinear principal component analysis by neural networks. Tellus 53A: 599–615

[13] Kim T, Adali T (2002) Fully complex multi-layer perceptron network for nonlinear signal processing. Journal of VLSI Signal Processing 32(1): 29–43

[14] Rattan S P, Hsieh W W (2005) Complex valued neural networks for nonlinear complex principal components analysis. Neural Networks 18(1): 61–69

[15] Nitta T (1997) An extension of the back-propagation algorithm to complex numbers. Neural Networks 10(8): 1391–1415

Chapter 5 Nonlinear ICA

5.1 Nonlinear Model and Source Separation

In the former chapters, it is assumed that the mixing model is linear; however, it is very different in practice since the mixture is always nonlinear mixture. Therefore, some error will be introduced into the solution. The difficulty in solving the nonlinear problem without any constraints is that it is exactly non-unique. The nonlinear ICA problem is much more complex than the linear model, so that a significant solution cannot be obtained by the assumption that the resources are statistically cannot mutually independent from each other. Thus, it is reasonable to make some restriction on the nonlinear function or mixing process. In recent years, many researchers have begun to study the nonlinear mixing model of ICA, such as Burel, 1992; Hermann and Yang, 1997; Lee *et al*, 1997; Lin and Cowan, 1997; Lin and Cowan, 1997; Table and Jutten, 1997; Hyvarinen and Oja, 1998; Hochreiter and Schemidhuber, 1998; Hyvarinen and Pajunen, 1999; Jutten, 2000; Jutten and Karhunen, 2003; Theis, 2003; Almeida, 2003, etc[1~5].

There are two different ways proposed for studying the nonlinear ICA . One is an expansion of linear ICA, that is, adding a nonlinear mixing model to the linear model, where the solution procedure is to get the inverse of the linear model and nonlinear model respectively. The other is to extract the nonlinear characteristic from data by using Self-Organizing Maps—SOM.

Since the basic linear ICA or BSS model is

$$\boldsymbol{x}=\boldsymbol{A}\boldsymbol{s} \tag{5.1.1}$$

The instantaneous nonlinear mixed model can be written as

$$\boldsymbol{x}=\boldsymbol{f}(\boldsymbol{s}) \tag{5.1.2}$$

where $\boldsymbol{x}$ is the observed M-dimensional mixed signal, $\boldsymbol{f}$ is the unknown real-valued mixing function with M components, and $\boldsymbol{s}$ is the N unknown statistically independent source signals. For simplification, let N=M; that is, suppose the number of source signals is equal to that of the mixed signals. Solving the nonlinear problem consists of finding a mapping $\boldsymbol{g}$: $\mathbf{R}^N \rightarrow \mathbf{R}^N$, to make

$$y = g(x) = \hat{s} \tag{5.1.3}$$

namely, to estimate the statistically independent sources.

The basic characteristics of nonlinear ICA is that although a solution exists, there are infinite solutions instead of a unique one. The simple reason is that if x and y are two independent random variables, for any measurable functions f and g, $f(x)$ and $g(y)$ are still mutually independent random variables. Finding an original source signal is different from finding the independent components in a nonlinear situation, since mutually independent output signals sometimes can't reflect the valuable information of the source signal themselves. The goal of BSS is to find the resource signals which are composed of the mixtures, but the independent components found from $\boldsymbol{x}$ may be quite different from the real ones. That is why ICA has the same solution as BSS in a linear model but not in a nonlinear model.

In a nonlinear model, the first model mentioned above is a model with real meaning named the Post Nonlinear-PNL-model[6], which can be solvable. The mixing models are listed as follows:

$$\begin{aligned} &\boldsymbol{x}=\boldsymbol{A}\boldsymbol{s} \\ &\boldsymbol{t}=\boldsymbol{f}(\boldsymbol{x}) \\ &\boldsymbol{z}=\boldsymbol{g}(\boldsymbol{t}) \\ &\boldsymbol{y}=\boldsymbol{W}\boldsymbol{z}=\boldsymbol{W}\boldsymbol{g}[\boldsymbol{f}(\boldsymbol{A}\boldsymbol{s})] \end{aligned} \tag{5.1.4}$$

where, $\boldsymbol{s}=[s_1, s_2,\cdots,s_N]^{\mathrm{T}}$ denotes source signal, $\boldsymbol{x}=[x_1, x_2, \cdots, x_N]^{\mathrm{T}}$ denotes the linear mixed signal, $\boldsymbol{t}=[t_1, t_2, \cdots, t_N]^{\mathrm{T}}$ denotes the nonlinear mixed observation signal, $\boldsymbol{z}=[z_1, z_2, \cdots, z_N]^{\mathrm{T}}$ is the linear mixed signal after nonlinear demixing, $\boldsymbol{y}=[y_1, y_2, \cdots, y_N]^{\mathrm{T}}$ is the estimation of the source signal obtained by linear demixing, $\boldsymbol{f}=[f_1(x_1), f_2(x_2), \cdots, f_N(x_N)]^{\mathrm{T}}$ is the nonlinear mixing function, $\boldsymbol{g}=[g_1(t_1), g_2(t_2), \cdots, g_N(t_N)]^{\mathrm{T}}$ is the nonlinear demixing function, $\boldsymbol{A}$ is the mixing matrix, and $\boldsymbol{W}$ is the demixing matirx. Figure 5.1 shows the whole mixing system, where the mixing part includes linear mixing matrix $\boldsymbol{A}$ and nonlinear function $\boldsymbol{f}(\cdot)$, and the demixing part includes nonlinear function $\boldsymbol{g}(\cdot)$ and demixing matrix $\boldsymbol{W}$.

Although the nonlinear mixed source cannot be separated only based on statistical independence of signals, one can separate the mixtures by choosing a suitable separation system for the post-nonlinear mixture. A natural method is minimizing the independence between signals; for instance, you can calculate the mutual information of the sources to estimate the parameters of the separation

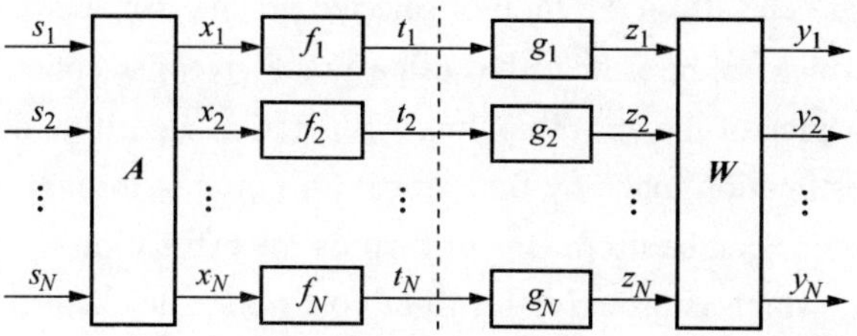

Fig. 5.1 The mixing model and demixing model of post-nonlinear ICA

system. The separation algorithm can be got by using weaker prior knowledge, such as the time correlation or nonstationarity of the source signal.

5.2 Learning Algorithm

Taleb and Jutten first presented the source separation algorithm for the PNL model[2], which is composed of two stages:

① Nonlinear stage: A nonlinear function g_i, which is the inversion of nonlinear function f_i, $i=1, 2, \cdots, N$, is needed. The parameter θ_i in $g_i(\theta_i)$ should be estimated in order to eliminate the distortion made by f_i.

② Linear stage: Separate the approximate linear mixture z obtained by the nonlinear stage to make each component of $\boldsymbol{y}=\boldsymbol{W}\boldsymbol{z}$ as independent as possible.

Mutual information among the components of output $y_1, y_2, \cdots, y_N$ is used as the evaluation function and the measure of independency. For the linear stage, minimizing the mutual information can give a component algorithm that is similar to the basic model.

$$\frac{\partial I(\boldsymbol{y})}{\partial \boldsymbol{W}}=-E\{\boldsymbol{\varphi}\boldsymbol{x}^{\mathrm{T}}\}-\{\boldsymbol{W}^{\mathrm{T}}\}^{-1} \tag{5.2.1}$$

where the component φ_i in vector $\boldsymbol{\varphi}$ is the evaluation function of each of the components y_i in output vector $\boldsymbol{y}$:

$$\varphi_i(y_i)=\frac{\mathrm{d}}{\mathrm{d}y}\lg p_i(y_i)=\frac{p'_i(y_i)}{p_i(y_i)} \tag{5.2.2}$$

here, $p_i(y_i)$ is the probability density function of y_i. In practice, we use the natural gradient method (3.6.12) instead of Eq. (5.2.1).

In the nonlinear stage, the rules of gradient learning are

$$\frac{\partial I(\boldsymbol{y})}{\partial \theta_k}=-E\left\{\frac{\partial \lg|g'_k(\theta_k,t_k)|}{\partial \theta_k}\right\}-E\left\{\sum_{i=1}^{N}\varphi_i(y_i)w_{ik}\frac{\partial g_k(\theta_k,t_k)}{\partial \theta_k}\right\} \tag{5.2.3}$$

here, t_k is the kth component of the observation vector, w_{ik} is the ikth element of separating matrix, and g'_k is the derivative of the kth nonlinear function g_k. The whole algorithm depends on the form of selected nonlinear mapping $g_k(\theta_k, t_k)$, the parameters of which can be estimated by using a multilayer perception network.

The performance of the above nonlinear separation algorithm depends on the self-adaptive estimation made by the estimation function through output vector $\boldsymbol{y}$. Several methods can be used. One of them is the estimation method based on Gram-Charlier, which is effective for light post nonlinear distortion, but under a severe nonlinear condition, it is more effective to directly estimate the score function.

The divisibility of PNL can be extended to a convolutive mixture; in this case, the instantaneous mixing matrix is replaced by the filter matrix $\boldsymbol{A}(z)$,

$$\boldsymbol{A}(z) = \sum_k \boldsymbol{A}_k z^{-k}$$

Define $\boldsymbol{s} \overset{\text{def}}{=} (\cdots, \boldsymbol{s}^{\mathrm{T}}(k-1), \boldsymbol{s}^{\mathrm{T}}(k), \boldsymbol{s}^{\mathrm{T}}(k+1), \cdots)^{\mathrm{T}}$ and $\boldsymbol{e} \overset{\text{def}}{=} (\cdots, \boldsymbol{e}^{\mathrm{T}}(k-1), \boldsymbol{e}^{\mathrm{T}}(k), \boldsymbol{e}^{\mathrm{T}}(k+1), \cdots)^{\mathrm{T}}$, then

$$\boldsymbol{e} = f(\overline{\boldsymbol{A}}\boldsymbol{s}) \tag{5.2.4}$$

here,

$$\overline{\boldsymbol{A}} = \begin{bmatrix} \cdots & \cdots & \cdots & \cdots & \cdots \\ \cdots & A_{k+1} & A_k & A_{k-1} & \cdots \\ \cdots & A_{k+2} & A_{k+1} & A_k & \cdots \\ \cdots & \cdots & \cdots & \cdots & \cdots \end{bmatrix}$$

The iid characteristic of the source, that is, the temporal independency of $s_i(k)$, $i=1, \cdots, n$, guarantees the space independency of $\boldsymbol{s}$. Moreover, a convolutive PNL can be considered as a mixture of special PNL. For the FIR mixing matrix $\boldsymbol{A}(z)$, Equation (5.2.4) corresponds to the PNL mixture of finite dimension with separability.On the other hand, for the IIR mixing matrix $\boldsymbol{A}(z)$, Equation (5.2.4) corresponds to the PNL mixture of infinite dimension, and the criterion of the algorithm is very complex.

5.3 Extended Gaussianization Method of Post Nonlinear Blind Separation

In real world applications of signal processing, nonlinear distortion introduced by sensors must be considered. The post nonlinear model discussed above is an important step, because by adding a reversible nonlinear transformation to the linear instantaneous mixture, a solotion results. The Gaussianization technology can be used as an approximation method to solve PNL mixing blind separation, which is based on the Gaussian and quasi-Gaussian distribution of the linear mixture of the independent sources.

Gaussianization technology is a method that transforms a random variable into a standard Gaussian random variable, and the approximation of the Gaussianization of the linear mixture to Gaussian distribution can directly recover linear mixed sources from observation signals[7,8]. In Fig. 5.1, nonlinear function g_i transforms variable x_i into a standard Gaussian variable z_i. Assuming the cumulative distribution function of x_i estimated from the observation sample $F_{xi}(x_i)$, and G is the cumulative distribution function of standard Gaussian distribution,$G(z) = \frac{1}{\sqrt{2\pi}}\int_{-\infty}^{z} e^{-t^2/2}\,dt$, then PNL separation referring to G is a two-step process[9].

Step 1: Construct every inverse nonlinear function according to Gaussianization, where $g_i = G^{-1} \circ F_{xi}$, the latent nonlinear mixture estimation, is $z_i = g_i(x_i)$.

Step 2: Extract independent component y_i from z using linear ICA algorithm, where $\boldsymbol{y} = \boldsymbol{W}z$, $\boldsymbol{y}$ is the estimation of $\boldsymbol{s}$. Although each linear ICA algorithm is effective, in some conditions, the latent linear mixture may be absent from G, so the residual distortion in the recovered linear mixture is still great and may lead to different results.

The Kernel estimation method is one kind of practical method that estimates cumulative distribution function F_{xi}. For simplification, the empirical cumulative distribution function $\hat{F}_{xi}(\boldsymbol{x})$, which has the same or fewer samples as x, is used.

$$\hat{F}(x) = \frac{1}{N+1}\sum_{i=1}^{N} I(x^{(i)} \leqslant x) \tag{5.3.1}$$

where $\boldsymbol{x}^{(i)}$ is the ith sample of variable $\boldsymbol{x}$, N is the total number of samples, and $I(\boldsymbol{x}^{(i)} \leqslant \boldsymbol{x})$ is an indicator function, which is 1 when $\boldsymbol{x}^{(i)} \leqslant \boldsymbol{x}$ and 0 under other conditions.

The inverse of the cumulative distribution function of standard G distribution, G^{-1}, can be approximated by using the numerical calculation method. For $0.5 \leqslant p < 1$, $G^{-1}(p)$ can be approximated as

$$z = G^{-1}(p) \approx u - \frac{c_0 + c_1 u + c_2 u}{1 + d_1 u + d_2 u + d_3 u} \tag{5.3.2}$$

where $u = \sqrt{\ln[1/(1-p)^2]}$ $c_0=2.515517$, $c_1=0.802853$, $c_2=0.010328$, $d_1=1.432788$, $d_2=0.189269$ and $d_3=0.001308$

The above Gaussianization method is a simple approximation of distribution, whose separating accuracy is not high enough, so one kind of extended Gaussianization method[9] is introduced in here, as shown in Fig. 5.2.This method looks at the nonlinear mapping constructed by Gaussianization as $\boldsymbol{g}^{(1)}$, and another nonlinear mapping $\boldsymbol{g}^{(2)}$ is construeted for the demixing system. In other words, the nonlinear part of demixing in the extended Gaussianization method is composed of two parts. First, one can obtain $\boldsymbol{v} = [v_1, v_2, \cdots, v_n]^{\mathrm{T}}$, $v_i = g_i^{(1)}(x_i) = G^{-1}(F_{xi}(x_i))$, from the observation vector $\boldsymbol{x}$ by using $\boldsymbol{g}^{(1)}$ and the Gaussianization method. Then $\boldsymbol{v}$, as a new observation, is used in recovering the latent linear mixture from $\boldsymbol{v}$ by using $\boldsymbol{g}^{(2)}$

$$g_i = g_i^{(2)} \circ g_i^{(1)} \tag{5.3.3}$$

$g_i^{(2)}$ transforms the standard Gaussian variable v_i into variable z_i. Since the distribution of z_i is assumed to approximate Gaussian, expanding it by using Comish-Fisher(C-F) is helpful to construct a simple and reasonable $g_i^{(2)}$. The whole nonlinear function can be constructed by using the equation above after $\boldsymbol{g}^{(2)}$is estimated, where all of g_i, $g_i^{(1)}$ and $g^{(2)}$ monotonously increases.

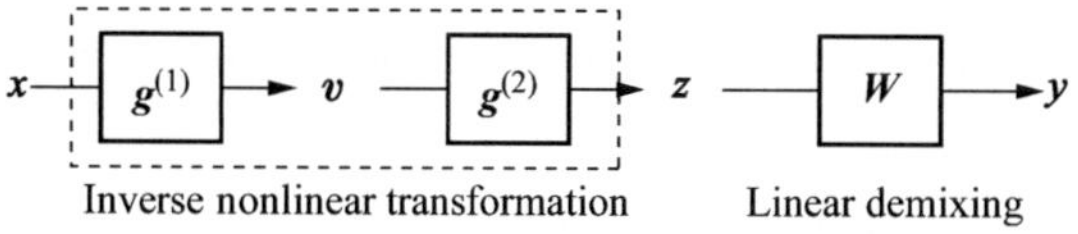

Fig. 5.2 Separation system using extended Gaussianization method[9]

C-F expansion is a cumulative distribution expansion form presented by Cornish and Fisher in 1937, Each item in the expanded equation is a multinomial of the standard Gaussian quantiles. The quadrinomial C-F expandedequation for α quantile is

$$x(v_\alpha) \approx v_\alpha + \frac{1}{6}(v_\alpha^2 - 1)k_3 + \frac{1}{24}(v_\alpha^3 - 3v_\alpha)k_4 - \frac{1}{36}(2v_\alpha^3 - 5v_\alpha)k_3^2 \tag{5.3.4}$$

where, k_3 and k_4 are third-order and forth-order cumulative amounts individually, v_α denotes α quantile of standard Gaussian distribution, and $x(v_\alpha)$ has zero mean value and unit variance.

Figure 5.3 shows the probability distribution function and the distribution from the quantile of standard Gaussian distribution with different k_3 and k_4 of $x(v_\alpha)$ obtained by Eq. (5.3.4).

The learning rules of extended Gaussianization can be obtained by using the common learning rules of separating the PNL mixture in the special mixing process ($g_i^{(2)}$, $\boldsymbol{W}$).

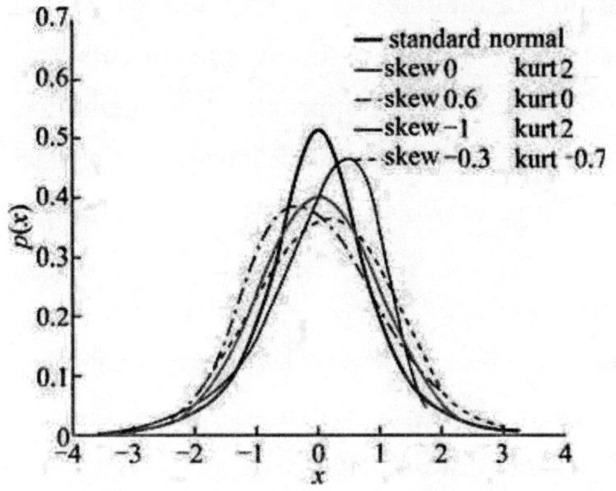

Fig. 5.3 x obtained by C-F extension[9] (courtesy of Zhang K and Chan L W. Used with permission)

Let D_i be the derivative of $g_i^{(2)}(v_i \mid \theta_i)$ depending on v_i,

$$D_i = g'^{(2)}_i(v_i \mid \theta_i) = 1 + \frac{1}{3}k_{3,i}v_i + \frac{1}{8}(v_i^2 - 1)k_{4,i} - \frac{1}{36}k_{3,i}^2(6v_i^2 - 5) \tag{5.3.5}$$

If the parameters of the nonlinear stage are $\theta_i = \pm\{k_3, i, k_4, i\}$, $i=1, 2, \cdots, n$, then the learning rules of updating all the unknown parameters are

$$\Delta k_{3,i} \propto E\left[\frac{v_i - \frac{1}{6}k_{3,i}(6v_i^2 - 5)}{3D_i} + \frac{\phi_i}{6}\left(v_i^2 - 1 - \frac{k_{3,i}(2v_i^3 - 5v_i)}{3}\right)\right] \tag{5.3.6}$$

and

$$\Delta k_{4,i} \propto E\left[\frac{v_i^2 - 1}{8D_i} + \frac{\phi_i}{24}(v_i^3 - 3v_i)\right] \tag{5.3.7}$$

5.4 Neural Network Method for Nonlinear ICA

The basic idea of the neural network method for nonlinear ICA is to establish a model for nonlinear mapping $f(\cdot)$ by using the NLP network[10~12], so the network can approximate any nonlinear mappings with precision as long as enough hidden neuron units are offered. But the learning process is based on non-supervised Bayesian general learning, which is quite different from the standard backpropagation algorithms. There is a problem of selecting the model in the learning process. If the selected model is too complex, overfitting will result that is, except for the real independent source, a meaningless solution will be obtained regarding the noises. On the other hand, the real source signals will be lost because of underfitting when the selected model is too simple. The solution for this problem is to consider all the possible explanations, and weight them according to their posterior probability, which is called Bayesian learning[13]. Of course, we cannot estimate such precise posterior probability for the model, so some appropriate approximation methods are needed. One of them is ensemble learning (or variational learning), the basic idea of which is to minimize the misfitting between the posterior probability and the parametric approximation.

Suppose $\boldsymbol{X}=\{\boldsymbol{x}(t)\mid t\}$, $\boldsymbol{S}=\{\boldsymbol{s}(t)\mid t\}$, and $\boldsymbol{\theta}$ represent the parameters of all the unknown models. Let $p(\boldsymbol{\theta}\mid\boldsymbol{X})$ represent the precise posterior probability and $q(\boldsymbol{\theta})$ denote its parametric approximation. Misfitting is weighed by the Kullback-Leibler(KL) divergence between p and q, which is defined as evaluation function

$$C_{\mathrm{KL}} = E_Q\left(\lg\frac{q}{p}\right) = \int \mathrm{d}\boldsymbol{\theta}\, q(\boldsymbol{\theta}) \lg\frac{q(\boldsymbol{\theta})}{p(\boldsymbol{\theta}\mid\boldsymbol{X})} \tag{5.4.1}$$

because the KL divergence includes the expectation of distribution, it is more sensitive to probability mass than probability density.

For nonlinear mixing mapping $f(\cdot)$, the model can be established by using the general approximate Property of the NLP network[14~16]:

$$\boldsymbol{x}(t) = \boldsymbol{B}[\boldsymbol{g}(\boldsymbol{A}\boldsymbol{s}(t)+\boldsymbol{b})]+\boldsymbol{n}(t) \tag{5.4.2}$$

where, $\boldsymbol{x}(t)$ denotes the observed data vector, matrix $\boldsymbol{B}$ and $\boldsymbol{A}$ represent the weight of the network output layer and hidden layer, $\boldsymbol{s}(t)$ denotes the source signal vector, $\boldsymbol{b}$ represents the offset vector of hidden layer, $\boldsymbol{n}(t)$ denotes the Gaussian noise vector, and $\boldsymbol{g}(t)$ represents the nonlinear activation function vector.

According to this model, only the network output $\boldsymbol{x}(t)$ is known and the input is unknown; therefore, the learning process must be an unsupervised learning method. First, use the linear PCA instead of a random selection method to find the initial useful value for the posterior mean of the source, then scan the whole dataset 50 times to find a significant mapping between the source and observation. Next, rotate the source with the linear ICA by using a simple nonlinear model of Gaussian distribution source (not Gaussian mixture), and compensate it by the inverse rotation of the weighting matrix $\boldsymbol{A}$ of the hidden layer. Finally, we can obtain the final representation of the source signal by the learning of the Gaussian mixting model of the source.

Posterior density $p(\boldsymbol{\theta}\mid X)$ and its parametric approximation $q(\boldsymbol{\theta})$ are needed to calculate the evaluation function. According to Bayesian rules, the posterior probability density of unknown variables $\boldsymbol{S}$ and $\boldsymbol{\theta}$ can be written as

$$p(\boldsymbol{S},\boldsymbol{\theta}\mid\boldsymbol{X}) = \frac{p(\boldsymbol{X}\mid\boldsymbol{S},\boldsymbol{\theta})\,p(\boldsymbol{S}\mid\boldsymbol{\theta})\,p(\boldsymbol{\theta})}{p(\boldsymbol{X})} \tag{5.4.3}$$

here, factor $p(\boldsymbol{X} \mid \boldsymbol{S}, \boldsymbol{\theta})$ is obtained by the neural network model of Eq. (5.4.2), since the components of noise are assumed independent, then $p(\boldsymbol{X} \mid \boldsymbol{S}, \boldsymbol{\theta})=\prod_{i} p(x_i(t) \mid s(t), \boldsymbol{\theta})$; $p(\boldsymbol{S} \mid \boldsymbol{\theta})$ and $p(\boldsymbol{\theta})$ is also the product of the Gaussian distribution, and can be obtained from the model structure. Since $p(\boldsymbol{X})$ does not relate to the model parameters, it can be ignored.

The calculation of parametric approximation $q(\boldsymbol{\theta})$ is based on the assumption that Gaussian density is a diagonal covariance matrix, so this approximation can be represented as the product of independent distribution $q(\boldsymbol{\theta}) = \prod_i q_i(\theta_i)$, and the parameters of each Gaussian component density $q_i(\theta_i)$are its mean $\overline{\boldsymbol{\theta}}_i$ and variance $\widetilde{\boldsymbol{\theta}}_i$. The ensemble learning method estimates the combined posterior probability of sources and network parameters instead of point-to-point estimation, so the evaluation function C_{KL} is a function of the posterior mean $\overline{\boldsymbol{\theta}}_i$ and variance $\widetilde{\boldsymbol{\theta}}_i$ of source signals and the network parameters. Divide C_{KL} into two parts, one $C_p=-E_p\{\lg p\}$and the other $C_q=E_q\{\lg q\}$, and then find its derivative on $\widetilde{\boldsymbol{\theta}}_i$

$$\frac{\partial C}{\partial \widetilde{\boldsymbol{\theta}}} = \frac{\partial C_p}{\partial \widetilde{\boldsymbol{\theta}}} + \frac{\partial C_q}{\partial \widetilde{\boldsymbol{\theta}}} = \frac{\partial C_p}{\partial \widetilde{\boldsymbol{\theta}}} - \frac{1}{2\widetilde{\boldsymbol{\theta}}} \tag{5.4.4}$$

Let the function above equal zero, and the update iterative equation of variance is given as follows:

$$\widetilde{\boldsymbol{\theta}} = \left[2\frac{\partial C_p}{\partial \widetilde{\boldsymbol{\theta}}}\right]^{-1} \tag{5.4.5}$$

and the mean $\overline{\boldsymbol{\theta}}_i$ can be estimated by the following approximate Newton iterative equation:

$$\overline{\boldsymbol{\theta}}_i = \overline{\boldsymbol{\theta}} - \frac{\partial C_p}{\partial \overline{\boldsymbol{\theta}}}\widetilde{\boldsymbol{\theta}} \tag{5.4.6}$$

5.5 Genetic Algorithm of Nonlinear ICA Solution

As a nonlinear processing method, the posterior PNL algorithm has received much attention. One of the main uses of this algorithm is that the mutual information of the estimated components serves as the contrast function. According to Section 2.3, mutual information I among each element of multidimensional variable $\boldsymbol{y}$ is defined as

$$I(y_1, y_2, \cdots, y_n) = \sum_{i=1}^{n} H(y_i) - H(y_1, y_2, \cdots, y_n) \tag{5.5.1}$$

When all the components y_1, y_2, $\cdots$, y_n are mutually independent, the mutual information is zero since the joint entropy is equal to the marginal entropy. On the other hand, the sum of marginal entropy is greater than joint entropy when they are dependent, which will lead to positive mutual information. In practice, the density function and even the estimation of the source are hard to get; therefore, some approximation methods for density function are needed to calculate the joint entropy and marginal entropy. The histogram approximation method is a simple one. This method divides the selected estimation space into m fields and calculates the number of the signal points in each area (B_i, i=1, 2, $\cdots$, m); thus, the marginal entropy can be approximated as

$$H(y) = -\sum_{i=1}^{n} p(y_i)\mathrm{lb}p(y_i) \approx -\sum_{j=1}^{m} \frac{\#B_j(y)}{n}\mathrm{lb}\frac{\#B_j(y)}{n} \tag{5.5.2}$$

where, $\#S$ represents the base of set S. The same method can also be used to calculate joint entropy.

$$\begin{aligned} H(y_1,\cdots,y_p) &= \sum_{i=1}^{p} H(y_i \mid y_{i-1},\cdots,y_1) \\ &\approx -\sum_{i_1=1}^{m}\sum_{i_2=1}^{m}\cdots\sum_{i_n=1}^{m} \frac{\#B_{i_1 i_2\cdots i_p}(y)}{n}\mathrm{lb}\frac{\#B_{i_1 i_2\cdots i_p}(y)}{n} \end{aligned} \tag{5.5.3}$$

where p is the approximated component number.

Using the above two equations to replace the entropy in Eq. (5.5.1), the contrast function which can minimize mutual information can be obtained. It is essential to avoid running into local minimum in the minimizing process.

Evolutionary calculation is the most popular optimization technology, and the genetic algorithm GA is one of them. The compendium about GA can be seen in Ref.[11]. For the linear mixture, the chromogene in GA algorithm is the corresponding parameter of the separating matrix $\boldsymbol{W}$. For the PNL mixture, the chromogene represents the parameters of the odd polynomial which approximates nonlinear g, so coding can be done by these parameters[17,18].

5.6 Application Examples of Nonlinear ICA

Since most actual mixing models of signals have nonlinear characteristics to some degrees, it is difficult to apply BSS/ICA algorithms without any prior assumption. However, the nonlinear ICA algorithms have had success in solving the mixing problems of actual signals.

The basic nonlinear ICA algorithms described in previous sections of this chapter have been developed into many similar algorithms. For example, the neural network-based nonlinear ICA method, methods based on a radial basis function neural network, a self-organizing feature map neural network, a spline neural network combined with high-order cumulants have been developed[16,19~22]. The genetic algorithm-based nonlinear ICA and designing methods of the evaluation function by using different statistical characteristics have also been developed[23,24]. Furthermore, the various extended algorithms based on kernel nonlinear ICA, etc. are proposed for different problems.

The application of nonlinear ICA methods on blind source separation of speech signals has been widely studied[17,20,25~28]. Ref.[18] presented a nonlinear blind separation method based on a high-order statistical characteristic and genetic algorithm. This method uses a neural network as the fitting solver of the demixing nonlinear function, high-order statistics to evaluate the independency between demixing output vectors, and a genetic algorithm instead of the gradient descent method to carry out global optimization on parameters of the neural network and evaluation functions, which can minimize nonlinear and nonconvex evaluation function. Research indicates that the method based on the genetic algorithm is adaptive to the blind signal separation of finite samples; therefore, global optimization can be realized in this algorithm, and it has the advantages of high accuracy, strong stability and fast convergence.

In the biomedical engineering area, nonlinear ICA has been used in signal separation of the ECG[23]. It is hot known if the components of the arrhythmia and normal electrocardiosignal in the ECG signal are mutually independent, but it is known that they are nonlinearly mixed according to the dynamic characteristics caused by arrhythmia signal. Ref.[23] introduced nonlinear dynamic control technology, which can estimate the unordered dynamic characteristics of a ventricular fibrillation electric signal in real time and separate or extract the arrhythmia signal by combining it with the ICA algorithm. The ICA algorithm based on nonlinear dynamic control technology can not only detect any variation in its systemic internal dynamic characteristics in a timely and quick way, but it has a robustness in estimating non-stationary signals, and can also hold a control action in the variation process of the dynamic characteristics of the arrhythmia signal, which has great significance for clinical application.

To separate the multiplicative noise from images, several researchers studied the nonlinear ICA blind separation method based on neural networks[21]. To simplify the image denoising problem, the noise is usually assumed to be additive noise, but the multiplicative noise through the nonlinear mixture can't be removed. The original noiseless signal and multiplicative noise can be separated by using the nonlinear ICA method based on the self-organizing mapping (SOM) neural network. SOM is an unsupervised nonlinear function mapping method based on a competitive learning criteria which uses a topological structure. Based on a post-nonlinear ICA structure, SOM can fit the inverse mapping g of nonlinear mixed function $f: \mathbf{R}^N \to \mathbf{R}^N$. Each nerve cell j, $1 \leqslant j \leqslant 1$, is connected to the input correlation through the connecting weight $\boldsymbol{W}_j=[w_{j1}, \cdots, w_{jm}]^T$. In each iteration process, SOM finds the optimum matching nerve cell v:

$$v(\boldsymbol{x}) = \arg\min_j \| \boldsymbol{x}(t) - \boldsymbol{W}_j \| , \quad j = 1,\cdots,l \tag{5.6.1}$$

Here $\boldsymbol{x}$ denotes m-dimensional input vector $\| \cdot \|$ and represents Euclidean distance. The iterative learning algorithm of connecting weight $\boldsymbol{W}_j$ is

$$\boldsymbol{W}_j(t+1) = \boldsymbol{W}_j(t) + \alpha(t) h_{j,v(x)}(t)[\boldsymbol{x}(t) - \boldsymbol{W}_j(t)], \quad j = 1,\cdots,l \tag{5.6.2}$$

where $\alpha(t)$ denotes the learning rate and $h_{j,v(x)}(t)$ is the adjacent neuron function.

Assume two noisy observation images I_1 and I_2, are nonlinear mixtures of the original noiseless signal and noise signal. The noisy observation signal can be separated into the original noiseless signal and noise signal for the purpose of denoising. The two observation images should be preprocessed before separation. First, divide each noisy image into several subblocks I_{1j} and I_{2j}, each of which should be rearranged in a one-dimensional vector, and pre-whitening should be done so that the mean is equal to zero and the variance is equal to 1. Then separate subimages of the two observation images by using the nonlinear ICA iteration algorithm based on SOM; then,the output separative signal is $y(t)=[\hat{s}_1(t), \hat{s}_2(t)]$, and $\hat{s}_1$and $\hat{s}_2$ are the estimation of the original noiseless signal and the noise signal respectively. Similarly, since the observation images have been divided into several images $\boldsymbol{X}_i$ in nonlinear blind separation, this meets the problem of indeterminacy. So, the correct recovery can be solved by using the correlation method:

$$\rho_{i,j} = C_{ij} / \sqrt{C_{ii} C_{jj}} \tag{5.6.3}$$

here, C_{ij} is the covariance of X_i and X_j, and $\rho_{i,j} \in [-1,1]$, which is equal to zero when the two vectors are mutually independent.

The experiment shows that the nonlinear blind separation algorithm can't separate the original image and the noise signal very precisely; this is because there is not any prior knowledge from the original image, the noise signal and the nonlinear mixing process, so the post processing for the separation output, such as image enhancement, is needed.

The nonlinear ICA method also has new applications on facial image processing. Since posture, light and expression are different, the actual facial image vectors always have complex nonlinear relationships. The nonlinear ICA based on the kernel method projects the facial image to a high dimensional feature space by nonlinear mapping, then separates the nonlinear independent components of the projection vector by using the ICA method to extract the nonlinear relationship in the facial image[19]. The method can offer a nice nonlinear expression for the actual facial images distorted by different factors.

The kernel ICA (KICA) method first projects the image input vectors into higher order feature space F by using the nonlinear kernel function, $\phi(\boldsymbol{x}) \in \mathrm{R}^N \rightarrow \phi(\boldsymbol{x}) \in F$. Suppose the image sequence is $\boldsymbol{x}=\{x_1, x_2, \cdots, x_n\}$; during this process, the calculation characteristic vectors are not needed; therefore, only the internal product of the two vectors in F are necessary by using the kernel function: $k(\boldsymbol{x},\boldsymbol{y})=(\phi(\boldsymbol{x}) \cdot \phi(\boldsymbol{y}))$. Then pre-whitening the characteristic vector in F space to find its eigenvectors and eigenvalues

$$(\boldsymbol{W}_p^\phi)^{\mathrm{T}}\phi(\boldsymbol{X})\phi(\boldsymbol{X})^{\mathrm{T}}\boldsymbol{W}_p^\phi = \boldsymbol{\Lambda}^\phi \tag{5.6.4}$$

where $\boldsymbol{W}_p^\phi$ denotes the eigenvectors vector matrix and $\boldsymbol{\Lambda}^\phi$ represents the diagonal matrix of eigenvalues. The eigenvectors vector matrix $\boldsymbol{W}_p^\phi$ can be represented as $\boldsymbol{W}_p^\phi=\phi(\boldsymbol{X})\boldsymbol{\alpha}$. Here $\boldsymbol{\alpha}$ denotes the coefficient matrix, $\boldsymbol{\alpha}\boldsymbol{\alpha}^{\mathrm{T}}=\boldsymbol{I}$, so the solution of eigenvectors and eigenvlaues in Eq. (5.6.4) can be represented as

$$\boldsymbol{\alpha}^{\mathrm{T}} k^2 \boldsymbol{\alpha} = \boldsymbol{\Lambda}^\phi \tag{5.6.5}$$

where $k=(\phi(\boldsymbol{x}_i) \cdot \phi(\boldsymbol{x}_j))$, $1 \leqslant i, j \leqslant n$. After finding the eigenvectors and eigenvalues, the prewhitening processing on the sample in F space can be done; at the same time, the dimension reduction is achieved:

$$\boldsymbol{X}_w^\phi = (\boldsymbol{\Lambda}_k^\phi)^{-1}\boldsymbol{\alpha}^{\mathrm{T}} k \tag{5.6.6}$$

Based on that, one can extract the independent components from the high dimensional samples found from nonlinear mapping by using the ICA method, such as informax algorithm. The iteration rules of the demixing matrix $\boldsymbol{W}_I^\phi$ can be represented as $\hat{\boldsymbol{W}}_I^\phi=\boldsymbol{W}_I^\phi+\rho\Delta\boldsymbol{W}_I^\phi$; here ρ is the learning rate.

$$(\Delta\boldsymbol{W}_I^\phi)^{\mathrm{T}} = \left[\boldsymbol{I}+\left(1-\frac{2}{1+\mathrm{e}^{-\boldsymbol{y}_I}}\right)(\boldsymbol{y}_I^\phi)^{\mathrm{T}}\right](\boldsymbol{W}_I^\phi)^{\mathrm{T}} \tag{5.6.7}$$

where $\boldsymbol{y}_I=\boldsymbol{W}_I^\phi\boldsymbol{x}$, the iteration can be completed when the convergence is achieved. Figure 5.4 shows the experimental results obtained by the SOM-based nonlinear ICA iteration algorithm[21], which represents a good denoising performance.

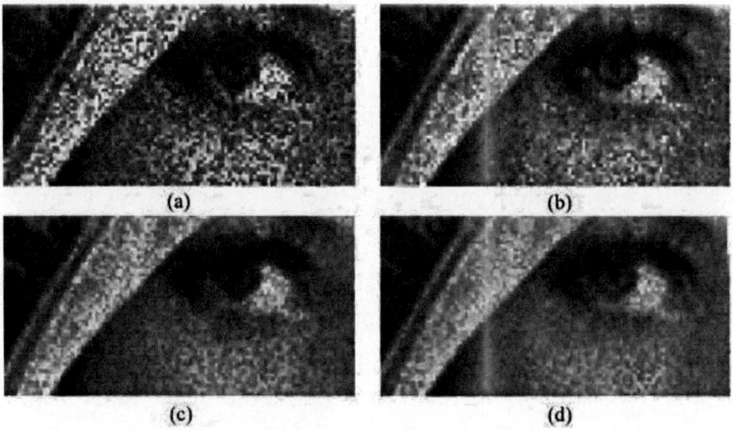

Fig. 5.4 Image denoising results by the SOM-ICA method
(a) Noisy image 1 (noise variance 0.05); (b) The denoising result of noisy image 1;
(c) Noisy image 2 (noise variance 0.01); (d) The denoising result of noisy image 2;

References

[1] Hyvarinen A, Pajunen P (1999) Nonlinear independent component analysis: Existence and uniqueness results. Neural Networks 12 (3): 429–439

[2] Taleb A, Jutten C (1999) Source separation in post-nonlinear mixtures. IEEE Transations on Signal Processing 47(10): 2807–2820

[3] Jutten C, Karhunen J (2003) Advances in nonlinear blind source separation. In: Proceedings of ICA2003, Nara, 2003, pp 245–256

[4] Lappalainen H, Honkela A (2000) Beyesian nonlinear independent component analysis by multi-layer perceptions. In: Mark Girolami (ed) Advances in Independent Component Analysis. Springer, London, pp 93–121

[5] Wang F S, Li H W, Shen Y T (2005) An overview on nonlinear blind source separation: Theory and algorithms. Signal Processing 21(3): 282–288 (Chinese)

[6] Sole-Casals J, Jutten C (2004) Post-nonlinear mixtures and beyond. In: Proceedings of World Automation Congress, Seville, 2004,16: 67–74

[7] Chen S S, Gopinath R A (2001) Gaussianization. In: Advances in Neural Information Processing Systems, Vancouver, 2001,13: 423–429

[8] Ziehe A, Kawanabe M, Harmiling S et al (2003) Blind separation of post-nonlinear mixtures using Gaussianizing transformations and temporal decorrelation. In: Proceedings of ICA2003, Nara, 2003, pp 269–274

[9] Zhang K, Chan L W (2005) Extended Gaussianization method for blind separation of post-nonlinear mixtures. Neural Computation 17(2): 425–452

[10] Haykin S (1998) Neural networks—A comprehensive foundation. 2nd edn. Prentice Hall, New Jersey

[11] Shi X Z (2003) Signal processing and soft-computing. Higher Education Press, Beijing

[12] Bishop C M (1995) Neural networks for pattern recognition. Oxford University Press, Oxford

[13] Lappalainen H, Giannakopoulos X, Karhunen J et al (2000) Nonlinear independent component analysis using ensemble learning theory. In: Proceedings of the 2nd International Workshop on Independent Component Analysis and Blind Signal Separation, Helsinki, 2000, pp 251–256

[14] Roberts S, Everson R (2001) Independent component analysis: Principles and Practice. Cambridge University Press, New York

[15] Tan Y, Wang J, Zurada J M (2001) Nonlinear blind source separation using a radial basis function network. IEEE Transactions on Neural Networks 12(1): 124–134

[16] Liu D, Zhao Y (2004) Nonlinear blind source separation based on radial basis function. In: Proceedings of the 5th World Congress on Intelligent Control and Automation, Hangzhou, 2004, pp 2002–2004

[17] Rojas F, Pintonet C G, Rojas I (2002) Genetic algorithm approach to nonlinear blind source separation. In: Proceedings of Congress on Evolutionary Computation. Honolulu, Hawaii, 2002, pp 1098–1102

[18] Tan Y,Wang J (2001) Nonlinear blind source separation using higher order statistics and a genetic algorithm. IEEE Transactions on Evolutionary Computation 5(6): 600–612

[19] Woo W L, Dlay S S (2005) Nonlinear blind source separation using a hybrid RBF-FMLP network. IEE Proceedings.-Vis. Image Signal Process 152(2): 173–183

[20] Wang G, Hu D W (2004) On nonlinear independent component analysis using self-organizing map. In: Proceedings of the 5th World Congress on Intelligent Control and Automation, Hangzhou, 2004, pp 91–95

[21] Haritopoulos M, Yin H J Allinson N M (2002) Image denoising using self-organizing map-based nonlinear independent component analysis.

Neural Networks 12 (5): 1085–1098

[22] Solazzi M, Piazza F, Uncini A (2001) Nonlinear blind source separation by spline neural networks. In: Proceedings of ICASSP, Salt Lake City, Utah, 2001, pp 2781–2784

[23] Bigan C (2003) Chaotic cardiac arrhythmia detection by ICA and nonlinear dynamic processing of ECG signal. In: IEEE International Symposium on Intelligent Signal Processing, Budapest, 2003, pp 117–120

[24] Nguyen T V, Patra J C, Das A (2005) Post nonlinear blind source separation geometric linearization. In: Proceedings of International joint Conference on Neural Networks, Montreal, 2005, pp 244–249

[25] Vigliano D, Uncini A (2003) Flexible ICA solution for nonlinear blind source separation problem. Electronics Letters 39(22): 1616–1617

[26] Martinez D, Bray A (2003) Nonlinear blind source separation using Kernels. IEEE Transaction on Neural Networks 14(1): 228–235

[27] Liu Q S, Cheng J, Lu H Q et al (2004) Modeling face appearance with nonlinear independent component analysis. In: Proceedings of the 6th IEEE International Conference on Automatic Face and Gesture Recognition, Seoul, 2004, pp 761–766

[28] Vigliano D, Parisi R,Uncini A (2005) An information theoretic approach to a novel nonlinearIndependent component analysis paradigm. Signal Processing 85 (4): 997–1028

Chapter 6 Convolutive Mixtures and Blind Deconvolution

6.1 Description of Issues

In Chapter 1 Section 1.2.2, the mathematical expression of convolutive mixtures was introduced in the convolutive mixtures separation problem. In order to facilitate reading, basic expressions are cited again here.

In convolutive mixtures, m received signals constitute an m dimensional received signal vector $\boldsymbol{x}(k)=[x_1(k), x_2(k), \cdots, x_m(k)]^{\mathrm{T}}$; n source signals constitute n dimensional source signal vector $\boldsymbol{s}(k)=[s_1(k), s_2(k), \cdots, s_n(k)]^{\mathrm{T}}$, and $n \leqslant m$, $\boldsymbol{x}(k)$ is obtained by applying a p-order linear convolution mixing model on $\boldsymbol{s}(k)$ with the mathematical expression

$$\boldsymbol{x}(k) = \sum_{p=0}^{P} \boldsymbol{H}(p)\boldsymbol{s}(k-p) \tag{6.1.1}$$

where

$$\boldsymbol{H}(p) = \begin{bmatrix} h_{11,p} & \cdots & h_{1n,p} \\ \vdots & \ddots & \vdots \\ h_{m1,p} & \cdots & h_{mn,p} \end{bmatrix}$$

is the mixing filter matrix with the p delay and h_{ij} is the impulse response between the jth source signal and the ith sensor. In a real environment, source signals arrive at receivers through different transmission paths with different delays. Therefore, the cocktail party problem is not an instantaneous linear mixtures problem but a convolutive mixtures problem. The blind separation of the convolution mixed signals is to seek a Q-order separating filter matrix $\boldsymbol{W}(q)$that makes

$$\boldsymbol{y}(k) = \sum_{q=0}^{Q-1} \boldsymbol{W}(q)\boldsymbol{x}(k-q) \tag{6.1.2}$$

is the estimate of the source signal $\boldsymbol{s}(k)$.

For a single-channel blind deconvolution problem, such as reverberation cancelling and earthquake deconvolution, Widrow[1] and Haykin[2] proposed good algorithms. A second-order decorrelation method can provide satisfactory solutions for a minimum-phase convolution system. But for a non-minimum phase system, nonlinear adaptive algorithms must be utilized to extract higher-order statistics. A class of nonlinear blind deconvolution methods is called the Bussgang algorithm, while another more direct method is to use fourth-order cumulants[3]. Bell and Sejnowski proved that Infomax can be used for deconvolution of independent sources[4].

Because time-domain convolution is equivalent to frequency-domain multiplication, the mixing process can be described by

$$\boldsymbol{x}(z) = \boldsymbol{H}(z)\boldsymbol{s}(z) \tag{6.1.3}$$

where $\boldsymbol{H}(z)$ is the FIR of a specific channel, $\boldsymbol{s}(z)$ and $\boldsymbol{x}(z)$are the z-domain expression of a source signal vector and mixed signal vector respectively. Similarly, Equation (6.1.2) can be written as

$$\boldsymbol{y}(z) = \boldsymbol{W}(z)\boldsymbol{x}(z) \tag{6.1.4}$$

it is clear that the convolutive mixtures blind separation in a time-domain is changed into instantaneous mixtures blind separation in the frequency-domain. Therefore, instantaneous mixtures blind separation methods in the frequency-domain can be used to solve the blind deconvolution problems in the time-domain. Blind separation algorithms of convolutive mixtures can be divided into two classes: separation algorithms in the time-domain and separation algorithms in the frequency-domain. The emphasis of this chapter is on frequency-domain methods because they have better computing efficiency and better separation effects when the mixing filter is longer.

6.2 Convolutive Mixtures in Time-Domain

In time-domain, the instantaneous mixtures scalar matrix is extended to the filter mixing matrix for convolution mixtures using the concept of independent component analysis. Ehlers[5] proposed a separation algorithm for the basic 2×2 convolutive mixtures system. The iterative algorithm of off-diagonal elements of the separating matrix is

$$\Delta w_{i,j}(p) = f_i(y_i(k))g_j(y_j(k-p)) \qquad i \neq j \in \{1,2\} \tag{6.2.1}$$

Van Gerven and Van Compernolle[6] proposed a blind separation algorithm of the 2×2 convolutive mixtures system based on second-order statistics:

$$\Delta w_{i,j}(p) = -y_i(k)y_j(k-p) \qquad i \neq j \in \{1,2\} \tag{6.2.2}$$

Kawamoto[7] further proposed another blind separation algorithm aiming at nonstationary signals:

$$\Delta w_{i,j}(p) = -\frac{1}{\beta_i}y_i(k)y_j(k-p) \qquad i \neq j \in \{1,2\} \tag{6.2.3}$$

where β_i is the energy of output signal y_j at a different period of time.

Other methods, such as the cross-cumulant minimum method and the exponential window method with tracking nonstationary signals, have the disadvantage of large computations and bad separation effects.

6.3 Convolutive Mixtures Algorithms in Frequency-Domain

Equations (6.1.3) and (6.1.4) indicate the equivalence of convolution in the time-domain and of the instantaneous mixtures in the frequency-domain. Many instantaneous blind separation methods have good convergence properties and separation effects; thus, an effort was made to solve the separation problem of convolutive mixtures in the time-domain through instantaneous mixture's separation in the frequency-domain.

The only difference with the instantaneous mixture's blind separation in the time-domain is that the data to be handled is complex when solving blind separation in the frequency-domain. Smaragdis[8] still adopts the natural gradient method to perform separation for complex in the frequency-domain, but in cases of super-Gaussian or sub-Gaussian signals, the following nonlinear evaluation functions are used respectively:

$$f(z) = \tanh(\gamma \mathrm{Re}\{z\}) + j\tanh(\gamma \mathrm{Re}\{z\}) \tag{6.3.1}$$

or

$$f(z) = \mathrm{Re}\{z\} \mid \mathrm{Re}\{z\} \mid^2 + j\mathrm{Re}\{z\} \mid \mathrm{Re}\{z\} \mid^2 \tag{6.3.2}$$

here γ is the scale coefficient.

Compared with the convolutive mixture's blind separation in the time-domain, the convolutive mixture's separation algorithms in the frequency-domain have many advantages, especially when handling higher-order mixing filters. The solution of each order of filters depends on the solution of other orders, which makes computation excessive and reduces the separation effect. Take Chan's algorithm for example[9]; when the order of filter is less than 5, it can recover the original signal fast. But when the order reaches longer than 6, the separation velocity will be slow with a bad separation effect. In frequency-domain algorithms, instantaneous blind separation is used for each fixed frequency ,and the computation for every frequency segment is mutually independent. Thus, the impact of filter orders to the algorithm is small. Smaragdis[8] experiments proved that for a system with two inputs and two outputs, if the filter length is 1,024 and sampling frequency is 44.1kHz, the computational amount of the convolutive mixture's separation algorithms in the frequency-domain is reduced by 70 times.

On the other hand, frequency-domain algorithms have two problems due to two indeterminacies of blind separation. The first is amplitude indeterminacy,which means amplitude deviations occur in signals for different frequency segments, resulting in the distortions of the signal frequency spectrum such as variations of speech quality in speech signal blind separations. The second is the indeterminacy of permutation which means signals separated from different frequency segments cannot be connected properly, leading to separation failure. Now, we further discuss these two problems.

Suppose the source signal is $s(\omega)$ at frequency ω and the separated signal is $y(\omega)$. The amplitude indeterminacy results in

$$y(\omega) = a(\omega)s(\omega) \tag{6.3.3}$$

here $a(\omega)$ is a scalar changing with ω, which indicates the amplitude indeterminacy. In the frequency-domain, Equation (6.3.3) is changed to the expression in the time-domain below

$$y(k) = a(k) * s(k) \tag{6.3.4}$$

From Eq. (6.3.4), we can learn that the separated signal is the convolution of the original signal and an unknown channel parameter rather than the real original signal due to amplitude indeterminacy.

In order to clarify the problem above, take a two-inputs two-outputs system for example, as shown in Fig. 6.1. The existing method[9] is based on the assumption that $H_{11}(\omega)=H_{22}(\omega)=1$ and $\boldsymbol{W}_{11}(\omega)=\boldsymbol{W}_{22}(\omega)=1$, that is

$$\begin{aligned}\begin{bmatrix} y_1(\omega) \\ y_2(\omega) \end{bmatrix} &= \begin{bmatrix} 1 & W_{12}(\omega) \\ W_{21}(\omega) & 1 \end{bmatrix} \begin{bmatrix} 1 & H_{12}(\omega) \\ H_{21}(\omega) & 1 \end{bmatrix} \begin{bmatrix} s_1(\omega) \\ s_2(\omega) \end{bmatrix} \\ &= \begin{bmatrix} 1+W_{12}(\omega)H_{21}(\omega) & W_{12}(\omega)+H_{12}(\omega) \\ W_{21}(\omega)+H_{21}(\omega) & 1+W_{21}(\omega)H_{12}(\omega) \end{bmatrix} \begin{bmatrix} s_1(\omega) \\ s_2(\omega) \end{bmatrix}\end{aligned} \tag{6.3.5}$$

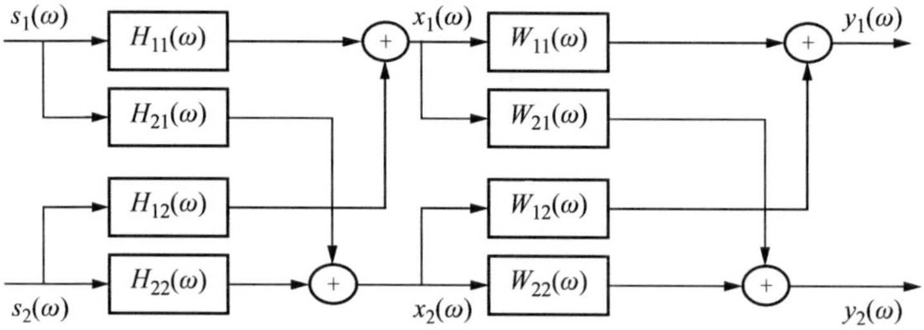

Fig. 6.1 Two-inputs two-outputs system for convolutive mixtures

If the convolutive mixtures will be separated successfully, there must be only one source signal contained in each output signal. Two cases are considered. In the first case, the first output only contains the first source signal and the second output only contains the second source signal. For this case, the off diagonal elements of Eq. (6.3.5) must be zero; that is

$$W_{12}(\omega) = -H_{12}(\omega), W_{21}(\omega) = -H_{21}(\omega) \tag{6.3.6}$$

substituting Eq. (6.3.6) into (6.3.5), we obtain

$$\begin{aligned} y_1(\omega) &= s_1(\omega)(1 - W_{12}(\omega)W_{21}(\omega)) \\ y_2(\omega) &= s_2(\omega)(1 - W_{12}(\omega)W_{21}(\omega)) \end{aligned} \tag{6.3.7}$$

The second case is that the first output only contains the second source signal and the second output only contains the first source signal. For this case, the off diagonal elements of Eq. (6.3.5) must also be zero; that is

$$W_1(\omega) = -\frac{1}{H_2(\omega)}, W_2(\omega) = -\frac{1}{H_1(\omega)} \tag{6.3.8}$$

Substituting Eq. (6.3.8) into (6.3.5), we obtain

$$\begin{aligned} y_1(\omega) &= s_2(\omega)\left(\frac{W_1(\omega)W_2(\omega) - 1}{W_2(\omega)}\right) \\ y_2(\omega) &= s_1(\omega)\left(\frac{W_1(\omega)W_2(\omega) - 1}{W_1(\omega)}\right) \end{aligned} \tag{6.3.9}$$

In order to eliminate the separation impact due to amplitude indeterminacy, a post-filtering process is added after the convolutive mixture's blind separation. For the first case, the post-filter is

$$Ps_1(\omega) = Ps_2(\omega) = \frac{1}{1 - W_1(\omega)W_2(\omega)} \tag{6.3.10}$$

and the second is

$$Ps_1(\omega) = \frac{W_1(\omega)}{W_1(\omega)W_2(\omega) - 1}, Ps_2(\omega) = \frac{W_2(\omega)}{W_1(\omega)W_2(\omega) - 1} \tag{6.3.11}$$

after doing so, the impact of the amplitude indeterminacy can be eliminated and the original signal is recovered. But such a method is based on the assumption that $H_{11}(\omega)=H_{22}(\omega)=1$, $W_{11}(\omega)=W_{22}(\omega)=1$, which does not accord with real applications; thus, such a method is just a theoretical discussion.

Schobben and Sommen proposed another solution for speech mixture's signals[10], which makes the energy of each frequency segment equal 1 after the convergence. Though speech signal energy is mainly distributed on 0~4 kHz, the distribution is not homogeneous, and the spectrum of speech for men is different from that for women. Therefore, this method still has a certain limitation.

Now we will discuss the indeterminacy of permutation. The indeterminacy of permutation has a huge impact on blind separation algorithms in the frequency-domain, which is the dominant key factor for the separation algorithm. The misalignment of separated signals in the adjacent frequency segment will occur due to the indeterminacy of permutation, and the failure of separation will result. Take the two-inputs two-outputs system for example; when at frequency segment ω_1 , the first recovered source is s_1 and the second is s_2. At frequency segment ω_2, the first recovered source is s_2 and the second is s_1. In such case, the connection of signals recovered from ω_1 and ω_2 will lead to a false original signal.

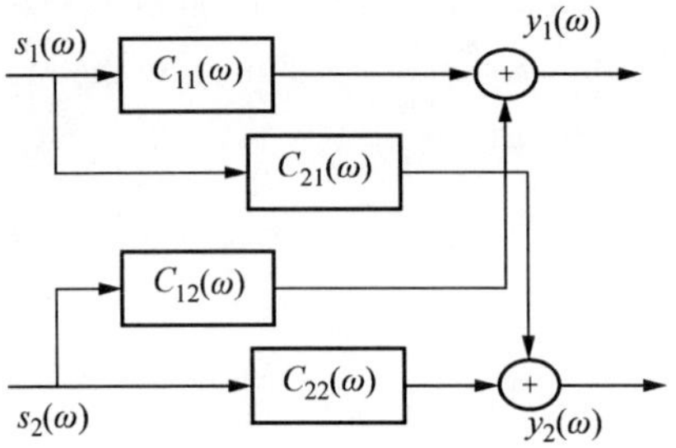

Fig. 6.2 Frequency-domain diagram of a two-inputs two-outputs system

Ikram[11] takes the two-inputs two-outputs system as an example and defines several parameters that are used as the standard to align the outputs of adjacent frequency segments. Figure 6.2 is the schematic diagram for a two-inputs two-outputs blind separation system. Here $S(\omega)$ and $Y(\omega)$ are the values of inputs and outputs in the frequency-domain respectively, $\boldsymbol{C}(\omega)$ is the frequency-domain matrix of the whole system,

$$\boldsymbol{C}(\omega) = \boldsymbol{W}(\omega)\boldsymbol{H}(\omega) \tag{6.3.12}$$

where $\boldsymbol{W}(\omega)$ and $\boldsymbol{H}(\omega)$ are the frequency-domain separating matrix and frequency-domain mixing matrix respectively.

When the first channel outputs the first source signal, that is to say, $y_1(\omega)$ is the estimation of the source signal $s_1(\omega)$, the Signal to Interference Ratio (SIR) is defined as

$$\mathrm{SIR}_{11}(\omega) = \frac{|C_{11}(\omega)|^2 |s_1(\omega)|^2}{|C_{12}(\omega)|^2 |s_2(\omega)|^2} \tag{6.3.13}$$

where we regard $s_1(\omega)$ as a signal and $s_2(\omega)$ as the interference.The SIR of the second output signal is

$$\mathrm{SIR}_{22}(\omega) = \frac{|C_{22}(\omega)|^2 |s_2(\omega)|^2}{|C_{21}(\omega)|^2 |s_1(\omega)|^2} \tag{6.3.14}$$

Similarly, we can also determine that the second channel outputs the first source signal, which means $y_1(\omega)$ is SIR_{12} when estimating the source signal $s_1(\omega)$. That is

$$\mathrm{SIR}_{12} = \frac{1}{\mathrm{SIR}_{22}} \tag{6.3.15}$$

Assuming the first channel outputs the first source signal and the second outputs the second after the separation algorithm convergence, it is required that

$$\mathrm{SIR}_{11} > \mathrm{SIR}_{12} \tag{6.3.16}$$

when $\mathrm{SIR}_{12}>\mathrm{SIR}_{11}$, if we exchange the positions of the first row and the second row of $\boldsymbol{W}(\omega)$ with each other, it can be assured that the first channel outputs the first source signal and the second channel outputs the second source; that is, $Y_1(\omega)$ is the estimate of source $S_1(\omega)$ and $Y_2(\omega)$ is the estimate of source $S_2(\omega)$.

From the above method we can see that the two original signals and the mixing matrix in the frequency-domain must be known while obtaining the SIR of each output signal. Therefore, such a method still has a long way to go before being applied to practical use.

In Ref.[12], a method for solving indeterminacy of the permutation problem was proposed; the principle and details are described as follows.

Suppose $Y(\omega)$ is the estimate at frequency ω and $Y(\omega)$ is expressed in the form of amplitude and phase

$$Y(\omega) = a(\omega)\mathrm{e}^{\mathrm{j}\phi(\omega)} \tag{6.3.17}$$

where $a(\omega)$ is the amplitude and $\phi(\omega)$ is the phase. For a nonstationary signal, amplitudes of the same signals at adjacent frequency segments have a certain correlation. But for different signals, their amplitudes are not correlated due to their mutual independence, i.e. $a_i(\omega_m)$ and $a_i(\omega_{m+1})$ are not correlated[13]. Reference[12] gives a new standard for solving the indeterminacy of permutation using this conclusion, and experiments prove the validity of such a conclusion.

In the developed alignment method, a parameter needs to be defined to measure the correlation of two random variables.

For random variables x and y, their cross-correlation coefficient is

$$r(x,y) = \frac{\mathrm{cov}(x,y)}{\sqrt{D(x)}\,\sqrt{D(y)}} \tag{6.3.18}$$

where $r(x, y)$ is a dimensionless value, $\mathrm{cov}(x, y)$ is the covariance of variables x and y, $D(x)$ and $D(y)$ are the variance of x and y respectively.

The cross-correlation coefficient $r(x, y)$ defined in Eq. (6.3.18) has the Property:

$$|r(x,y)| \leqslant 1 \tag{6.3.19}$$

when x and y are linearly correlated, $|r(x, y)|=1$, and when they are not correlated, $|r(x, y)|=0$.

Similarly, the correlation coefficient of amplitudes of two signal spectrums can be defined. Take the cross-correlation coefficient between amplitude $a_i(\omega)$ and $a_j(\omega)$ of output signals $Y_i(\omega)$ and $Y_j(\omega)$, for example,

$$r(a_i(\omega), a_j(\omega)) = \frac{\operatorname{cov}(a_i(\omega), a_j(\omega))}{\sqrt{D(a_i(\omega))}\sqrt{D(a_j(\omega))}} \tag{6.3.20}$$

where the cross covariance is

$$\operatorname{cov}(a_i(\omega), a_j(\omega)) = \frac{1}{M}\sum_{m=1}^{M} a_i(\omega, m) a_j(\omega, m) - \frac{1}{M}\sum_{m=1}^{M} a_i(\omega, m) \cdot \frac{1}{M}\sum_{m=1}^{M} a_j(\omega, m) \tag{6.3.21}$$

where $a_i(\omega, m)$ is the amplitude at frequency ω in the mth window and M is the number of windows. In computations, the mean of the signals are first eliminated and Eq. (6.3.21) becomes

$$\operatorname{cov}(a_i(\omega), a_j(\omega)) = \frac{1}{M}\sum_{m=1}^{M} a_i(\omega, m) a_j(\omega, m) \tag{6.3.22}$$

The variance of amplitudes $a_i(\omega, m)$ and $a_j(\omega, m)$ is

$$D(a_i(\omega, m)) = \frac{1}{M}\sum_{m=1}^{M} \left\{ a_i(\omega, m) - \frac{1}{M}\sum_{m=1}^{M} a_i(\omega, m) \right\}^2 \tag{6.3.23}$$

For the signals with zero mean, the above equation is simplified as

$$D(a_i(\omega, m)) = \frac{1}{M}\sum_{m=1}^{M} (a_i(\omega, m))^2 \tag{6.3.24}$$

$r(a_i(\omega), a_j(\omega))$ defined in Eq. (6.3.20) is just the parameters used to measure the correlation between $a_i(\omega, m)$ and $a_j(\omega, m)$.

Taking a two-inputs two-outputs system, for example, the detailed algorithm is shown below. For the two-inputs two-outputs system, its output will have two cases:

① If the outputs of channel 1 at the adjacent frequency segments ω_m and ω_{m+1} are the same, then the outputs of channel 2 must be the same. So it is easy to obtain that: $|r(a_1(\omega_m), a_1(\omega_{m+1}))|>0$, $|r(a_2(\omega_m), a_2(\omega_{m+1}))|>0$, and $|r(a_1(\omega_m), a_2(\omega_{m+1}))|=0$, $|r(a_2(\omega_m), a_1(\omega_{m+1}))|=0$.

② If the outputs of channel 1 at the adjacent frequency segments ω_m and ω_{m+1} are not the same, then the outputs of channel 2 must be different. So it is obtained that: $|r(a_1(\omega_m), a_1(\omega_{m+1}))|=0$, $|r(a_2(\omega_m), a_2(\omega_{m+1}))|=0$, and $|r(a_1(\omega_m), a_2(\omega_{m+1}))|>0$, $|r(a_2(\omega_m), a_1(\omega_{m+1}))|>0$.

Synthesizing the above two cases, we can consider if

$$\frac{|r(a_1(\omega_m), a_1(\omega_{m+1}))|}{|r(a_1(\omega_m), a_2(\omega_{m+1}))|} > \frac{|r(a_2(\omega_m), a_1(\omega_{m+1}))|}{|r(a_2(\omega_m), a_2(\omega_{m+1}))|} \tag{6.3.25}$$

then the outputs of channel 1 at the adjacent frequency segments ω_m and ω_{m+1} are from the same source and so are the outputs of channel 2. In this case, we can use the same separating matrix $\boldsymbol{W}(\omega_{m+1})$ for both ω_m and ω_{m+1}.

If Eq. (6.3.25) is not satisfied, namely

$$\frac{|r(a_1(\omega_m), a_1(\omega_{m+1}))|}{|r(a_1(\omega_m), a_2(\omega_{m+1}))|} > \frac{|r(a_2(\omega_m), a_1(\omega_{m+1}))|}{|r(a_2(\omega_m), a_2(\omega_{m+1}))|} \tag{6.3.26}$$

then the outputs of channel 1 at the adjacent frequency segments ω_m and ω_{m+1} are different and so are the outputs of channel 2. Now it is necessary to exchange the positions of the two rows of the separating matrix $\boldsymbol{W}(\omega_{m+1})$in order to assure that the same signal at different frequencies can be correctly connected.

From the above discussion, we can conclude that after the separation algorithm converged and the four parameters for each frequency, i.e. $|r(a_1(\omega_m), a_1(\omega_{m+1}))|$, $|r(a_1(\omega_m), a_2(\omega_{m+1}))|$, $|r(a_2(\omega_m), a_1(\omega_{m+1}))|$ and $|r(a_2(\omega_m), a_2(\omega_{m+1}))|$, have been computed, according to both sides of Eq.(6.3.26), it is needed to determine if the position of the two rows in the separating matrix must be exchanged to assure the correct connection between the signals at different frequency segments.

Two source signals are now used to verify the correlation of the adjacent frequency segments in the alignment method: One is the Chinese reading signal and the other is light music. The window length is set to $T=500$, overlapping coefficient $\beta=1$, and the parameters computed at different frequency segments are shown in Table 6.1.

Table 6.1 The correlation coefficients at different frequency segments

	γ_{11}	γ_{12}	γ_{21}	γ_{22}	γ_{11}/γ_{12}	γ_{21}/γ_{22}
$\omega_{50}\sim\omega_{51}$	0.5165	0.0071	0.0168	0.1728	72.7465	0.0972
$\omega_{150}\sim\omega_{151}$	0.9821	0.0062	0.0070	0.9102	158.4032	0.0077
$\omega_{200}\sim\omega_{201}$	0.9966	0.0037	0.0039	0.9816	269.3514	0.0040
$\omega_{250}\sim\omega_{251}$	1.0000	0.0078	0.0028	0.9999	128.2051	0.0028
$\omega_{300}\sim\omega_{301}$	0.9978	0.0034	0.0033	0.9880	293.4706	0.0033
$\omega_{350}\sim\omega_{351}$	0.9869	0.0061	0.0050	0.9329	161.7869	0.0053
$\omega_{400}\sim\omega_{401}$	0.7757	0.0068	0.0091	0.5990	114.0735	0.0152
$\omega_{450}\sim\omega_{451}$	0.5165	0.0168	0.0071	0.0728	30.7440	0.0975

Parameters in the table γ_{11}, γ_{12}, γ_{21} and γ_{22} are $|r(a_1(\omega_m), a_1(\omega_{m+1}))|$, $|r(a_1(\omega_m), a_2(\omega_{m+1}))|$, $|r(a_2(\omega_m), a_1(\omega_{m+1}))|$ and $|r(a_2(\omega_m), a_2(\omega_{m+1}))|$ respectively. From Table 6.1, we can find that $\gamma_{11}>\gamma_{12}$, $\gamma_{22}>\gamma_{21}$, and the experiment verified that there is a stronger amplitude correlation of same nonstationary signals at adjacent frequency segments; however,there is not amplitude correlation for different signals.

6.4 Frequency-Domain Blind Separation of Speech Convolutive Mixtures

Defining the frequency-domain correlation matrix of mixed signals vector $\boldsymbol{x}(t)$ at frequency ω is

$$\boldsymbol{R}_x(\omega) = E\{\boldsymbol{x}(\omega)\boldsymbol{x}^{\mathrm{H}}(\omega)\} \tag{6.4.1}$$

then the frequency-domain correlation matrix $\boldsymbol{R}_y(\omega)$ of output $\boldsymbol{y}(t)$is

$$\boldsymbol{R}_y(\omega) = \boldsymbol{W}(\omega)\boldsymbol{R}_x(\omega)\boldsymbol{W}^{\mathrm{H}}(\omega) \tag{6.4.2}$$

where the superscript H denotes the complex conjugated transposed operation, and $\boldsymbol{W}(\omega)$ is the frequency-domain separating matrix at the frequency ω.

For nonstationary signals such as speech, by first making a windowing process, the signals within each window can be considered as stationary; then perform FFT on the data within every window. The mixed signal within the mth window is $\underline{\boldsymbol{x}}$ (m) and T sampling points exist within each window, then

$$\underline{\boldsymbol{x}}(m) = [\boldsymbol{x}(\beta mT), \cdots, \boldsymbol{x}(\beta mT + T - 1)] \tag{6.4.3}$$

where β is the overlapping coefficient.Transforming $\underline{\boldsymbol{x}}$ (m) to frequency-domain, we have

$$\underline{\boldsymbol{x}}(\omega, m) = \sum_{\tau=0}^{T-1} \mathrm{e}^{-\mathrm{j}2\pi\omega\tau/T} \boldsymbol{x}(\beta mT + \tau) \tag{6.4.4}$$

Because of colored characteristics of speech, the correlation function with delays must be computed when calculating the correlation function. Therefore, the data with delays is taken into account when $\underline{\boldsymbol{x}}(\omega, m)$is obtained.We use shift $\underline{\boldsymbol{x}}(\omega, m, \tau)$ to express $\underline{\boldsymbol{x}}(\omega, m)$ with delay τ.The frequency-domain correlation matrix of $\boldsymbol{x}(t)$ at the mth window with frequency ω is

$$\boldsymbol{R}_{\underline{x}}(\omega, m, \tau) = E\{\underline{\boldsymbol{x}}(\omega, m)\,\mathrm{shift}\,\underline{\boldsymbol{x}}^{\mathrm{H}}(\omega, m, \tau)\} \tag{6.4.5}$$

the frequency-domain correlation matrix $\boldsymbol{R}_y(\omega$, m, $\tau)$ of $\boldsymbol{y}(t)$ at the mth window with frequency ω is

$$\boldsymbol{R}_y(\omega, m, \tau) = \boldsymbol{W}(\omega)\boldsymbol{R}_x(\omega, m, \tau)\boldsymbol{W}^{\mathrm{H}}(\omega) \tag{6.4.6}$$

When $\boldsymbol{y}(t)$ is the perfect estimation of $\boldsymbol{s}(t)$, according to the independence of the signal and the colored characteristics of speech, $\boldsymbol{R}_y(\omega, m, \tau)$ is always a diagonal matrix, which is shown by unit matrix $\boldsymbol{I}$ as follows.

$$\boldsymbol{R}_y(\omega, m, \tau) - \boldsymbol{I} = 0 \tag{6.4.7}$$

Suppose M windows and J delays exist, then

$$\sum_{m=1}^{M}\sum_{\tau=0}^{J-1} \boldsymbol{R}_y(\omega, m, \tau) - \boldsymbol{I} = 0 \tag{6.4.8}$$

Use the cost function based on the matrix Frobenius norm:

$$\phi = \frac{1}{4}\sum_{m=1}^{M}\sum_{\tau=0}^{J-1} \| \boldsymbol{W}\boldsymbol{R}_{\underline{x}}(\omega, m, \tau)\boldsymbol{W}^{\mathrm{H}} - \boldsymbol{I} \|_{\mathrm{F}} \tag{6.4.9}$$

Compute its negative gradient:

$$\begin{aligned} -\nabla\phi &= -\frac{1}{4}\frac{\partial}{\partial \boldsymbol{W}}\sum_{m=1}^{M}\sum_{\tau=0}^{J-1} \| \boldsymbol{W}\boldsymbol{R}_{\underline{x}}(\omega, m, \tau)\boldsymbol{W}^{\mathrm{H}} - \boldsymbol{I} \|_{\mathrm{F}} \\ &= -\frac{1}{4}\frac{\partial}{\partial \boldsymbol{W}}\sum_{m=1}^{M}\sum_{\tau=0}^{J-1} \mathrm{Trace}\{[\boldsymbol{W}\boldsymbol{R}_{\underline{x}}(\omega, m, \tau)\boldsymbol{W}^{\mathrm{H}} - \boldsymbol{I}][\boldsymbol{W}\boldsymbol{R}_{\underline{x}}(\omega, m, \tau)\boldsymbol{W}^{\mathrm{H}} - \boldsymbol{I}]^{\mathrm{H}}\} \\ &= \sum_{m=1}^{M}\sum_{\tau=0}^{J-1} (\boldsymbol{W}\boldsymbol{R}_{\underline{x}}(\omega, m, \tau)\boldsymbol{W}^{\mathrm{H}} - \boldsymbol{I})\boldsymbol{W}\boldsymbol{R}_{\underline{x}}(\omega, m, \tau) \end{aligned} \tag{6.4.10}$$

Then the frequency-domain blind separation algorithm based on second-order statistics for speech is obtained:

$$\begin{aligned} \boldsymbol{W}(\omega, k+1) = {} & \boldsymbol{W}(\omega, k) + \mu\sum_{m=1}^{M}\sum_{\tau=0}^{J-1} (\boldsymbol{W}(\omega, k)\boldsymbol{R}_{\underline{x}}(\omega, m, \tau)\boldsymbol{W}(\omega, k)^{\mathrm{H}} - \boldsymbol{I}) \cdot \\ & \boldsymbol{W}(\omega, k)\boldsymbol{R}_{\underline{x}}(\omega, m, \tau) \end{aligned} \tag{6.4.11}$$

Detailed implementation of Eq. (6.4.11) is shown as follows:

① The mean is removed from the original mixed signal and the result is expressed by $\boldsymbol{x}$.

② Divide $\boldsymbol{x}$ into segments with T points in each, and the mth segment of the signal is

$$\underline{\boldsymbol{x}}(m) = [\boldsymbol{x}(\beta mT), \cdots, \boldsymbol{x}(\beta mT + T - 1)] \tag{6.4.12}$$

where $0 < \beta \leqslant 1$, when $\beta=1$, no overlap exists in adjacent signal segments; when $\beta=0.5$, half of the data in the adjacent signal segments overlaps; $m=1; 2, \cdots, M$, $M=N/(\beta T)-1$, N is the length of mixed signal $\boldsymbol{x}$.

③ Perform the FFT of data in the mth window:

$$\underline{\boldsymbol{x}}(\omega, m) = \mathrm{FFT}\{\underline{\boldsymbol{x}}(m)\} \qquad \omega = 1, 2, \cdots, T \tag{6.4.13}$$

④ Compute $\underline{\boldsymbol{x}}(\omega, m)$ with delay τ. Shift $\underline{\boldsymbol{x}}(\omega, m)$ to the right by τ bits and add zero in the left to obtain shift $\boldsymbol{x}(\omega, \mathrm{m}, \tau)$, $\tau=0, 1, \cdots, J-1$.

⑤ Compute the frequency-domain correlation matrix at the mth window with frequency ω:

$$\boldsymbol{R}_{\underline{x}}(\omega, m, \tau) = \frac{1}{T-\tau}[\underline{\boldsymbol{x}}(\omega, m)\,\mathrm{shift}\,\underline{\boldsymbol{x}}(\omega, m, \tau)]$$

$$m=1,2,\cdots,T,\quad \omega=1,2,\cdots,T,\quad \tau=0,1,\cdots,J-1 \tag{6.4.14}$$

⑥ Select data of Q continuous windows to compute the average of the correlation matrix:

$$\boldsymbol{R}_{\underline{x}}(\omega,k,\tau)=\frac{1}{Q}\sum_{m=1}^{Q}\boldsymbol{R}(\omega,(k-1)Q+m,\tau)$$

$$k=1,2,\cdots,K,\quad K=\frac{N}{\beta TQ} \tag{6.4.15}$$

⑦ To initialize the separating matrix, set

$$\boldsymbol{W}=\begin{bmatrix}1 & 0\\ 0 & 1\end{bmatrix}$$

⑧ Iteratively compute the separating matrix at $\omega=\omega_l$, and in the lth iteration

• Compute the error matrix between the correlation matrix and unit matrix

$$\boldsymbol{V}^{(l)}(\omega_l,k,\tau)=\boldsymbol{W}^{(l)}\boldsymbol{R}_{\underline{x}}(\omega_l,k,\tau)\boldsymbol{W}^{(l)\mathrm{H}}-\boldsymbol{I} \tag{6.4.16}$$

• Calculate the random gradient of the Frobenius norm of the error matrix $\boldsymbol{V}^{(l)}(\omega_l, k, \tau)$

$$\frac{\partial}{\partial \boldsymbol{W}}\left\{\sum_{k=1}^{K}\sum_{\tau=0}^{J-1}\|\boldsymbol{V}^{(l)}(\omega_l,k,\tau)\|_{\mathrm{F}}\right\}$$

$$=\sum_{k=1}^{K}\sum_{\tau=0}^{J-1}\boldsymbol{V}^{(l)}(\omega_l,k,\tau)\boldsymbol{W}^{(l)}\boldsymbol{R}_{\underline{x}}(\omega_l,k,\tau) \tag{6.4.17}$$

• Iteratively compute separating matrix $\boldsymbol{W}^{(l+1)}$

$$\boldsymbol{W}^{(l+1)}=\boldsymbol{W}^{(l)}+\mu\sum_{k=1}^{K}\sum_{\tau=0}^{J-1}\boldsymbol{V}^{(l)}(\omega,k,\tau)\boldsymbol{W}^{(l)}\boldsymbol{R}_{\underline{x}}^{(l)}(\omega,k,\tau) \tag{6.4.18}$$

⑨ Adopt the frequency alignment method mentioned in the last section and repeat steps 7 and 8 until all the separating matrix for all frequency segments are obtained.

⑩ Transform the frequency-domain matrix into the time-domain:

$$\boldsymbol{W}=\mathrm{IFFT}\{\boldsymbol{W}(\omega)\}\qquad \omega=1,2,\cdots,T \tag{6.4.19}$$

⑪ Recover the original signal using separating filters in the time-domain:

$$\boldsymbol{y}=\boldsymbol{W}\boldsymbol{x} \tag{6.4.20}$$

What must be pointed out is that the energy at different frequency segments is different to a large extent. In order to obtain a good convergence Property, different learning rates should be applied at different frequency segments. The learning rate is selected according to Ref. [14],

$$\mu(\omega)=\frac{\alpha}{\sqrt{\sum_{m=1}^{M}|\boldsymbol{R}_{\underline{x}}(\omega,m,0)|^2}} \tag{6.4.21}$$

where α is a scale factor and can be set to 0.6.

A certain relationship among the order number p, the filter length q, and window length T exists. Generally, it is satisfied by the criterion,

$$p < q \ll T \tag{6.4.22}$$

The following is a simulation of acoustic signals.

The same sources as in the previous section are used here. The first is a Chinese reading voice, and the second is light music. They are recorded independently with a sampling frequency of 20 kHz and a duration of 7 seconds with 140,000 sampling points. Figure 6.3 shows the waveforms of the two source signals.

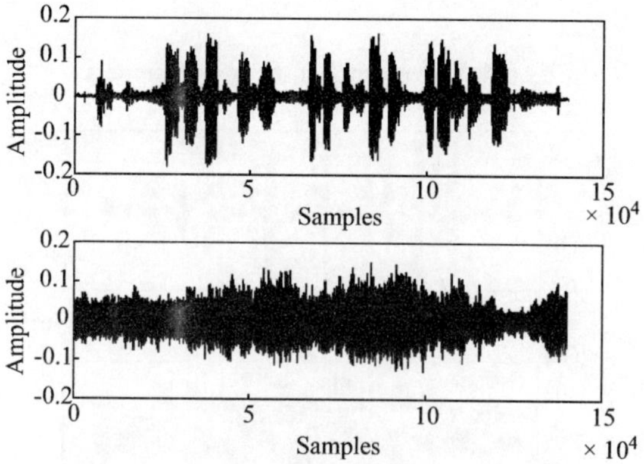

Fig. 6.3 Waveforms of the source signals. (The top is the Chinese reading speech signal, and the bottom is the music signal)

Two source signals are handled first by mixing matrix $\boldsymbol{H}$; every element in the matrix is a filter. In this experiment,

$$\boldsymbol{H} = \begin{bmatrix} h_{11} & h_{12} \\ h_{21} & h_{22} \end{bmatrix}$$

where

$$h_{11}=[1, 0.8, 0.7, 0.4, 0.3, 0.25, 0.2, 0.15, 0.1]$$

$$h_{12}=[0.6, 0.5, 0.5, 0.4, 0.3, 0.2, 0.25, 0.1,0.05]$$

$$h_{21}=[0.5, 0.5, 0.4, 0.35, 0.3, 0.3, 0.2,0.1, 0.05]$$

$$h_{22}=[1, 0.9, 0.8, 0.6, 0.4, 0.35, 0.3, 0.15, 0.08]$$

and the waveforms of the two mixed signals generated from the signals in Fig. 6.3 are shown in Fig. 6.4.

Separate the mixed signal using Eq. (6.4.11) and the alignment method in the previous section. According to the criterion of $p<q\ll T$, let T=500, β=1, J=60, and q=20, determine the learning rate with Eq. (6.4.21). After 3,000 iterations, the waveforms of separated signals are shown in Fig. 6.5.

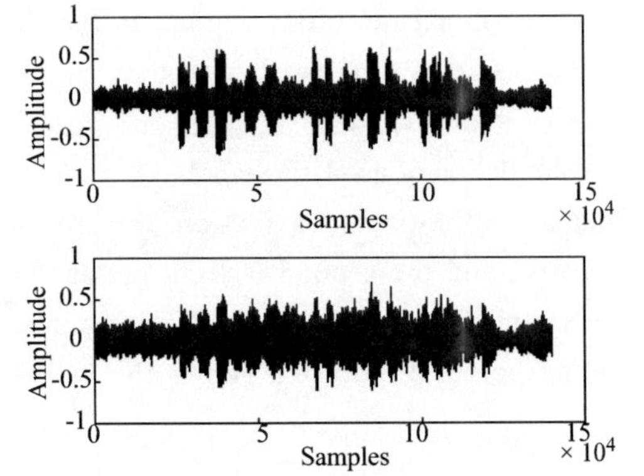

Fig. 6.4 Waveforms of the mixed signals

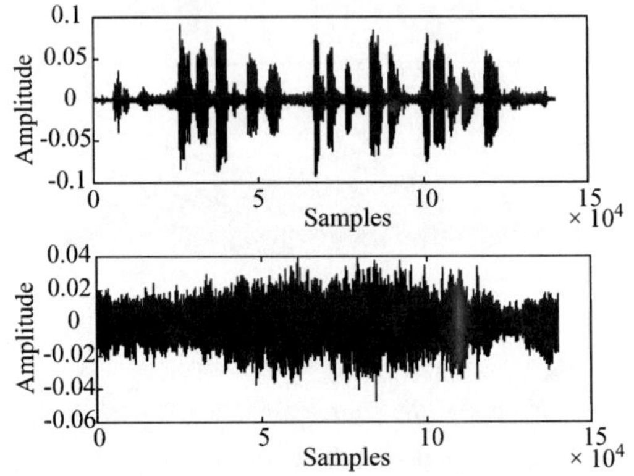

Fig. 6.5 Waveforms of separated signals

Comparing the waveforms of the separated signals with the original one, we can find that the waveforms of the original signals are well recovered through blind separation. In order to evaluate the separation effect quantitatively, we compute the SIR before and after separation. The former is 1.5459 and the latter 7.1740, which indicates a large improvement of SIR.

In Fig. 6.6, the amplitudes of C_{11} and C_{22} are larger than that of C_{12} and C_{21} by three times. It means that the first output mainly contains the first source signal and the second contains the second source signal.

6.5 Bussgang Method

The Bussgang algorithm is a blind deconvolution method developed by Bellini (1994) for the blind equalization of communication channels. It was extended

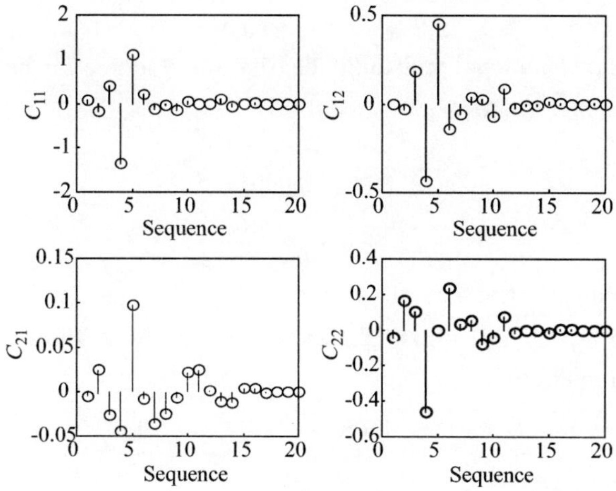

Fig. 6.6 Schematic of the system filter

into a multi-channel blind deconvolution method by Lambert (1996). Further researches by Bell, Sejnowski (1995), Lambert (1997), Girolami, Fyfe (1997) and Lee (1997) *et al* showed that the Bussgang algorithm is equivalent to an information theory-based learning algorithm.

A white zero-mean random process with a Bussgang Property must satisfy the condition[15,16]

$$E(u(t)u(t+k)) = E(f[u(t)]u(t+k)) \tag{6.5.1}$$

where k is the delay label and $f(\cdot)$ is the Bussgang nonlinearity in the form of the non-linear monotone functions. This nonlinearity shows that the autocorrelation function of $u(t)$ is equal to the cross-correlation function of $u(t)$ and the nonlinear output $f[u(t)]$. The above Property can be extended from time samples to space samples, that is

$$E(u_i u_j) = E(f(u_i)u_j) \tag{6.5.2}$$

here subscripts i, j indicate that it is an independent white random process. The nonlinear function $f(\cdot)$ can be regarded as the combination of a higher-order Taylor series expansions. The left side of Eq. (6.5.2) is a second-order cross-correlation of two estimated sources, and the right side is the higher-order cross-correlation of two estimated sources.

The derivation of learning rules still adopts the MSE method between estimated u_i and real source s_i. In order to minimize the MSE between the estimated sources and Bussgang nonlinearity $f(u_i)$, we define the evaluation function C as follows:

$$C = E((u_i - f(u_i))^2) \tag{6.5.3}$$

Now, define an error variable

$$e_i = s_i - f(u_i) \tag{6.5.4}$$

for given e_i, the conditional probability density of the source can be described by maximum posterior (MAP):

$$p(s_i \mid e_i) = p(e_i \mid s_i)p(s_i) \tag{6.5.5}$$

Suppose $p(e_i \mid s_i)$ is subjected to Gaussian distribution,

$$p(e_i \mid s_i) = K\exp\left(-\frac{(e_i - s_i)^2}{2\sigma_u^2}\right) \tag{6.5.6}$$

where K is a constant and σ_u^2 is the variance of u_i.

Substitute Eq. (6.5.6) into (6.5.5) and take the logarithm; the logarithm conditional probability is

$$\lg(p(s_i \mid e_i)) = \lg(K) - \frac{(e_i - s_i)^2}{2\sigma_u^2} + \lg(p(s_i)) \tag{6.5.7}$$

Take the derivative

$$\frac{\partial \lg(p(s_i \mid e_i))}{\partial s_i} = \frac{(e_i - s_i)}{\sigma_u^2} + \frac{\partial \lg(p(s_i))}{\partial s_i} \tag{6.5.8}$$

when the estimation error is minimized, the above formula equals zero and the equation for e_i is obtained as

$$e_i = s_i - \sigma_u^2 \frac{\partial \lg(p(s_i))}{\partial s_i} \tag{6.5.9}$$

Compare Eq. (6.5.9) with (6.5.4) and suppose unit variance; the expression of the nonlinear Bussgang estimator is

$$f(s_i) = \frac{\dfrac{\partial p(s_i)}{\partial s_i}}{p(s_i)} \tag{6.5.10}$$

it is proportional to the derivative of the logarithm density of the real source distribution.

Applying Eq. (6.5.10) to (6.5.2) and rewriting in matrix form gives

$$E\{\boldsymbol{u}\boldsymbol{u}^{\mathrm{T}}\} = E\{f(\boldsymbol{u})\boldsymbol{u}^{\mathrm{T}}\}$$

$$E\{[\boldsymbol{W}\boldsymbol{A}\boldsymbol{s}\boldsymbol{s}^{\mathrm{T}}\boldsymbol{A}^{\mathrm{T}}\boldsymbol{W}^{\mathrm{T}}]\} = E(f(\boldsymbol{u})\boldsymbol{u}^{\mathrm{T}}) \tag{6.5.11}$$

Suppose $\boldsymbol{W}=\boldsymbol{A}^{-1}$, the left of Eq. (6.5.11) is a unit matrix. Multiplying by $\boldsymbol{W}$ gives

$$[\boldsymbol{I} - E\{f(\boldsymbol{u})\boldsymbol{u}^{\mathrm{T}}\}]\boldsymbol{W} = 0 \tag{6.5.12}$$

Compared with ICA learning rules, we can see that the optimal nonlinear Bussgang function $f(\boldsymbol{u})$ is the evaluation function $\varphi(\boldsymbol{u})$ of the Infomax learning rules.

6.6 Multi-channel Blind Deconvolution

The purpose of multi-channel blind deconvolution is to design $\boldsymbol{W}(z, t)$ to satisfy the following conditions of a composite impulse response[17,18] without any knowledge of $\boldsymbol{H}(z)$.

$$\lim_{t\to\infty} \boldsymbol{C}(z,t) = \boldsymbol{PD\Lambda}(z) \tag{6.6.1}$$

where $\boldsymbol{C}$ is the impulse response of the system, $\boldsymbol{P}$ is an $m\times m$ permutation matrix, $\boldsymbol{D}$ is an $m\times m$ nonsingular diagonal matrix, $\boldsymbol{\Lambda}$ is a delay matrix and m is the number of channels. The multi-channel blind deconvolution is to recover source signals under the indeterminacy conditions of permutation and the scale of source signals, as well as delays. We can consider that the whole system is transparent and channel $\boldsymbol{H}(z)$ is equalizable. When Eq. (6.6.1) is satisfied, the ith estimated signal $y_i(t)$ is the estimation of the waveform of the jth source signal $s_i(t)$ with the delay Δ_i, i. e. $y_i(t)=d_{ii}s_j(t-\Delta_i)$.

The on-line estimation model of multi-channel blind deconvolution is

$$\boldsymbol{y}(t) = \sum_{p=-\infty}^{+\infty} \boldsymbol{W}_p(t)\boldsymbol{x}(t-p) \tag{6.6.2}$$

For off-line or batch processing learning, then

$$\boldsymbol{y}^{(l)}(t) = \sum_{p=-\infty}^{+\infty} \boldsymbol{W}_p(l)\boldsymbol{x}(t-p) \tag{6.6.3}$$

where $\boldsymbol{y}^{(l)}(t)=[y_1^{(l)}(t), y_2^{(l)}(t), \cdots, y_m^{(l)}(t)]^{\mathrm{T}}$ is the vector of m estimated signals in the lth iteration, $\{\boldsymbol{W}_p(l), -\infty \leqslant p \leqslant +\infty\}$ is the separating filter matrix with the delay p, when it is expressed with the operator z,

$$\boldsymbol{y}^{(l)}(t) = \boldsymbol{W}(z,l)[\boldsymbol{x}(t)] = \boldsymbol{C}(z,l)[\boldsymbol{s}(t)] + \boldsymbol{W}(z,l)\boldsymbol{n}(t) \tag{6.6.4}$$

where

$$\boldsymbol{W}(z,l) = \sum_{p=-\infty}^{+\infty} \boldsymbol{W}_p(l) z^{-p}$$

$$\boldsymbol{C}(z,l) = \boldsymbol{W}(z,l)\boldsymbol{H}(z)$$

are the z-transform of the equalizer and impulse response of whole system channel respectively.

The system diagram for multi-channel blind deconvolution is illustrated in Fig. 6.7.

Blind deconvolution algorithms have intrinsic indeterminacies [16~18], which

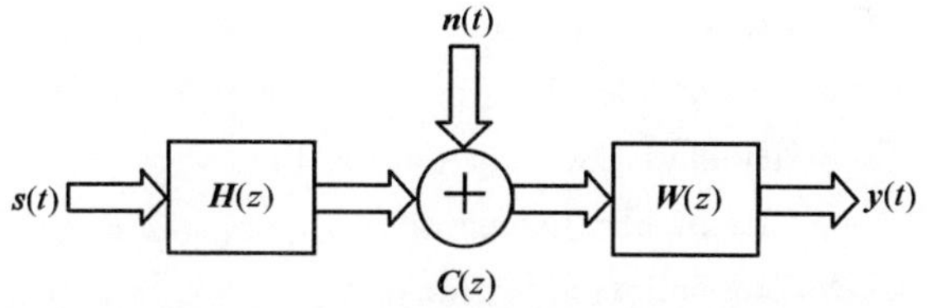

Fig. 6.7 System diagram of multichannel blind deconvolution

mainly contain the indeterminacy of amplitude, delays and permutation.

Discussion of the multi-channel blind deconvolution algorithm as related to nonstationary signals and noisy signals will be expanded in details in Chapter 9 of this book.

References

[1] Widrow B, Stearns S (1985) Adaptive signal processing. Prentice Hall, New Jersey

[2] Haykin S (1994) Blind deconvolution. Prentice Hall, New Jersey

[3] Common P (1996) Contrasts for multichannel blind deconvolution.Signal Processing Letters 3(7): 209–211

[4] Bell A J, Sejnowski T J (1995) An information-maximization approach to blind separation and blind deconvolution. Neural Computation 7(6): 1129–1159

[5] Ehlers F, Schuster H G (1997) Blind separation of convolutive mixturess and an application in automatic speech recognition in noisy environment. IEEE Transactions on Signal Processing 45(10): 2609–2619.

[6] Gerven S V, Compernolle D V (1995) Signal separation by symmetric adaptive decorrelation: Stability, converge, and uniqueness. IEEE Transactions Signal Processing 43(7): 1602–1612

[7] Kawamoto M, Barros A K, Mansour A et al (1999) Real world blind separation of convolved non-stationary signals. In: Proceedings of ICA99. Aussois, 1999, pp 365–370

[8] Smaragdis P (1998) Blind separation of convolved mixturess in the frequency domain. Neurocomputing 22(1): 21–34

[9] Chan D C B (1997) Blind signal separation. Dissertation, Cambridge University

[10] Schobben D W E, Sommen P C W (1998) A new blind signal separation algorithm based on second order statistics. In: Proceedings of IASTED International Conference Signal Image Processing, Las Vegas, 1998

[11] Ikram M Z, Morgan D R (2000) Exploring permutation inconsistency in blind separation of speech signals in a reverberant environment. In: ICASSP '2000, Istanbul, 2000, pp1041–1044

[12] Jia P (2003) Study of blind separation of acoustic signal. Dissertation, Shanghai Jiao Tong University (Chinese)

[13] Murata N, Ikeda S, Ziche A (2001) An approach to blind source separation

based on temporal structure of speech signals. Neurocomputing 41(1): 1–24

[14] Ikram M Z (2001) Multichannel blind separation of speech signals in a reverberant environment. Dissertation, Georgia Institute of Technology

[15] Lee T W (1998) ICA theory and applications. Kluwer Academic, Boston

[16] Haykin S (2000) Unsupervised adaptive filtering, Vol. Ⅱ, Blind deconvolution. Wiley, New York

[17] Inouye Y (1997) Criteria for blind deconvolution of multichannel linear time-invariant systems of nonminimum phase. In: Katayama T. Sugimoto S (eds) Statistical methods in control and signal processing. Dekker, New York, pp 375–397

[18] Inouye Y, Liu R W (2002) A system-theoretic foundation for blind equalization of an FIR MIMO channel system.Circuits and Systems Ⅰ: Fundamental Theory and Applications. IEEE Transactions on Signal Processing 49(4): 425–436

Chapter 7 Blind Processing Algorithm Based on Probability Density Estimation

7.1 Advancing the Problem

Most algorithms for resolving the problems of blind source separation (BSS) and blind deconvolution are mainly focused on blind separation for mixed signals using higher order statistics of random signals, including third-order (for sources with dissymmetrical distribution) and fourth-order cumulant (for sources with symmetrical distribution). Usually, this kind of method directly employs an algebraic structure of a mixed signal cumulant matrix. The common method makes eigenvalues decomposition to the cumulant matrix that is estimated by the mixed signal samples. It can also perform joint diagonalization for the cumulant matrix through complex matrix transformation to estimate the mixing matrix in order to solve the BSS problem.

Another method of resolving the problem of blind source separation is the independent component analysis methods based on neural networks. It processes mixed signals using the neural network $\boldsymbol{W}$ to optimize a contrast function through unsupervised adaptive learning. It also adjusts the weights constantly to make $\boldsymbol{W}(z)\boldsymbol{H}(z)$ converge gradually to the product of a generalized permutation matrix $\boldsymbol{P}$, diagonal matrix $\boldsymbol{D}$ and delayed matrix $\boldsymbol{\Lambda}(z)$; that is to say $\boldsymbol{W}(z)\boldsymbol{H}(z)=\boldsymbol{P}\boldsymbol{D}\boldsymbol{\Lambda}(z)$, and provides the estimation of the source signal waveform at the output of the network at the same time.

Contrasting this with blind source separation based on a high cumulant, We find a remarkable advantage of blind source separation based on neural networks because its algorithm realization is comparatively simpler.A blind source separation algorithm with equivariant Property is achieved if using a relative gradient and natural gradient to optimize the evaluation function of the network (i.e. the contrast function in the blind separation problem). This means that the Property of the algorithm is not relying on a specific mixing matrix, and the component with relatively little power in signals can be recovered. In addition, for a neural networks based independent component analysis algorithm, which optimizes the contrast functions using relative gradient, the convergence speed is also very fast.

The main disadvantage of most blind source separation algorithms based on neural networks is that only homologous mixed signals can be separated. That is to say, corresponding to different source signals, the sub-Gaussian signals with cumulant less than zero or the super-Gaussian signals with cumulant more than zero result in different nonlinear functions being selected as activation functions. Because the nonlinear function fitting both sub-Gaussian and super-Gaussian signals does not exist, the blind separation of the hybrid mixed signals is not realized effectively by the neural network algorithm.

A representative algorithm in the blind source separation algorithms based on neural networks is the equivariant adaptive separation algorithm[1]. It is named the EASI algorithm (Equivariant Adaptive Separation via Independence). Most other neural network algorithms developed after it are seen as special cases or extensions of the EASI algorithm.

With regard to most algorithms, the key problem is the probability density function (PDF) estimation of the source signals, where the nonparametric estimation method is also rooted in the basic statistical theory. the nonparametric estimation method estimates directly the probability distribution using signal samples on the condition that it is ignorant of a distribution form of random signals or closed formula. Convergence to any unknown density is ensured only if there are enough samples that properly watch the condition of unknown source singals in the blind separation problem. So the nonparametric estimation method is very useful in solving the estimation problem of the probability density function in blind separation.

7.2 Nonparametric Estimation of Probability Density Function

The probability density function of random variables is a basic and important concept of statistics. After determining the probability density function $f(x)$ of random variable x, the probability about x located in the interval (a,b) can be calculated expediently for any $a<b$:

$$p(a<x<b)=\int_a^b f(x)\mathrm{d}x \tag{7.2.1}$$

Assume that there is one group sample $\{x_1,x_2,\cdots,x_N\}$ of random variable x

whose probability density function is unknown. The probability density function can be estimated by using this group of samples in practice.

The parametric estimation is a general method of probability density function estimation in statistics[2]. First assume that x is subjected to a certain known statistical distribution according to some prior knowledge of random variable x, and the parametric model $f(x;\boldsymbol{\theta})$ of its probability density function is provided; then the unknown parameter $\boldsymbol{\theta}$ of the parametric model of the probability density function is estimated by using the samples $\{x_1, x_2,\cdots, x_N\}$ of x and a certain estimation method, such as a maximum likelihood method. At last, the estimation $f(x;\hat{\boldsymbol{\theta}})$ of the probability density function to x is obtained by replacing $\boldsymbol{\theta}$ of $f(x;\boldsymbol{\theta})$ with $\hat{\boldsymbol{\theta}}$. For example, assuming that random variable x has a Gaussian distribution, because the probability density function is determined entirely by mathematical expectation and variance, the estimation of the probability density function to x is equivalent to the estimation of its mean value μ and variance σ^2, i.e. $\boldsymbol{\theta}=(\mu,\sigma^2)$.

The advantage of the parametric estimation method of the probability density function of the random variable is that its calculation is relatively simple and the theory is more mature, but the statistical distribution of the random variable must be assumed in advance in this approach. This approach is not applied effectively when the distribution of the random variable is not inferred rationally because of insufficient prior knowledge. In addition, when the probability density function of the random variable does not belong to any known distribution family, the result of using the parametric modeling method may not be satisfactory.

Besides the parametric estimation method of the probability density function, there is the nonparametric method[3] that does not make any prior assumption on the probability density function of random variable x; thus, the data "speak for themselves."

7.2.1 Kernel Function Method of Probability Density Estimation

There is some relationship between the nonparametric estimation of the probability density function and a histogram. According to the definition of the probability density function of a random variable, if $f(x)$ is the probability density function of random variable x, there is

$$f(x) = \lim_{h\to 0}\frac{1}{h}P(x-h<x<x+h) \tag{7.2.2}$$

for any h, the proportion between samples is located in the interval $(x-h,x+h)$

and all samples can be calculated conveniently. Therefore, one natural nonparametric estimation method of the probability density function is to select one minor number h at first, then make an estimation of $f(x)$ by using the following formula:

$$\hat{f}(x)=\frac{1}{2hN}[\text{samples of } \{x_1,x_2,\cdots,x_N\} \text{ located in the interval}(x-h,x+h)] \tag{7.2.3}$$

define the weighting function as

$$w(x)=\begin{cases}\frac{1}{2} & |x|<1\\ 0 & |x|\geqslant 1\end{cases} \tag{7.2.4}$$

then Eq. (7.2.3) can be expressed in the following more convenient form:

$$\hat{f}(x)=\frac{1}{N}\sum_{m=1}^{N}\frac{1}{h}w\left(\frac{x-x_m}{h}\right) \tag{7.2.5}$$

The form of the probability density function can now be estimated conveniently by using Eq. (7.2.5). Figure 7.1 is the graph obtained from the estimation of the probability density function of certain Gaussian random variables using Eq. (7.2.5). Apparently, the basic form of the probability density function estimated is correctly reflected by Fig. 7.1.

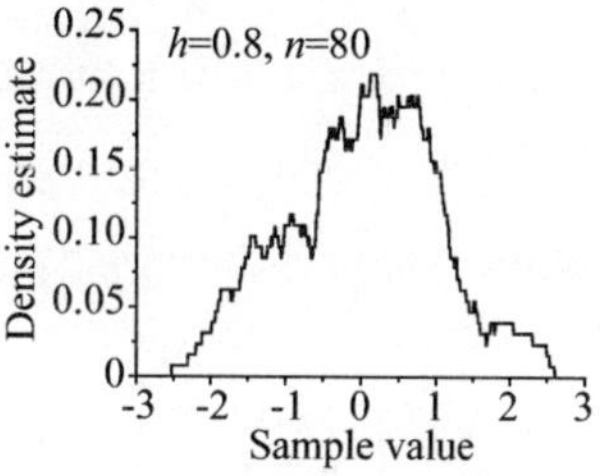

Fig. 7.1 Probability density function obtained from Eq. (7.2.5)

Viewing Fig. 7.1, we see that the estimation of the probability density function by Eq. (7.2.5) is a discontinuous function, and the estimate has discontinuities at $x_m \pm h$, while some derivatives are zero at other values.These cause the estimation of the probability density function to have a sawtooth form fluctuation.

Apparently, the sawtooth form fluctuation of the estimation of the probability density function in Fig. 7.1 is owing to the discontinuous weighting function in Eq. (7.2.5). A suitable method to get a continuous estimation of the probability density function is to replace the weighting function $w(x)$ in Eq. (7.2.4) with continuous function $K(x)$. Generally the continuous weighting function is called the kernel function. Except for a continuous Property condition, the kernel function $K(x)$ used in the nonparametric estimation method of the probability density function must satisfy the following several assumptions[4]:

① $K(x)$ is symmetrical, i.e. $K(x)=K(-x)$.

② $\int_{-\infty}^{+\infty} K(x)\mathrm{d}x=1$.

③ $\int_{-\infty}^{+\infty} x^j K(x)\mathrm{d}x = 0$ for $j=1,2,\cdots,k-1$, and $\int_{-\infty}^{+\infty} x^k K(x)\mathrm{d}x \neq 0$.

Usually the continuous function $K(x)$ satisfying the above three assumptions is called the kernel of order k. Replacing the weighting function defined by Eq. (7.2.4) in Eq. (7.2.5) with $K(x)$, the estimation of the probability density function by the kernel function method is given by[5~8]

$$\hat{f}(x) = \frac{1}{N}\sum_{m=1}^{N}\frac{1}{h}K\left(\frac{x-x_m}{h}\right) \tag{7.2.6}$$

$K(x)$ is an even function due to the symmetric assumption of the kernel function, and $\hat{f}(x)$ is a density function for the second assumption of the kernel function, namely $\int_{-\infty}^{+\infty}\hat{f}(x)\mathrm{d}x = 1$. When $k=2$, one may choose $K(x)\geqslant 0$, i.e. a nonnegative kernel function, and there is $\hat{f}(x)\geqslant 0$, $K(x)$ and $\hat{f}(x)$ themselves are probability density functions. When $k>2$, $K(x)$ must take negative values in some places, which can make the ultimate estimation $\hat{f}(x)$ negative, therefore, second order kernels are most frequently used in kernel density function estimation. One of the most popular second-order kernels is the Gaussian kernel $\phi(x)=\frac{1}{\sqrt{2\pi}}\mathrm{e}^{-(1/2)x^2}$.

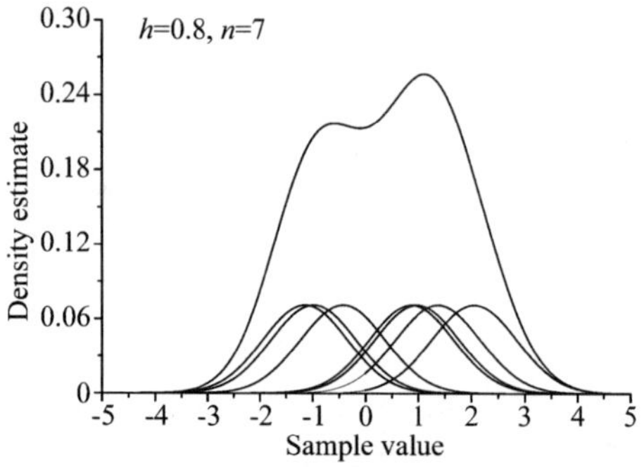

Fig. 7.2 Sketch of estimation of probability density function by the kernel function method

From the discussion above, the kernel function method of estimation of the probability density function using second-order kernel weights smoothly for every sample of random variable x through a series of continuous function windows, It takes the average for all weighting windows and obtains the estimation for the probability density function of x. In Eq. (7.2.6) of the probability density estimation by the kernel function method, the parameter h

is window width, which is usually called a smoothing parameter or bandwidth. A sketch of the kernel function method of estimation of the probability density function is shown in Fig. 7.2. For the estimation of the probability density function by the kernel function method, the choice of the bandwidth h is much more important than the choice of the kernel function $K(x)$ [3].

In order to quantitatively analyze the accuracy of the probability density function estimation by the kernel function method, namely the difference between $\hat{f}(x)$ and the real probability density function $f(x)$, more than one measurement method can be used. The estimation of the probability density function depends on samples $\{x_1,x_2,\cdots,x_N\}$ of random variable x; therefore, there also is one random variable for every x, $\hat{f}(x)$. When the estimation quality of $\hat{f}(x)$ at the single point x is considered, one of the natural measures is the mean square error (MSE):

$$\mathrm{MSE}(h) = E\{[\hat{f}(x) - f(x)]^2\} \tag{7.2.7}$$

When estimating accuracy for $\hat{f}(x)$ on the real number axis, one can use the mean integrated square error, namely

$$\mathrm{MISE}(h) = E\left\{\int_{-\infty}^{+\infty} [\hat{f}(x) - f(x)]^2 \mathrm{d}x\right\} \tag{7.2.8}$$

apparently, after the kernel function $K(x)$ is selected, MISE of $\hat{f}(x)$ is determined by the bandwidth of kernel function h, and an ideal bandwidth is determined by the following equation:

$$\mathrm{MISE}(h_*) \leqslant \mathrm{MISE}(h), \quad \forall h > 0 \tag{7.2.9}$$

except for some special cases (for example, if the kernel function $K(x)$ is standard Gaussian function and the real probability density function $f(x)$ is also Gaussian function), the optimal bandwidth h_* satisfying Eq. (7.2.9) does not generally have a closed formula.

From Eq. (7.2.8), we have

$$\begin{aligned}
\mathrm{MISE}(h) &= \int_{-\infty}^{+\infty} E\{[\hat{f}(x) - f(x)]^2\}\mathrm{d}x \\
&= \int_{-\infty}^{+\infty} E\{\hat{f}(x) - E[\hat{f}(x)] + E[\hat{f}(x)] - f(x)\}^2 \mathrm{d}x \\
&= \int_{-\infty}^{+\infty} (E\{\hat{f}(x) - E[\hat{f}(x)]\}^2 + E\{E[\hat{f}(x)] - f(x)\}^2)\mathrm{d}x \\
&= \int_{-\infty}^{+\infty} \mathrm{Var}\hat{f}(x)\mathrm{d}x + \int_{-\infty}^{+\infty} \mathrm{bias}^2 \hat{f}(x)\mathrm{d}x \\
&= \mathrm{IV}(h) + \mathrm{IB}(h)
\end{aligned} \tag{7.2.10}$$

That is, the mean integrated square error of the kernel function density estimation method can be decomposed to the sum of integrated variance IV(h)= $\int_{-\infty}^{+\infty} \mathrm{Var}\hat{f}(x)\mathrm{d}x$ and integrated deviation square $\mathrm{IB}(h) = \int_{-\infty}^{+\infty} \mathrm{bias}^2 \hat{f}(x)\mathrm{d}x$ of $f(x)$. If bandwidth h is very small, sample number N of random variable x is very large, and the samples $\{x_1,x_2,\cdots,x_N\}$ are i.i.d. for second-order kernels, using the Taylor series expansion for both the integrated variance and integrated deviation square of Eq. (7.2.10); Ignoring the high order component of h, and then taking a summation, the following approximate mean integrated square error (AMISE) can be obtained:

$$\mathrm{AMISE}(h) = \frac{1}{4}h^4 \left(\int_{-\infty}^{+\infty} t^2 K(t)\mathrm{d}t\right)^2 \int_{-\infty}^{+\infty} f''(x)^2 \mathrm{d}x + \frac{1}{Nh}\int_{-\infty}^{+\infty} K(t)^2 \mathrm{d}t \tag{7.2.11}$$

where $f''(x)$ is the second-order derivative of the probability density function $f(x)$. The AMISE(h) is asymptotically the same as the MISE(h) when $N\to\infty$ and $h=h(N)\to 0$ in such a way that $Nh(N)\to\infty$.

The expression of ideal bandwidth h_{opt}, which makes the approximate integral mean square error AMISE(h) minimum, is achieved through a fairly simple calculation, as follows:

$$h_{\mathrm{opt}} = \left\{\int_{-\infty}^{+\infty} t^2 K(t)\mathrm{d}t\right\}^{-2/5} \left\{\int_{-\infty}^{+\infty} K(t)^2 \mathrm{d}t\right\}^{1/5} \left\{\int_{-\infty}^{+\infty} f''(x)^2 \mathrm{d}x\right\}^{-1/5} N^{-1/5} \tag{7.2.12}$$

In literature about kerual function bandwictth selection, when an explicit expression (7.2.12) of ideal bandwidth h_{opt} is gotten, AMISE(h) usually is used as the measure of accuracy of the probability density fouction estimation.

Two functionals are defined as follows:

$$R(f) = \int_{-\infty}^{+\infty} f^2(x)\mathrm{d}x \tag{7.2.13}$$

and

$$\sigma_K^2 = \int_{-\infty}^{+\infty} x^2 K(x)\mathrm{d}x \tag{7.2.14}$$

then Eq. (7.2.12) can be written as

$$h_{\mathrm{opt}} = \left\{\frac{R(K)}{\sigma_K^4 R(f'')}\right\}^{1/5} N^{-1/5} \tag{7.2.15}$$

in Eq. (7.2.15), the only unknown is $R(f'')$, which depends on the estimated real probability density function $f(x)$.

One of the many bandwidth selection methods is called the inserting method. The unknown variable $R(f'')$ of Eq. (7.2.15) is estimated at first, and then the estimated $\hat{R}(f'')$ is substituted into Eq. (7.2.15) to obtain the "optimal" bandwidth for the kernel function method. Different bandwidth selection methods are obtained by using different estimation methods for $R(f'')$.

7.2.2 Experiential Bandwidth Selection of Kernel Function Method

As a rule of thumb in bandwidth, the unknown probability density function $f(x)$ is supposed to have some standard distribution family $f(x;\boldsymbol{\theta})$ that is called the reference family of distribution; then the estimated value $\hat{\boldsymbol{\theta}}$ of $\boldsymbol{\theta}$ is obtained by the parametric estimation method.The $f(x;\hat{\boldsymbol{\theta}})$ is used in Eq. (7.2.15) for the ideal bandwidth estimation.

In general, the reference family of distribution of the experiential selection method is a standard normal distribution $\phi(x)=\frac{1}{\sqrt{2\pi}}\mathrm{e}^{-(1/2)x^2}$. If variance of standard normal distribution is σ^2, then

$$\int_{-\infty}^{+\infty} f''(x)^2\mathrm{d}x=\sigma^{-5}\int_{-\infty}^{+\infty}\phi''(x)^2\mathrm{d}x$$
$$=\frac{3}{8}\pi^{-1/2}\sigma^{-5}\approx 0.212\sigma^{-5} \quad (7.2.16)$$

When using the Gaussian kernel function, the optimal bandwidth estimation is achieved by substituting Eq. (7.2.16) into Eq. (7.2.15) (denoted h_{rot1})[3]:

$$h_{\mathrm{rot1}}=1.06\hat{\sigma}N^{-1/5} \quad (7.2.17)$$

where $\hat{\sigma}$ is the estimated value of σ calculated by using samples $\{x_1, x_2,\cdots, x_N\}$.

If $\hat{\sigma}$ is replaced by interquartile range $\hat{\lambda}$ as the measure of sample distribution, then Eq. (7.2.17) is changed to[3,4]

$$h_{\mathrm{rot2}}=0.79\hat{\lambda}N^{-1/5} \quad (7.2.18)$$

Normal distribution PDF can give good results by using the bandwidth selection method of Eqs. (7.2.17) and (7.2.18). These two equations can also provide satisfactory results for the case of unimodal PDF. But for bimodal or multimodal PDF, both bandwidth estimation h_{rot1} and h_{rot2} are bigger than normal, which makes the last PDF estimation result too smooth hiding some useful structure information, such as a bimodal Property, multimodal Property and so on. One eclectic method is to use the following equation to estimate bandwidth:

$$h_{\mathrm{rot}}=0.9\min(\hat{\sigma},\hat{\lambda}/1.34)N^{-1/5} \quad (7.2.19)$$

The bandwidth h_{rot} obtained from Eq. (7.2.19) can be applied to the estimation of the unimodal PDF better, and as to the estimation effect of bimodal PDF, it is also better than using bandwidth h_{rot1} and h_{rot2}.

7.2.3 Sheather-Jones Bandwidth Selection of Kernel Function Method

The experiential bandwidth selection method was attributed to the first generation bandwidth selection method, whose advantage is convenience of calculation, but this method is only suitable to estimation of the unimodal PDF, because the bandwidth obtained from the experiential bandwidth selection method is too large in general.

Theory proved that the nonparametric kernel function method used to estimate the unknown variable $R(f'')$ of Eq. (7.2.15) is more ascendant than the parametric estimation method. When using the kernel function method to estimate $R(f'')$, a suitable bandwidth denoted by g is needed. Apparently, the optimal bandwidth used to estimate $R(f'')$ is different from the bandwidth used to estimate PDF $f(x)$. In general, the selection of bandwidth g has a smaller influence on $\hat{f}(x)$ estimation quality than on the selection of bandwidth h; therefore, it is usual to use the parametric estimation method to select a bandwidth g. According to different calculation methods for $R(f'')$, we can have several second-generation bandwidth selection methods whose properties are more ascendant than the experiential bandwidth selection method. In the following, the calculation steps of the Sheather-Jones bandwidth selection method[5~7] (SJ bandwidth selection method) advanced by sheather and Jones are introduced briefly.

The SJ bandwidth selection method calculates the optimal bandwidth h_{opt} by solving for the root of the following equation:

$$H-\left[\frac{R(K)}{\sigma_K^4 R(\hat{f}''(\cdot\,;g(H)))}\right]^{1/5} N^{-1/5}=0 \tag{7.2.20}$$

where $R(\hat{f}''(\cdot\,; g(H)))$ denotes that the kernel function method uses the bandwidth $g(H)$ depending on H to obtain the estimation of $R(f'')$. Because the estimation of $R(f'')$ is involved in the estimation of $R(f''')$, the calculation process is complicated.

When the standard normal distribution function $\phi(x)=\dfrac{1}{\sqrt{2\pi}}\mathrm{e}^{-(1/2)x^2}$ is selected as a PDF estimated kernel function, the whole calculation process of the SJ bandwidth selection method is shown as follows:

Step 1: Calculate the interquartile range $\hat{\lambda}$ of samples $\{x_1, x_2, \cdots, x_N\}$ distribution of random variable x;

Step 2: Calculate $a=0.920\hat{\lambda}N^{-1/7}$, $b=0.912\hat{\lambda}N^{-1/9}$;

Step 3: Calculate $\hat{R}_a(f'') = \frac{1}{(N(N-1))}\sum_{i=1}^{N}\sum_{j=1}^{N}\phi^{(4)}\left(\frac{x_i - x_j}{a}\right)$;

Step 4: Calculate $\hat{R}_b(f''') = \frac{-1}{(N(N-1))}\sum_{i=1}^{N}\sum_{j=1}^{N}\phi^{(6)}\left(\frac{x_i - x_j}{b}\right)$;

Step 5: Calculate $g(H) = (6\sqrt{2})^{1/7}\left[\frac{\hat{R}_a(f'')}{\hat{R}_b(f''')}\right]^{1/7} H^{5/7}$;

Step 6: Calculate . $\hat{R}_{g(H)}(f'') = \frac{-1}{(N(N-1))}\sum_{i=1}^{N}\sum_{j=1}^{N}\phi^{(4)}\left(\frac{x_i - x_j}{g(H)}\right)$;

Step 7: Calculate $\mathrm{SJ}(H) = \left(\frac{1}{2\sqrt{\pi}}\right)^{1/5} R_{g(H)}(f'')^{-1/5} N^{-1/5} - H$;

Step 8: Solve the root of equation SJ(H)=0 and regard the result as the optimal bandwidth h_{opt} of the SJ bandwidth selection method.

In step 8, the Newton method is used to solve the equation SJ (H)=0.

7.3 Estimation of Evaluation Function

In order to estimate the evaluation function of the BSS algorithm based on the natural gradient method, it is also necessary to estimate the derivative of the signal PDF. Consider using samples $\{x_1, x_2, \cdots, x_N\}$ of the random signal x to estimate the derivative $f'(x)$ of PDF $f(x)$. In general, the k-order derivative of $f(x)$ can be estimated by the k-order derivative of $\hat{f}(X)$, which is the kernel function method of estimation of $f(x)$[8,9], namely

$$\hat{f}^{(k)}(x) = \frac{1}{Nh^{k+1}}\sum_{i=1}^{N} K^{(k)}\left(\frac{x - x_i}{h}\right) \tag{7.3.1}$$

The estimation of the kernel function method for the first derivative $f'(x)$ of the PDF is achieved when k=1 in Eq. (7.3.1).

Using the same kernel function $K(x)$, we can obtain the estimation of the evaluation function by substituting estimation value $\hat{f}(x)$ and $\hat{f}'(x)$ into the definition of the evaluation function, as

$$\hat{\varphi}(x) = -\frac{\hat{f}'(x)}{\hat{f}(x)} = -\frac{\sum_{i=1}^{N} K'\left(\frac{x-x_i}{h}\right)}{h\sum_{i=1}^{N} K\left(\frac{x-x_i}{h}\right)} \tag{7.3.2}$$

When selecting the Gaussian kernel function $K(x) = \frac{1}{\sqrt{2\pi}} e^{-0.5x^2}$, we can obtain the estimation of the evaluation function:

$$\hat{\varphi}(x) = \frac{\sum_{i=1}^{N} \left(\frac{x-x_i}{h}\right) e^{-0.5\left(\frac{x-x_i}{h}\right)^2}}{h\sum_{i=1}^{N} e^{-0.5\left(\frac{x-x_i}{h}\right)^2}} \tag{7.3.3}$$

7.4 Blind Separation Algorithm Based on Probability Density Estimation

Using the kernel function method of PDF estimation discussed in the last section, we can estimate the evaluation function of output signals of the separating matrix for each channel and obtain $\hat{\varphi}_i(y)$, $i=1,2,\cdots,n$. Substituting

$$\hat{\varphi}(\mathbf{y}) = \begin{bmatrix} \hat{\varphi}(y_1) \\ \hat{\varphi}(y_2) \\ \vdots \\ \hat{\varphi}(y_n) \end{bmatrix} \tag{7.4.1}$$

into Eq. (3.6.15) to obtain the following BSS algorithm:

$$\boldsymbol{W}(k+1) = \boldsymbol{W}(k) + \mu_k\{\boldsymbol{I} - E[\mathbf{y}(k)\mathbf{y}(k)^{\mathrm{T}} - \hat{\varphi}(\mathbf{y}(k))\mathbf{y}(k)^{\mathrm{T}} + \mathbf{y}(k)\hat{\varphi}(\mathbf{y}(k))^{\mathrm{T}}]\}\boldsymbol{W}(k) \tag{7.4.2}$$

We must estimate the evaluation function of Eq. (7.4.2) again after every iterative step. Equation (7.4.2) is called the density estimation-based blind source separation (DEBBSS).Because the DEBBSS algorithm is achieved through the improvement of EASI algorithm, the following examples will be given to verify and show the availability of DEBBSS compared with EASI.

Some new improved algorithms were proposed to solve the problem that hybrid mixed signals cannot be separated by the artificial neural network BSS algorithm, which uses a single nonlinear function in place of a single PDF or evaluation function. These algorithms first estimate each output cumulant of the separating matrix in every iterative step; and depending on whether the cumulant estimated is greater than zero or smaller than zero, different nonlinear functions suitable to sub-and super-Gaussian signals respectively as a network activation function of each node to realize the separation of hybrid mixed signals is selected. The representative algorithm in this kind of algorithm is called the extended infomax ICA (EXTICA), which uses the following iterative algorithm to realize blind separation of the source signal[10]:

$$\Delta \boldsymbol{W} \propto [\boldsymbol{I} - \boldsymbol{K}\tanh(\boldsymbol{y})\boldsymbol{y}^{\mathrm{T}} - \boldsymbol{y}\boldsymbol{y}^{\mathrm{T}}]\boldsymbol{W} \tag{7.4.3}$$

where $\boldsymbol{K}$=diag{sign[$k(y_1)$], sign[$k(y_2)$],···, sign [$k(y_n)$]} is a diagonal matrix whose diagonal elements are the cumulant symbol of each output signal of the separating matrix.Depending on whether the cumulant of the separating matrix output signal is greater or smaller than zero, the algorithm of Eq. (7.4.3) selects different nonlinear functions accordingly.

In order to illustrate the advantage of the DEBBSS algorithm, it and the EXTICA algorithm are compared in the following examples.

7.4.1 Mixtures Case of Hybrid Signals

Whereas there is a clear analysis on sub-Gaussian mixed signals and super-Gaussian mixed signals[11~13], in this subsection, The EASI blind source separation algorithm and DEBBSS algorithm will be compared in their separation capability for mixed signals, which include sub-Gaussian and super-Gaussian signals in the source signals at the same time.

Three sub-Gaussian signals and three super-Gaussian signals are selected as source signals as follows:

$$\boldsymbol{s}(t) = \begin{bmatrix} \boldsymbol{s}^{>}(t) \\ \boldsymbol{s}^{<}(t) \end{bmatrix} \tag{7.4.4}$$

where

$$\boldsymbol{s}^{<} = \begin{bmatrix} s_1^{<}(t) \\ s_2^{<}(t) \\ s_3^{<}(t) \end{bmatrix} = \begin{bmatrix} \sin(1\,600\pi\Delta t) \\ \sin(600\pi\Delta t + 6\cos(120\pi\Delta t)) \\ \sin(180\pi\Delta t) \end{bmatrix},$$

$$\boldsymbol{s}^{>} = \begin{bmatrix} s_1^{>}(t) \\ s_2^{>}(t) \\ s_3^{>}(t) \end{bmatrix} = \begin{bmatrix} [(\mathrm{mod}(t,23) - 11)/9]^5 + 3 \\ X(t)\lg(Y(t)) + 1.4 \\ \cos(300\pi\Delta t)\sin(200\pi\Delta t)^7 \end{bmatrix}.$$

the superscript "<" denotes that the kurtosis of signals is smaller than zero, ">"denotes that the kurtosis of signals is greater than zero, t=1,2,···,1,000, and $\Delta = 2 \times 10^{-4}$. From the top down, the kurtosis of six signals are $k(s_1^{<}) = -1.5030$、$k(s_2^{<}) = -1.0492$, and $k(s_3^{<}) = -1.5030$, $k(s_1^{>}) = 2.4037$, $k(s_2^{>}) = 2.4310$, and $k(s_3^{>}) = 1.1889$, respectively.

The mixing matrix is one 6 × 6 random matrix as follows:

$$\boldsymbol{A} = \begin{bmatrix} 0.0291 & 0.0555 & 0.5162 & 0.1818 & 0.6896 & 0.7237 \\ 0.5962 & 0.1520 & 0.7679 & 0.4610 & 0.0797 & 0.5482 \\ 0.5010 & 0.4678 & 0.9567 & 0.6993 & 0.4450 & 0.3496 \\ 0.3964 & 0.7412 & 0.2844 & 0.0547 & 0.3627 & 0.0134 \\ 0.9928 & 0.6492 & 0.4114 & 0.9129 & 0.6701 & 0.9772 \\ 0.2837 & 0.7268 & 0.7202 & 0.5019 & 0.0853 & 0.3286 \end{bmatrix} \tag{7.4.5}$$

Figure 7.3 shows six mixed signals obtained from source signals mixed by the above mixing matrix.

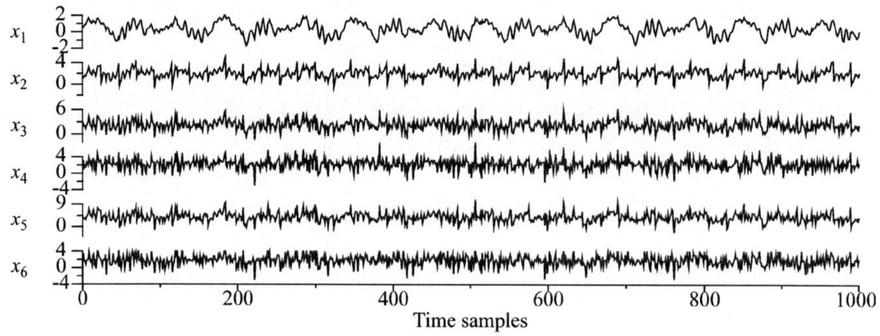

Fig. 7.3 Hybrid mixed signals

All mixed signals are passed through preprocessing, including removal of the average and whitening before blind separation.

The convergence of performance index PI_{DEBBSS} based on the DEBBSS algorithm, performance index PI_{SUB} calculated using nonlinear function $\varphi(t)=t^3$ with EASI algorithm and performance index PI_{SUPER} calculated using nonlinear function $\varphi(t)=\tanh(10t)$ with EASI algorithm are now compared. Owing to having the whitening processing before the separation of the mixtur's signals, The whitening matrix $\boldsymbol{T}$ must be considered in calculating the performance index. Figure 7.4 shows the convergent situation of PI_{DEBBSS}, PI_{SUB} and PI_{SUPER} in the process of the first to the 800th iterative step, from which performance index PI_{SUB} begins to converge to a stable value at about the 200th iterative step. This stable value is very large, about 13, and PI_{SUPER} begins to converge to stable value at about the 400th iterative step; this stable value is also large, at about 8.7. It means that both EASI blind separation algorithms using nonlinear function $\varphi(t)=t^3$, as well as nonlinear function $\varphi(t)=\tanh(10t)$, cannot separate hybrid mixed signals well. Figure 7.4 also shows that the performance index PI_{DEBBSS} using the DEBBSS algorithm begins to stably decrease after the beginning of iteration and to converge to one very small stable value of about 0.58 at about the 280th iterative step. It means that the hybrid mixed signals of Fig. 7.3 can be separated successfully by the DEBBSS algorithm.

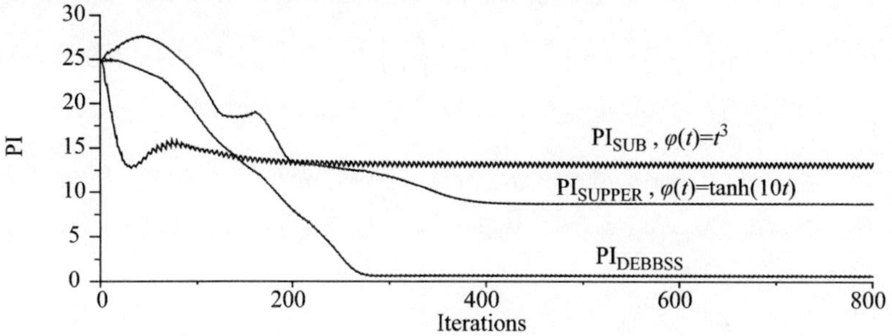

Fig. 7.4 The convergence of the performance index by using the DEBBSS algorithm and the EASI algorithm during the separation of hybrid mixed signals

Figure 7.5 shows that the hybrid mixed signals of Fig. 7.3 are separated by the DEBBSS algorithm at the 2000th iterative step. Comparing Fig. 7.5 with the source signals waveform of Fig. 7.3, it is shown that this algorithm realizes the separation of hybrid mixed signals successfully. As a comparison, Figure 7.6 shows the separated hybrid mixed signals of Fig. 7.3 by the EASI algorithm using nonlinear function $\varphi(t)$=tanh(10t) at the 2000th iterative step. Apparently, the EASI algorithm cannot succeed in realizing blind separation of the hybrid mixed signals.

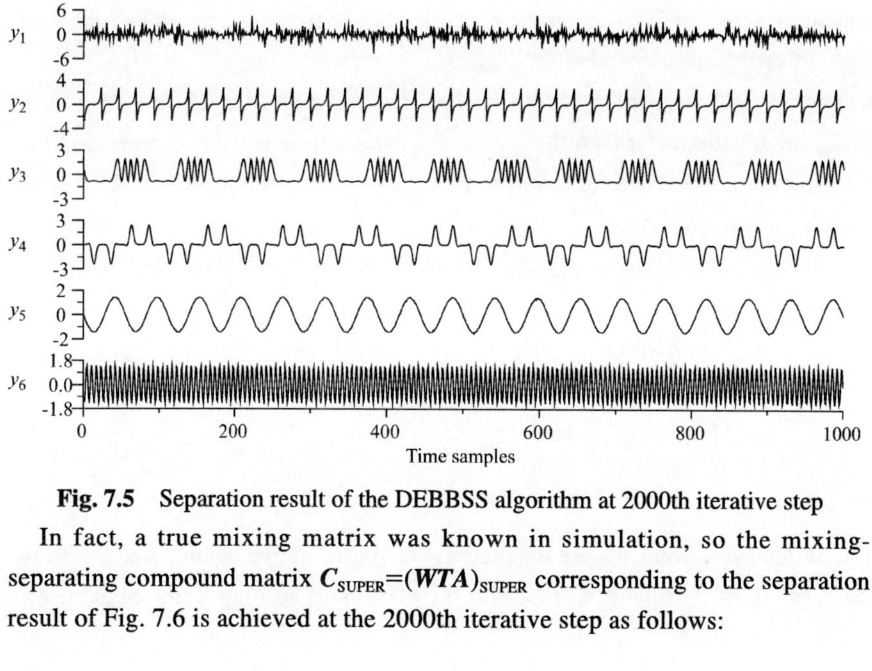

Fig. 7.5 Separation result of the DEBBSS algorithm at 2000th iterative step

In fact, a true mixing matrix was known in simulation, so the mixing-separating compound matrix $\boldsymbol{C}_{\mathrm{SUPER}}=(\boldsymbol{WTA})_{\mathrm{SUPER}}$ corresponding to the separation result of Fig. 7.6 is achieved at the 2000th iterative step as follows:

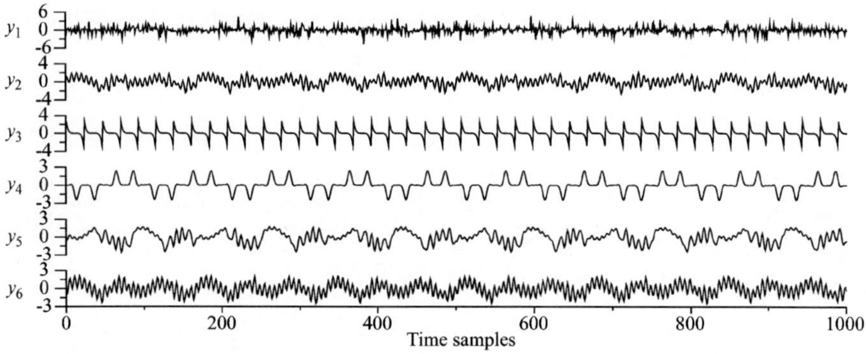

Fig. 7.6 Separation result of the EASI algorithm using $\varphi(t)=\tanh(10t)$ at the 2000th iterative step

$$\boldsymbol{C}_{\text{SUPER}} = \begin{bmatrix} 0.0285 & \underline{-0.6808} & 0.0929 & 0.0036 & -0.0009 & -0.0134 \\ 0.0091 & -0.0195 & -0.0244 & \underline{0.9484} & 0.7585 & 0.7060 \\ \underline{-0.9900} & 0.0025 & -0.0034 & 0.0094 & -0.0077 & -0.0025 \\ 0.0074 & 0.0031 & \underline{3.0835} & 0.0153 & 0.0056 & -0.0080 \\ 0.0086 & 0.0255 & -0.0281 & -0.2390 & \underline{1.0588} & -0.8752 \\ 0.0021 & -0.0253 & 0.0287 & \underline{-1.0211} & 0.4611 & 0.8607 \end{bmatrix} \tag{7.4.6}$$

in the above matrix, the elements underlined are the maximal ones of absolute value in every row of the mixing-separating compound matrix $\boldsymbol{C}_{\text{SUPER}}$. The separation result of Fig. 7.6 can be explained easily from comparison of absolute value of every row elements in $\boldsymbol{C}_{\text{SUPER}}$.

The maximum absolute value in the first, third and fourth row of matrix $\boldsymbol{C}_{\text{SUPER}}$ are much greater than the second largest absolute value of elements in the corresponding row.Therefore, output $\boldsymbol{y}=\boldsymbol{C}_{\text{SUPER}}\, s$ of separating matrix $\boldsymbol{W}$ can be obtained as follows:

$$\boldsymbol{y} = \begin{bmatrix} y_1 \\ y_2 \\ y_3 \\ y_4 \\ y_5 \\ y_6 \end{bmatrix} = \begin{bmatrix} 0.0285s_1 \underline{-0.6808s_2} + 0.0929s_3 + 0.036s_4 - 0.0009s_5 - 0.0134s_6 \\ \# \\ \underline{-0.99s_1} + 0.0025s_2 - 0.0034s_3 + 0.0094s_4 - 0.0077s_5 - 0.0025s_6 \\ 0.0074s_1 + 0.0031s_2 + \underline{3.0835s_3} + 0.0153s_4 + 0.0056s_5 - 0.008s_6 \\ \# \\ \# \end{bmatrix} \tag{7.4.7}$$

The uninterested elements are denoted by "#" in the above column vector. They are $y_1 \approx -s_2$, $y_3 \approx -s_1$ and $y_4 \approx s_3$. And the maximum absolute value of elements in the second, fifth and sixth rows of matrix $\boldsymbol{C}_{\text{SUPER}}$ is not much greater than the second largest absolute value in the corresponding row; therefore, there are no separated source signals in the output corresponding to those channels of the separating matrix. Besides, both element 0.9484, whose absolute value is maximal in the second row of matrix $\boldsymbol{C}_{\text{SUPER}}$ and element -1.0211, whose absolute value is maximal in the sixth row, appear in the fourth column of the matrix; thus, the signal waveform of the second channel is similar to the signal waveform of the sixth channel in Fig. 7.6.

After convergence of the EASI blind separation algorithm using the nonlinear function $\varphi(t)=t^3$ (used in separation of sub-Gaussian mixtures signals), none of the source signals of the hybrid mixed signals in Fig. 7.3 is recovered. In fact, it is only a special case that the super-Gaussian source signal components of the hybrid mixed signals is separated after convergence of the EASI blind separation algorithm using the nonlinear function $\varphi(t)=\tanh(10t)$, and it seldom appears in other simulation calculations.

The corresponding mixing-separating composite matrix $\boldsymbol{C}_{\text{DEBBSS}}=(\boldsymbol{WTA})_{\text{DEBBSS}}$ at the 2000th iterative step is as follows:

$$\boldsymbol{C}_{\text{DEBBSS}}=\begin{bmatrix} 0.0292 & \underline{-0.6818} & 0.0626 & 0.0236 & -0.0034 & 0.0348 \\ \underline{0.9898} & -0.0019 & 0.0533 & -0.0069 & 0.0164 & 0.0008 \\ -0.0059 & 0.0009 & 0.0030 & 0.0032 & \underline{-1.3816} & -0.0029 \\ -0.0083 & -0.0038 & \underline{3.0842} & -0.0010 & -0.0052 & 0.0094 \\ -0.0010 & 0.0178 & 0.0183 & 0.0003 & -0.0018 & \underline{-1.4154} \\ -0.0009 & -0.0107 & -0.0023 & \underline{-1.4138} & 0.0002 & 0.0003 \end{bmatrix} \tag{7.4.8}$$

Similarly, the element whose absolute value is maximal in every row of $\boldsymbol{C}_{\text{DEBBSS}}$ is underlined. Apparently, there is only one element whose absolute value is maximal in its row in every row and column of matrix $\boldsymbol{C}_{\text{DEBBSS}}$, and the absolute value of the element whose absolute value is second greater in the corresponding row is much less it. Namely, matrix $\boldsymbol{C}_{\text{DEBBSS}}$ approximates a generalized permutation matrix.

The above three calculation examples show that the EASI algorithm of BSS can separate only homogeneous mixed signals that include only sub-Gaussian signals or super-Gaussian signals, and must select different nonlinear functions corresponding to whether the source signals are super-Gaussian signals or sub-Gaussian signals. Source signals are actually unknown for the blind separation problem in the real world. What is more serious is that the EASI algorithm cannot separate hybrid mixed signals, and apparently limit its application. It is a common problem for many BSS algorithms at the present. The algorithm based on the probability density function estimation solves this problem successfully by directly estimating the score function of the signals and automatically making an adjustment according to statistical characteristics of each channel signal.

7.4.2 Comparison of Density Estimation-Based Algorithm and EXTICA Algorithm

The density estimation-based algorithm and EXTICA blind source separation algorithm are compared in this subsection. Because the EXTICA algorithm can also separate hybrid mixed signals with sub-Gaussian source signals and super-Gaussian source signals existing at the same time, it is useful to show the advantage of the density estimation-based algorithm by comparing it with the EXTICA algorithm.

In order to be consistent, we use the hybrid mixed signals of Fig. 7.3 in subsection 7.4.1.Through simulation and comparison, we discover that the EXTICA algorithm has one serious shortcoming in that its separation capability is quite sensitive to the learning rate of the algorithm, and the density estimation-based algorithm does not have this problem.

First, we select different learning rate μ_1=0.01, μ_2=0.02, μ_3=0.04, μ_4=0.08, μ_5=0.1, μ_6=0.2, μ_7=0.4 and μ_8=0.6, and use this group learning rate to test the density estimation-based blind separation algorithm. We use this group of the learning rate to separate the hybrid mixed signals of Fig. 7.3 and obtain a group with a performance index of the DEBBSS algorithm, which is shown in Fig. 7.7(a) and Fig. 7.7(b).

Comparing the four figures in Fig. 7.7(a), we learn that $\mathrm{PI}^1_{\mathrm{DEBBSS}}$ converges to the minimum at about the 1600th iterative step, and $\mathrm{PI}^2_{\mathrm{DEBBSS}}$, $\mathrm{PI}^3_{\mathrm{DEBBSS}}$ and $\mathrm{PI}^4_{\mathrm{DEBBSS}}$ converge to the minimum at about the 800th, 400th and 200th iterative steps respectively; namely, with the increase of the learning rate in germination, the convergence speed also increases in germination, and the four performance indexes are maintained at the same level after convergence. The same conclusion can be obtained by comparing the four figures in Fig. 7.7(b). Figure 7.7(b) also shows that with the increase of the learning rate, the performance index of the DEBBSS algorithm appears to have a small magnitude oscillation after converging to its minimum, but even including this small magnitude oscillation, the performance index is also maintained at a very low level. In fact, the

DEBBSS algorithm using the maximal learning rate μ_8=0.6 also succeeds in separating all source signals after convergence, and has converged quickly in fewer than 30 iterative steps.

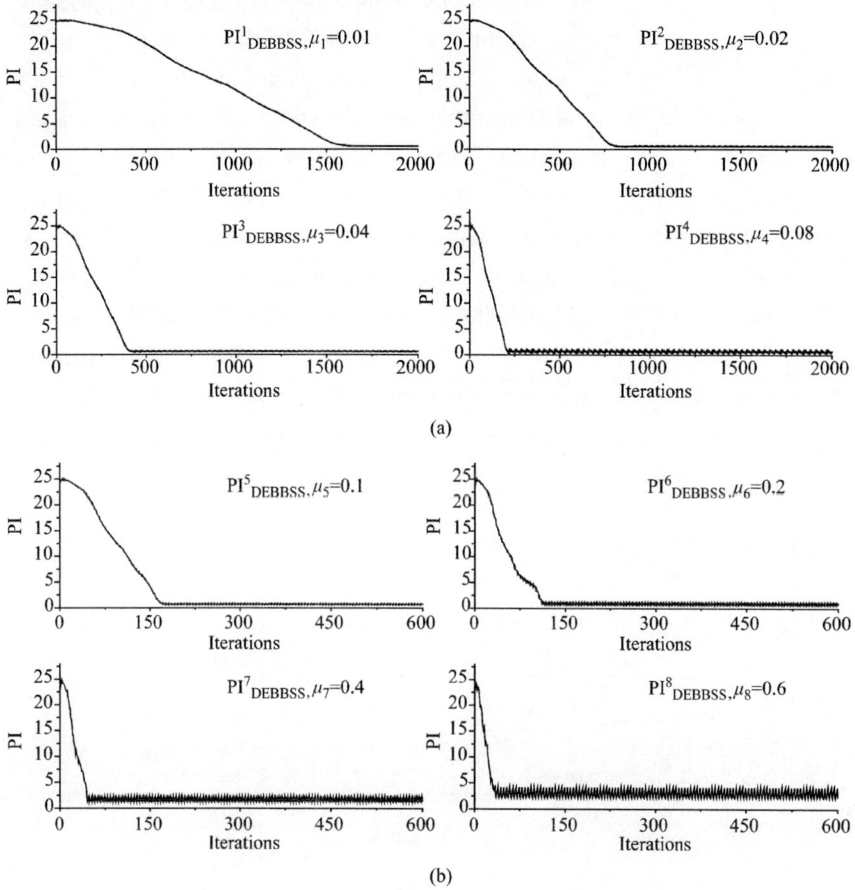

Fig. 7.7 The convergence of performance index of DEBBSS algorithm using different learning rates

From Fig. 7.7(a) and Fig. 7.7(b), we can obtain the very obvious conclusion that the learning rate influences the convergence speed but not the separation capability of the DEBBSS algorithm.

When we select another group of different learning rates μ_1=0.0001, μ_2=0.0002, μ_3=0.0004, μ_4=0.0005, μ_5=0.0007, μ_6=0.001 and μ_7=0.0014 to test EXTICA algorithm, we use this group learning rate to separate the hybrid mixed signals of Fig. 7.3 and obtain a group with a performance index of the EXTICA algorithm, which is shown in Fig. 7.8. The broken line in Fig. 7.8 denotes the performance index of the EXTICA algorithm using the learning rate μ_7=0.0014;

the EXTICA algorithm has diverged after 6 iterative steps and does not realize blind source separation. Figure 7.8 shows that the separation capability of the EXTICA algorithm is dependent on the learning rate and the dependency does not follow any rule. From the convergence of PI^1_{EXTICA}, PI^2_{EXTICA} and PI^3_{EXTICA}, it seems that if the learning rate of the EXTICA algorithm is greater, its performance index can converge to a smaller value, but the performance index PI^4_{EXTIC} using the smaller learning rate μ_4=0.0005 converges to a much lower level than the performance index PI^5_{EXTICA} using the larger learning rate μ_5=0.0007 in Fig. 7.8. Figure 7.8 shows the test result of only one group learning rate, but the same phenomenon is observed frequently in other tests. The EXTICA algorithm successfully separated all source signals using only the learning rate μ_4=0.0005 and μ_6=0.0001 in seven separation tests of Fig. 7.8. The separation capability of the EXTICA algorithm is very sensitive to the learning rate, and there is no certain rule relating separation capability and the learning rate of the algorithm; therefore, the EXTICA algorithm can separate hybrid mixed signals but its actual performance is not ideal.

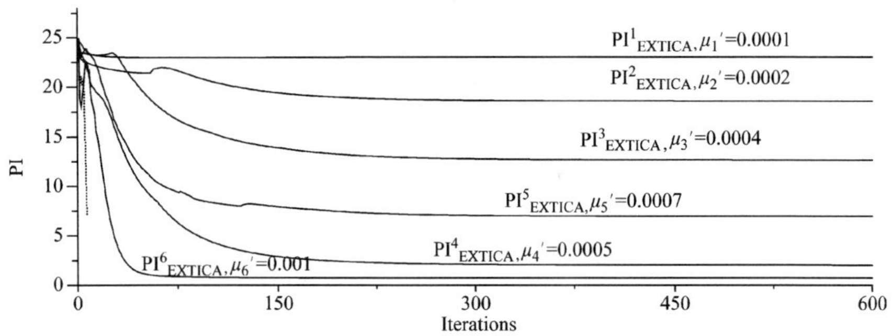

Fig. 7.8 The convergence of the performance index of the EXTICA algorithm using different learning rates

7.5 Probability Density Estimation of Gaussian Mixtures Model

From the above algorithms criteria of blind separation, we can conclude that no matter whether the blind separation algorithm is based on mutual information minimization or entropy maximization or on maximizing the nongaussianity of the output vectors or the maximum negentropy criteria, it is necessary to estimate the probability distribution of source signals, which plays a very important role in the convergence of the blind signal separation algorithm during the optimization process of the algorithm.The more accurate the probability distribution of the source signals is estimated, the better effect there is on separation[14].

The common blind separation algorithm, such as the infomax algorithm, has a good separation effect on the speech signal with super-Gaussian distribution. The derivative function $g'(y)=\exp(-y)/(1+\exp(-y))^2$ of the nonlinear function $g(y)=(1+\exp(-y))^{-1}$ in the algorithm has a probability density function of super-Gaussian distribution, which can approximate the distribution of the speech signal well; therefore, such an algorithm has a good effect on the blind separation of super-Gaussian signals such as speech and so on. For non-Gaussian and non-stationary noise signals, such as radiated ship and sea noise signals or other signals whose distribution are more complex, the super-Gaussian function does not fit anymore. So the essence of the blind separation algorithm is the estimation problem of the probability density function of source signals. When the PDF used in the algorithm is one of true signals, the blind separation algorithm can provide optimal solution. Therefore, the aim to improve the accuracy of the blind separation algorithm is that no matter how complex the distribution form of the signals is, the estimated PDF should approach the PDF of the source signals as closely as possible.

As a probability density model of non-Gaussian signals, the generalized Gaussian model (GGM)[15], with distribution being symmetrical and unimodal, was advanced by Miller, *et al* in 1972. A family of distribution functions that are symmetrical with respect to the peak at the mean value of the function is built by the generalized Gaussian model, whose general mathematic expression is[16]

$$p(s \mid \boldsymbol{\theta}) = \frac{\omega(q)}{\sigma} \exp\left[-c(q) \left| \frac{s-\mu}{\sigma} \right|^{q} \right], \qquad \boldsymbol{\theta} = \{\mu, \theta, q\} \tag{7.5.1}$$

where

$$c(q) = \left[\frac{\Gamma[3/q]}{\Gamma[1/q]} \right]^{q/2} \tag{7.5.2}$$

$$\omega(q) = \frac{\Gamma[3/q]^{\frac{1}{2}}}{(2/q)\Gamma[1/q]^{\frac{3}{2}}} \tag{7.5.3}$$

$\mu=E[s]$, $\sigma=\sqrt{E[(s-\mu)^2]}$ are mean value and standard deviation of the signal respectively. [$\cdot$] is the Gamma function in Eqs. (7.5.2) and (7.5.3). A different distribution function model can be obtained by adjusting the index value q. For example, when $q=2,q=1$ and $q<1$, the functions denote Gaussian and the Laplacian and strong Laplacian (such as a speech signal) distribution model respectively. Its simplified model can be expressed as follows:

$$p(s) \propto \exp\left(-\frac{1}{2}|s|^q\right) \tag{7.5.4}$$

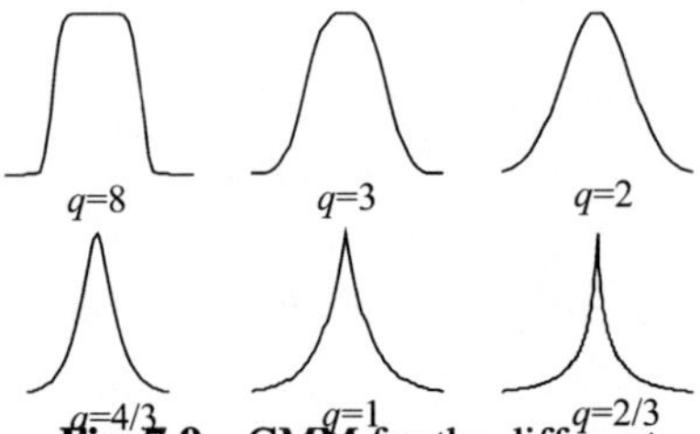

Fig. 7.9 GMM for the different q value

The GGM distribution situation with the different value q is shown in Fig. 7.9. From the figure we can see that the difference of the distribution function peak value and its deviation extent from normal distribution are determined by the different index parameter value q. GGM denotes standard Gaussian distribution when $q=2$, super-Gaussian distribution when $q<2$, and sub-Gaussian distribution when $q>2$.

The key problem using GGM as the probability density estimator of the signal source is how to estimate the value q of the model, which can be converted to estimate the maximum posteriori value of q. It can be obtained from the distribution of the observation vector $x=[x_1,x_2,\cdots,x_n]$:

$$p(q \mid \boldsymbol{x}) \propto p(\boldsymbol{x} \mid q)p(q) \tag{7.5.5}$$

where $p(\boldsymbol{x}|q)$ is the likelihood function:

$$p(\boldsymbol{x} \mid q) = \prod_n \omega(q)\exp[-c(q) \mid x_n \mid^q] \tag{7.5.6}$$

$p(q)$ is prior PDF of index q, which is replaced by Gamma function $\Gamma[\cdot]$ here. The index value q can be inferred according to maximum posteriori value. Figure 7.10 shows examples of several acoustic signals with different probability distributions that are estimated by the GGM, which shows that speech and music signals belong to typical super-Gaussian distribution, sea ambient noise signals approach Gaussian distribution, and radiated ship noise is sub-Gaussian distribution.

In the process of blind separation solved by ICA, its likelihood function can be obtained as follows, according to the definition of observation signals $\boldsymbol{x}=\boldsymbol{A}\boldsymbol{s}$:

$$p(\boldsymbol{x}|\mathbf{A}) = \frac{p(\boldsymbol{s})}{|\det\mathbf{A}|} \tag{7.5.7}$$

Then we can obtain an updated iterative learning rule based on the maximum likelihood method for the unmixing matrix $\boldsymbol{W}$ as follows:

$$\Delta\boldsymbol{W} \propto \frac{\partial \lg p(\boldsymbol{x}|\mathbf{A})}{\partial \mathbf{W}}\mathbf{W}^{\mathrm{T}}\mathbf{W} = \eta[\boldsymbol{I} - \varphi(\boldsymbol{s})\boldsymbol{s}^{\mathrm{T}}]\boldsymbol{W} \tag{7.5.8}$$

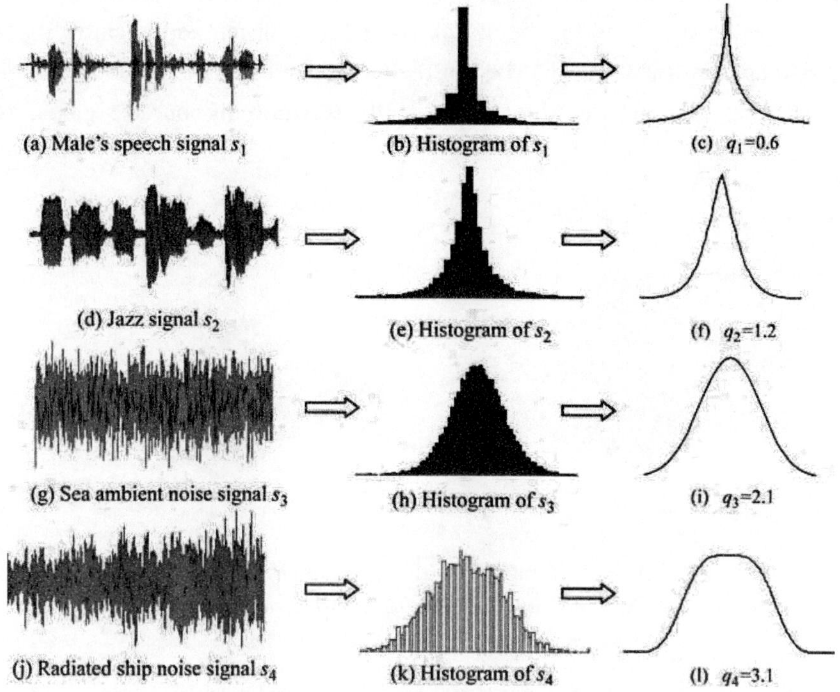

Fig. 7.10 The probability density function estimated by the GMM

where η is the learning rate, and $\boldsymbol{W}^{\mathrm{T}}\boldsymbol{W}$ denotes natural gradient algorithm. The vector $\varphi(\boldsymbol{s})$ is defined as $\varphi(s)=\frac{\partial \lg p(s)}{\partial \boldsymbol{s}}$, and $p(\boldsymbol{s})$ is the generalized Gaussian PDF. Substituting $\varphi(\boldsymbol{s})$ we obtain:

$$\varphi_i(s_i) = -qc\sigma_i^{-q} \mid s_i - \mu_i \mid^{q-1} \mathrm{sign}(s_i - \mu_i) \tag{7.5.9}$$

where $\mu_i = E[s_i]$, $\sigma_i = \sqrt{E[(s_i - \mu_i)^2]}$ are mean value and standard deviations of the signals respectively.

In order to verify the performance of this algorithm, it is compared with the popular FastICA algorithm in the separation of two kinds of mixed signals.

The first kind of mixed signal is a homogeneous mixture, namely the mixtures of six kinds of speech signals with super-Gaussian distribution. The nonlinear function of the FastICA algorithm is $g(u)=\tanh(u)$. The separation performance of two kinds of algorithms is shown by performance matrix $\boldsymbol{G}=\boldsymbol{WA}=\boldsymbol{PDA}^{-1}\boldsymbol{A}=\boldsymbol{PD}$, in which $\boldsymbol{D}$ is a nonsingular diagonal matrix and $\boldsymbol{P}$ is a permutation matrix. After separation, $\boldsymbol{W}$ is equal to $\boldsymbol{PDA}^{-1}$ and at the same time $\boldsymbol{G}$ is equal to a unit matrix with changed amplitude and order. Two clear performance matrices can be obtained by adjusting columns of matrix

G, as shown in Fig. 7.11. From the figure we can see that two source signals cannot be separated by the fixed nonlinear function method because its *G* matrix cannot form a unit matrix; but the GGM can estimate different source signals correctly and therefore successfully separate all source signals, as shown in Fig. 7.11(b).

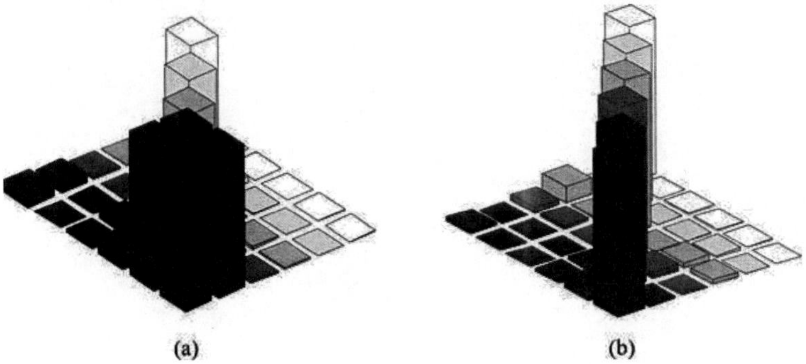

Fig. 7.11 Performance matrix of two kinds of algorithms
(a) Performance matrix *G* of FastICA algorithm;
(b) Performance matrix *G* of GGM algorithm

The second kind of mixed signal is a hybrid mixture, namely mixtures of two sub-Gaussian distribution signals and one super-Gaussian distribution signal; the separation result is shown in Fig. 7.12.

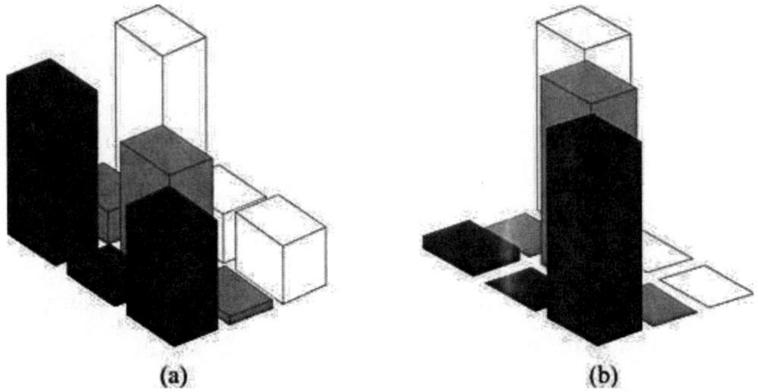

Fig. 7.12 Performance matrix of two kinds of algorithms
(a) Performance matrix *G* of FastICA algorithm;
(b) Performance matrix *G* of GGM algorithm

Corresponding to the GGM algorithm, the Gaussian mixture's model can estimate the real PDF of the source signal more accurately, and its output signal-to-noise ratio is higher. But its convergence speed is lower, because it is necessary to estimate more model parameters for GGM.The model is more complex for more accurate estimation, and the calculation load increases exponentially with the time index. Therefore, we need to give attention to both the separation effect and convergence speed for the blind separation of signals with complex distribution in order to select a proper separation method.

The mathematical model of GMM can be denoted as

$$p(y_i) = \sum_{j=1}^{M} \frac{c_j}{\sqrt{2\pi}\sigma_j} \exp\left(-\frac{(y_i - \mu_j)^2}{2\sigma_j^2}\right) \tag{7.5.10}$$

where $\boldsymbol{\mu}_j=(\mu_1, \mu_2, \cdots, \mu_M)$ is mean value, $\boldsymbol{\sigma}_j=(\sigma_1, \sigma_2, \cdots, \sigma_M)$ is the standard deviations, $\boldsymbol{c}=(c_1, c_2, \cdots, c_M)$ is the mixed-weight coefficient of the Gaussian function, and $c_j \geqslant 0$, $\sum_{j=1}^{M} c_j = 1$. When $M=1$, Equation (7.5.10) is an ordinary Gaussian density function.

Based on the maximum entropy principle and the PDF estimation of output signals using GMM, the learning rule can be solved by the gradient descent algorithm. Similarly, we can use the natural gradient $\frac{\partial H(z;\boldsymbol{W})}{\partial \boldsymbol{W}}\boldsymbol{W}^{\mathrm{T}}\boldsymbol{W}$ to replace ordinary random gradient $\frac{\partial H(z;\boldsymbol{W})}{\partial \boldsymbol{W}}$ and obtain:

$$\frac{\mathrm{d}\boldsymbol{W}}{\mathrm{d}t} = \eta \frac{\partial H(z;\boldsymbol{W})}{\partial \boldsymbol{W}}\boldsymbol{W}^{\mathrm{T}}\boldsymbol{W} = \eta(\boldsymbol{I} - \varphi(\boldsymbol{y})\boldsymbol{y}^{\mathrm{T}})\boldsymbol{W} \tag{7.5.11}$$

the updating iterative equation for the unmixing matrix $\boldsymbol{W}$ is

$$\boldsymbol{W}(t+1) = \boldsymbol{W}(t) + \eta(t)(\boldsymbol{I} - \varphi(\boldsymbol{y}(t))\boldsymbol{y}^{\mathrm{T}}(t))\boldsymbol{W}(t) \tag{7.5.12}$$

where $\varphi(\boldsymbol{y})$ is the function of *GMM*:

$$\varphi_i(y) = -\frac{p'_i(y_i)}{p_i(y_i)} = \frac{\sum_{j=1}^{M} -\left(\frac{y_i - \mu_i}{\sigma_j^2} \cdot \frac{c_j}{\sqrt{2\pi}\sigma_j} \exp\left(-\frac{(y_i - \mu_j)^2}{2\sigma_j^2}\right)\right)}{\sum_{j=1}^{M} \frac{c_j}{\sqrt{2\pi}\sigma_j} \exp\left(-\frac{(y_i - \mu_j)^2}{2\sigma_j^2}\right)} \tag{7.5.13}$$

where each parameter of the GMM, such as mean value μ_j, variance σ_j and mixed-weight coefficient c_j are obtained by iteration of the expectation maximum (EM) method[17,18].

7.6 Blind Deconvolution Algorithm Based on Probability Density Function Estimation

In the simulation of deconvolution mixtures, we can compare the separation capability of the traditional blind deconvolution algorithm and the improved blind deconvolution algorithm using the nonparametric density estimation method for sub-Gaussian signals, super-Gaussian signals, and hybrid mixed signals. In blind deconvolution, besides different signals with mutual independence, we consider the case that all signals are independent identically distributed (i.i.d.) processes. So in this section we cannot use the same signals as in the last section. The signals used in this section are independent identically distributed random processes generated using the Matlab program. In order to estimate the result of blind deconvolution better, the evaluation criterion "intersymbol interference (ISI)" frequently used in blind deconvolution is introduced first.

7.6.1 Evaluation Criterion in Blind Deconvolution

Some signals used in blind deconvolution are random processes whose time-domain characteristics are not very obvious. The method by which the separation effect is judged by comparison of time-domain wave forms of signals separated before and after is not a suitable criterion. So the concept of ISI is introduced.

ISI is a common concept in the communication field, and decreasing it is the ultimate aim of channel equalization. ISI is not sensitive to amplitude uncertainty and phase uncertainty that exists in blind deconvolution. So it is often used as an evaluation criterion of blind deconvolution to assure whether the transfer function of the entire system satisfies the separation condition through blind deconvolution[11].

ISI of a single channel system in communication is defined as

$$\mathrm{ISI} = \frac{\sum_{k} |c(k)|^2 - \max_k |c(k)|^2}{\max_k |c(k)|^2} \tag{7.6.1}$$

where $\boldsymbol{c}$ denotes the impulse response of the system $\boldsymbol{c}=\boldsymbol{w}*\boldsymbol{h}$, $\boldsymbol{w}$ is the impulse response of the separation channel, and $\boldsymbol{h}$ is the impulse response of the mixing system. It is obvious from the above equation, when and only when $\boldsymbol{c}=[\cdots, 0, 0, a, 0, 0, \cdots]$, ISI=0, where a is a constant, and the output signals $y(k)=\boldsymbol{c}*s(k)=as(k-L)$, L is the time delay of the estimated signal relative to a real signal. When ISI is zero, we can say that signals are recovered completely. On this basis, Lambert had expanded single channel ISI to the multiple input multiple output (MIMO) system[12]. In an N input N output system, the ISI of the ith channel is defined as

$$\mathrm{ISI}_i = \frac{\sum_{j}\sum_{k} |\boldsymbol{c}_{ij}(k)|^2 - \max_{j,k} |\boldsymbol{c}_{ij}(k)|^2}{\max_{j,k} |\boldsymbol{c}_{ij}(k)|^2} \tag{7.6.2}$$

If $\boldsymbol{c}_{ij}$ is the (i, j)th term of the whole system filter matrix, through the analysis of Eq. (7.6.2) we can see that when every row and every column of the system filter matrix has only one filter, such as $c_{ij}=[\cdots, 0, 0, a_{ij}, 0, 0, \cdots]$ and all other filters are zero, the ISI of every channel could be zero. Then output signals of the system are $y_i(k)=a_{ij}s(k-L)$, which is the result required by blind deconvolution. Therefore, in multi-channel blind deconvolution, when the separating filter matrix can make the ISI of every channel zero, we can say that the system is deconvolved successfully.

We can also use filter algebra[12] to calculate the ideal inverse filter of the mixing filter and compare it with the separating filter calculated by the blind deconvolution algorithm. Taking a system with two inputs and two outputs as an example, when the mixing filter matrix is

$$\boldsymbol{H}=\begin{bmatrix}\boldsymbol{h}_{11} & \boldsymbol{h}_{12}\\ \boldsymbol{h}_{21} & \boldsymbol{h}_{22}\end{bmatrix} \tag{7.6.3}$$

its inverse matrix (namely ideal separating filter matrix) can be solved using filter matrix algebra as follows:

$$\boldsymbol{W}=\frac{1}{(\boldsymbol{h}_{11}*\boldsymbol{h}_{22}-\boldsymbol{h}_{12}*\boldsymbol{h}_{21})}\begin{bmatrix}\boldsymbol{h}_{22} & -\boldsymbol{h}_{12}\\ -\boldsymbol{h}_{21} & \boldsymbol{h}_{11}\end{bmatrix} \tag{7.6.4}$$

The product of them is

$$\begin{aligned}\boldsymbol{I}=\boldsymbol{W}\boldsymbol{H}&=\frac{1}{\boldsymbol{h}_{11}*\boldsymbol{h}_{22}-\boldsymbol{h}_{12}*\boldsymbol{h}_{21}}\begin{bmatrix}\boldsymbol{h}_{11}*\boldsymbol{h}_{22}-\boldsymbol{h}_{12}*\boldsymbol{h}_{21} & \boldsymbol{h}_{11}*\boldsymbol{h}_{12}-\boldsymbol{h}_{11}*\boldsymbol{h}_{12}\\ \boldsymbol{h}_{21}*\boldsymbol{h}_{22}-\boldsymbol{h}_{21}*\boldsymbol{h}_{22} & \boldsymbol{h}_{11}*\boldsymbol{h}_{22}-\boldsymbol{h}_{12}*\boldsymbol{h}_{21}\end{bmatrix}\\ &=\begin{bmatrix}\cdots,0,0,1,0,0,\cdots & \cdots,0,0,0,\cdots\\ \cdots,0,0,0,\cdots & \cdots,0,0,1,0,0,\cdots\end{bmatrix}\end{aligned} \tag{7.6.5}$$

The ideal separating filter, the separating filter calculated by blind deconvolution, and the product $\boldsymbol{C}=\boldsymbol{W}\boldsymbol{H}$ (namely the transfer function of the whole system) of the filter calculated by blind deconvolution and the mixing filter can be compared to verify the correctness of the blind deconvolution algorithm.

7.6.2 Blind Deconvolution Simulation

Corresponding to the blind separation algorithm of Eq. (7.4.2), the iterative equation of the blind deconvolution algorithm based on the PDF estimation is

$$\begin{aligned}\boldsymbol{W}(k+1)=&\boldsymbol{W}(k)+\mu_k\{\boldsymbol{I}-E[\boldsymbol{y}(k)\boldsymbol{y}(k)^{\mathrm{T}}-\\ &\hat{\varphi}(\boldsymbol{y}(k))\boldsymbol{y}(k)^{\mathrm{T}}+\boldsymbol{y}(k)\hat{\varphi}(\boldsymbol{y}(k))^{\mathrm{T}}]\}\boldsymbol{W}(k)\end{aligned} \tag{7.6.6}$$

For simplicity, the system with two inputs and two outputs and the same mixing channels is used in the simulation of blind deconvolution. The parameters of the mixing channels are

$$\boldsymbol{h}_{11}=[0.5, 0.2, 0.2] \qquad \boldsymbol{h}_{12}=[0.2, 0.2, 0.1]$$

$$h_{21} = [0.1, 0.2, -0.1] \qquad h_{22} = [0.5, 0.2, 0.1] \tag{7.6.7}$$

Its inverse channels are solved by Eq. (7.6.4) as shown in Fig. 7.13, in which the horizontal coordinate denotes the order of the inverse channel filter. In the following simulation, the inverse channels of the mixing channels can be compared with the separating filter matrix obtained using the blind deconvolution algorithm based on the nonparametric density estimation to show the validity of this algorithm.

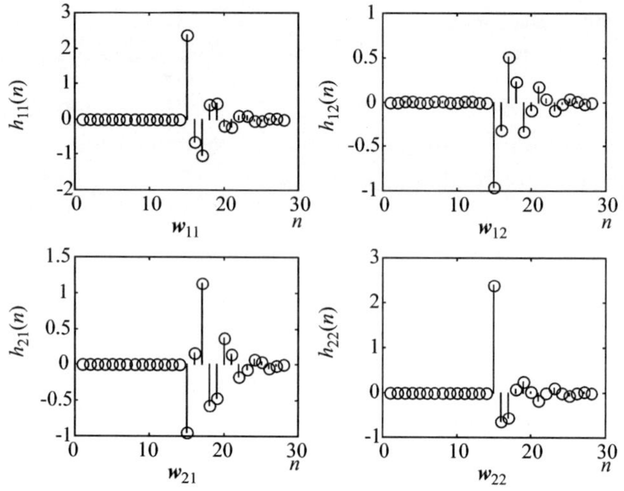

Fig. 7.13 Ideal inverse channels of mixing channels

7.6.3 Sub-Gaussian Mixtures Case

Both source signals are generated by Matlab.The first one is s_1=rand (60,000, 1)−0.5; the second one is s_2=rand (60,000, 1).∧2. The kurtosis of these two signals are −1.2098 and −0.8710 respectively, both are sub-Gaussian signals. Through the channel of Eq. (7.6.7), two mixed signals can be obtained and then separated using the blind deconvolution algorithm based on the nonparametric density estimation. In the algorithm, the learning rate is 0.0003. After 60,000 iterations, the ISI of the two channels is shown in Fig. 7.14, and ISI approaches zero gradually after about 10,000 iterations.

Figure 7.15 shows a sketch of the separating filter solved by blind deconvolution algorithm Eq. (7.6.6). Comparing Fig. 7.13 and Fig. 7.15, the separating filter obtained after blind deconvolution and the ideal inverse filter are basically the same. Figure 7.16 shows the sketch of the global filter after blind deconvolution, from which we can see that the system filter basically satisfies the condition of $[\cdots, 0, 0, a, 0, 0, \cdots]$. That is to say, the output signals passed

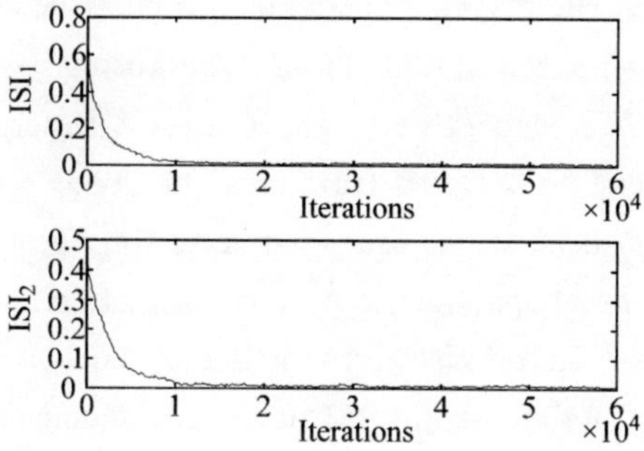

Fig. 7.14 Change of ISI between two channels

through the system filter satisfy the condition of blind deconvolution.

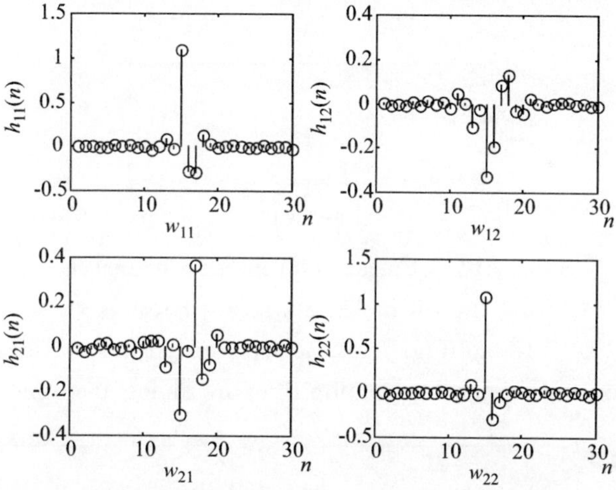

Fig. 7.15 Sketch of the separating filter after blind deconvolution

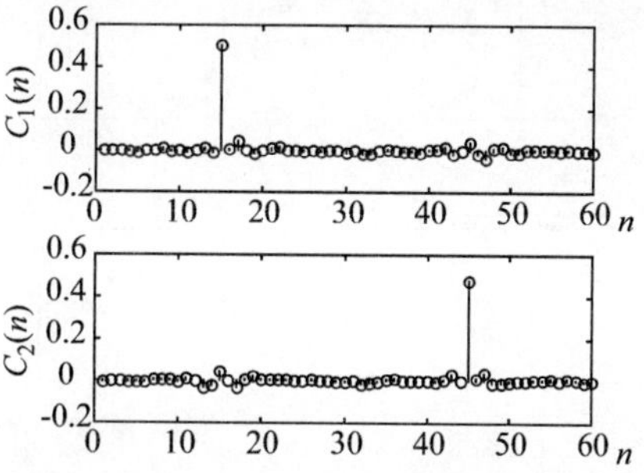

Fig. 7.16 Sketch of the global filter after blind deconvolution

7.6.4 Super-Gaussian Mixtures Case

In the simulation of super-Gaussian mixtures, source signals are also generated by Matlab. The first one is s_1=randn(60,000, 1); s_1=sign(s_1).* s_1.∧2; the second one is s_2=randn (60,000, 1).∧3. The kurtosis of these two signals is 8.9651 and 44.1795 respectively, and both are super-Gaussian signals. Two mixed signals can be obtained from these two source signals through the system of equation (7.6.7), which are separated using the blind deconvolution algorithm based on the nonparametric density estimation. In the algorithm, the learning rate is 0.0008. The ISI change of the two channels is shown in Fig. 7.17, and after about 3,000 iterations, the ISI of the two channels approaches zero.

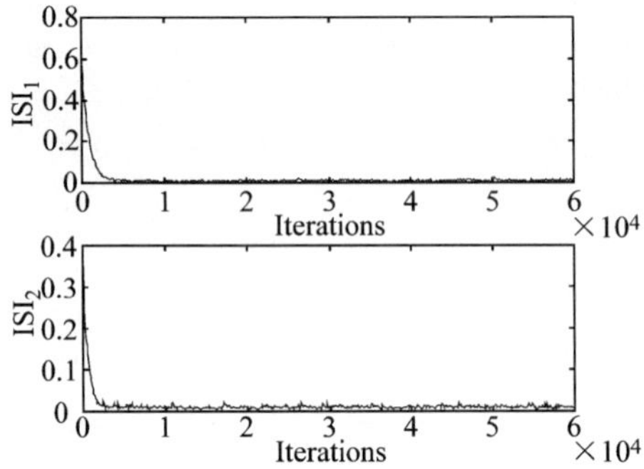

Fig. 7.17 Change of ISI between two channels

after deconvolution, the ultimate separating filter is shown in Fig.7.18. Comparing Fig. 7.18 and Fig. 7.13, the separating filter obtained after blind deconvolution and the ideal separation filter are almost the same. Figure 7.19

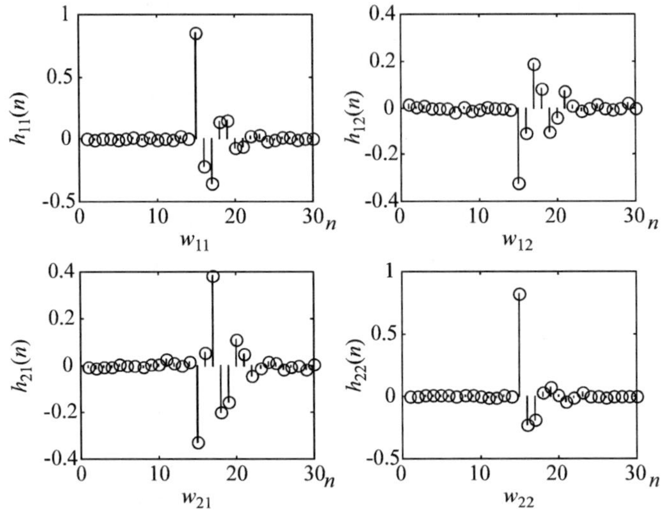

Fig. 7.18 Sketch of the separating filter after blind deconvolution

shows the sketch of the global filter after blind deconvolution; the global filter basically satisfies the condition of [···,0, 0, *a*, 0, 0, ···] . That is to say that the system satisfies the condition of blind deconvolution $y_j(k)=as_j(k-L)$.

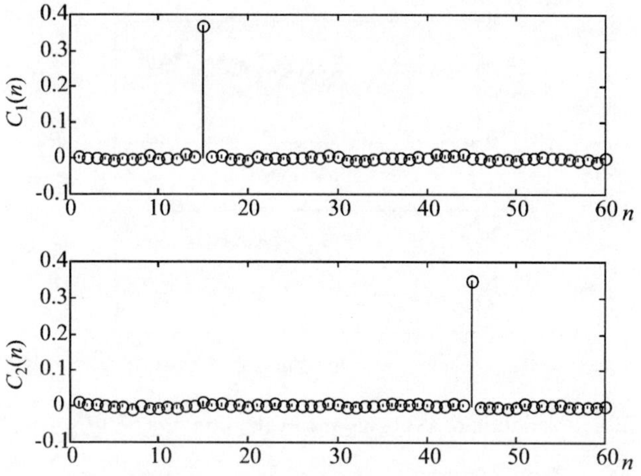

Fig. 7.19 Sketch of the global filter after blind deconvolution

7.6.5 Hybrid Mixed Signals Case

From the above two experiments we can see that the improved blind deconvolution algorithm using the nonparametric density estimation can solve deconvolution for homogeneous mixture's signals as with the traditional blind deconvolution algorithm. We can compare the deconvolution capability to hybrid mixtures signals by both as follows. Here the source signals are made up of two signals that are selected from the above two simulations. The first signal is s_1=rand (60,000, 1)−0.5; and one s_2=rand *n* (60,000,1); s_2=sign (s_2).* s_2.∧2. Their kurtosis are −1.1983 and 8.9137 respectively, and obviously they are hybrid mixed signals. First, we separate the hybrid mixed signals using the typical blind deconvolution algorithm, then select the learning rate as 0.0008 and the evaluation functions as $f(y)=\tanh(10y)$ and $f(y)=y^3$ respectively. The variation of the ISI in the separation process is shown in Fig. 7.20 and Fig. 7.21. Obviously, whether the evaluation function is $f(y)=\tanh(10y)$ or $f(y)=y^3$, its ISI does not approach zero, so the hybrid mixed signals cannot be restored by using the traditional blind deconvolution algorithm. Figure 7.22 shows the variation of the ISI when the blind deconvolution algorithm based on the nonparametric density estimation solved the hybrid mixed signals. The sketch of the separating filter and system filter obtained by using the above algorithm is shown in Fig. 7.23 and Fig. 7.24 respectively. We can see that the blind deconvolution algorithm based on the nonparametric density estimation can recover hybrid mixed signals well.

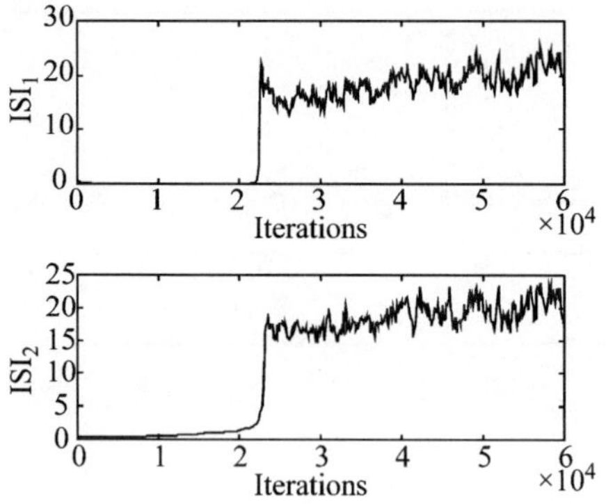

Fig. 7.20 Variation of ISI using the evaluation function $f(y)$=tanh(10y)

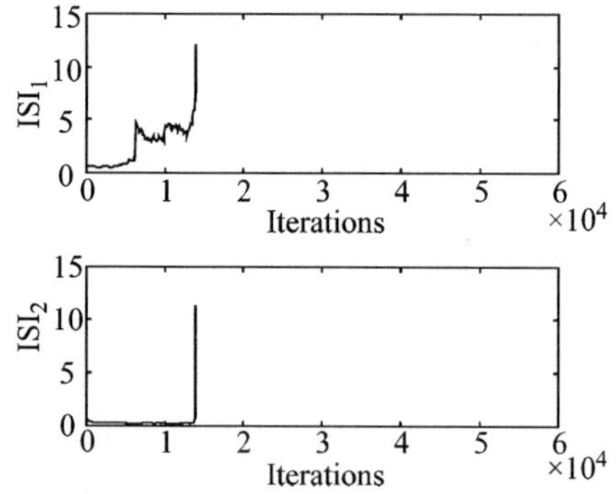

Fig. 7.21 Variation of ISI using the evaluation function $f(y)=y^3$

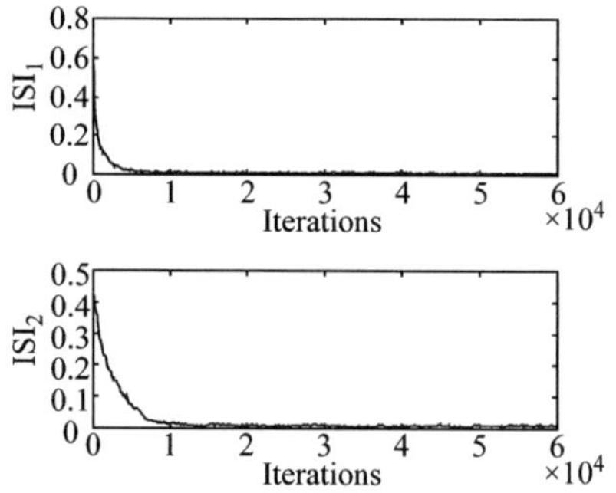

Fig. 7.22 Variation of ISI using the nonparametric density estimation algorithm

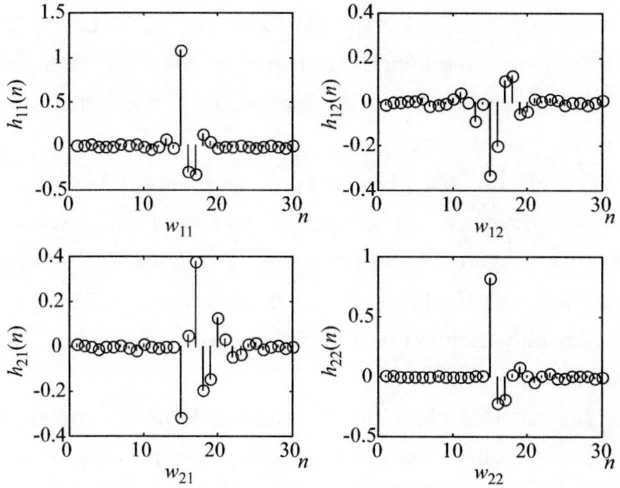

Fig. 7.23 Separating filter using the nonparametric density estimation algorithm

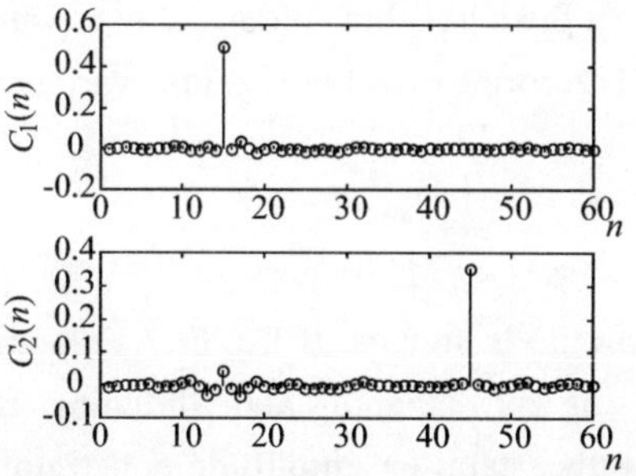

Fig. 7.24 System filter using the nonparametric density estimation algorithm

7.7 On-line Algorithm of Nonparametric Density Estimation

The algorithm based on the probability density function estimation is discussed in the last two sections. The basic idea is that every sample of random variable y_i is weighted smoothly through a series of continuous Gaussian windows; then the average is taken for all weighted windows as the estimation of the probability density function of y_i and the score function is calculated for the blind separation algorithm. The estimation equation of the PDF includes one kernel function. The most common one is the Gaussian kernel function, in which parameter h is window width and is usually also called the smoothing parameter or bandwidth. The selection of bandwidth h has an important influence on the ultimate estimation result. If the bandwidth is too narrow, the estimated PDF graph would have pseudo peaks, and if the bandwidth is too wide, the graph would be too smooth to hide some important characteristics. Therefore, the algorithm is closely related to whether a rational bandwidth h is selected; otherwise, the algorithm even diverges at times.

The approximation of the PDF based on the Gram-Charlier expansion is introduced in this section; it can be suitable for hybrid mixed signals and has a good convergence Property.Compared with the estimation algorithm based on the kernel function, its calculation is simple, with good real-time Property.The problem of bandwidth selection in the PDF estimation is avoided, which makes the algorithm more practical[19].

The basic idea of this algorithm is that the PDF is estimated using the interceptive Gram-Charlier expansion; then the score function is calculated and a new blind separation algorithm is obtained.Aspects considered with existing algorithms include the deduction of the score function and calculation formula as well as the Property of the algorithm. We make the assumption that the source is a random signal with zero mean value and unit variance before deducing the algorithm, namely:

$$E(\boldsymbol{s}) = \boldsymbol{0}; \tag{7.7.1}$$

$$E(\boldsymbol{s}\boldsymbol{s}^{\mathrm{T}}) = \boldsymbol{I} \tag{7.7.2}$$

it is trivial for many mixture's signals. If Eq. (7.7.1) is not satisfied, the source can be preprocessed using zero meaning and whitening. Because the amplitude of the output signal is uncertain, an amplitude constraint item is added in the evaluation function for the requirement of Eq. (7.7.2),

$$E[\boldsymbol{y}\boldsymbol{y}^{\mathrm{T}}] = \boldsymbol{I} \tag{7.7.3}$$

according to the above analysis and assumption, we can further make the following inference:

$$E[\boldsymbol{y}] = E[\boldsymbol{WAs}] = \boldsymbol{0}, E[\boldsymbol{y}\boldsymbol{y}^{\mathrm{T}}] = \boldsymbol{I} \tag{7.7.4}$$

The concrete method using the Gram-Charlier expansion to approach the PDF uses the interceptive Gram-Charlier expansion to estimate the PDF of output signals y_i[20,21], which is given by

$$p_i(y_i) \approx \alpha(y_i)\left\{1 + \frac{\kappa_{i,3}}{3!}H_3(y_i) + \frac{\kappa_{i,4}}{4!}H_4(y_i)\right\} \tag{7.7.5}$$

three items are considered in the above equation, where $\kappa_{i,3}$, $\kappa_{i,4}$ are third-order and fourth-order cumulants respectively

$$\kappa_{i,3} = E(y_i^3) \tag{7.7.6}$$

$$\kappa_{i,4} = E(y_i^4) - 3E(y_i^2) \tag{7.7.7}$$

In Eq. (7.7.5), $\alpha(y_i)$ is the PDF of normal distribution with a zero mean value and unit variance, i.e. $\alpha(y_i)=\frac{1}{\sqrt{2\pi}}\cdot\exp\left(-\frac{y_i^2}{2}\right)$, $H_k(y_i)$ is the Chebyshev-Hermite polynomial, which is defined by the following formula:

$$(-1)^k\frac{\mathrm{d}^k\alpha(y_i)}{\mathrm{d}y_i^{\,k}}=H_k(y_i)\alpha(y_i) \tag{7.7.8}$$

when k=3, 4, from Eq. (7.7.8) we can obtain

$$H_3(y_i)=y_i^3-3y_i \tag{7.7.9}$$

$$H_4(y_i)=y_i^4-6y_i^2+3 \tag{7.7.10}$$

Substituting Eq. (7.7.5) into a score function:

$$f_i(y_i)=-\frac{\mathrm{d}}{\mathrm{d}y_i}\lg p_i(y_i)=-\frac{p'_i(y_i)}{p_i(y_i)} \tag{7.7.11}$$

using the above relation, we can obtain

$$f_i(y_i)=-\frac{-y_i+\frac{1}{6}\kappa_{i,3}g_1(y_i)+\frac{1}{24}\kappa_{i,4}g_2(y_i)}{1+\frac{\kappa_{i,3}}{3!}H_3(y_i)+\frac{\kappa_{i,4}}{4!}H_4(y_i)} \tag{7.7.12}$$

where $g_1(y_i)=-y_i^4+6y_i^2-3$, and $g_2(y_i)=-y_i^5+10y_i^3-15y_i$.

Substituting Eq. (7.7.12) into the following relation:

$$\boldsymbol{W}(t+1)=\boldsymbol{W}(t)+\mu(t)(\boldsymbol{\Lambda}(t)-\boldsymbol{f}(\boldsymbol{y}(t))\boldsymbol{y}^{\mathrm{T}}(t))\boldsymbol{W}(t) \tag{7.7.13}$$

where $\mu(t)$ denotes learning rate, $\boldsymbol{\Lambda}(t)$ denotes any positive definite diagonal matrix,in general, $\boldsymbol{\Lambda}(t)=\boldsymbol{I}$. This is the newly developed algorithm.

The concrete steps of this algorithm are as follows:

Step 1: Initialize the separating matrix $\boldsymbol{W}(0)$, which in general is a unit matrix $\boldsymbol{I}$; the present iterative step (i. e. present time) $t=0$;

Step 2: Adjust the time $t+1$ as t;

Step 3: Calculate the output $\boldsymbol{y}(t)=\boldsymbol{W}(t)\boldsymbol{x}(t)$ of the whole system;

Step 4: Calculate the cumulant on line according to Eqs. (7.6.6) and (7.6.7) respectively:

$$\hat{\kappa}_{i,3}(t)=(1-\alpha_{\kappa3})\hat{\kappa}_{i,3}(t-1)+\alpha_{\kappa3}y_i^3 \tag{7.7.14}$$

$$\hat{\kappa}_{i,4}(t)=(1-\alpha_{\kappa4})\hat{\kappa}_{i,4}(t-1)+\alpha_{\kappa4}(y_i^4-3y_i^2) \tag{7.7.15}$$

where $\alpha_{\kappa3}$, $\alpha_{\kappa4}$ are estimation coefficients, which are selected as 0.1~0.001 in general. $\hat{\kappa}_{i,3}(t)$, $\hat{\kappa}_{i,4}(t)$ is the cumulant at the time t;

Step 5: Calculate the Chebyshev-Hermite polynomial $H_3(y_i)$, $H_4(y_i)$ according to Eqs. (7.6.9) and (7.6.10);

Step 6: Calculate the score function (7.7.11) according to Eq. (7.7.12);

Step 7: Update $\boldsymbol{W}$ on line according to Eq. (7.7.13) as follows:

$$W(t+1)=W(t)+\mu(t)(\Lambda(t)-f(y(t))y^{T}(t))W(t)$$

where $\Lambda(t)=I$, the value range for $\mu(t)$ is 0.1~ 0.001;

Step 8: Judge whether it reaches a certain iterative number. If it does, calculate an output estimation for the sources; otherwise continue with step 2.

We should point out that when the above algorithm is realized in batches, we only need to select certain data, size L, and calculate third-order and fourth-order cumulants respectively, according to Eqs. (7.7.6) and (7.7.7). In practice, we can select data size as $L=300\sim500$.

In order to further develop the algorithm, we apply the traditional conjugate gradient idea to the Riemannian manifold optimization problem defined on the nonsingular matrix, i.e., create the conjugate gradient method of the instantaneous blind separation model.

Therefore, we introduce the concept of geodesic first.The geodesic is the shortest path between two points W_1 and W_2 defined in the Riemannian manifold[22,23], whose value can be obtained by using a variational method of optimization problems:

$$\mathrm{dist}(W_1,W_2)=\min_{W(u)}\int_0^1 \langle W'(u),W'(u)\rangle_{W(u)}^{\frac{1}{2}}\,du \tag{7.7.16}$$

$$\text{s. t. } W(0)=W_1,\quad W(1)=W_2 \tag{7.7.17}$$

its solution corresponding to path u is

$$W(u)=(W_2W_1^{-1})^{u}W_1 \tag{7.7.18}$$

In the blind source separation algorithm, the parameter space W described by the basic model is the imbedded manifold of the nonsingular matrix, in which the geodesic is defined to minimize the score function. The variant of Eq. (7.7.17) is adopted in general in the blind source separation algorithm, providing the following initial points and their tangent vectors:

$$W(0)=W_1,\quad W'(0)=X_1 \tag{7.7.19}$$

Solve the objective function using the variational method and obtain the variant of Eq. (7.7.18):

$$W(u)=\exp(uX_1W_1^{-1})^{u}W_1 \tag{7.7.20}$$

when minimizing the score function, the unknown parameter matrix W is defined on the manifold of a nonsingular matrix, so the steepest descent direction is the natural gradient given by

$$\nabla J(W(t))=\Lambda(t)-f(y(t))y^{T}(t) \tag{7.7.21}$$

for notation, the natural gradient is denoted as $\boldsymbol{G}(t)$. Along its steepest descent direction $\boldsymbol{G}(t)$ on the geodesic, we adopt a line search method to solve optimal step size u and obtain the minimal value of the score function at time t, as

$$\boldsymbol{W}(t+1) = \arg\min_{u} J(\boldsymbol{W}(t,u)) \tag{7.7.22}$$

where $\boldsymbol{W}(t, u)$ is the position u on the geodesic for parameter matrix $\boldsymbol{W}$ at time t in the on-line algorithm, as Eq. (7.7.20) concretely shows.

The line search method solving the optimal step size along the steepest descent direction on the geodesic is added in the expansion algorithm; thus, a better effect is obtained.

Next, the shift of the tangent vector along the geodesic is introduced and the conjugate is defined on the matrix manifold.

The shift $\tau\boldsymbol{H}(t_1)$ of tangent vector $\boldsymbol{H}(t_1)$ along the geodesic $\boldsymbol{W}(t, u)$ can be stated as follows[22,23]: Suppose that $\boldsymbol{H}(t_1)$ $\boldsymbol{H}(t_2)$ are vectors of point $\boldsymbol{W}(t_1)$, $\boldsymbol{W}(t_2)$ defined on the matrix manifold in the tangent plane $T_{W_1} M$, $T_{W_2} M$ respectively, and the tangent vector $\boldsymbol{H}(t_1)$ moves from point $\boldsymbol{W}(t_1)$ to point $\boldsymbol{W}(t_2)$ of the tangent plane $T_{W_2} M$ along the geodesic, then

$$\tau\boldsymbol{H}(t_1) = \boldsymbol{H}(t_1)\boldsymbol{W}(t_1)^{-1}\boldsymbol{W}(t_2) \tag{7.7.23}$$

The conjugate on the matrix manifold means that $\boldsymbol{H}(t_1)$, $\boldsymbol{H}(t_2)$ are tangent vectors of point $\boldsymbol{W}(t_1)$, $\boldsymbol{W}(t_2)$ defined on the matrix manifold in the tangent plane and satisfy

$$\boldsymbol{H}\mathrm{e}(\boldsymbol{H}(t_2), \tau\boldsymbol{H}(t_1)) = 0 \tag{7.7.24}$$

where $\tau\boldsymbol{H}(t_1)$ is the shift of tangent vector $\boldsymbol{H}(t_1)$ moving from point $\boldsymbol{W}(t_1)$ to point $\boldsymbol{W}(t_2)$ along the geodesic $\boldsymbol{W}(t, u)$; $\boldsymbol{H}\mathrm{e}(\boldsymbol{H}, \boldsymbol{H})$ is the Hessian matrix and is defined as the second derivative of the evaluation function along the geodesic:

$$\boldsymbol{H}\mathrm{e}(\boldsymbol{H}, \boldsymbol{H}) = \frac{\mathrm{d}^2}{\mathrm{d}u^2} J(\boldsymbol{W}(t,u))\big|_{u=0} \tag{7.7.25}$$

The conjugate gradient method performed on the nonsingular matrix manifold includes two steps[22,23]:

The first step is to find a search direction $\boldsymbol{H}(t)$, which is conjugate with the former direction ,namely

$$\boldsymbol{H}(t) = \boldsymbol{G}(t) + \gamma(t)\tau\boldsymbol{H}(t-1) \tag{7.7.26}$$

where $\boldsymbol{G}(t)$ is the natural gradient

$$\boldsymbol{G}(t) = (\boldsymbol{\Lambda}(t) - \boldsymbol{f}(\boldsymbol{y}(t))\boldsymbol{y}^{\mathrm{T}}(t))\boldsymbol{W}(t) \tag{7.7.27}$$

The selection of parameter $\gamma(t)$ in Eq. (7.7.26) must be insured to satisfy Eq. (7.7.24), i. e. that the conjugate gradient direction is conjugate with the former search direction.

In the Polak and Ribiere algorithm[23], every iterative step is

$$\gamma(t)=\frac{\langle \boldsymbol{G}(t)-\tau\boldsymbol{G}(t-1),\boldsymbol{G}(t)\rangle}{\langle \tau\boldsymbol{G}(t-1),\tau\boldsymbol{G}(t-1)\rangle} \tag{7.7.28}$$

where the interior product $\langle \boldsymbol{W}_1, \boldsymbol{W}_2\rangle = \text{trace}\,(\boldsymbol{W}_1\boldsymbol{W}_2^{\mathrm{T}})$ for any matrix $\boldsymbol{W}_1$ and $\boldsymbol{W}_2$.

The second step is to perform the line search optimization algorithm along the conjugate direction $\boldsymbol{H}(t)$ on the geodesic. The optimal step size $u(t)$ is determined according to Eq. (7.6.22), and the parameter matrix $\boldsymbol{W}(t+1)$ is determined at time $(t+1)$.

The stability of the algorithm, that is the separation solution using the learning algorithm is stable, is judged according to the stability theory of the algorithm. The criterion is when and only when

$$m_i = E[y_i^2 f'_i(y_i)] > -1\ ; \tag{7.7.29}$$

$$k_i = E[f'_i(y_i)] > 0\ ; \tag{7.7.30}$$

$$\sigma_i^2\sigma_j^2 k_i k_j > 1,\ \forall\, i,j\,(i\neq j) \tag{7.7.31}$$

where $\sigma_i^2 = E[y_i^2]$, $f'_i(y_i) = \dfrac{\mathrm{d}f_i(y_i)}{\mathrm{d}y_i}$, $i,j=1,2,\cdots,m$.

The first-order derivative of the score function is calculated according to Eq. (7.7.12):

$$f'_i(y_i) = \frac{1+\boldsymbol{\kappa}^{\mathrm{T}}\boldsymbol{g}+\boldsymbol{\kappa}^{\mathrm{T}}\boldsymbol{Q}\boldsymbol{\kappa}}{\left[1+\frac{\kappa_{i,3}}{3!}H_3(y_i)+\frac{\kappa_{i,4}}{4!}H_4(y_i)\right]^2} \tag{7.7.32}$$

where

$$\boldsymbol{\kappa} = \left(\frac{\kappa_{i,3}}{6},\frac{\kappa_{i,4}}{24}\right)^{\mathrm{T}} \tag{7.7.33}$$

$$\boldsymbol{g} = (2y_i^3 - 12y, 2y_i^4 - 24y_i^2 + 18)^{\mathrm{T}} \tag{7.7.34}$$

$$\boldsymbol{Q} = \begin{bmatrix} y_i^6 - 3y_i^4 + 9y_i^2 + 9 & y_i^7 - 6y_i^5 + 15y_i^3 \\ y_i^7 - 6y_i^5 + 15y_i^3 & y_i^8 - 8y_i^6 + 30y_i^4 + 45 \end{bmatrix} \tag{7.7.35}$$

Because the separation output signal $y(t)$ is the estimate of source $s(t)$, from inference formula (7.7.4), we can see that

$$E[y_i(t)] = 0, \qquad E[y_i^2(t)] = 1 \tag{7.7.36}$$

from Eq. (7.7.34) and noticing Eqs. (7.7.6) and (7.7.7), we can obtain

$$E(\boldsymbol{g}) = E[(2y_i^3, 2y_i^4 - 6)^{\mathrm{T}}] = [2\kappa_{i,3}, 2\kappa_{i,4}]^{\mathrm{T}} \tag{7.7.37}$$

then

$$E[\boldsymbol{\kappa}^{\mathrm{T}}\boldsymbol{g}] > 0 \tag{7.7.38}$$

From Eq. (7.7.35), we can easily validate that despite any value of y_i, the principal minor of $\boldsymbol{Q}$ is greater than zero.Thus,

$$Q > 0 \tag{7.7.39}$$

From Eqs. (7.7.38) and (7.7.39), we can obtain

$$E[\kappa^{\mathrm{T}} \boldsymbol{g} + \kappa^{\mathrm{T}} Q \kappa] > 0 \tag{7.7.40}$$

We can validate this from Eqs. (7.7.40) and (7.7.32):

$$m_i + 1 = E[y_i^2 f'_i(y_i) + 1] > 0 \tag{7.7.41}$$

and

$$k_i = E\left[\frac{1 + \kappa^{\mathrm{T}} \boldsymbol{g} + \kappa^{\mathrm{T}} Q \kappa}{\left(1 + \frac{\kappa_{i,3}}{3!} H_3(y_i) + \frac{\kappa_{i,4}}{4!} H_4(y_i)\right]^2}\right] > 0; \tag{7.7.42}$$

Next we consider the numerator of $(k_i - 1)$:

$$\begin{aligned} \text{Numerator}(k_i - 1) &= E[\boldsymbol{\kappa}^{\mathrm{T}} \Delta \boldsymbol{g} + \boldsymbol{\kappa}^{\mathrm{T}} \Delta Q \kappa] \\ &= \boldsymbol{\kappa}^{\mathrm{T}} E[\Delta \boldsymbol{g}] + \boldsymbol{\kappa}^{\mathrm{T}} E[\Delta Q] \boldsymbol{\kappa} \end{aligned} \tag{7.7.43}$$

where

$$\Delta \boldsymbol{g} = [-6y_i, -12y_i^2 + 12]^{\mathrm{T}} \tag{7.7.44}$$

from Eq. (7.7.36), we have

$$\mathrm{E}[\Delta \boldsymbol{g}] = [0, 0]^{\mathrm{T}} \tag{7.7.45}$$

Besides,

$$\Delta Q = \begin{bmatrix} 3y_i^4 + 9 & 3y_i^5 - 6y_i^3 + 9y_i \\ 3y_i^5 - 6y_i^3 + 9y_i & 4y_i^6 - 12y_i^4 + 36y_i^2 + 36 \end{bmatrix}$$

ΔQ has positive first-order and second-order principal minors, so it is positive definite, namely

$$\Delta Q > 0 \tag{7.7.46}$$

Considering Eqs. (7.7.45) and (7.7.46) at the same time, we can obtain from Eq. (7.7.43):

$$\text{Numerator}(k_i - 1) > 0 \tag{7.7.47}$$

The denominator term of $(k_i - 1)$ is greater than zero

$$k_i > 1 \tag{7.7.48}$$

it satisfies Eq. (7.7.31) with regard to the stability criterion, and it is proved that the algorithm is finally stable.

In order to further show the convergence Property of the algorithm, it is performed 30 times by the Monte Carlo calculation, in each of which the source signal is generated according to the following equation, and the mixing matrix is generated by random number.

$$\boldsymbol{s}(t) = \begin{bmatrix} s_1(t) \\ s_2(t) \\ s_3(t) \end{bmatrix} = \begin{bmatrix} n(t) \\ 0.1\sin(400t)\cos(30t) \\ 0.01\mathrm{sgn}\{\sin[500t + 9\cos(40t)]\} \end{bmatrix}$$

where $n(t)$ is the noise source uniformly distributed in the interval $[-1,$

+1], and the, other two signals are deterministic sources whose waveforms are shown in Fig. 7.25.

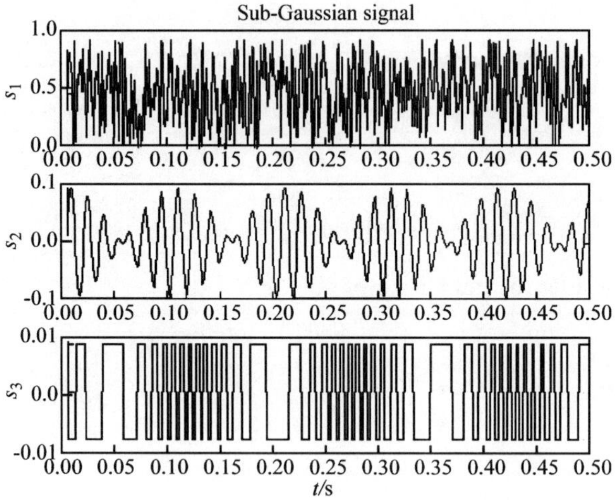

Fig. 7.25 Waveforms of the source signals

The result of running the Monte Carlo 30 times is shown in Fig. 7.26. The difference of the performance index of the algorithm is not great, though the mixing matrix is not the same every time. The result shows that the algorithm is robust and stable.

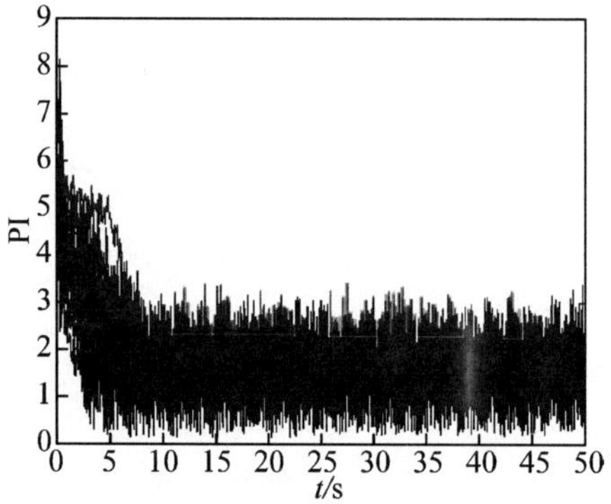

Fig. 7.26 Result of running the Monte Carlo 30 times

In addition, the on-line function of the probability density estimation algorithm is also tested using real acoustic signals. The sources are real acoustic signals, and the mixing matrix is generated by random numbers. We selected four acoustic signals recorded in a semi-anechoic room, with sampling frequency of 10.24 kHz. The data size is close to 100 seconds with approximately 1,000,000 samples. The first source is a Chinese speech signal, and the second, and third ones are English speech signals, while the fourth is a piece of light music. Their waveforms are shown in Fig. 7.27. The mixing matrix $\boldsymbol{A}$ is selected as 4×4 matrix, whose elements are random numbers distributed uniformly over the interval 0~1:

$$\boldsymbol{A} = \begin{bmatrix} 0.9355 & 0.0579 & 0.1389 & 0.2722 \\ 0.9169 & 0.3529 & 0.2028 & 0.1988 \\ 0.4103 & 0.8132 & 0.1987 & 0.0153 \\ 0.8936 & 0.0099 & 0.6038 & 0.7468 \end{bmatrix}$$

The waveforms of the mixed signals are shown in Fig. 7.28.

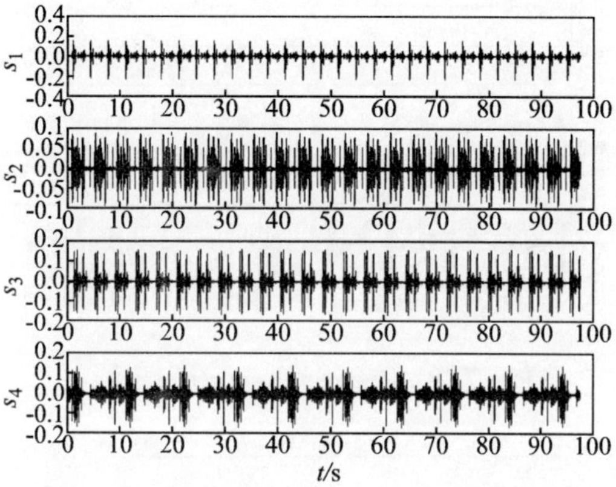

Fig. 7.27 Waveforms of the source signals

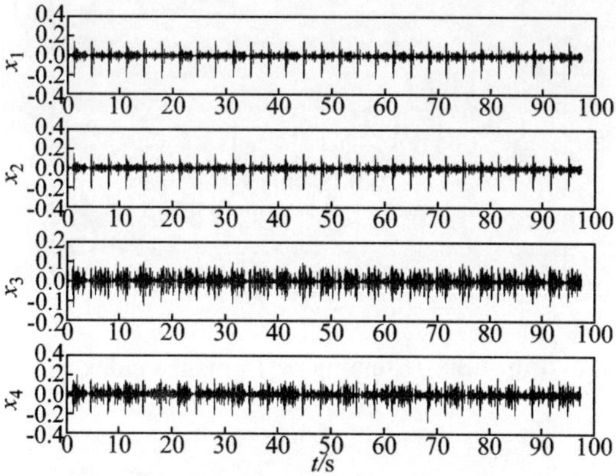

Fig. 7.28 Waveforms of the mixed signals

Considering that acoustic signals are non-stationary random signals, in learning the iterative equation, we select

$$\boldsymbol{\Lambda}(t) = \mathrm{diag}[\boldsymbol{f}(\boldsymbol{y}(t))\boldsymbol{y}^{\mathrm{T}}(t)]$$

At the same time, we select mixed signals with a learning rate of $\mu(t)$=0.02, a parametric learning rate of $\alpha_{k2}=\alpha_{k3}=\alpha_{k4}=0.1$, and an on-line separation length of 10 s. The signal waveforms after separation are shown in Fig. 7.29, and the running performance index of the algorithm is shown in Fig. 7. 30. From the figure we can see that the content can be heard clearly after 5 s, and the separation algorithm is basically stable after 10 s.

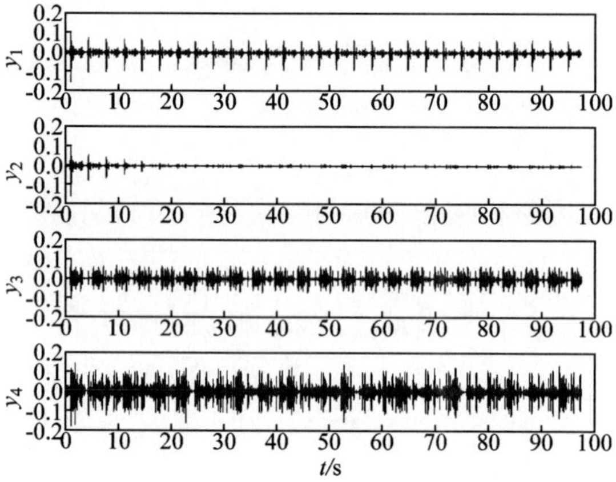

Fig. 7.29 Recovered source signals

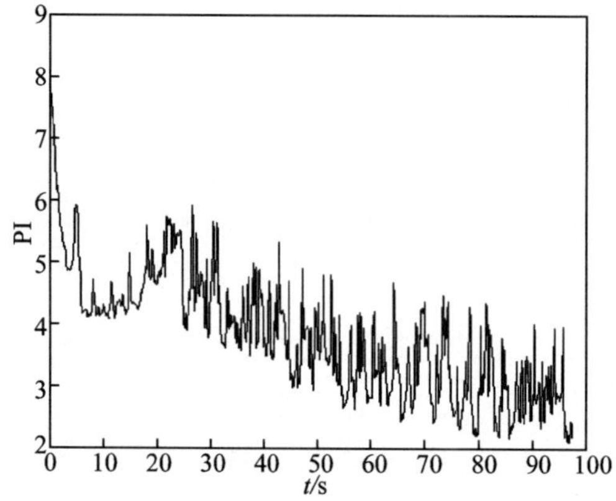

Fig. 7.30 Running performance index

References

[1] Cardoso J F, Laheld B (1996) Equivariant adaptive source separation. IEEE Transactions on Signal Processing 44(12): 3017–3030

[2] Yu Y (1995) Advanced engineering mathematics, 2nd edn. Huazhong University of Science and Technology Press, Wuhan

[3] Silverman B W (1986) Density estimation for statistics and data analysis. Chapman and Hall, London

[4] Turlach B A (1993) Bandwidth selection in kernel density estimation: A review. C.O.R.E. and Institut de Statistique, Universite' Catholique de Louvain, B-1348 Louvain-la-Neuve, Belgium

[5] Sheather S J, Jones M C (1991) A reliable data-based bandwidth selection method for kernel density estimation.Journal of Royal Statistical Society, Series B 53(3): 683–690

[6] Jones M C, Sheather S J (1996) A brief survey of bandwidth selection for density estimation. Journal of the American Statistical Association 91(433): 401–407

[7] Jones M C, Marron J S, Sheather S J (1996) Progress in data-based bandwidth selection for kernel density estimation. Computational Statistics 11: 337–381

[8] Hädel W, Marron J S, Wand M P (1990) Bandwidth choice for density derivatives. Journal of the Royal Statistical Society, Series B 52: 223–232

[9] Fukunaga K, Hostetler L D (1975) The estimation of the gradient of a density function, with applications in pattern recognition. IEEE Transaction on Information Theory 21(1): 32–40

[10] Lee T W, Girolami M, Sejnowski T J (1999) Independent component analysis using an extended infomax algorithm for mixed sub-gaussian and super-gaussian sources. Neural Computation 11: 417–441

[11] Haykin S (2000) Unsupervised adaptive filter, Vol.2, Blind source separation. Wiley, New York

[12] Lambert R H, Bell A J (1996) Multichannel blind deconvolution: FIR matrix algebra and separation of multipath mixturess. Dissertation, University of Southern California

[13] Zhang H Y (2001) Theoretical and experimental studies of blind source separation. Dissertation, Shanghai Jiao Tong University (Chinese)

[14] Jia P (2001) Study of blind separation of acoustic signal. Dissertation, Shanghai Jiao Tong University (Chinese)

[15] Miller J H, Thomas J B (1972) Detectors for discrete-time signals in non-gaussian noise. IEEE Transaction on Information Theory IT-18(2): 241–250

[16] Lee T W, Lewicki M S (2000) The generalized gaussian mixtures model using ICA. In: International Workshop on Independent Component

Analysis (ICA'00), Helsinki, 2000, pp 239–244

[17] Zhao Y X, Zhang X H (1995) Gaussian mixtures density modeling of non-gaussian source for autoregressive process. IEEE Transactions on Signal Processing 43(4): 894–903

[18] Kong W (2005) Research on blind source separation algorithm and its application in acoustic signals. Dissertation, Shanghai Jiao Tong University (Chinese)

[19] Shen X Z (2005) Study of algorithms of second and higher order blind signal processing. Dissertation, Shanghai Jiao Tong University (Chinese)

[20] Stuart A, Ord J K (1994) Kendall's advanced theory of statistics. Edward Arnold, Lodon

[21] Papoulis A (1985) Probability, random variables, and stochastic processes, 2nd edn. McGraw-Hill, New York

[22] Zhang L Q (2004) Conjugate gradient approach to blind separation of temporally correlated signals. In: Proceedings of International Conference on Communications, Circuits and Systems, Chengdu, 2004, 2: 1008–1012

[23] Edelman A, Tomas A, Smith S T (1999) The geometry of algorithms with orthogonality constraints. SIAM Journal on Matrix Analysis & Applications 20(2): 303–353

[24] Amari S I, Chen T P, Cichocki A (1997) Stability analysis of adaptive blind source separation. Neural Networks 10(8): 1345–1351

Chapter 8 Joint Approximate Diagonalization Method

8.1 Introduction

The joint approximate diagonalization (JAD) method is a means to find an orthogonal matrix $\boldsymbol{U}$ according to the least squares or weighted least squares criteria to diagonalize a data matrix, which can be a covariance matrix or cumulant matrix. Since the 1990s, as the solution of the statistical identification problem, JAD has been used to extract information from the feature structure of the data matrix. For example, the forth-order joint cumulant matrix presented by Cardoso[1], the cumulant matrix generalized by Moreau[2] is greater or equal to the third-order cumulant matrix, SOBI (Second-Order Blind Identification) algorithm developed by Belouchrani,[3,4] and second derivative matrix of the logarithmic characteristic matrix[5], etc. The major advantage of the JAD method is its high efficiency, Jacobi-like technology alternative least square, parallel factor analysis, and subspace fitting, etc. For infinite samples, JAD can calculate the common feature structure from any part of matrix set A_∞ or linear combination of matrix in A_∞; but for finite samples T, the matrices in set A_T can't accurately share the same feature structure. The viewpoint of statistics, an "averaging" processing should be used to improve accuracy and robustness, then an appropriate weighting can obtain a mean structure. Details will be discussed in Section 5 of this chapter.

In blind source separation, if source signals have non-overlapping spectrums, they can be separated only by using a second-order cumulant. A second-order separation algorithm includes two parts. First, after whitening the data, it can remove the correlation from the data but cannot achieve independence at this time. Then, make the Jacobi rotation on the whitened covariance to achieve the unitary diagonalization to minimize the elements beyond the diagonal. The processing of the unitary matrix and whitening are strong enough to estimate the mixing matrix and separate mixtures.

X. Shi, *Blind Signal Processing*

8.2 JAD Algorithm of Frequency-Domain Feature

Based on the mutual uncorrelation of source signals, the JAD blind separation algorithm defines a second-order statistical contrast function related to the power spectrum. This algorithm estimates the mixing matrix by applying JAD on the power spectrums of the whitened mixed signals to separate spectrums of the source signals. Figure 8.1 shows the process of this method.

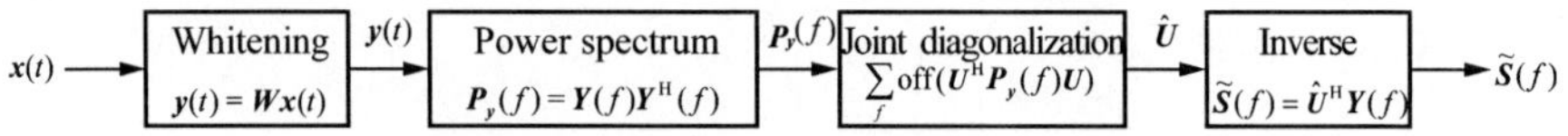

Fig. 8.1 JAD blind separation algorithm

When the Fourier transformation of the whitened signal $\boldsymbol{y}(t)$ can obtain $\boldsymbol{Y}(f)$, then the power spectrum matrix $\boldsymbol{P}_y(f)$ of whitened signals is shown as follows:

$$\boldsymbol{P}_y(f) = \boldsymbol{Y}(f)\boldsymbol{Y}^{\mathrm{H}}(f) = \boldsymbol{W}\boldsymbol{P}_x(f)\boldsymbol{W}^{\mathrm{H}} = \boldsymbol{U}\boldsymbol{P}_s(f)\boldsymbol{U}^{\mathrm{H}} \tag{8.2.1}$$

where the power spectrum matrix $\boldsymbol{P}_s(f)$ is a diagonal matrix, and matrix $\boldsymbol{U}$ is a unitary matrix, so the following equation can be obtained:

$$\boldsymbol{U}^{\mathrm{H}}\boldsymbol{P}_y(f)\boldsymbol{U} = \boldsymbol{P}_s(f) = \mathrm{diag}(P_{s1}(f), \cdots, P_{sN}(f)) \tag{8.2.2}$$

The above equation uses a unitary matrix to trans form the power spectrom matrix of the mixed signal into a diagonal matrix. Therefore, blind identification will be realized as long as a unitary matrix is found that can diagonalize the function matrix $\boldsymbol{P}(f)$. For this, the following frequency spectrum diagonalization contrast function is given:

$$C(\boldsymbol{U}) = \sum_f \mathrm{trace}(\boldsymbol{U}^{\mathrm{H}}\boldsymbol{P}_y(f)\boldsymbol{U}) \tag{8.2.3}$$

where trace (•) represents the matrix trace. The diagonalization contrast function of the frequency spectrum is an object function related to the power spectrum features; when the unitary matrix is a mixing one to be identified, the object function gets the maximum value. In this case, the power spectrum matrix is a diagonal matrix. Since frequency-domain transformation has been performed, the power spectrum feature of the signals will be characterized, and the instantaneous noise is spread over the whole frequency domain; therefore, the signal to noise ratio (SNR) increases. So the diagonalization contrast function in the frequency domain not only guarantees mutual independence of power spectrum features of the source signals to the utmost; but also reduces the sensitivity to noise. Actually, when the source signals are mutually uncorrelated, the power spectrum matrix is a diagonal matrix. Therefore, the frequency spectrum diagonalization contrast function still can be used for blind separation of uncorrelated source signals, which relaxes the assumption of mutual independence.

In actual calculation, because of the existence of numerical calculation error and interference noise, the complete diagonalization of Eq. (8.2.3) can't be obtained. The approximate diagonalization instead of the complete diagonalization was used, relative to the diagonal elements, Let the non-diagonal elements of the matrix after unitary transformation be as small as possible. To measure the deviation between matrix $\boldsymbol{M}=(m_{ij})_{n\times n}$ and the diagonal matrix, the following index can be used,

$$\mathrm{off}(\boldsymbol{M}) = \sum_{1\leqslant i\neq j\leqslant n} (m_{ij})^2 \tag{8.2.4}$$

in order to improve robustness of the algorithm, the power spectrum set $\{\boldsymbol{P}_y(i)|i=1, \cdots, K\}$ should be of joint diagonalization. Define the contrast function of the spectrum approximate diagonalization as follows:

$$C(\boldsymbol{U}) = \sum_{i=1}^{K} \mathrm{off}(\boldsymbol{U}^{\mathrm{H}}\boldsymbol{P}_y(i)\boldsymbol{U}) \tag{8.2.5}$$

It can be shown from the above equation that minimizing the sum of the squares of the non-diagonal elements is used as the separation target of the contrast function in spectrum approximate diagonalization, so the algorithm allows the existence of noise, and it still has separating ability for the mixed signals with a certain SNR. Furthermore, since the algorithm allows non-diagonal elements having minor values, it also can separate some weakly correlated source signals. In this case, in order to guarantee that the mixing matrix can be transferred to a unitary matrix and guarantee the validity of algorithm, the variance matrix of the source signal vector must be a unit matrix. The contrast function of spectrum approximate diagonalization takes the unique global minimum value when and only when the searched unitary matrix is a mixing unitary matrix to be identified. In the process of searching the unitary matrix for optimization, the search does not converge at a local minimum value, which guarantees the reliability of the separated signals. Finally, for choosing appropriate spectrum set $\{\boldsymbol{P}_y(i)|i=1, \cdots, K\}$ for JAD, $\boldsymbol{P}_y(f)$ should be selected in the range of a characteristic frequency to guarantee the selected $\boldsymbol{P}_y(i)$ with high SNR, so the robustness and reliability of the algorithm can be improved.

First define the contrast function of spectrum approximate diagonalization, then search for a unitary matrix to optimize the power spectrum matrix of the mixed signal, and the unitary matrix $\hat{\boldsymbol{U}}$ which minimizes the contrast function value of the mixed unitary matrix to be identified. The search process can be implemented by using the Jacobi-likelihood method[6,7]. After obtaining the mixed unitary matrix, the power spectrum matrix of the source signal vector can be obtained by using the following function:

$$\boldsymbol{P}_s(f) = \hat{\boldsymbol{U}}^{\mathrm{H}} \boldsymbol{P}_y(f) \hat{\boldsymbol{U}} \tag{8.2.6}$$

According to the unit orthogonal Property of column (row) vectors in the unitary matrix, the unitary matrix set can be composed of a unit vector with rotary parameters, and different unitary matrices will be created by adjusting the parameters. The formation method of the second-order unitary matrix is

$$\boldsymbol{U} = \begin{bmatrix} \cos\theta & \mathrm{e}^{\mathrm{j}\varphi}\sin\theta \\ -\mathrm{e}^{-\mathrm{j}\varphi}\sin\theta & \cos\theta \end{bmatrix} \tag{8.2.7}$$

in the above function, adjusting parameter θ and φ can search different unitary matrices in the second-order unitary matrix set.

To check the validity of the JAD blind separation algorithm, we make a simulation and the results are as follows.

Four source signals in the first simulation are

$$\boldsymbol{s}(t) = \begin{pmatrix} s_1(t) \\ s_2(t) \\ s_3(t) \\ s_4(t) \end{pmatrix} = \begin{pmatrix} \sin(2\pi 475t) + 0.8\sin\left(2\pi 1000t + \frac{\pi}{6}\right) \\ 0.8\,\mathrm{sgn}(\sin(2\pi 175t)) \\ \sin(2\pi 560t + \sin(2\pi 215t)) \\ 1.25\sin(2\pi 150t)\cos(2\pi 345t) \end{pmatrix}$$

the signal sampling frequency is 3 kHz, and the sampling length is 6,000 points. Performing 2,048 point FFT on the signals by sections, we obtain their power spectrums after being averaged. The source signal waveforms and their power spectrums are shown in Fig. 8.2(a). The mixing matrix is generated randomly by the computer, with the elements distributed between (0,1), and the 4×4 mixing matrix in this example is shown as:

$$\boldsymbol{A} = \begin{pmatrix} 0.9501 & 0.8913 & 0.8214 & 0.9218 \\ 0.2311 & 0.7621 & 0.4447 & 0.7382 \\ 0.6068 & 0.4565 & 0.6154 & 0.1763 \\ 0.4860 & 0.0185 & 0.7919 & 0.4057 \end{pmatrix}$$

four mixtures can be obtained by mixing source signals with the mixing matrix; the waveforms of mixed signals and their power spectrums are shown in Fig. 8.2(b). Using the spectrum joint approximate diagonalization algorithm to separate the mixed signals, the result is shown in Fig. 8.2(c). Comparing the results in Fig. 8.2(c) with Fig. 8.2(a), it can be seen that the source signals made a satisfied recovery. At the same time, it also can be seen that the absolute

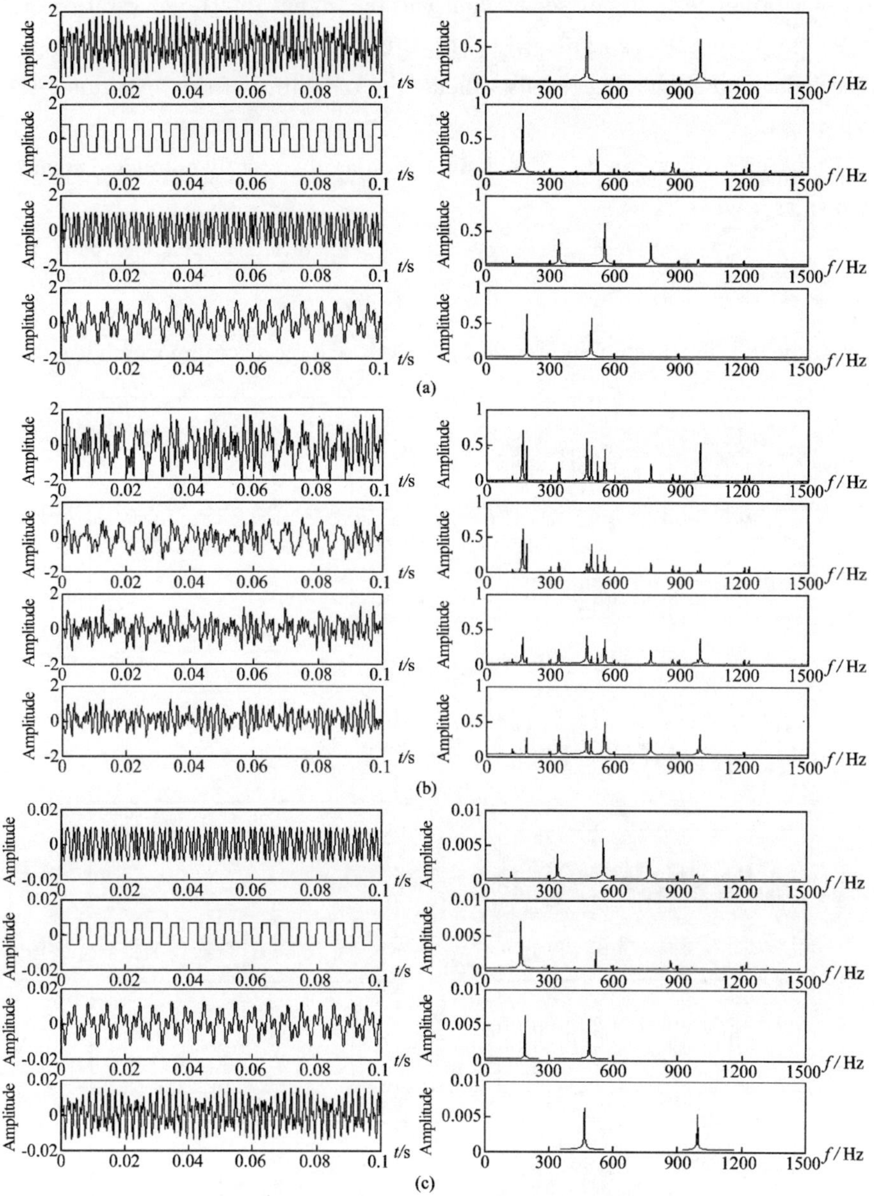

Fig. 8.2 Example 1 of spectrum separation
(a) Waveforms of source signals and their power spectrums
(b) Waveforms of mixed signals and their power spectrums
(c) Waveforms of separated signals and their power spectrums

amplitude values and the permutation of the separated signals are different from the original sources. This indicates there is a problem with indeterminacy of scale and permutation. $PI(\boldsymbol{C})=0.1038$ can be obtained by calculating the performance index of separation, and the signal interference ratios are SIR_1=18.4 dB, SIR_2=20.1 dB, SIR_3=18.7 dB, and SIR_4=21.3 dB respectively. The index and simulating results validate the reliability of the spectrum JAD algorithm.

The source signals in the second simulation are two partly correlated signals shown as follows:

$$\boldsymbol{s}(t)=\begin{pmatrix} s_1(t) \\ s_2(t) \end{pmatrix}=\begin{pmatrix} 0.5\sin(2\pi 225t)+0.8\sin(2\pi 600t+\frac{\pi}{3})+\sin(2\pi 875t) \\ \mathrm{sgn}(\sin(2\pi 225t)) \end{pmatrix}$$

similarly, the signal sampling frequency is 3 kHz and the sampling length is 6,000

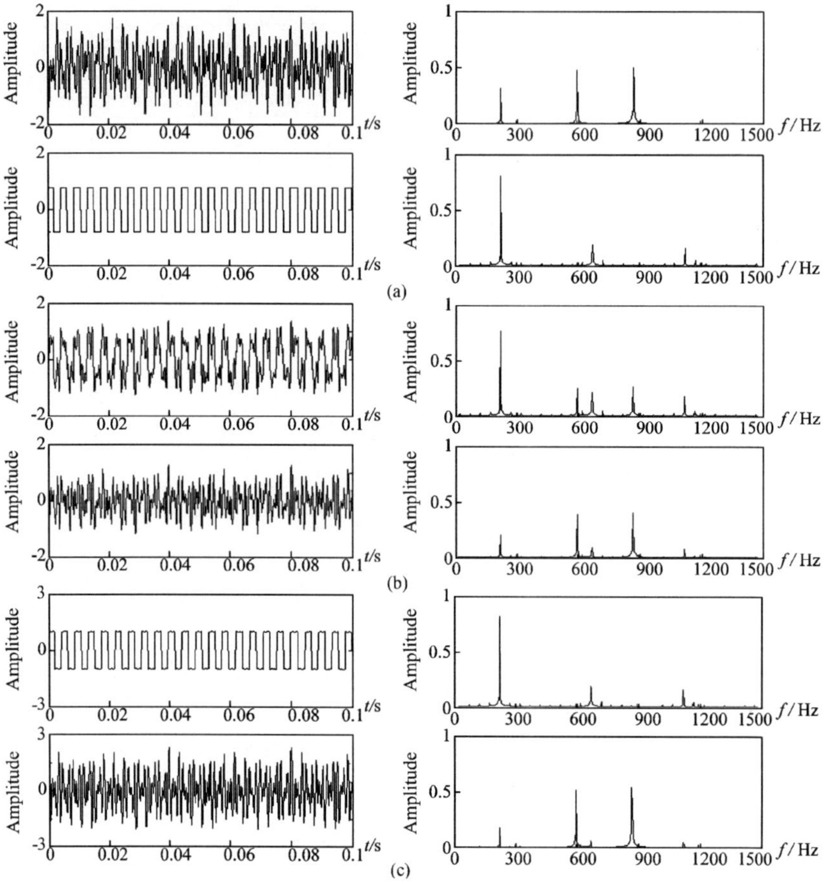

Fig. 8.3 Example 2 of spectrum separation

(a) Waveforms of source signals and their power spectrums

(b) Waveforms of mixed signals and their power spectrums

(c) Waveforms of separated signals and their power spectrums points. Performing a 2,048 point FFT on the signals by sections, the waveforms of source signals and their power spectrums are shown in Fig. 8.3(a), where the two signals both have components with 225 Hz frequency.The waveforms of mixed signals and their power spectrums are shown in Fig. 8.3(b). Using the spectrum joint approximate diagonalization algorithm to separate the mixed signals, the result of the separation is shown in Fig. 8.3(c). Comparing the results in Fig. 8.3(c) with Fig. 8.3(a), it can be seen that the source signals were successfolly recovered. PI (*C*)=0.2072 can be obtained by calculating the performance index of separation, with the interference ratios of signals of $\mathrm{SIR}_1=13.2$ dB and $\mathrm{SIR}_2=15.2$ dB. The performance index and simulation indicate that the spectrum JAD algorithm can be used in the blind separation of partly correlated signals.

8.3 JAD Algorithm of Time-Frequency Feature

The spectrum analysis discussed in the above section is only applicable for extracting the energy of a periodic component of signals. A time-frequency joint analysis method is more effective for the instantaneous features of a non-stationary signal. The blind separation algorithm for time-frequency features is very similar to that for spectrum features. The difference between them is that the blind separation algorithm for time-frequency features uses a time-frequency distribution function instead of the power spectrum function matrix in the latter algorithm.

For source signal vector $\boldsymbol{s}(t)$, the Cohen class time-frequency distribution function matrix $\boldsymbol{D}_s(t,f)$ is defined as follows:

$$\boldsymbol{D}_s(t,f)=\sum_{l=-\infty}^{+\infty}\sum_{m=-\infty}^{+\infty}\phi(m,l)\boldsymbol{x}(t+m+l)\boldsymbol{x}^{\mathrm{H}}(t+m-l)\mathrm{e}^{-\mathrm{j}4\pi ft}$$
$$=\begin{bmatrix} d_{11}(t,f) & \cdots & d_{1n}(t,f) \\ \vdots & \ddots & \vdots \\ d_{n1}(t,f) & \cdots & d_{nn}(t,f) \end{bmatrix} \tag{8.3.1}$$

where $\varphi(m,l)$ is the kernel function, and $d_{ij}(t,f)$ is the mutual time-frequency distribution function between *i*th and *j*th source signal, i.e.

$$d_{ij}(t,f)=\sum_{l=-\infty}^{+\infty}\sum_{m=-\infty}^{+\infty}\phi(m,l)x_i(t+m+l)x_j^*(t+m-l)\mathrm{e}^{-\mathrm{j}4\pi ft} \tag{8.3.2}$$

Since the source signals are mutually independent, the mutual time-frequency distribution function of them is equal to zero. Therefore the time-frequency distribution function matrix of the source signal vector is a diagonal matrix, and the diagonal elements of this are the time-frequency distribution functions of each source signal, i.e.

$$\boldsymbol{D}_s(t,f) = \mathrm{diag}(d_{11}(t,f),\cdots,d_{mm}(t,f))$$

According to the indeterminacy of amplitude in blind separation, a reference signal $\tilde{s}_i(t) = \sigma_i^{-1} s_i(t)$ with the same waveform can be obtained by multiplying a real source signal $s_i(t)$ with a coefficient σ_i^{-1}. We can use the two signals without any distinction and consider the covariance matrix $\boldsymbol{R}_s(0)$ of the source signal vector to be a unit matrix. Using the reference signal instead of the real source signal will cause uncertainty of the absolute amplitude value of the time-frequency distribution function, but the relative amplitude value of each time-frequency component stays invariant, and this assumption is acceptable for the extraction of time-frequency features.

Whitening on the mixed signal, assuming the whitening matrix and whitened signals are $\boldsymbol{W}$ and $\boldsymbol{y}(t)$ respectively, the correlation matrix of the whitened signal satisfies the following equation:

$$\boldsymbol{R}_y(0) = \boldsymbol{W}\boldsymbol{R}_x(0)\boldsymbol{W}^{\mathrm{H}} = \boldsymbol{U}\boldsymbol{R}_s(0)\boldsymbol{U}^{\mathrm{H}} = \boldsymbol{U}\boldsymbol{U}^{\mathrm{H}} = \boldsymbol{I} \tag{8.3.3}$$

indicating that matrix $\boldsymbol{U}$ is an orthogonal unitary matrix.

Performing this time-frequency transformation on the whitened signals, we can obtain this time-frequency distribution function matrix:

$$\boldsymbol{D}_y(t,f) = \boldsymbol{W}\boldsymbol{D}_x(t,f)\boldsymbol{W}^{\mathrm{H}} = \boldsymbol{U}\boldsymbol{D}_s(t,f)\boldsymbol{U}^{\mathrm{H}} \tag{8.3.4}$$

since the time-frequency distribution function matrix $\boldsymbol{D}_s(t,f)$ of the source signal is a diagonal matrix and matrix $\boldsymbol{U}$ is a unitary matrix, the following relation can be obtained:

$$\boldsymbol{U}^{\mathrm{H}}\boldsymbol{D}_y(t,f)\boldsymbol{U} = \boldsymbol{D}_s(t,f) = \mathrm{diag}(d_{11}(t,f),\cdots,d_{mm}(t,f)) \tag{8.3.5}$$

the above equation indicates that the time-frequency distribution function matrix of the whitened signals is transformed into a diagonal matrix by a unitary matrix; and the blind identification of the mixed unitary matrix can be realized as long as a unitary matrix that can diagonalize the time-frequency distribution function $\boldsymbol{D}_y(t,f)$ is found.

Considering the influence of a calculation error and interference noise, the complete diagonalization of the time-frequency distribution function cannot be realized; only JAD can be carried out. Make the non-diagonal elements of the unitary transform matrix as small as possible compared to the diagonal elements. The following time-frequency diagonalization contrast function is given for the time-frequency distribution function matrix set $\{\boldsymbol{D}_y(t_i,f_i)\,|\,i=1,\cdots,K\}$

$$C(\boldsymbol{U}) = \sum_{i=1}^{K} \mathrm{off}(\boldsymbol{U}^{\mathrm{H}}\boldsymbol{D}_y(t_i,f_i)\boldsymbol{U}) \tag{8.3.6}$$

The time-frequency diagonalization contrast function is an objective function related to time-frequency distribution, and it takes the minimum value when the unitary matrix is the mixed unitary matrix to be identified. At this point, the mutual independence of the time-frequency features will be guaranteed to be maximized. Similarly, the time-frequency contrast function is also effective for mutually uncorrelated source signals. In the process of searching and optimizing the contrast function, the learning rules will not stop when they converge to the local minimum, which guarantees the reliability of separation of signals.

Searching the unitary set to optimize the time-frequency diagonalization contrast function, the unitary matrix $\hat{\boldsymbol{U}}$ that minimizes the contrast function is the mixed unitary to be identified. Then the time-frequency distribution function matrix of the source signal vector can be obtained by using the following equation:

$$\boldsymbol{D}_s(t,f) = \hat{\boldsymbol{U}}^{\mathrm{H}} \boldsymbol{D}_y(t,f) \hat{\boldsymbol{U}} \tag{8.3.7}$$

the diagonal element $d_{ij}(t, f)$ of this matrix is the time-frequency distribution function of the ith source signal.

There are two examples for the time-frequency JAD algorithm. Two source signals of the first simulation are

$$\boldsymbol{s}(t) = \begin{pmatrix} s_1(t) \\ s_2(t) \end{pmatrix} = \begin{pmatrix} \sin(250\pi t^2) \\ \sin(100\pi t) + 0.8\sin(250\pi t) + \delta(t-0.5) \end{pmatrix}$$

where signal $s_1(t)$ is the linear frequency modulation signal. Its instantaneous frequency f=250t , the sampling frequency, is 1 kHz, and the sampling length is 1,000 points. Since the exponential distribution (CWD, Choï-Williams Distribution) can reject cross-terms in the Cohen class time-frequency distribution of signals very well, CWD are used to describe the time-frequency distribution of signals. The CWD of source signals are shown in Fig. 8.4(a). The mixing matrix is generated randomly by a computer where the elements are distributed between (0,1). The 2×2 mixing matrix in this example is shown as follows:

$$\boldsymbol{A} = \begin{pmatrix} 0.5634 & 0.5576 \\ 0.4498 & 0.6734 \end{pmatrix}$$

After the source signals are mixed by the mixing matrix, two mixed signals can be obtained. The CWD of the mixed signals are shown in Fig. 8.4(b). Carrying out blind separation on mixed signals by using the time-frequency JAD algorithm, the CWD of separated signals are shown in Fig. 8.4(c). Comparing Fig. 8.4(c) with Fig. 8.4(a), it can be seen that the time-frequency feature of the source signals is recovered satisfactorily. At the same time, it can also be seen that there is indeterminacy of amplitude and permutation. Calculating the performance index of separation gives $PI(C)=0.1938$, and the interference ratios of signals are $SIR_1=14.9$ dB, $SIR_2=15.7$ dB respectively.

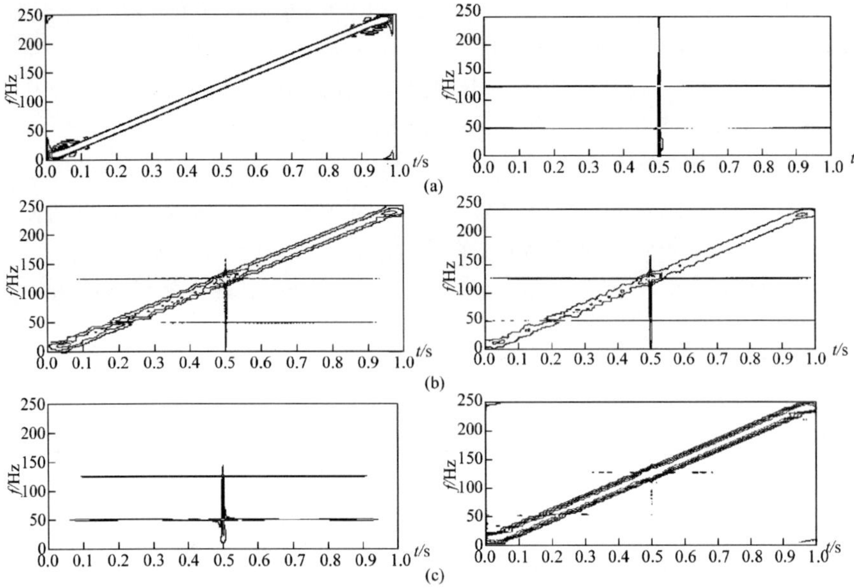

Fig. 8.4 Simulation example of time-frequency separation

(a) Time-frequency distribution of source signals;

(b) Time-frequency distribution of mixed signals;

(c) Time-frequency distribution of separated signals

8.4 Joint Approximate Block Diagonalization Algorithm of Convolutive Mixtures

Based on the feature blind separation algorithm on instantaneous mixtures, we can study deconvolution algorithms on the convolution mixing model. The basic idea is that after transforming the convolution mixing model into a high dimensional instantaneous model, We can apply the joint approximate block diagonalization algorithm to get blind deconvolution of source signals on the spectrum feature or time-frequency feature.

Referring to the convolutive mixture's problem discussed in Chapter 6, the discrete form of the convolution mixing model is shown as follows:

$$\begin{cases} \boldsymbol{x}(z) = \boldsymbol{H}(z)\boldsymbol{s}(z) \\ \boldsymbol{x}(n) = \sum_{i=0}^{M} \boldsymbol{H}(i)\boldsymbol{s}(n-i) \end{cases} \tag{8.4.1}$$

where $\boldsymbol{x}(z), \boldsymbol{H}(z), \boldsymbol{s}(z)$ are the z transformation of their corresponding function, M is the order number of the filter of the function matrix $\boldsymbol{H}(z)=(h_{ij}(z))_{m\times n}$; here M is the maximum order number of filters among all the matrix elements $h_{ij}(z)$. Define the signal vector $\boldsymbol{x}_{M'}(n)$ and $\boldsymbol{s}_{M'}(n)$ as

$$\begin{aligned} \boldsymbol{x}_{M'}(n) =& [x_1(n), \cdots, x_1(n-M'+1), x_2(n), \cdots, \\ & x_2(n-M'+1), \cdots, x_m(n-M'+1)]^{\mathrm{T}} \end{aligned} \tag{8.4.2}$$

$$\begin{aligned} \boldsymbol{s}_{M'}(n) =& [(s_1(n), \cdots, s_1(n-(M+M')+1), s_2(n), \cdots, \\ & s_2(n-(M+M')+1), \cdots, s_n(n-(M+M')+1)]^{\mathrm{T}} \end{aligned} \tag{8.4.3}$$

where M' is a positive integer, $\boldsymbol{x}_{M'}(n)$ is the column vector of $(m\times M')\times 1$, $\boldsymbol{s}_{M'}(n)$ is the row vector of $(m+M')\times 1$. According to the above definition, the following equation can be obtained:

$$\boldsymbol{x}_{M'}(n) = \overline{\boldsymbol{H}}_{M'}\boldsymbol{s}_{M'}(n) \tag{8.4.4}$$

where $\overline{\boldsymbol{H}}_{M'}$ is the following block matrix.

$$\overline{\boldsymbol{H}}_{M'} = \begin{pmatrix} \overline{\boldsymbol{H}}_{11} & \cdots & \overline{\boldsymbol{H}}_{1n} \\ \vdots & \ddots & \vdots \\ \overline{\boldsymbol{H}}_{m1} & \cdots & \overline{\boldsymbol{H}}_{mn} \end{pmatrix} \tag{8.4.5}$$

its sub-matrix $\overline{\boldsymbol{H}}_{ij}$ is

$$\overline{\boldsymbol{H}}_{ij} = \begin{pmatrix} h_{ij}(0) & h_{ij}(1) & \cdots & h_{ij}(M) & \cdots & 0 \\ & \ddots & \ddots & \ddots & \ddots & \\ 0 & \cdots & h_{ij}(0) & h_{ij}(1) & \cdots & h_{ij}(M) \end{pmatrix} \tag{8.4.6}$$

where $\overline{\boldsymbol{H}}_{M'}$ is a matrix with the order of $mM'\times n(M+M')$ and $\overline{\boldsymbol{H}}_{ij}$ is a matrix with the order of $M'\times(M+M')$.

It can be seen from Eq. (8.4.4) that the convolution mixing model is transformed into a high dimensional instantaneous mixing model. In order to satisfy the condition that the number of mixed signals isn't less than that of the source signals, the selected M' in Eq. (8.4.2) and Eq. (8.4.3) must satisfy $mM' \geqslant n(M+M')$.

For signal vector $\boldsymbol{s}_{M'}(n)$, its Fourier transformation is $\boldsymbol{S}_{M'}(k)$, and the corresponding power spectrum function matrix $\boldsymbol{P}_{sM'}(k)$ is defined as

$$\boldsymbol{P}_{sM'}(k) = \boldsymbol{S}_{M'}(k)\boldsymbol{S}_{M'}^{\mathrm{H}}(k) = \begin{pmatrix} P_{s1}(k) & \cdots & 0 \\ \vdots & \ddots & \vdots \\ 0 & \cdots & P_{sn}(k) \end{pmatrix} \tag{8.4.7}$$

in the above equation, since source signals $s_i(n)$ and $s_j(n)(j\neq i)$ are independent of each other, its power spectrum function matrix is a block diagonal matrix, and the sub-matrices $\boldsymbol{P}_{s_i}(k)$ are all $(M+M') \times (M+M')$ order matrices.

The real source signals are replaced by the unit variance reference signals with the same waveforms, that whiten the mixed signals. Denoting the whitening matrix and whitening signal with $\boldsymbol{W}$ and $\boldsymbol{y}_{M'}(n)$ respectively, then the following expression can be obtained:

$$\boldsymbol{R}_{y_{M'}}(0) = \boldsymbol{W}\boldsymbol{R}_{x_{M'}}(0)\boldsymbol{W}^{\mathrm{H}} = \boldsymbol{U}\boldsymbol{R}_{s_{M'}}(0)\boldsymbol{U}^{\mathrm{H}} = \boldsymbol{U}\boldsymbol{U}^{\mathrm{H}} = \boldsymbol{I} \tag{8.4.8}$$

It indicates that matrix $\boldsymbol{U}$ is an orthogonal unitary matrix.

According to the relationship between whitened signals and source signals, their power spectrum function matrix is

$$\boldsymbol{P}_{y_{M'}}(k) = \boldsymbol{W}\boldsymbol{P}_{x_{M'}}(k)\boldsymbol{W}^{\mathrm{H}} = \boldsymbol{U}\boldsymbol{P}_{s_{M'}}(k)\boldsymbol{U}^{\mathrm{H}} \tag{8.4.9}$$

Since the power spectrum function matrix of the source signal $\boldsymbol{P}_{sM'}(k)$ is a block diagonal matrix, and matrix $\boldsymbol{U}$ is a unitary matrix, the following equation can be obtained:

$$\boldsymbol{U}^{\mathrm{H}}\boldsymbol{P}_{y_{M'}}(k)\boldsymbol{U} = \boldsymbol{P}_{s_{M'}}(k) \tag{8.4.10}$$

It means that the power spectrum matrix of the whitened signal is transformed into a block diagonal matrix by a unitary matrix, so blind identification can be achieved so long as we can find a unitary matrix which can block diagonalize the power spectrum matrix $\boldsymbol{P}_{y_{M'}}(k)$.

Considering the effect of interference noise, the complete block diagonalization of the power spectrum matrix can't be implemented. Only joint approximate block diagonalization can be carried out here, i.e. the non-block diagonal elements of the matrix after unitary transformation should be as few as possible compared to the block diagonal elements, so the following block diagonalization contrast function is given for the power spectrum matrix set $\{\boldsymbol{P}_{y_{M'}}(k) \mid k=1, \cdots, K\}$

$$C(\boldsymbol{U}) = \sum_{k=1}^{K} \overline{\mathrm{off}}(\boldsymbol{U}^{\mathrm{H}}\boldsymbol{P}_{y_{M'}}(k)\boldsymbol{U}) \tag{8.4.11}$$

where $\overline{\mathrm{off}}(\boldsymbol{M})$ means the square sum of the non-block diagonal matrix elements, i.e.

$$\overline{\mathrm{off}}(\boldsymbol{M}) = \|\boldsymbol{M}\| - \sum_{l=1}^{n} \sum_{i,j=(M'+M)(l-1)+1}^{(M'+M)l} (m_{ij})^2 \tag{8.4.12}$$

where $\|\boldsymbol{M}\|$ is the norm of the matrix (i.e., $\|\boldsymbol{M}\| = \sum_{i,j} m_{ij}^2$). Compared with the JAD algorithm, this block diagonalization contrast function only calculates the norms sum of the non-block diagonal matrix; and, in fact, it also displays the blind separating ability for weakly correlated source signals.

Searching for the unitary matrix to optimize the diagonalization contrast function in the unitary matrices set, the unitary matrix $\hat{\boldsymbol{U}}$ that makes the contrast function minimum is the mixed unitary matrix to be identified. When the mixed unitary matrix is obtained, the power spectrum matrix of the source signal vector can be obtained by using the following formula:

$$\boldsymbol{P}_{s_{M'}}(k) = \hat{\boldsymbol{U}}^{\mathrm{H}} \boldsymbol{P}_{y_{M'}}(k)\, \hat{\boldsymbol{U}} \tag{8.4.13}$$

diagonal elements of this matrix are the power spectrum functions of the source signals. It needs to be pointed out that the above equation will recover the power spectrums of $n(M+M')$ source signals, in which every $(M+M')$ power spectrum is the power spectrum of one source signal. The separation can be achieved only by selecting n mutually uncorrelated power spectrums from these $n(M+M')$ spectrums.

The above steps show the process of the spectrum feature blind separation algorithm of the convolution mixing model. The time-frequency feature blind separation algorithm of the convolution mixing model has the same process and can be done only by replacing the power spectrum function matrix with the time-frequency distribution function matrix.

It needs to be clear that the blind deconvolution of the convolutive mixtures is more complex than the blind separation of instantaneous mixtures. The main problem is that the computation workload of the blind deconvolution algorithm is great, and the ill-conditioned problem may result in severe distortion in the separation results that can't satisfy the actual requirement. It usually needs a multi-order transfer function matrix to describe the convolution mixing model, and the identification of its transfer function matrix becomes more complex. Usually, there may be non-convergence and ill-conditioned problems in the algorithms that result in severe distortion of identification and separation results. Even for the classical deconvolution problem of a known transfer function matrix, the reliability problem in the deconvolution process can't be ignored. Miyoshi and Kaneda pointed out[8] that the deconvolution problem of the multi-input multi-output system is reliable only when the order number of the filter is limited to 1,000 or less; the reliability of the deconvolution result can't be guaranteed if its order number exceeds 1,000. For blind deconvolution, since the transfer function matrix is estimated, the order number of the filter should be lower in order to guarantee the reliability of the separation results.

8.5 JAD Method Based on Cayley Transformation

Theoretically, the covariance matrices set is a group of regular exchange matrices, that has a unique joint unitary diagonalization matrix. Because of estimation error and noise, whitened covariance $\boldsymbol{R}_k$ is not a complete Hermit. Therefore, the diagonalization matrix of a single matrix is almost not a unitary matrix, and the single unitary matrix which can make diagonalization of all the covariance matrixces does not exist. The solution is to compel transform $\boldsymbol{R}_k$ into a Hermit by replacing the non-diagonal elements of $\boldsymbol{R}_k$ with their means, and at the same time, to find a class of average unitary matrices to approximate each covariance matrix with a different diagonalization.

In order to solve the problem that the linear combination of unitary matrices doesn't keep to the unitary matrix, two diagonalization algorithms based on the Cayley transformation are proposed. The linear combination diagonalization (LCD) algorithm can obtain a kind of "mean" unitary matrix by linearly combining single unitary matrices in the Cayley domain. The Cayley-based joint unitary diagonaliser algorithm searches the optimum diagonalization matrix for all the covariance matrixes directly[9]. The generalized Cayley transformation is defined as

$$C(\boldsymbol{S}):\boldsymbol{U}=(\boldsymbol{I}-\boldsymbol{S})(\boldsymbol{I}+\boldsymbol{S})^{-1} \tag{8.5.1}$$

and the generalized inverse Cayley transformation is

$$C^{-1}(\boldsymbol{U}):\boldsymbol{S}=(\boldsymbol{I}-\boldsymbol{U})(\boldsymbol{I}+\boldsymbol{U})^{-1} \tag{8.5.2}$$

where $\boldsymbol{U}$ and $\boldsymbol{S}$ are a unitary matrix and skew the Hermitian matrix respectively. If neither $(\boldsymbol{I}+\boldsymbol{S})$ nor $(\boldsymbol{I}+\boldsymbol{U})$ is singular, Equation (8.5.1) will transform a unitary matrix into a skew Hermitian matrix. Thus, transform the unitary matrix into the Cayley domain, average it, and inverse it back to the regular domain; then the diagonalization mean corresponding to the single matrix is completed.

In order to find a unitary matrix $\boldsymbol{U}$ that diagonalizes the matrix set $\mathfrak{R}$ to minimize the non-diagonal elements of the whitened covariance matrix, we define the evaluation function

$$\phi(\boldsymbol{U},\mathfrak{R})=\sum_{k=1}^{K}\mathrm{off}(\boldsymbol{U}^{\mathrm{H}}\boldsymbol{R}_k\boldsymbol{U}) \tag{8.5.3}$$

here, define the non-diagonal element component off (•) as

$$\mathrm{off}(\boldsymbol{A})=\sum_{i\neq j}|a_{ij}|^2 \tag{8.5.4}$$

Therefore, the matrix$\boldsymbol{U}$ that minimizes Eq. (8.5.3) is the approximate joint unitary diagonalization matrix. Minimizing Eq. (8.5.3) is equivalent to maximizing energy of diagonal elements $\boldsymbol{U}^{\mathrm{H}}\boldsymbol{R}_k\boldsymbol{U}, k=1, \cdots, K$. Therefore, the evaluation function of Eq. (8.5.3) can be written as

$$\phi(\boldsymbol{U}, \mathfrak{R}) = \sum_{k=1}^{K} \{\mathrm{diag}^{\mathrm{H}}[\boldsymbol{U}^{\mathrm{H}}\boldsymbol{R}_k\boldsymbol{U}]\mathrm{diag}[\boldsymbol{U}^{\mathrm{H}}\boldsymbol{R}_k\boldsymbol{U}]\} \tag{8.5.5}$$

where diag[•] represents the vector composed by the matrix diagonal elements.

If skew-Hermitian matrix $\boldsymbol{C}_k (k=1, \cdots, K)$ is the Cayley transform of matrix $\boldsymbol{U}_k$ that only diagonalizes on a single matrix $\boldsymbol{R}_k$ of set $\mathfrak{R}$, its weighted mean is

$$\overline{\boldsymbol{C}} = \sum_{k=1}^{K} w_k \boldsymbol{C}_k \tag{8.5.6}$$

the weight $w_k (k=1, \cdots, K)$ in the equation is to be determined. Similarly,

$$\overline{\boldsymbol{U}} = (\boldsymbol{I} - \overline{\boldsymbol{C}})(\boldsymbol{I} + \overline{\boldsymbol{C}})^{-1} \tag{8.5.7}$$

where $\overline{\boldsymbol{U}}$ is the Cayley inverse transform of the weighted mean of the single unitary matrix in the Cayley domain. Combining Eqs. (8.5.5), (8.5.6) and (8.5.7), the optimization problem is summed up as follows:

$$\min_{w} \phi(\overline{\boldsymbol{U}}, \mathfrak{R}) \tag{8.5.8}$$

where $\boldsymbol{w}$ is the $K \times 1$ weight vector of $\boldsymbol{C}_k$.

Differentiate the evaluation function of Eq. (8.5.5)

$$\frac{\partial \phi}{\partial w_p} = 2\sum_{k=1}^{K} \mathrm{diag}^{\mathrm{H}}[\overline{\boldsymbol{U}}^{\mathrm{H}}\boldsymbol{R}_k\overline{\boldsymbol{U}}]\mathrm{diag}\left[\overline{\boldsymbol{U}}^{\mathrm{H}}\boldsymbol{R}_k\frac{\partial \overline{\boldsymbol{U}}}{\partial w_p} + \left(\frac{\partial \overline{\boldsymbol{U}}}{\partial w_p}\right)^{\mathrm{H}}\boldsymbol{R}_k\overline{\boldsymbol{U}}\right] \tag{8.5.9}$$

here, w_p represents the weight of $\boldsymbol{R}_p$. Since

$$\mathrm{diag}\left[\left(\frac{\partial \overline{\boldsymbol{U}}}{\partial w_p}\right)^{\mathrm{H}}\boldsymbol{R}_k\overline{\boldsymbol{U}}\right] = \mathrm{diag}\left[\overline{\boldsymbol{U}}^{\mathrm{H}}\boldsymbol{R}_k^{\mathrm{H}}\frac{\partial \overline{\boldsymbol{U}}}{\partial w_p}\right] \tag{8.5.10}$$

Equation (8.5.9) can be simplified as

$$\frac{\partial \phi}{\partial w_p} = 2\sum_{k=1}^{K} \mathrm{diag}^{\mathrm{H}}[\overline{\boldsymbol{U}}^{\mathrm{H}}\boldsymbol{R}_k\overline{\boldsymbol{U}}]\mathrm{diag}\left[\overline{\boldsymbol{U}}^{\mathrm{H}}(\boldsymbol{R}_k + \boldsymbol{R}_k^{\mathrm{H}})\frac{\partial \overline{\boldsymbol{U}}}{\partial w_p}\right] \tag{8.5.11}$$

the next step is to calculate the derivative of averaging unitary diagonal matrix $\overline{\boldsymbol{U}}$ with respect to the weight.

$$\begin{aligned}
\frac{\partial \overline{\boldsymbol{U}}}{\partial w_p} &= -\frac{\partial \overline{\boldsymbol{C}}}{\partial w_p}(\boldsymbol{I}+\overline{\boldsymbol{C}})^{-1} - (\boldsymbol{I}+\overline{\boldsymbol{C}})(\boldsymbol{I}+\overline{\boldsymbol{C}})^{-1}\frac{\partial \overline{\boldsymbol{C}}}{\partial w_p}(\boldsymbol{I}+\overline{\boldsymbol{C}})^{-1} \\
&= -\{\boldsymbol{I} + (\boldsymbol{I}+\overline{\boldsymbol{C}})(\boldsymbol{I}+\overline{\boldsymbol{C}})^{-1}\}\frac{\partial \overline{\boldsymbol{C}}}{\partial w_p}(\boldsymbol{I}+\overline{\boldsymbol{C}})^{-1} \\
&= (\boldsymbol{I}+\overline{\boldsymbol{U}})\frac{\partial \overline{\boldsymbol{C}}}{\partial w_p}(\boldsymbol{I}+\overline{\boldsymbol{C}})^{-1} \\
&= (\boldsymbol{I}+\overline{\boldsymbol{U}})\boldsymbol{C}_p(\boldsymbol{I}+\overline{\boldsymbol{C}})^{-1}
\end{aligned} \tag{8.5.12}$$

where, $\boldsymbol{C}_p$ is the Cayley transform of the unitary diagonal matrix of $\boldsymbol{R}_p$. By using Eqs. (8.5.11) and (8.5.12), and K nonlinear equations sets, including K unknown numbers can be constructed. Then we can maximize the weights of Eq. (8.5.5) by using the steepest ascent algorithm:

$$W(k+1) = W(k) + \mu\nabla \mid_{w(k)} \tag{8.5.13}$$

this is a linear combination diagonalization. This algorithm is attractive in concept; however, it has two shortcomings: one is the heavy calculation workload in Cayley domain, the other is that the eigenvalues must have great differences from 1 for the purpose of the Cayley transform. Therefore,another improved diagonalization method based on the Cayley transform is developed.

$\boldsymbol{H}$ is a real $n \times n$ skew Hermitian matrix, h_{ij} denotes the element, if $i=j$, $h_{ij}=0$, and if $i \neq j$, $h_{ij}=-h_{ji}$. Similarly, define ∂h_{ij} as δ variation on h_{ij} and $-\delta$ variation on h_{ji}, then

$$\frac{\partial \boldsymbol{H}}{\partial h_{ij}} = \boldsymbol{T}_{ij} \tag{8.5.14}$$

except for $t_{ij}=1$ and $t_{ji}=-1$, all the elements of $\boldsymbol{T}_{ij}$ are equal to zero. Replace $\overline{C}$ by $\boldsymbol{H}$ and replace w_p by h_{ij}, then Eq. (8.5.12) can be rewritten as

$$\frac{\partial \overline{\boldsymbol{U}}}{\partial w_p} = -(\boldsymbol{I}+\overline{\boldsymbol{U}})\frac{\partial \boldsymbol{H}}{\partial h_{ij}}(\boldsymbol{I}+\boldsymbol{H})^{-1} \tag{8.5.15}$$

Using Eq. (8.5.14), the above equation can be simplified as

$$\frac{\partial \overline{\boldsymbol{U}}}{\partial h_{ij}} = -(\boldsymbol{I}+\overline{\boldsymbol{U}})T_{ij}(\boldsymbol{I}+\boldsymbol{H})^{-1} \tag{8.5.16}$$

Similar to Eq. (8.5.13), a new gradient can be written as

$$\frac{\partial \phi}{\partial h_{ij}} = 2\sum_{k=1}^{K} \mathrm{diag}^{\mathrm{H}}[\overline{\boldsymbol{U}}^{\mathrm{H}}\boldsymbol{R}_k\overline{\boldsymbol{U}}]\mathrm{diag}\left[\overline{\boldsymbol{U}}^{\mathrm{H}}(\boldsymbol{R}_k+\boldsymbol{R}_k^{\mathrm{H}})\frac{\partial \overline{\boldsymbol{U}}}{\partial h_{ij}}\right] \tag{8.5.17}$$

Use steepest ascent algorithm of Eq. (8.5.13) and Eq. (8.5.17) to maximize Eq. (8.5.5). The merit of this algorithm is that the evaluation function can change along with the sample length, noise or prior information, and all the elements of the matrix can be continuously adjusted to the optimum unitary matrix.

8.6 Joint Diagonalization and Joint Non-Diagonalization Method

For linear mixtures of non-stationary sources, most blind separation algorithms are based on time-frequency distributions[10]. These methods only consider the auto-terms of signal time-frequency distributions, not the cross-terms in order to guarantee the diagonal structure of the time-frequency distribution matrix of the source signal, which is the necessary condition of a joint diagonalization program. A new method different from the existing ones relaxes the previous conditions and uses both diagonal structural and non-diagonal structural time-frequency distributions of the source signal, i.e. select time-frequency points from both the auto-term and the cross-term at the same time. Since both of the two domains include the important information of signal separation, establishing a unified frame can increase the quantity of the space time-frequency distribution matrix to improve separation performance[10,21].

As mentioned above, the basic model of the separation problem is

$$x(t) = As(t) + n(t) \tag{8.6.1}$$

where A is a full-rank $m \times n$ mixing matrix and $n(t)$ is a non-specific distributed stationary white noise.

The time-frequency distribution of the discrete space of the output signals is defined as

$$D_{xx}(t,f) = \sum_{l=-\infty}^{\infty} \sum_{m=-\infty}^{\infty} \phi(m,l) x(t+m+l) x^{\mathrm{H}}(t+m-l) \mathrm{e}^{-\mathrm{j}4\pi fl} \tag{8.6.2}$$

where $\phi(m, l)$ is the distribution kernel function.

For a noise-free model, the time-frequency distribution matrix of the source is

$$D_{xx}(t,f) = AD_{ss}(t,f)A^{\mathrm{H}} \tag{8.6.3}$$

to distinguish them, define the estimated auto-source space time-frequency distribution with the auto-term time-frequency point as $D^{\mathrm{a}}_{xx}(t,f)$, and define the cross-source space time-frequency distribution estimated with the cross-term time-frequency point as $D^{\mathrm{c}}_{xx}(t,f)$.

Denote the $n \times m$ whitening matrix as W, $(WA)(WA)^{\mathrm{H}}=UU^{\mathrm{H}}=I$. Premultiply and postmultiply the time-frequency distribution matrix to obtain the whitened time-frequency distribution matrix:

$$\breve{D}_{xx}(t,f) = WD_{ss}(t,f)W^{\mathrm{H}} = UD_{ss}(t,f)U^{\mathrm{H}} \tag{8.6.4}$$

When the time-frequency points of the auto-term are selected, the structure of the whitened time-frequency distribution is

$$\breve{D}_{xx}(t,f) = UD^{\mathrm{a}}_{ss}(t,f)U^{\mathrm{H}} \tag{8.6.5}$$

where $D^{\mathrm{a}}_{ss}(t,f)$ is diagonal. The unitary matrix U can be obtained through the joint diagonalization for the set of the p auto-term time-frequency distribution matrices $\{\breve{D}_{xx}(t_i,f_i) \mid i=1,\cdots,p\}$. Multiple auto-term time-frequency points can decrease the probability of eigenvalue degeneration, consequently enhancing the noise immune capacity.

The joint diagonalization of the p $n \times n$ matrixes set $\{M_k \mid k=1,\cdots,p\}$ is defined as maximizing the unitary matrices set $\{V \mid v_1,\cdots,v_n\}$ according to the following criteria:

$$C_{\mathrm{JD}}(\boldsymbol{V}) = \sum_{k=1}^{p}\sum_{i=1}^{n} |\boldsymbol{v}_i^{\mathrm{H}}\boldsymbol{M}_k\boldsymbol{v}_i|^2 \tag{8.6.6}$$

the details of the effective algorithms for achieving the above maximization can be found in Ref.[10]

When selecting the cross-term time-frequency point, the structure of the whitened time-frequency distribution is

$$\breve{\boldsymbol{D}}_{xx}(t,f) = \boldsymbol{U}\boldsymbol{D}_{ss}^{\mathrm{c}}(t,f)\boldsymbol{U}^{\mathrm{H}} \tag{8.6.7}$$

where $\boldsymbol{D}_{ss}^{\mathrm{c}}(t,f)$ is non-diagonal. We can obtain the unitary matrix $\boldsymbol{U}$ through the joint diagonalization for the q cross-term time-frequency distribution matrices set $\{\breve{\boldsymbol{D}}_{xx}(t_i,f_i)\,|\,i=1,\cdots,q\}$.

The non-diagonalization of an $n\times n$ matrix $\boldsymbol{N}$ is equivalent to the maximization of the following criteria on the whole unitary matrices set $\boldsymbol{V}=[\boldsymbol{v}_1,\cdots,\boldsymbol{v}_n]$:

$$C(\boldsymbol{N},\boldsymbol{V}) = -\sum_{i=1}^{n} |\boldsymbol{v}_i^{\mathrm{H}}\boldsymbol{N}\boldsymbol{v}_i|^2 \tag{8.6.8}$$

in the unitary transform, the Frobenius norm $\|\boldsymbol{N}\|_{\mathrm{F}} = \|\boldsymbol{V}^{\mathrm{H}}\boldsymbol{N}\boldsymbol{V}\|_{\mathrm{F}}$ is constant. Therefore, the joint anti-diagonalization of the q $n\times n$ matrices set $\{\boldsymbol{N}_k|k=1,\cdots,q\}$ can be defined as maximizing the joint anti-diagonalization criteria under the same unitary constraint.

$$C_{\mathrm{JOD}}(\boldsymbol{V}) = \sum_{k=1}^{q} C(\boldsymbol{N}_k,\boldsymbol{V}) = -\sum_{k=1}^{q}\sum_{i=1}^{n} |\boldsymbol{v}_i^{\mathrm{H}}\boldsymbol{N}_k\boldsymbol{v}_i|^2 \tag{8.6.9}$$

then further define that the simultaneous joint diagonalization and non-diagonalization of the two sets $\{\boldsymbol{M}_k|k=1,\cdots,p\}$ and $\{\boldsymbol{N}_k|k=1,\cdots,q\}$ of the $n\times n$ matrices is the joint criteria maximization on the unitary matrix set $\boldsymbol{V}=[\boldsymbol{v}_1,\cdots,\boldsymbol{v}_n]$.

$$C_{\mathrm{JD-JOD}}(\boldsymbol{V}) = \sum_{i=1}^{n}\left(\sum_{k=1}^{p} |\boldsymbol{v}_i^{\mathrm{H}}\boldsymbol{M}_k\boldsymbol{v}_i|^2 - \sum_{k=1}^{q} |\boldsymbol{v}_i^{\mathrm{H}}\boldsymbol{N}_k\boldsymbol{v}_i|^2\right) \tag{8.6.10}$$

the Givens rotation method can be adopted to maximize this equation (see the appendix).

Actually, since the auto-term main-lobe and side-lobe leakage caused by windowing may enter into the cross-term area, the time-frequency distribution matrix of the cross-source is not a complete non-diagonal, but the approximate Property of the simultaneous joint diagonalization decreases this interference.

The steps of the simultaneous joint diagonalization algorithm are described as follows:

Step 1. Determine the estimation of the whitening matrix $\hat{\boldsymbol{W}}$ from the eigenvalues decomposition of the covariance matrix of data;

Step 2. Calculate the time-frequency distribution of the array outputs by Eq. (8.6.2);

Step 3. Select a group of auto-term time-frequency points and cross-term time-frequency points;

Step 4. Maximize the whitened time-frequency distribution matrix $\breve{D}_{xx}(t, f)$ by using joint diagonalization and joint non-diagonalization criteria on the selected time-frequency points (t_i, f_i), $i=1, \cdots, l$, $l=p+q$, and estimate unitary matrix $\hat{U}$;

Step 5. Estimate $\hat{A}$ by $\hat{A}=\hat{W}^{\#}\hat{U}$, where $\hat{W}^{\#}$ is pseudo inverse of $\hat{W}$.

Step 6. Estimate source signals $\hat{s}(t)=\hat{U}^{\#}\hat{W}x(t)$.

8.7 Nonparametric Density Estimating Separating Method Based on Time-Frequency Analysis

The basic ICA algorithm in blind separation assumes that the source are stationary independent stationary random signals, but it is not sufficient for nonstationary signals, such as a voice signal, biomedical signals, sonar signals, etc. The time-frequency algorithm doesn't have any limitation on Gaussianity of the source, which is specially applicable for nonstationary signals. Therefore, more attention was paid to research based on time-frequency analysis, and three representative algorithms are presented.

The first type of algorithm has a closer relationship to the basic blind separation algorithm. Belouchrani[3] carried out joint diagonalization on a series of space time-frequency distributed matrices by utilizing the nonstationary Property. Then Holobar[22], represented the TFBSS algorithm to find the single non-zero diagonal elements matrix of the space time-frequency distributed matrix after joint diagonalization, which corresponds to the space time-frequency distributed matrix of a single source auto-term. Selecting the right time-frequency point in this algorithm is a key point for the recovery of source signals, which is achieved by a threshold defined by a signal and has some requirement on the time-frequency distribution. The second type of algorithm is based on the time-frequency transform ratio of observed signals[23,24],which requires no overlap among the time-frequency distribution of sources; this is a strict requirement. The third type of algorithm[25] has no constraint on the localization Property of the source, and it doesn't matter whether the mixing system is linear or not, as each independent signal has independent time-frequency transforms and time-frequency distributions. The methods presented in this section don't need to select the right time-frequency points, so they have the potential of separating real mixed signals.

The real signal usually contains a noise component. When the singal-to-noise ratio (SNR) is higher, a blind separation algorithm can be used directly; but when the SNR is lower, noise must first be removed or reduced. For noisy nonstationary signals, time-frequency analysis extends the noise energy to the whole time-frequency domain, which corresponds to increasing the SNR of each time-frequency point. Therefore, confirming if the signal feature can be reflected in the time-frequency plane is the main starting point when considering the time-frequency plane filter. In fact, the time-frequency plane Wiener filtering has been used to reduce the noise from an electrocardiogram containing period deterministic time-frequency structure signal[26~28].

Traditional nonparametric density estimation blind separation[3] focuses on the independent identical distribution of signals in time or frequency domain and needs more sample points to get an accurate estimation. Thus, utilizing limited signal data appropriately and effectively becomes key. The time-frequency transformation of a nonstationary signal can reflect the changes of spectrum values along with time; therefore, the nonparametric density estimation of it can adequately use the limited sample points to get more information in the time and frequency domain.

1. Mixing model

Consider N source signals observed by M sensors, where the observed signal $\boldsymbol{x}(t)$ is a linear instantaneous mixing model of sources:

$$\boldsymbol{x}(t) = \boldsymbol{A}\boldsymbol{s}(t) + \boldsymbol{n}(t) \tag{8.7.1}$$

where $\boldsymbol{A}=(a_{mn}) \in R^{M\times N}$ is a full rank instantaneous mixing matrix to guarantee the difference in signal space. Taking K periodic samples from the mth sensor, the ensemble average signal is $\overline{x_m}(t) = \frac{1}{K}\sum_{i=1}^{K} x_m^{(i)}(t)$ (i is the serial number of sample). Perform the short-time Fourier transform on Eq. (8.7.1), and get

$$\overline{\mathbf{X}_m}(t,f) = \sum_{n=1}^{N} a_{mn}\boldsymbol{S}_n(t,f) + \overline{\mathbf{N}_m}(t,f) \qquad 1 \leqslant m \leqslant M \tag{8.7.2}$$

where $\overline{\mathbf{X}_m}(t,f)$, $\boldsymbol{S}_n(t,f)$ are the time-frequency transform matrix of $\overline{\boldsymbol{x}_m}(t)$ and $\boldsymbol{s}_n(t)$ respectively, and $\overline{\mathbf{N}_m}(t,f)$ is the time-frequency transform matrix of the ensemble's average noise $\overline{\boldsymbol{n}}_m(t)$.

For the condition of a lower SNR ratio, blind separation can be carried out only after the noise energy is reduced and the SNR is increased. The basic idea of this section is to show how to use time-frequency wiener filtering to reduce noise and then to run a TFNPICA algorithm.

2. Noise reduction in the time-frequency domain

The time-frequency Wiener filtering is based on the posterior Wiener filter theory, which improves its performance through an ensemble average. The ensemble average of posterior statistics makes the filter match the time-frequency structure of the signals to be estimated. Here the application of the time-frequency plane makes the algorithm more robust for nonstationary signals. Compared with the normal ensemble average, the credibility of the signals to be estimated and the SNR can be increased significantly, and it is applicable to the noise reduction of signals with a definite time-frequency structure and ensemble additive noise[28~30]. Since the optimizing criteria of Wiener filtering is the minimization of the mean-square error, one can obtain its optimum filter parameters by solving the Wiener-Hopf equation. We design the time-frequency Wiener filtering in the time-frequency plane, filter the transformed time-frequency signal, and reconstruct the denoising signals in the time-frequency domain. In order to obtain the time-frequency energy distribution form of signals, square Eq. (8.7.2):

$$\overline{\boldsymbol{X}_m}^2(t,f) = \| \mathrm{STFT}(\overline{x_m}(t)) \|^2 \tag{8.7.3}$$

Similarly,the ensemble average of the square of the time-frequency transform is

$$\overline{\boldsymbol{X}_m(t,f)^2} = \frac{\sum_{i=1}^{K} \| \mathrm{STFT}(\overline{x_m}(t)) \|^2}{K} \tag{8.7.4}$$

where i is the sample serial number of the mth sensor, the expansion of Eqs. (8.7.3) and (8.7.4) are

$$\begin{aligned} \overline{\boldsymbol{X}_m}^2(t,f) = {} & \sum_{i=1}^{N} a_{nm}^2 \boldsymbol{S}_n^2(t,f) + \overline{\boldsymbol{N}_m}^2(t,f) + \\ & 2\sum_{i=1}^{N} a_{nm} \mathrm{COV}(\boldsymbol{S}_n(t,f), \overline{\boldsymbol{N}_m}(t,f)) + \\ & \mathrm{IF}(\overline{\boldsymbol{X}_m}^2(t,f)) \end{aligned} \tag{8.7.5}$$

$$\begin{aligned} \overline{\boldsymbol{X}_m(t,f)^2} = {} & \sum_{n=1}^{N} a_{nm}^2 \boldsymbol{S}_n^2(t,f) + \overline{\boldsymbol{N}_m(t,f)^2} + \\ & 2\sum_{i=1}^{K}\sum_{n=1}^{N} a_{nm} \mathrm{COV}(\boldsymbol{S}_n^2(t,f), \boldsymbol{N}_m^{(i)}(t,f))/K + \end{aligned}$$

$$\sum_{i=1}^{K} \mathrm{IF}(\boldsymbol{X}_m^{2,(i)}(t,f))/K \tag{8.7.6}$$

where $\overline{\boldsymbol{X}_m}^2(t,f)$ and $\overline{\boldsymbol{X}_m(t,f)^2}$ denote the square spectrum of the ensemble average of the time-frequency transform and the ensemble average of the square spectrum of the time-frequency transform of the noise signal for the mth sensor respectively, a_{mn}^2 denotes the square of mixing matrix element, $\boldsymbol{N}_m^{(i)}(t,f)$ denotes the time-frequency transform of the ith sample sequence, $\boldsymbol{X}_m^{2,(i)}(t,f)$ denotes the square spectrum of time-frequency transform of ith sample for the mth sensor, $\boldsymbol{S}_n^2(t,f)$ denotes the square spectrum of time-frequency transform of source n, COV, and IF denotes the correlated term and interference term. Since the source signals are statistically independent of each other and of noise, the ideal expression is

$$\overline{\boldsymbol{X}_m}^2(t,f) = \sum_{n=1}^{N} a_{mn}^2 \boldsymbol{S}_n^2(t,f) + \overline{\boldsymbol{N}_m(t,f)}^2/K \tag{8.7.7}$$

$$\overline{\boldsymbol{X}_m(t,f)^2} = \sum_{n=1}^{N} a_{mn}^2 \boldsymbol{S}_n^2(t,f) + \overline{\boldsymbol{N}_m(t,f)^2}/K \tag{8.7.8}$$

Actually, the cross-term and correlated term on the time-frequency plane will not be zero, so it is necessary to design a smooth function or kernel function in the time-frequency plane to reduce the existing noise variance, interference term and correlated terms; then we can design a time-frequency Wiener filter.

First, it can be considered that the time-frequency transform spectrum is the filtering form of Wigner-Ville distribution[29], namely

$$\overline{\boldsymbol{X}_m}^2(t,f) = \| \mathrm{STFT}(\overline{x_m}(t)) \|^2 = \mathrm{WD}[\overline{x_m}(t)] * \mathrm{WD}[w(-t)] \tag{8.7.9}$$

where w is the window function of the short-time Fourier transform. This time-frequency spectrum can be further smoothed by a two-dimension lowpass filter to increase the centrality of the signal in the time-frequency plane and attenuate the cross-term, correlated term and noise variance. Here, select a Gaussian window as the smoothing function in the time-frequency plane:

$$\gamma(t) = \mathrm{e}^{-\alpha t} \tag{8.7.10}$$

where α is determined by signals. The smoothed time-frequency distribution can be represented as

$$\begin{aligned}\overline{\boldsymbol{X}_m}^2(t,f) &= \| \mathrm{STFT}(\overline{x_m}(t)) \|^2 \\ &= \mathrm{WD}[\overline{x_m}(t)] * \mathrm{WD}[w(-t)] * \mathrm{WD}[\gamma(t)]\end{aligned} \tag{8.7.11}$$

Similarly, perform the above processing on Eq. (8.7.4):

$$\overline{\boldsymbol{X}_m(t,f)^2} = \sum_{i=1}^{K} \{\mathrm{WD}[x_m^{(i)}(t)] * \mathrm{WD}[w(-t)] * \mathrm{WD}[\gamma(t)]\}/K \tag{8.7.12}$$

According to the expression of the posterior Wiener filtering in time-frequency plane[28] and considering Eqs. (8.7.5) and (8.7.6), the weight function of the time-frequency Wiener filtering can be obtained:

$$\boldsymbol{h}_m(t,f) = \frac{\sum_{n=1}^{N} a_{mn}^2 \boldsymbol{S}_n^2(t,f)}{\sum_{n-1}^{N} a_{mn}^2 \boldsymbol{S}_n^2(t,f) + \overline{\boldsymbol{N}_m(t,f)}^2 / K} \tag{8.7.13}$$

By multiplying weight parameters of the time-frequency Wiener filtering with the short-time Fourier transform of the ensemble average of the signal, the estimated denoising receiving signal can be obtained as follows:

$$\hat{\boldsymbol{X}}_m(t,f) = \boldsymbol{h}_m(t,f) \cdot \mathrm{STFT}\left(\frac{1}{K}\left(\sum_{i=1}^{K} x_m^{(i)}(t)\right)\right) \tag{8.7.14}$$

3. Time-frequency nonparametric density estimation ICA algorithm (TFNPICA)

Since the time-frequency transfor of the signal keeps the linear relationship among signals, the time-frequency transfor of the reconstructed signal can be written as $\boldsymbol{Y}(t,f) = \boldsymbol{W}\hat{\boldsymbol{X}}(t,f)$ $(\boldsymbol{W} \in R^{N \times M})$. Here, we assume $\boldsymbol{W}$ is a square matrix, i.e. positive definite. The algorithm of the signal reconstruction based on the minimum mutual information is

$$\boldsymbol{W}_{\mathrm{opt}} = \arg\min_{W} I(\boldsymbol{Y}_1(t,f), \cdots, \boldsymbol{Y}_N(t,f)) \tag{8.7.15}$$

where $I(\boldsymbol{Y}_1(t,f), \cdots, \boldsymbol{Y}_N(t,f))$ is the mutual information of the time-frequency transformation of the reconstructed signal, the evaluation function is finally obtained:

$$C(\boldsymbol{W}) = -H(\hat{\boldsymbol{X}}(t,f)) - \lg|\det(\boldsymbol{W})| - \sum_{i=1}^{N} E\{\lg(p_i(\boldsymbol{Y}_i(t,f)))\} \tag{8.7.16}$$

where $p_i(\boldsymbol{Y}_i(t,f))$ denotes ith marginal density function in time-frequency domain, $H(\hat{X}(t,f))$ denotes entropy; it can be omitted since it is irrelevant to the separating matrix $\boldsymbol{W}$. Differentiating on $\boldsymbol{W}$, we can obtain:

$$\nabla C(\boldsymbol{W}) = \boldsymbol{W}^{-\mathrm{T}} - f(\boldsymbol{Y}(t,f))\hat{\boldsymbol{X}}(t,f)^{\mathrm{T}} \tag{8.7.17}$$

where $f(\boldsymbol{Y}(t,f)) = [f_1(\boldsymbol{Y}_1(t,f)), \cdots, f_L(\boldsymbol{Y}_L(t,f))]^{\mathrm{T}}$ is a nonlinear function, its ith component is

$$f_i(\boldsymbol{Y}_i(t,f)) = -\frac{p'_i(\boldsymbol{Y}_i(t,f))}{p_i(\boldsymbol{Y}_i(t,f))} \tag{8.7.18}$$

(1) Time-frequency domain nonparametric density estimation

In actual problems, the marginal density function and nonlinear function can be estimated directly by using nonparametric density estimation from data. The discrete expression of the nonparametric density estimation function is as follows:

$$\hat{p}(y_i(t_k, f_g)) = \sum_{j=1,l=1}^{J,L} \phi\left(\frac{(y_i(t_j, f_l) - y_i(t_k, f_g))}{h}\right) \Big/ (JLh) \quad (8.7.19)$$

where j, l is the serial umber for the time-frequency point, $\hat{p}(y_i(t_k, f_g))$ denotes the spectrum density estimation for each time-frequency point, ϕ is the Gaussian kernel function, the experience formula of the kernel function bandwidth[25] is $h_{\text{opt}} = 1.06\,\hat{\sigma} N^{-1/5}$, and $\hat{\sigma}$ is the variance of the estimation samples. This estimation is unbiased and effective, and it can converge to the real probability density function in a definite range.

Because of the Gaussian kernel function Property, the density of each time-frequency point is defined by the spectrum value of its surrounding time-frequency point, so more time-varying information can be obtained without selecting a special time-frequency point. In the strict sense, the density estimation should be a two-dimension density estimation, but it can be considered as one-dimensional data in calculations because of the characteristic of the Gaussian function. The spectrum in the time-frequency domain is time-varying, but the linear relationship among independent components remains invariant in this linear time-frequency transform, and the time-frequency transform of each independent component is independent in the time-frequency domain.

Its derivative formula is

$$\nabla p_i(y_i(t_k, f_g)) = \sum_{j=1,l=1}^{J,L} (y_i(t_j, f_l) - y_i(t_k, f_g)) \cdot \phi((y_i(t_k, f_l) - y_i(t_k, f_g))/h)/(Lh^2)$$
$$i = 1, \cdots, N \quad (8.7.20)$$

The Quasi-Newton method can be used for optimization of the evaluation function. By using the information of the evaluation function and first-order derivative to establish the curvature approximation of the object function, it does not need to form an obvious Hessian matrix and has the merit of fast convergence. The BFGS (Broyden-Fletcher-Goldfarb-Shanno) correction formula is used in this algorithm to obtain the Hessian matrix.

(2) Signal reconstruction

Since $\boldsymbol{Y}(t,f) = \boldsymbol{W}\hat{\boldsymbol{X}}(t,f)$, the time domain expression of the separated signal can be written as follows:

$$\boldsymbol{y}_m(t) = \text{STFT}^{-1}(\boldsymbol{Y}_m(t,f)) \quad (8.7.21)$$

the analysis window used in the short-time Fourier transform is a Hamming window with a 50% overlap ratio, and adds an adequate time shift on the inverse

Fourier transform for signal reconstruction. Because of the application of the filter smoothing function in the time-frequency plane, there will be some small signals peak at the edge of the analyzing window, but we can use medium filter to smooth it.

4. Simulation experiment

Record three kinds of signals, which are speech, music and noise signals generated by a noise generator in a semianechoic room. The sampling frequency is 11,025 Hz. A randomly generated matrix $\boldsymbol{A} \in \boldsymbol{R}^{5\times 2}$ is used to mix the speech and music signals. There are five observed signals, and each signal has added Gaussian white noise with SNR of 25, 35 and 45 dB respectively. After whitening, perform the short-time Fourier transform on the ensemble average of the whitened signals. The length of the Hamming window is 45 ms, and the overlap ratio between windows is 50%. The time interval includes 512 points and the frequency section includes 128 points in the time-frequency plane; the two source signals are shown in Fig. 8.5. Time-frequency Wiener filtering is done as preprocessing to reduce the noise of signals in each time-frequency section. When SNR=25 dB, the SNR of the spectrum of the signal after time-frequency Wiener filtering (Fig. 8.8) is improved about 7 dB by comparison to the original spectrum (Fig. 8.7). After preprocessing, we separate the mixed signals by using the time-frequency nonparametric density estimation, and the separation results under SNR=45 dB are shown in Fig. 8.6.

The signal interference ratio (SIR) used to evaluate the separation effect is defined as

$$10\,\lg\left(\frac{\sum\limits_{m=1}^{M} s_m^2}{\sum\limits_{m=1}^{M} (\hat{s}_m - s_m)^2}\right)$$

where s is the source signal, and $\hat{s}$ is the reconstructed source signal; the interference signal of the reconstructed source signal means the other reconstructed signals except for this recovered source signal. figures 8.9 and figures 8.10 respectively represent the comparison results of this algorithm with the time-frequency domain blind separation algorithm (TFBSS[22]) under 45 dB SNR for the samples from 5,000 to 10,000. It can be seen that the SIR of the time-frequency domain nonparametric density estimation is 2 dB higher than the TFBSS. Along with increasing the number of samples, the separation results of the time-frequency nonparametric density estimation ICA (TFNPICA) is better than the TFBSS algorithm.

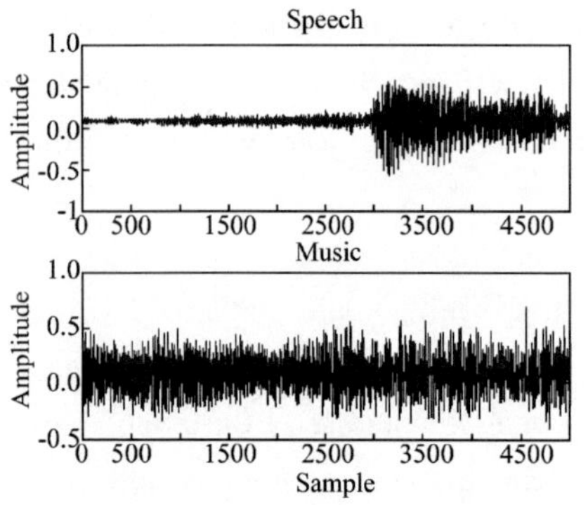

Fig. 8.5 Source signals

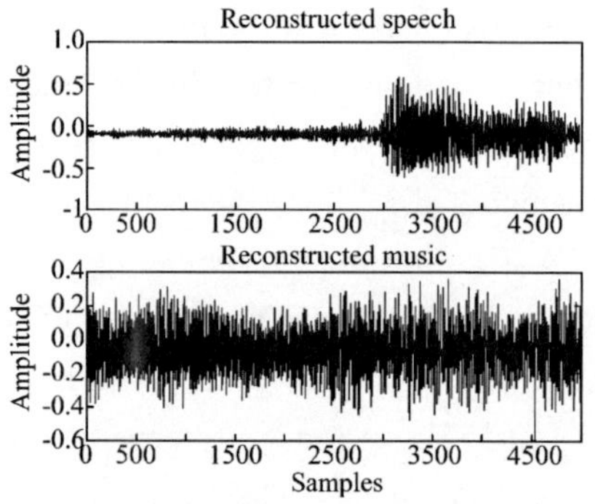

Fig. 8.6 Recovered signal SNR=45 dB

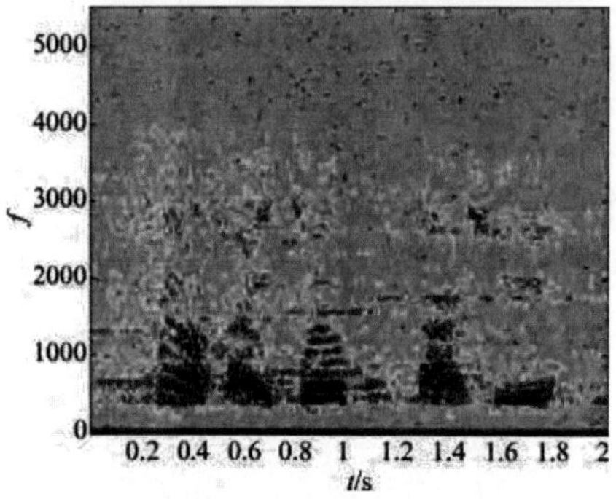

Fig. 8.7 SNR=25 dB signal spectrum (signal length is 2.04 s)

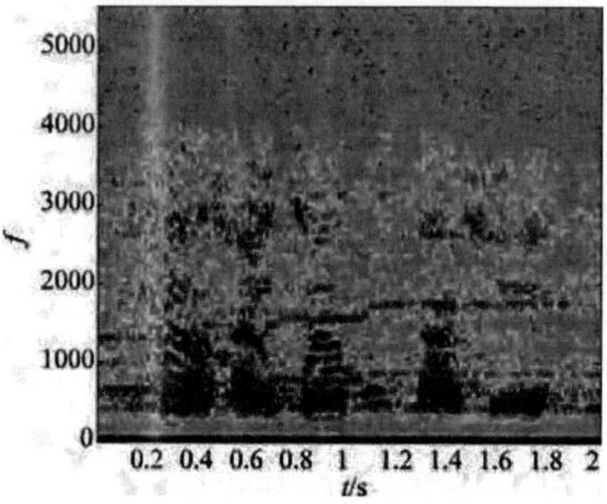

Fig. 8.8 Spectrum after time-frequency filtering SNR ≈ 32.52 dB (signal length is 2.04 s)

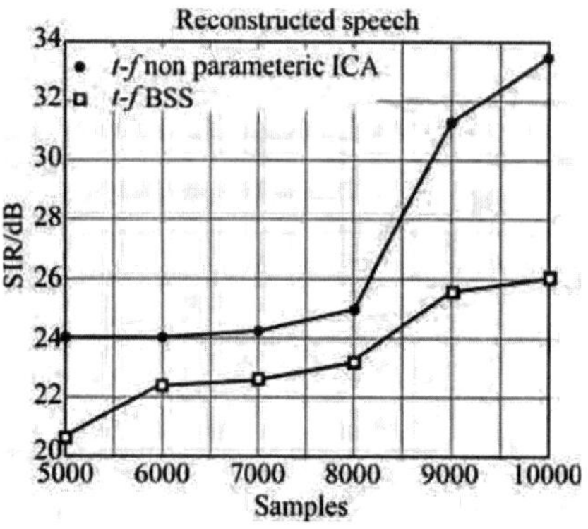

Fig. 8.9 SIR of the recovered speech source signal

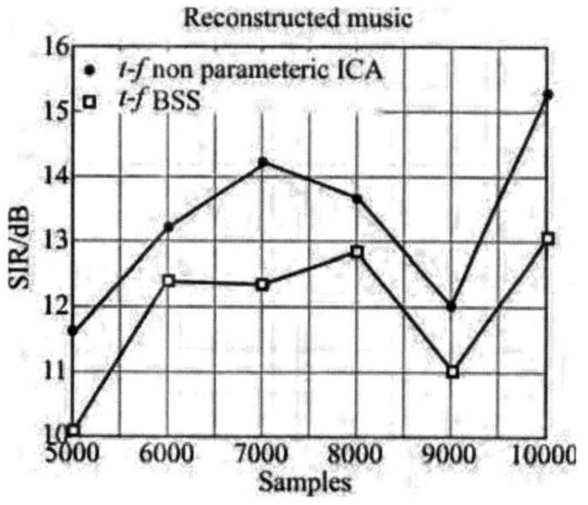

Fig. 8.10 SIR of the recovered music source signal

Appendix:

Givens rotation is a kind of method that selectively eliminates some elements (set them to zero) of a matrix to transform it into a diagonal type; it is also called a Jacobi rotation. These matrices are the rank-2 correction forms for a unit matrix.

$$J(i,k,\theta) = \begin{bmatrix} 1 & \vdots & & \vdots & \\ \cdots & c & \cdots & s & \cdots \\ & \vdots & & \vdots & \\ \cdots & -s & \cdots & c & \cdots \\ & \vdots & & \vdots & \end{bmatrix}$$

where $c=\cos\theta$, $s=\sin\theta$ so Givens rotation is an orthogonal transformation. Left-multiplying vector $\boldsymbol{x}$ with $J(i, k, \theta)$ is equivalent to rotating for θ degree in (i, k) coordinate plane.

Assume $\boldsymbol{v} \in \boldsymbol{R}^n$ is nonzero. An $n \times n$ matrix $\boldsymbol{P}$, if its form is

$$\boldsymbol{P} = \boldsymbol{I} - 2\boldsymbol{v}\boldsymbol{v}^{\mathrm{T}}/(\boldsymbol{v}^{\mathrm{T}}\boldsymbol{v})$$

it is an Householder matrix or transformation and is a symmetric orthogonal matrix.

For $n=2$, $\boldsymbol{Q}=\boldsymbol{I}-2\boldsymbol{v}\boldsymbol{v}^{\mathrm{T}}/\boldsymbol{v}^{\mathrm{T}}\boldsymbol{v}$ is a 2×2 Householder matrix. Assume $v_1=\cos\theta$ and $v_2=\sin\theta$, we can obtain

$$\boldsymbol{Q} = \begin{bmatrix} 1-2v_1^2 & -2v_1v_2 \\ -2v_1v_2 & 1-2v_2^2 \end{bmatrix} = \begin{bmatrix} -\cos2\theta & -\sin2\theta \\ -\sin2\theta & \cos2\theta \end{bmatrix}$$

if $\boldsymbol{Q} \in \boldsymbol{R}^{2\times 2}$ is an orthogonal matrix, then $\boldsymbol{Q} = \begin{bmatrix} c & s \\ s & -c \end{bmatrix}$ is a Householder matrix, or $\boldsymbol{Q}=\begin{bmatrix} c & s \\ -s & c \end{bmatrix}$ is a Givens matrix, the relationship between the two matrices can be obtained, and the numerical properties of the two matrices are both very useful. For a more common situation, if $\boldsymbol{Q} \in \boldsymbol{R}^{n\times n}$ is an orthogonal matrix and this matrix is the same as a unit matrix except for a 2×2 sub-matrix, then $\boldsymbol{Q}$ is a Givens rotation.

$$\mathrm{off}(\boldsymbol{A}) = \sum_{i=1}^{n} \sum_{\substack{j=1 \\ j\neq i}}^{n} a_{ij}^2$$

in this given process, the object is to determine the Givens rotation $\boldsymbol{J}(i, k, \theta)$ to minimize off $(\boldsymbol{B})$, here

$$\boldsymbol{B} = \boldsymbol{J}^{\mathrm{T}}\boldsymbol{A}\boldsymbol{J}$$

then correct $\boldsymbol{A}$ into $\boldsymbol{A}:=\boldsymbol{J}^{\mathrm{T}}\boldsymbol{A}\boldsymbol{J}$, and repeat this process until the requirement is met.

References

[1] Cardoso J F (1989) Source separation using higher order moments. In: Proceedings of ICASSP, Glasgow, 1989, pp 2109–2112

[2] Moreau E (2001) A generalization of joint-diagonalization criteria for source separation. IEEE Transactions on Signal Processing 49(3): 530–541

[3] Belouchrani A, Amin M G (1998). Blind source separation based on time-frequency signal representation. IEEE Transactions on Signal Processing 46(12): 2888–2898

[4] Belouchrani A, Abed M K, Cardoso J F et al (1997) A blind source separation technique using second-order statistics. IEEE Transactions on Signal Processing 45(2): 434–444

[5] Yeredor A (2000) Blind source separation via the second characteristic function. Signal Processing 80(5): 897–902

[6] Cardoso J F, Souloumiac A (1996) Jacobi angles for simultaneous diagonalization. SIAM Journal of Matrix Analysis and Application 17(1): 161–164

[7] Yeredor A (2002) Non-orthogonal joint diagnalization in the least-squares sense with application in blind source separation. IEEE Transactions on Signal Processing 50(7): 1545–1553

[8] Miyoshi M, Kaneda Y (1988) Inverse filtering of room acoustics. IEEE Transactions on Acoustics, Speech and Signal Processing 36(1): 145–152

[9] Klajman M, Chambers J A (2002) A novel approximate joint diagonalization algorithm. In: Mathematics in signal processing V, Clarendon Press, Oxford

[10] Belouchrani A, Aded-Meraim K, Amin M G (2004) Blind separation of nonstationary sources. IEEE Signal Processing Letter 11(7): 605–608

[11] Tong L, Xu G H, Hassibi B et al (1995) Blind Channel identification based on second order statistics: A frequency domain approach. IEEE Transactions on Information Theory 41(1): 329–334

[12] Schobben D W, Sommen P (2002). A frequency domain blind signal separation method based on decorrelation. IEEE Transactions on Signal Processing 149(5): 253–262

[13] Bousbia H, Belouchrani A, Abed K (2001) Jacobi-like algorithm for blind signal separation of convolutive mixturess. Electronics Letters 37(16): 1049–1050

[14] Hild K E, Erdogmus D, Principe J C (2002) Blind source separation

of time-varying, instantaneous mixturess using an on-line algorithm. In: IEEE International Conference on Acoustics, Speech and Signal Processing-Proceedings, Orlando, 2002, 1: 993–996

[15] Wu J B (2003) Study on blind separation of noise signals and method of acoustical feature extraction. Dissertation, Shanghai Jiao Tong University (Chinese)

[16] Jia P (2003) Study of blind separation of acoustic signal. Dissertertion, Shanghai Jiao Tong University (Chinese)

[17] Wu J B, Chen J, Zhong P (2003) Time frequency-based blind source separation for elimination of cross-terms in Wigner distribution. Electronics Letters 39(5): 475–477

[18] Nuzillard D, Nuzillard J M (2003) Second-order blind source separation in the Fourier space of data. Signal Processing 83(3): 627–631

[19] Shi X Z (2003) Signal processing and soft computing. Higher Education Press, Beijing (Chinese)

[20] Golub G H, Van Loan C F (1998) Matrix computation. Translated by Lian Q R, et al. Dalian University of Technology Press, Dalian

[21] Xu H X,Chen C C, Shi X Z (2005) Independent component analysis based on nonparametric density estimation on time-frequency domain. In: IEEE International Workshop on Machine Learning for Signal Processing, Mystic, 2005, pp 171–176

[22] Holobar A, F'evotte C, Doncarli C et al (2002) Single autoterms selection for blind source separation in time frequency plane. In: Proceedings of EUSIPCO'2002 , Toulouse, 2002

[23] Balan R, Rosca J (2000) Statistical properties of STFT ratios for two channel systems and applications to blind source separation. In: Proceedings. ICA'2000, Helsinki, 2000

[24] Jourjine, Rickard S, Yilmaz O (2000). Blind separation of disjoint orthogonal signals: Demixing *N* sources from 2 mixtures. In: Proceedings of ICASSP'2000, Istanbul, 2000, 5: 2985–2988

[25] Fahmy M F,Osama M H, Amin (2004) Blind source separation using time-frequency distribution. In: Proceedings of 21st National Radio Science Conference (NRSC2004), Cairo, 2004

[26] Doyle D J (1977) A proposed methodology for evaluation of the Wiener filtering method of evoked potential estimation. Electroencephalogr. Cli. Neurophysiol., 43 : 749–751

[27] de Weerd J P C, Kap J I (1981) A posteriori time-varying filtering of averaged evoked potentials-I: Introduction and conceptual basis. Biol. Cybern., 41: 223–234

[28] Lander P (1997) Time-frequency plane Wiener filtering of the high-resolution ECG: Development and application. IEEE Transactions on Biomedical Engineering 44(4): 247–255

[29] Boashash B (2004) Signal enhancement by time-frequency peak filtering. IEEE Transactions on Signal Processing 52(4): 929–937

[30] Kirchauer H (1995) Time-frequency formulation and design of nonstationary Wiener filters. In: IEEE Processing ICASSP-95, Detroit, 1995, pp 1549–1552

Chapter 9 Extension of Blind Signal Processing

9.1 Blind Signal Extraction

We discussed the blind source separation algorithm of linear instantaneous mixtures, which are based on the simultaneous separation of all sources, above. Actually, in many situations, we only want to separate several interesting sources from many sensor signals, such as the EEG signals in a head electrode array (64~128 electrodes) of brain research, a speech signal in a microphone array of speech recognition, and a sonar signal in sonar stations.The method based on sequential blind extraction, which extracts the required signal from the mixed signal successively, has been very important. Compared with the simultaneous separation of all sources, blind extraction has the following advantages. We can extract signals with random characteristics (kurtosis, sparsity, non-Gaussian and smooth Property). The extraction methods are very flexible. We can propose different criteria in each phase according to the characteristics of the signals to be extracted but can extract only a few signals, which can reduce the computing time, so the extraction algorithm is easier than the separation algorithm.

The classical extraction algorithm includes two steps: extraction and deflation in a cascade, as shown in Fig. 9. 1[1]. Extraction extracts one source with each step; however, deflation subtracts the extracted signal from the mixed signal. This process is similar to extracting the principal component in principal component analysis.

Assume the interesting signals are independent and non-Gaussian. We can now explain the whole process. First, mixed signals are pre-whitened, satisfying

$$E(\boldsymbol{x}_1 \boldsymbol{x}_1^{\mathrm{T}}) = \boldsymbol{I}_n \tag{9.1.1}$$

where $x_1 = \bar{x} = Qx$, which means to construct a new $n \times n$ orthogonal matrix $\boldsymbol{A}=\boldsymbol{QH}$. This unit is to extract a zero-mean source signal:

$$y_1(k) = \boldsymbol{w}_1^{\mathrm{T}} \boldsymbol{x}_1 = \sum_{i=1}^{m} w_{1i} x_{1i} \tag{9.1.2}$$

which satisfies $\boldsymbol{w}_1^{\mathrm{T}}\boldsymbol{A}=\boldsymbol{e}_j^{\mathrm{T}}$, and $\boldsymbol{e}_j$ is the jth column of a nonsingular diagonal matrix.

X. Shi, *Blind Signal Processing*

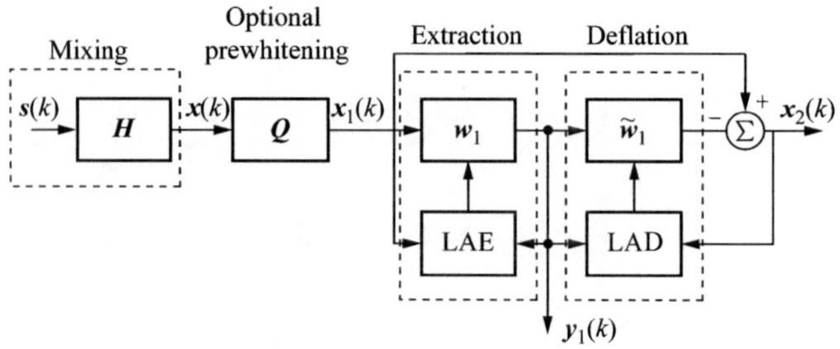

Fig. 9.1 Flowchart of blind extraction

As the minimum cost function, we use a normalized kurtosis to define[1]

$$J_1(w_1) = -\frac{1}{4}|k_4(y_1)| = -\frac{\beta}{4}k_4(y_1) \tag{9.1.3}$$

where β is used to determine the sign of kurtosis of the extracted signal, where for sub-Gaussian source signals, $\beta = -1$, and for super-Gaussian $\beta = +1$.

Using the gradient descent method to obtain the minimum cost function, we have

$$\frac{\mathrm{d}\boldsymbol{w}_1}{\mathrm{d}t} = -\mu_1 \frac{\partial J_1(\boldsymbol{w}_1)}{\partial \boldsymbol{w}_1} = \mu_1(t)\varphi(y_1(t))\boldsymbol{x}_1(t) \tag{9.1.4}$$

where the learning rate $\mu_1(t)>0$, non-linear activation function is

$$\varphi_1(y_1) = \beta\left[\frac{\hat{m}_2(y_1)}{\hat{m}_4(y_1)}y_1^3 - y_1\right] \tag{9.1.5}$$

or

$$\varphi_2(y_1) = \beta\left[\frac{1}{\hat{m}_4(y_1)}y_1^3 - \frac{1}{m_2(y_1)}y_1\right] \tag{9.1.6}$$

where moment $m_q(y_1)=E(y_1(t)^q)$, $q=2$ and 4; we can use the moving average model on-line. Generally, the non-linear activation function is not constant, but can be changed with the statistical characteristic of the extracted signals.

Besides the above algorithm, there are also the generalized kurtosis-based learning algorithm, constrainted generalized kurtosis-based learning algorithm, and the KuicNet and Hyvärinen-Oja fixed-point algorithm[2].

After the first source signal is extracted, then we compress,

$$\boldsymbol{x}_{j+1}(k) = \boldsymbol{x}_j(k) - \tilde{\boldsymbol{w}}_j y_j(k), \qquad j = 1,2,\cdots,n \tag{9.1.7}$$

where $\tilde{w}_j$ is estimated to minimize the below cost function:

$$J_j(\tilde{\boldsymbol{w}}_j) = \frac{1}{2}E\left(\sum_{p=1}^{m} x_{j+1,p}^2\right) \tag{9.1.8}$$

the learning rule is

$$\tilde{\boldsymbol{w}}_j(k+1) = \tilde{\boldsymbol{w}}_j(k) + \tilde{\eta}_j(k)y_j(k)\boldsymbol{x}_{j+1}(k), \qquad j = 1,2,\cdots,n \tag{9.1.9}$$

The cost function is based on energy, which means that as the signals are extracted, the energy of the mixed signal minimizes. This process can be called again and again to extract useful signals. The details of the above algorithm are available in Appendix A and B in section 5 of Ref.[1].

Excepting the assumption of independence and non-Gaussian, sequential blind extraction also can be used in sources of a different time structure and arbitrary distribution, including a time-correlated source and a convolutive mixed source, an especially independent identical distribution source, and a color source.

The above methods are based on the number of sources being equal to the number of sensors, $n=m$, and $\boldsymbol{A}$ is nonsingular or full column rank. However, in many situations, there are 4 kinds of ill-conditioned conditions. The mixing matrix is square but singular; the number of sensors is less than the number of sources; the number of sensors is more than the number of sources, but the mixing matrix is not full column rank; and the number of sources is unknown. The solvable condition of the sequential blind extraction is weaker than in blind separation, and the key to solving the ill-conditioned cases is to process the nonfull column rank of $\boldsymbol{A}$.

For blind extraction, the general model is

$$z = \boldsymbol{B}y = \boldsymbol{C}x \tag{9.1.10}$$

where z is an m-dimension output vector an $\boldsymbol{B}$ is an $m \times m$ blind extraction matrix, $\boldsymbol{C} \overset{\text{def}}{=} \boldsymbol{B}\boldsymbol{A}$. The objective of blind extraction is to obtain $\boldsymbol{B}$, making sure one component corresponds to one source. Although this model is similar to blind separation, only the extracted source is pairwise independent with the other sources in blind extraction, not all the output pairwise independent. The solvability analysis in the ill-conditioned situation was shown in Refs.[3, 4]. If the total number of the $m \times (n-1)$ sub-matrix $(\overline{\boldsymbol{A}_1}, \cdots, \overline{\boldsymbol{A}_L})$ is L in $\boldsymbol{A}$, which lets the rank of $\overline{\boldsymbol{A}_i}$ be smaller than that of $\boldsymbol{A}$ for $i=1, \cdots, L$, then we can extract L sources according to Eq. (9.1.10).

The fourth-order accumulative evaluation function as extraction criteria[3] follows:

$$J = \sum_{j=2}^{m} \mathrm{Cum}_{2,2}^{2}(z_1, z_j) \tag{9.1.11}$$

obviously, if z_1 and z_j are pairwise independent, $\mathrm{Cum}_{2,2}(z_1,z_j)=0$, $j=2, \cdots, m$, then $J=0$.

in Eq. (9.1.10) .

$$z_k = \sum_{j=1}^{m} b_{kj} y_j = \sum_{l=1}^{n} c_{kl} x_l, \qquad k = 1, \cdots, m \tag{9.1.12}$$

As $x_1, \cdots, x_n$ are pairwise independent, the following formula is tenable:

$$\mathrm{Cum}_{2,2}(z_1, z_j) = \sum_{l=1}^{n} c_{1l}^2 c_{jl}^2 \beta_l \tag{9.1.13}$$

where $\beta_l = \mathrm{Cum}_4(z_l)$.

From Eq. (9.1.11) to Eq. (9.1.13), for $k=1, \cdots, n$ and $j=2, \cdots, m$, there is

$$\begin{aligned}\frac{\partial J}{\partial c_{1k}} &= \sum_{j=2}^{m} \frac{\partial \mathrm{Cum}_{2,2}^2(z_1, z_j)}{\partial c_{1k}} \\ &= 4\sum_{j=2}^{m} \Big(\sum_{l=1}^{n} c_{1l}^2 c_{jl}^2 \beta_l\Big) c_{1k} c_{jk}^2 \beta_k \\ &= 4\Big[\sum_{j=2}^{m} \Big(\sum_{l=1}^{n} c_{1l}^2 c_{jl}^2 \beta_l\Big) c_{jk}^2\Big] c_{1k} \beta_k,\end{aligned}$$

$$\frac{\partial J}{\partial c_{jk}} = 4\Big(\sum_{l=1}^{n} c_{1l}^2 c_{jl}^2 \beta_l\Big) c_{1k}^2 c_{jk} \beta_k \tag{9.1.14}$$

let $\frac{\partial J}{\partial c_{1k}} = 0$ and $\frac{\partial J}{\partial c_{jk}} = 0$, then we can present the constrained minimization problem as

$$\min_{|\boldsymbol{B}| \neq 0} J \tag{9.1.15}$$

this optimization problem can be solved with a gradient algorithm:

$$\mathrm{vec}(\boldsymbol{B}(k+1)) = \mathrm{vec}(\boldsymbol{B}(k)) - \Big(\sum_{j=2}^{m} 2 \frac{\partial \mathrm{Cum}_{2,2}(z_1, z_j)}{\partial \mathrm{vec}(\boldsymbol{B})} \cdot \frac{\partial \mathrm{Cum}_{2,2}(z_1, z_j)^{\mathrm{T}}}{\partial \mathrm{vec}(\boldsymbol{B})} + \mu \beta^k \boldsymbol{I}\Big)^{-1} \frac{\partial J}{\partial \mathrm{vec}(\boldsymbol{B})} \tag{9.1.16}$$

where $\mathrm{vec}(\boldsymbol{B}) = [b_{11}, \cdots, b_{1m}, b_{21}, \cdots, b_{mm}]^{\mathrm{T}}$, μ and β are positive constants, $\beta \leqslant 1$, and $\boldsymbol{I}$ is a $m^2 \times m^2$ unit matrix. In order to make sure of the nonsingular constraint condition $|\boldsymbol{B}| \neq 0$, in the minimization process, we need to frequently check whether $|\boldsymbol{B}|=0$ or not. If $|\boldsymbol{B}|=0$, then we should provide some perturbation to avoid singular $\boldsymbol{B}$, which is also used in some blind separation algorithms, and iterate again.

9.2 From Projection Pursuit Technology to Nonparametric Density Estimation-Based ICA

9.2.1 Introduction

ICA is one method of multivariate data analysis from the view of mathematical statistics[4]. Its thought can be traced back to the exploratory projection pursuit[5] in the early 1970s. In the mathematical statistics field, there is much development in the theory and application of multivariate data analysis and projection pursuit technology. Quite a lot of good results rarely can be used in engineering application, however, because of the gulf between fields and professions. This section makes a schematic review of ICA, and based on this it proposes a nonparametric density estimation-based ICA algorithm.

Projection pursuit is a method for exploratory statistics analysis of higher-dimensional data. Its objective is based on a functional projection index to pursue "interesting" low-dimensional linear orthogonal projections[6~8].Compared with thoughts and functions of classical PCA, projection pursuit technology becomes one of effective multivariate analysis methods to avoid "dimension's malediction" in a general sense, but the most interesting fact is that it has the ability of neglecting a noise variable and non-information variable; the only cost is that it greatly increases computing time. The projection pursuit technology-based regression model, classification, and density estimation methods emerge, in which the principal components found with projection pursuit technology are taken as the basis function. It is worth pointing out that early projection pursuit technology and computer tomography technology were mutually benificial. The latter mainly solves the inverse of the former, which is to effectively reconstruct a higher-dimensional structure from some lower-dimensional projections. What is more, if the variable of the regression model is based on non-linear transformation, the projection pursuit technology-based regression model has a relation with the famous Hilbert 13th problem[6]. Although the proposed ICA method in the engineering field has realized that the blind separation problem has a relationship with some projection pursuit algorithms[9], it is usually used to verify the rationality of the contrast function selection; however, it neglects the further meaning. This can be found from two aspects below:

① Actually, all the existing ICA algorithms can be seen as special cases of projection pursuit technology, which is an order-one entropy index function-based projection pursuit algorithm. The process to find the demixing matrix is the same as the process to find the "interesting" projection direction.

② No matter what function (maximum likelihood method, stochastic gradient method etc.) is based on a demixing matrix evaluation algorithm, choosing a proper model of the density distribution of the source signal is the key to solving the blind separation problem. It does not matter which criteria (information maximization, redundancy reduction, etc.) is used on the objective function, these algorithms need an evaluation method for differential entropy of the signal after projection. Much research has been on several non-linear contrast functions for special source signal modeling, and most is based on the evaluation of fourth-order moment of the source signal (or other higher-order or parameter model), but such simplification can lead to problems. Algorithms cannot flexibly adapt to the probability density distribution of the source signal, and cannot do universal modeling of source signals with arbitrary distribution; an algorithm is more sensitive to outliers, and may lead to failure of the source separation, thus losing many advantages of the former projection pursuit technology.

Based on the relation between projection pursuit and the ICA algorithm and their theoretical background, as well as practical experience, we can see that universal modeling of source signals with arbitrary distribution is the key. However, the nonparametric method, such as the normal kernel function density evaluation method[7,10], directly models the probability density function of the signal source with arbitrary distribution and can provide the algorithm full adaptability in actual application.

This section first establishes the theoretical framework of projection pursuit for blind separation and then, under this framework, derives and designs a new ICA algorithm with the robust kernel function density evaluation method[11,12]. The simulation results show this algorithm can reliably and exactly separate the source signal of arbitrary marginal distribution; it does not need to predefine any contrast functionor choose any adjustment parameter.

9.2.2 Projection Pursuit and Blind Separation

As mentioned above for N independent signal sources, linear instantaneous mixtures is $\boldsymbol{x}=\boldsymbol{As}$, and the task of separation is to find an unmixing matrix $\boldsymbol{W}_{N\times N}$. After linear projection $\boldsymbol{y}=\boldsymbol{Wx}$, we can obtain a set of random variables $\boldsymbol{y}$ that approach the original signal source $\boldsymbol{s}$. Suppose each mixed signal has M samples, then the observation data can be denoted as $\boldsymbol{X}_{N\times M}$. If signal sources are uncorrelated, $\boldsymbol{W}$ must be orthogonal and is solved based on prewhitening $\boldsymbol{X}$.

Jones proposed a method that pursues N-dimension data and projects to one-dimension[7]. The first-order entropy index function-based projection pursuit algorithm recognizes the unknown projection vector w through solving the following optimization problem:

$$\max_{w}(-H(y)) = \max_{w}\int p(y)\lg p(y)\mathrm{d}y = \max_{w}\int p_y(\boldsymbol{wx})\lg p_y(\boldsymbol{wx})\mathrm{d}(\boldsymbol{wx})$$

$$\text{s. t.} \quad \|\boldsymbol{w}\| = 1 \tag{9.2.1}$$

where constraint condition $\|\boldsymbol{w}\| = 1$ makes the searching space restricted to the linear projection in which the original data variance is invariant. At the same time, it makes sure that $H(y)$ is bounded, which poses the above optimization problem well.

Therefore, we can extend the one-dimension projection pursuit to the N linear independent projection vectors:

$$\max_{W}\left(-\sum_{i=1}^{N} H(y_i)\right), \quad \text{s. t.} \quad \det \boldsymbol{W} \neq 0,$$

$$\|\boldsymbol{w}_i\| = 1, \qquad i = 1,\cdots,N \tag{9.2.2}$$

where $\boldsymbol{w}_i$ is the row vector of matrix $\boldsymbol{W}$, that is $y_i=\boldsymbol{w}_i\boldsymbol{x}$. The constraint condition $\det \boldsymbol{W} \neq 0$ guarantees that N projections are linear independent. This constraint is asymptotically equivalent to solving the following relaxation problem:

$$\max_{W}\left(-\sum_{i=1}^{N} H(y_i) + \lg|\det \boldsymbol{W}|\right)$$

$$\text{s. t.} \ \|\boldsymbol{w}_i\| = 1, \qquad i = 1,\cdots,N \tag{9.2.3}$$

where $\lg|\det \boldsymbol{W}|$ is to ensure the full-rank of $\boldsymbol{W}$. The constraint condition is $\|\boldsymbol{w}_i\| = 1$, $i=1,\cdots,N$, and it also considers the Hadamard inequality:

$$|\det \boldsymbol{W}| \leqslant \prod_{i=1}^{N} \|\boldsymbol{w}_i\| = 1 \tag{9.2.4}$$

Obviously

$$\lg|\det \boldsymbol{W}| \leqslant 0 \tag{9.2.5}$$

when $\boldsymbol{W}$ is normalized and orthogonal, the equality is true. So, the optimization problems given by Eqs. (9.2.2) and (9.2.3) are asymptotically equal, if the estimation of the unmixing matrix is based on prewhitening data. Actually, in Eq. (9.2.3), $\lg|\det \boldsymbol{W}|$ plays a constraint role to force iterative solution in the feasible solution set, and when the iterative solution approaches optimal solution, $\lg|\det \boldsymbol{W}|$ becomes zero.

There is one basic object for all ICA methods: to minimize the statistical interdependence between the reconstructed y_i. It is easy to see that we can get the same result by using the Eq. (9.2.3) based projection pursuit method. If we add a constant term $H(\boldsymbol{x})$, which is the entropy of the observation data, on the right side of Eq. (9.2.3), with the consideration $H(\boldsymbol{y})=H(\boldsymbol{x})+\lg|\det \boldsymbol{W}|$, we can modify Eq. (9.2.3) into the following form:

$$\max_{W}\left(-\sum_{i=1}^{N} H(y_i) + \lg|\det \boldsymbol{W}| + H(\boldsymbol{x})\right)$$

$$=\min_{W} I(y_1, \cdots, y_N) \tag{9.2.6}$$

this shows that Eq. (9.2.3) obeys the principle of decreased redundancy and minimizes the interdependence between reconstructed y_i. If the probability density distribution of the source signal is known, this principle is equivalent to the maximum likelihood principle. Because $H(\boldsymbol{x})$ is a constant term of $\boldsymbol{W}$, the objective function of the optimization process can be simplified as

$$\begin{aligned} L(\boldsymbol{W}) &= \sum_{i=1}^{N} H(y_i) - \lg|\det \boldsymbol{W}| \\ &= -\sum_{i=1}^{N} E[\lg p_{y_i}(\boldsymbol{w}_i \boldsymbol{x})] - \lg|\det \boldsymbol{W}| \end{aligned} \tag{9.2.7}$$

9.2.3 Signal Source Probability Density and Joint Estimation of Demixing Matrix

In order to estimate marginal entropy $H(y_i)$ in Eq. (9.2.7), we need the modeling marginal distribution of the source signal. Cardoso[13] proved that if the marginal distribution assumption is wrong, then the estimation of $\boldsymbol{W}$ will become worse and even lead to failure of the blind separation. For the solution to this key problem, we use kernel function estimation technology[7,11] to do nonparametric modeling for the marginal probability density function p_{y_i} from observed signal data. Meanwhile, it can estimate the objective function and its derivative, avoiding an extra estimate of the objective function beyond the iterative process[14,15].

Supposeing each mixed signal has M samples, then any marginal distribution of the reconstructed signal can be approached by the following model:

$$p_{y_i}(y_i) = \frac{1}{Mh}\sum_{m=1}^{M}\phi\left(\frac{y_i - Y_{im}}{h}\right), \qquad i = 1, \cdots, N \tag{9.2.8}$$

where h is the kernel bandwidth, ϕ is the normal kernel function:

$$\phi(u) \stackrel{\text{def}}{=} \frac{1}{\sqrt{2\pi}} \mathrm{e}^{-u^2/2} \tag{9.2.9}$$

Centroid Y_m of the kernel function depends on a given mixed signal and an unknown unmixing matrix

$$Y_{im} = \boldsymbol{w}_i \boldsymbol{x}^{(m)} = \sum_{n=1}^{N} w_{in} \boldsymbol{X}_{nm} \tag{9.2.10}$$

where $\boldsymbol{x}^{(m)}$ is the mth column vector of the mixed signal matrix. It has been proven[10] that this estimation is asymptotically unbiased and effective; it can converge to a real distribution of sources under kinds of error index. Meanwhile, the kernel function approximation model of marginal distribution is a continuous derivable function of the unmixing matrix unit:

$$p_{y_i}(w_i x^{(k)}) = \frac{1}{Mh}\sum_{m=1}^{M}\phi\left(\frac{w_i(x^{(k)} - x^{(m)})}{h}\right) \tag{9.2.11}$$

the gradient function is

$$\nabla p(y_i) = \frac{1}{Mh^2}\sum_{m=1}^{M} x^{(m)}(y_i - w_i x^{(m)})\phi\left(\frac{y_i - w_i x^{(m)}}{h}\right) \tag{9.2.12}$$

in practical application, the expectation in objective function Eq. (9.2.7) can be approximated by ergodic averaging:

$$L(W) \approx -\frac{1}{M}\sum_{i=1}^{N}\sum_{k=1}^{M}\lg p_{y_i}(w_i x^{(k)}) - \lg|\det W| \tag{9.2.13}$$

So, the objective function can be rewritten as

$$L(W) = -L_0(W) - \lg|\det W| \tag{9.2.14}$$

where the marginal probability density function p_{y_i} is replaced with the kernel density function estimation:

$$\begin{aligned} L_0(W) &= \sum_{i=1}^{N} E\lg\left[\frac{1}{Mh}\sum_{m=1}^{M}\phi\left(\frac{y_i - Y_{im}}{h}\right)\right] \\ &\approx \frac{1}{M}\sum_{i=1}^{N}\sum_{k=1}^{M}\lg\left[\frac{1}{Mh}\sum_{m=1}^{M}\phi\left(\frac{w_i(x^{(k)} - x^{(m)})}{h}\right)\right] \end{aligned} \tag{9.2.15}$$

Therefore, the whole optimization problem can be expressed as

$$\begin{aligned} &\min_{W} -\frac{1}{M}\sum_{i=1}^{N}\sum_{k=1}^{M}\lg\left[\frac{1}{Mh}\sum_{m=1}^{M}\phi\left(\frac{w_i(x^{(k)} - x^{(m)})}{h}\right)\right] - \lg|\det W| \\ &\text{s. t. } \|w_i\| = 1, i = 1, \cdots, N \end{aligned} \tag{9.2.16}$$

Given sample data $x^{(k)}$,k=1, ⋯, M, the objective function in Eq. (9.2.16) is the non-linear function of the W matrix unit and is restricted by additional constraints. As the mixed signal has been pre-whitened, along with the restriction, all the reconstructed signals are random variable with a zero mean and unit variance, which can help to simplify the kernel density estimation. In this situation, the optimal value of the kernel function bandwidth h, which is used to control objective functional smoothing, is only a function of sample size (h =1.06 $M^{-1/5}$) [10]. The simulation results show the Eq. (9.2.16) based algorithm is relatively insensitive to h, even if h offsets the empirical optimal value to 50%, it still does not significantly affect the algorithm.

9.2.4 Optimization Algorithm and Global Convergence

The objective function in Eq. (9.2.16) is the smoothing non-linear function of the unmixing matrix W, and we can compute its gradient in Eq. (9.2.12):

$$\nabla L(W) = -\nabla L_0(W) - \nabla\lg|\det(W)|$$

$$= -\nabla L_0(\boldsymbol{W}) - (\boldsymbol{W}^{\mathrm{T}})^{-1} \tag{9.2.17}$$

If we define

$$Z_i(k,m) \stackrel{\text{def}}{=} \frac{\boldsymbol{w}_i(\boldsymbol{x}^{(k)} - \boldsymbol{x}^{(m)})}{h} = \frac{1}{h}\sum_{j=1}^{N} w_{ij}(\boldsymbol{X}_{jk} - \boldsymbol{X}_{jm}) \tag{9.2.18}$$

it is easy to see $\frac{\partial Z_i(k,m)}{\partial w_{ij}} = \frac{\boldsymbol{X}_{jk} - \boldsymbol{X}_{jm}}{h}$, so then the component of $\nabla L_0(\boldsymbol{W})$ is

$$\frac{\partial L_0(\boldsymbol{W})}{\partial w_{ij}} = \frac{1}{M}\sum_{k=1}^{M} \frac{\sum_{m=1}^{M} \frac{\partial Z_i(k,m)}{\partial w_{ij}} \phi'[Z_i(k,m)]}{\sum_{m=1}^{M} \phi[Z_i(k,m)]}$$

$$= -\sum_{k=1}^{M} \frac{\sum_{m=1}^{M} (\boldsymbol{X}_{jk} - \boldsymbol{X}_{jm}) Z_i(k,m) \phi[Z_i(k,m)]}{h \cdot \sum_{m=1}^{M} \phi[Z_i(k,m)]} \tag{9.2.19}$$

In Eq. (9.2.16), there is constraint $\| \boldsymbol{w}_i \| = 1$, $i=1,\cdots,N$. If we can eliminate the constraints, then the constrained optimization problem can be changed into a non-constrained optimization problem, which can be solved with the Newton method or the conjugate gradient method. Therefore, we make the following substitution in Eq. (9.2.16):

$$\boldsymbol{w}_i = \frac{\tilde{\boldsymbol{w}}_i}{\| \tilde{\boldsymbol{w}}_i \|}, \qquad i = 1,\cdots,N \tag{9.2.20}$$

Matrix $\boldsymbol{W}$ can be written as, $\boldsymbol{W} = \tilde{\boldsymbol{D}}^{-1}\tilde{\boldsymbol{W}}$,

$$\tilde{\boldsymbol{D}} = \begin{bmatrix} \| \tilde{\boldsymbol{w}}_1 \| & & 0 \\ & \ddots & \\ 0 & & \| \tilde{\boldsymbol{w}}_N \| \end{bmatrix} \tag{9.2.21}$$

then $\tilde{\boldsymbol{W}} = \tilde{\boldsymbol{D}}\boldsymbol{W}$,

$$\lg|\det \boldsymbol{W}| = -\sum_{i=1}^{N} \lg \| \tilde{\boldsymbol{w}}_i \| + \lg|\det \tilde{\boldsymbol{W}}| \tag{9.2.22}$$

The partial derivative with respect to $\tilde{w}_{ij}$ is

$$\frac{\partial(\lg|\det \boldsymbol{W}|)}{\partial \tilde{w}_{ij}} = -\frac{\tilde{w}_{ij}}{\| \tilde{\boldsymbol{w}}_i \|^2} + [(\tilde{\boldsymbol{W}}^{\mathrm{T}})^{-1}]_{ij} \tag{9.2.23}$$

When $\boldsymbol{W}$ is orthogonal, we have

$$(\tilde{\boldsymbol{W}}^{\mathrm{T}})^{-1} = \tilde{\boldsymbol{D}}^{-1}(\boldsymbol{W}^{\mathrm{T}})^{-1} = \tilde{\boldsymbol{D}}^{-2}\tilde{\boldsymbol{W}} \tag{9.2.24}$$

so all gradient coefficients in Eq. (9.2.23) are zero. Therefore, when $\boldsymbol{W}$ approaches the orthogonal matrix in the optimization process, the second term in Eq. (9.2.7) has no effect on the optimization process. According to Eq. (9.2.23), the component $\nabla L_0(\tilde{\boldsymbol{W}})$ can be computed as follows:

$$\frac{\partial L_0(\widetilde{W})}{\partial \widetilde{w}_{ij}} = \frac{1}{M}\sum_{k=1}^{M} \frac{\sum_{m=1}^{M} \frac{\partial \widetilde{Z}_i(k,m)}{\partial \widetilde{w}_{ij}} \phi'[\widetilde{Z}_i(k,m)]}{\sum_{m=1}^{M} \phi[\widetilde{Z}_i(k,m)]} \tag{9.2.25}$$

where

$$\frac{\partial \widetilde{Z}_i(k,m)}{\partial \widetilde{w}_{ij}} = \frac{1}{h}(\boldsymbol{X}_{jk} - \boldsymbol{X}_{jm} - \widetilde{Z}_i(k,m)\widetilde{w}_{ij}) \tag{9.2.26}$$

$$\phi'[\widetilde{Z}_i(k,m)] = -\widetilde{Z}_i(k,m)\phi[\widetilde{Z}_i(k,m)] \tag{9.2.27}$$

$$\widetilde{Z}_i(k,m) \stackrel{\text{def}}{=} \frac{\widetilde{\boldsymbol{w}}_i(\boldsymbol{x}^{(k)} - \boldsymbol{x}^{(m)})}{h} = \frac{1}{h}\sum_{j=1}^{N} \widetilde{w}_{ij}(\boldsymbol{X}_{jk} - \boldsymbol{X}_{jm}) \tag{9.2.28}$$

The main steps of the entire nonparametric ICA algorithm are shown in Table 9.1. By selecting the quasi-Newton method as the optimal algorithm,using tracking method to select step size, even if the objective function is nonconvex, it can also converge to the nearest minimum.

Table 9.1 Main steps of whole nonparametric ICA algorithm [12]

Initialize $\boldsymbol{W}, \alpha, \beta$

Initialize Hessian estimation $\boldsymbol{H} := \boldsymbol{I}$

Repeat

① Compute the search direction: $\boldsymbol{V} := -\boldsymbol{H}^{-1}\nabla L(\boldsymbol{W})$

② Compute the step size by tracking method:

$\mu := 1$

while $L(\boldsymbol{W}+\mu\boldsymbol{V}) > L(\boldsymbol{W}) + \alpha\mu\nabla L(\boldsymbol{W})^{\mathrm{T}}\boldsymbol{V}$

$\mu := \beta\mu$

③ Update $\boldsymbol{H}^{-1}$

④ Update $\boldsymbol{W}$: $\boldsymbol{W} := \boldsymbol{W} + \mu\boldsymbol{V}$

Until $\sqrt{-\boldsymbol{V}^{\mathrm{T}}\nabla L(\boldsymbol{W})} \leqslant \varepsilon$

9.3 Second-Order Statistics Based Convolutive Mixtures Separation Algorithm

9.3.1 Introduction

For the stationary signal, if it has independent identical distribution and non-Gaussian signal, we can use higher-order statistics (HOS) of the output signal to separate source signals from mixed signals. Based on this premise, many blind separation algorithms called HOS-based algorithms have been developed. Because there is no prior knowledge for probability distribution of the source signal, many algorithms use a score function to adapt to the type of the source signal, or to directly estimate the probability density function[16], thus completing blind separation. Generally, the score function estimation is quite complex, leading to difficulties in its application.

There are two reasons for the proposed SOS algorithm. One is the complexity of HOS-based algorithm, and the other is that many real signals are uncorrelated in nature and are not independent, so it is possible and necessary to develop a simple but fast convergent SOS-based algorithm.

For research on the SOS-based algorithm, in the case of instantaneous mixtures, Tong *et al* proposed AMOUSE algorithm[17], and Choi *et al* used the natural gradient method resulting in a better algorithm[18,19]. In this section, we emphasize the convolutive mixture's separation algorithm of a multiple-input / multiple-output (MIMO) system.

In recent years, there have been achievements[20,21] in the blind processing of the MIMO system, and it has been proved that using SOS alone can realize blind deconvolution of the linear convolutive mixing model[22,23]. For example, Hua and Tugnait[22] proved that if the FIR system cannot be simplified and the source signals are spatially uncorrelated, each source signal has a different power spectrum, and the system can be recognized by SOS in the sense that there are indeterminacies of permutation, arbitrary scaling and some delay. However, Ref. [22] did not give a specific algorithm. On this basis, Kawamoto and Inouye[23] gave a necessary and sufficient condition of blind equalization, deducing the blind equalization algorithm in the frequency domain under a weak condition.

Our research shows that in the situation of a nonstationary source signal, the power spectrum matrix of the mixed signal is not necessarily nonnegative; therefore, a new evaluation function for the nonstationary source signal is constructed and an algorithm with better performance both in the time-domain and frequency-domain is developed for more extensive applications.

9.3.2 Identification Condition

First of all, we present the conditions for identifiability: assumption 9.1, assumption 9.2, assumption 9.3 and Theorem 9.1.

Assumption 9.1 Input signal $s(t)$ is a zero-mean stationary random process, and its components are temporally non-white and spatially uncorrelated.

Assumption 9.2 Input signal $\boldsymbol{s}(t)$ has a different normalized power spectrum element $S_{s,\,ii}(f)\,/\,S_{s,\,ii}(0)$, $i=1, \cdots, m$, where $S_{s,\,ii}(f)$ is the (i, i) element of power spectrum matrix $\boldsymbol{S}_s(f)$.

Assumption 9.3 Channel transfer function $\boldsymbol{H}(z)$ is equalizable[24].

Theorem 9.1 Assume that $\boldsymbol{y}(t)$ is the output sequence of a compound system without noise (i.e., $\boldsymbol{n}(t)=\boldsymbol{0}$), and the transfer function of the equalizer cannot be simplified. Then, under the assumptions 9.4, 9.5 and 9.6, the compound system is transparent if and only if the power spectrum matrix $\boldsymbol{S}_y(f)$ of $\boldsymbol{y}(t)$ is a diagonal matrix.

In order to present the SOS-based algorithm, we give the assumption 9.4 modified on assumption 9.1, then add time-domain assumption 9.6 and frequency-domain assumption 9.7. Meanwhile, in order to show that SOS does not fit the equalization of the stationary signal, we provide assumption 9.5.

Assumption 9.4 Source signal $\boldsymbol{s}(t)$ is a zero-mean non-stationary random process; its correlation function matrix $\boldsymbol{R}_s(t, \tau)$ is a function of (t, τ), not the single variable function of τ; and its vector components $s_i(t), i=1, \cdots, m$ are mutually statistically uncorrelated.

Suppose assumption 9.4 is the generalization of assumption 9.1. This assumption means that its correlation function $\boldsymbol{R}_s(t, \tau)$ is a diagonal matrix. When time t is fixed, its power spectrum matrix $\boldsymbol{S}_s(f)$ is also a diagonal matrix. Actually, the two-dimensional Fourier transforms of $\boldsymbol{R}_s(t, \tau)$, which is $\boldsymbol{S}_s(f_1, f_2)$, are also diagonal matrices.

Assumption 9.5 Source signals are mutually independent, and each source signal is a temporarily independent and identically distributed non-Gaussian signal.

Assumption 9.5 is used to explain that SOS cannot do the blind deconvolution of the stationary signal in the following section.

According to Theorem 9.1, we can easily reach this conclusion:

Corollary 9.1 Assume that $\boldsymbol{y}(t)$ is the output sequence of a compound system without noise (i.e. $\boldsymbol{n}(t)=\boldsymbol{0}$), and the transfer function of the equalizer cannot be simplified. Then under assumptions 9.2, 9.3 and 9.4, the compound system is transparent if and only if the correlation function $\boldsymbol{R}_y(t, \tau)$ and power spectrum matrix $\boldsymbol{S}_y(f)$ of $\boldsymbol{y}(t)$ is a diagonal matrix.

The only difference between Theorem 9.1 and Corollary 9.1 is that the assomptions range from 9.1 to 9.4, and its proof can be referred to in Ref.[23].

Because of the introduction of assumption 9.4, the power spectrum matrix $\boldsymbol{S}_y(f)$ is no longer a nonnegative definite matrix, and the evaluation function given by Ref.[23] should be modified.

Assumption 9.6 Assume $(m-1)\times(m-1)$ dimension matrix $\boldsymbol{R}_{y,ii}(t,\tau)\boldsymbol{I}_{m-1}+\boldsymbol{R}_i(t,\tau)$, $\forall i=1,2,\cdots,m$ is reversible, and $\boldsymbol{I}_{m-1}$ is $(m-1)\times(m-1)$ unit matrix, $\boldsymbol{R}_i(t,\tau)$, $\forall i=1,2,\cdots,m$ is $(m-1)\times(m-1)$ matrix defined as a matrix that removes the ith row and ith column from $\boldsymbol{R}_y(t,\tau)$,

$$\boldsymbol{R}_i(t,\tau)=\begin{bmatrix} \cdots & \cdots & \cdots & \cdots \\ \cdots & R_{y,(i-1)(i-1)}(t,\tau) & R_{y,(i-1)(i+1)}(t,\tau) & \cdots \\ \cdots & R_{y,(i+1)(i-1)}(t,\tau) & R_{y,(i+1)(i+1)}(t,\tau) & \cdots \\ \cdots & \cdots & \cdots & \cdots \end{bmatrix}$$

Assumption 9.7 Assume $(m-1)\times(m-1)$ dimension matrix $\boldsymbol{S}_{y,ii}(f)\boldsymbol{I}_{m-1}+\boldsymbol{S}_i^*(f)$, $i=1,\cdots,m$ is reversible, and $\boldsymbol{I}_{m-1}$ is $(m-1)\times(m-1)$ unit matrix, superscript $*$ means complex conjugate, $\boldsymbol{S}_i(f)$ is $(m-1)\times(m-1)$ matrix and defined as a matrix that removes the ith row and ith column from $\boldsymbol{S}_y(f)$,

$$\boldsymbol{S}_i(f)=\begin{bmatrix} \cdots & \cdots & \cdots & \cdots \\ \cdots & S_{y,(i-1)(i-1)}(f) & S_{y,(i-1)(i+1)}(f) & \cdots \\ \cdots & S_{y,(i+1)(i-1)}(f) & S_{y,(i+1)(i+1)}(f) & \cdots \\ \cdots & \cdots & \cdots & \cdots \end{bmatrix}$$

Time-domain assumption 9.6 and frequency-domain assumption 9.7 are used to establish the evaluation function in the time-domain and frequency-domain respectively; they supplement assumption 9.2. It should be pointed out that it appears difficult to satisfy assumption 9.6 and assumption 9.7, whereas, simulation results show that it is easy unless an ill-conditioned matrix apperars.

Property 9.1 Under assumption 9.4, correlations function $\boldsymbol{R}_y(t,\tau)$ and power spectrum matrix $\boldsymbol{S}_y(f_1,f_2)$ are not always a nonnegative definite matrix.

Considering output vector $\boldsymbol{y}(t)$, according to the definition of the correlation function, we can get

$$\begin{aligned}\boldsymbol{R}_y(t_1,t_2)&=\langle \boldsymbol{C}(z)\boldsymbol{s}(t_1)\boldsymbol{s}^{\mathrm{H}}(t_2)\boldsymbol{C}^{\mathrm{H}}(z)\rangle \\ &=\boldsymbol{C}(z)\boldsymbol{R}_s(t_1,t_2)\boldsymbol{C}^{\mathrm{H}}(z)\end{aligned} \tag{9.3.1}$$

Its power spectrum matrix $\boldsymbol{S}_y(f_1,f_2)$ is

$$\boldsymbol{S}_y(f_1,f_2)=\boldsymbol{C}(f_1)\boldsymbol{S}_s(f_1,f_2)\boldsymbol{C}^{\mathrm{H}}(f_2) \tag{9.3.2}$$

Because f_1 and f_2 are independent variables and cannot guarantee $\boldsymbol{S}_s(f_1,f_2)=\boldsymbol{I}$ as a nonnegative definite matrix when the source signal is nonstationary, then $S_y(f_1, f_2)$ is not always a nonnegative definite matrix. For example, in Eq. (9.2.2), if $S_s(f_1,f_2)=I$, and $C(f)$ is a diagonal matrix, its diagonal elements are $C_{11}(f)=\frac{\sin f}{f}$ respectively, $C_{ii}(f)=1$, $\forall i \neq 1$, when $f_1=0$, $f_2=\frac{3\pi}{2}$, $S_y(f_1,f_2)$ is also a diagonal matrix, with its diagonal element $S_{y,11}(f_1,f_2)=-\frac{2}{3\pi}$, $S_{y,ii}(f_1,f_2)=1$, $\forall i \neq 1$, i.e., $S_y(f_1,f_2)=\mathrm{diag}\begin{bmatrix}-\frac{2}{3\pi} & 1 & \cdots & 1\end{bmatrix}$. Obviously, the power spectrum matrix $S_y(f_1,f_2)$ constructed in this way is not always a nonnegative definite matrix. Similarly, the correlation's function matrix $R_y(t,\tau)$ is not always a nonnegative definite matrix.

Meanwhile, under assumption 9.5, because source signals are mutually independent and each source signal is a temporary independent identically distributed non-Gaussian signal. When $t_1=t_2=t$, correlation function matrix $R_s(t_1,t_2)$ is a diagonal matrix, however, then $t_1 \neq t_2$, $R_s(t,\tau)$, is zero matrix. In Eq. (9.3.2), supposing $f_1=f_2=f$, we can get

$$S_y(f)=C(f)S_s(f)C^{H}(f) \tag{9.3.3}$$

Its diagonal elements $S_{y,ii}(f)$ *is*

$$S_{y,ii}(f)=\sum_{j=1}^{m} S_{s,ii}(f)\,|C_{ij}(f)|^2 \tag{9.3.4}$$

this equation indicates that the power spectrum matrix $S_y(f)$ loses phase information of a compound system under assumption 9.5, so that the SOS can only recognize the minimum phase system and not the non-minimum phase system. However, under assumption 9.4, we can see that the power spectrum matrix $S_y(f_1,f_2)$ in Eq. (9.3.2) includes phase information about the channel and can recognize the minimum phase system.

In practical application of the SOS blind processing algorithm, the most popular is the correlation function matrix $R_y(t_1, t_2)$ and power spectrum matrix $S_y(f)$ at time t, and the corresponding Eqs. (9.3.1) and (9.3.2) can be simplified as

$$R_y(t,\tau)=C(z)R_s(t,\tau)C^{H}(z) \tag{9.3.5}$$

$$S_y(f)=C(f)S_s(f)C^{H}(f) \tag{9.3.6}$$

Equations (9.3.5) and (9.3.6) are deduced under assumption 9.4 for a nonstationary signal, and $R_y(t,\tau)$ and $S_y(f)$ have conjugate symmetry for non-diagonal elements, so it has the following Property 9.2.

Property 9.2 Under assumption 9.4,

$$R_{y,ij}(t,\tau) = R_{y,ij}^{*}(t,\tau), \qquad \forall i \neq j$$
$$S_{y,ij}(f) = S_{y,ji}^{*}(f), \qquad \forall i \neq j \tag{9.3.7}$$

When $\boldsymbol{y}(t)$ is the real function, its correlation function matrix $\boldsymbol{R}_y(t,\tau)$ is a real symmetric matrix, namely

$$R_{y,ij}(t,\tau) = R_{y,ji}(t,\tau), \qquad \forall i \neq j \tag{9.3.8}$$

9.3.3 Time-Domain Algorithm

First, look at the following nonnegative definite matrix function:

$$\boldsymbol{M}_{\boldsymbol{R}}(t,\tau) = \boldsymbol{R}_y(t,\tau)\boldsymbol{R}_y^{\mathrm{T}}(t,\tau) \tag{9.3.9}$$

it has the properties of Property 9.3 and Theorem 9.2.

Property 9.3 $\boldsymbol{M}_{\boldsymbol{R}}(t,\tau) \geqslant 0$, i.e. matrix $\boldsymbol{M}_{\boldsymbol{R}}(t,\tau)$ is a nonnegative definite.

Theorem 9.2 Under assumption 9.4 and assumption 9.6, if and only if correlation function matrix $\boldsymbol{R}_y(t,\tau)$ is a diagonal matrix, $\boldsymbol{M}_{\boldsymbol{R}}(t,\tau)$ is also a diagonal matrix.

We can know from Property 9.2 that we must diagonalize $\boldsymbol{R}_y(t,\tau)$ if we want to make the system transparent for blind deconvolution. According to Theorem 9.1, if nonnegative definite matrix $\boldsymbol{M}_{\boldsymbol{R}}(t,\tau)$ is diagonal, the system can become transparent. Therefore, we apply the Hadamard inequality:

$$\prod_{i=1}^{m} a_{ii} \geqslant \det\boldsymbol{A} \tag{9.3.10}$$

where $\boldsymbol{A}=(a_{ij})_{m\times m}$ is defined as a nonnegative definite matrix. If and only if $i \neq j$, $a_{ij}=0$, then the above equation exists. Therefore, the evaluation function is constructed as the function in parameter space:

$$J_R = \sum_{\tau}\left(\frac{1}{2}\sum_{i=1}^{m}\lg\sum_{j=1}^{m}R_{y,ij}^{2}(t,\tau) - \lg|\det\boldsymbol{R}_y(t,\tau)|\right) \geqslant 0 \tag{9.3.11}$$

The evaluation function constructed in this way has the following Theorem:

Theorem 9.3 Under assumptions 9.2, 9.3, 9.4 and 9.6, if and only if $\boldsymbol{R}_y(t,\tau)$ is diagonal matrix, then evaluation function J_R can reach minimum.

The current object is to obtain parameter matrix $\boldsymbol{W}_p(t)$ and minimize the evaluation function Eq. (9.3.11).

Compute the partial derivative of evaluation function J_R with respect to $\boldsymbol{W}_p(t)$:

$$\begin{aligned}\frac{\partial J_R}{\partial \boldsymbol{W}_p(t)} = &-\sum_{\tau}[\boldsymbol{A}_R(t,\tau)\langle \boldsymbol{y}(t-\tau)\boldsymbol{x}^{\mathrm{T}}(t-p)\rangle + \\ &\boldsymbol{A}_{\boldsymbol{R}}^{\mathrm{T}}(t,\tau)\langle \boldsymbol{y}(t)\boldsymbol{x}^{\mathrm{T}}(t-\tau-p)\rangle]\end{aligned} \tag{9.3.12}$$

where

$$\boldsymbol{A}_R(t,\tau) = \boldsymbol{R}_y^{-1}(t,\tau) - \boldsymbol{\Lambda}_R^{-1}\boldsymbol{R}_y(t,\tau) \tag{9.3.13}$$

Substituting Eq. (9.3.13) into a conventional stochastic gradient learning algorithm:

$$\Delta \boldsymbol{W}(k) = \boldsymbol{W}(k+1) - \boldsymbol{W}(k) = -\mu(k)\frac{\partial J[\boldsymbol{y}(t), \boldsymbol{W}(k)]}{\partial \boldsymbol{W}(k)}$$

We now have the following deconvolution algorithm:

$$\begin{aligned}\boldsymbol{W}_p(t+1) = {} & \boldsymbol{W}_p(t) + \mu(t) \cdot \\ & \sum_{\tau} [\boldsymbol{A}_R(t,\tau)\langle \boldsymbol{y}(t-\tau)\boldsymbol{x}^{\mathrm{T}}(t-p)\rangle + \\ & \boldsymbol{A}_R^{\mathrm{T}}(t,\tau)\langle \boldsymbol{y}(t)\boldsymbol{x}^{\mathrm{T}}(t-\tau-p)\rangle]\end{aligned} \tag{9.3.14}$$

In Eq. (9.3.14), the estimation of the correlation function matrix directly affects the convergence of the algorithm and fluctuation in the convergence process. The on-line estimation is as follows:

$$\hat{\boldsymbol{R}}_y(t,\tau) = (1-\alpha_R)\hat{\boldsymbol{R}}_y(t-1,\tau) + \alpha_R \boldsymbol{y}(t)\boldsymbol{y}^{\mathrm{T}}(t-\tau) \tag{9.3.15}$$

where α_R is the influence coefficient such as α_R=0.1~0.001. Similarly, we can follow the same procedure for the ensemble average in Eq. (9.3.14).

The parametric model on which the evaluation function Eq. (9.3.11) is defined, is a curve manifold and we can use the natural gradient algorithm to deduce an effective on-line learning algorithm.

Similarly, compute the partial derivative of evaluation function J_R in Eq. (9.3.11) with respect to parameter matrix $\boldsymbol{W}_p(t)$:

$$\begin{aligned}\frac{\mathrm{d}J_R}{\mathrm{d}\boldsymbol{X}_p(t)} = {} & -\sum_{\tau}(\boldsymbol{A}_R(t,\tau)\langle \boldsymbol{y}(t-\tau)\boldsymbol{y}^{\mathrm{T}}(t-p)\rangle + \\ & \boldsymbol{A}_R^{\mathrm{T}}(t,\tau)\langle \boldsymbol{y}(t)\boldsymbol{y}^{\mathrm{T}}(t-\tau-p)\rangle)\end{aligned} \tag{9.3.16}$$

where the definition of $\boldsymbol{A}_R(t,\tau)$ is the same as Eq. (9.3.13). Substitute it into a natural gradient algorithm,

$$\Delta \boldsymbol{W}_p = -\mu(t)\left[\frac{\mathrm{d}J(\boldsymbol{W}(z,t))}{\mathrm{d}\boldsymbol{X}_p(t)}\right]\boldsymbol{W}(z,t)$$

we obtain

$$\begin{aligned}\boldsymbol{W}_p(t+1) = {} & \boldsymbol{W}_p(t) + \mu(t) \cdot \\ & \sum_{\tau}(\boldsymbol{A}_R(t,\tau)\langle \boldsymbol{y}(t-\tau)\boldsymbol{u}_p^{\mathrm{T}}(t)\rangle + \boldsymbol{A}_R^{\mathrm{T}}(t,\tau)\langle \boldsymbol{y}(t)\boldsymbol{u}_p^{\mathrm{T}}(t-\tau)\rangle)\end{aligned} \tag{9.3.17}$$

where

$$\boldsymbol{u}_p(t) = \sum_{q=-\infty}^{+\infty} \boldsymbol{W}_q^{\mathrm{T}}(t)\boldsymbol{y}(t-p+q) \tag{9.3.18}$$

the algorithm in Eq. (9.3.17) has equivariant Property, i.e. the algorithm does not simply depend on channel parameters.

When p=0, the updating of the pth coefficient matrix depends on the output of demixing $y(t-p+q), 0 \leqslant q \leqslant L$. In order to realize on-line algorithm, delay L

samples for Eq. (9.3.17). This delay does not change the statistical relation of the signal; therefore, provide the same algorithm performance under the same learning rate[25,26].

In the lower learning rate,

$$\boldsymbol{W}_p(t) \approx \boldsymbol{W}_p(t-1) \approx \boldsymbol{W}_p(t-2L) \tag{9.3.19}$$

from Eq. (9. 3. 18), we obtain

$$\boldsymbol{u}_p(t) \approx \boldsymbol{u}_0(t-p), \qquad p = 0,1,\cdots,L \tag{9.3.20}$$

the iterative updating of coefficient matrix $\boldsymbol{W}_p(t)$ is modified as follows:

$$\begin{aligned}\boldsymbol{W}_p(t+1) = \boldsymbol{W}_p(t) + \mu(t)\sum_{\tau}[&\boldsymbol{A}_R(t-L,\tau)\langle \boldsymbol{y}(t-\tau-L)\boldsymbol{u}^{\mathrm{T}}(t-p)\rangle + \\ &\boldsymbol{A}_R^{\mathrm{T}}(t-L,\tau)\langle \boldsymbol{y}(t-L)\boldsymbol{u}^{\mathrm{T}}(t-\tau-p)\rangle]\end{aligned} \tag{9.3.21}$$

where

$$\boldsymbol{u}(t) = \sum_{q=0}^{L}\boldsymbol{W}_{L-q}^{\mathrm{T}}(t)\boldsymbol{y}(t-p-q) \tag{9.3.22}$$

In order to avoid computing the inverse of matrix $\boldsymbol{R}_y(t,\tau)$, by simplifying the evaluation function, the iterative updating for coefficient matrix $\boldsymbol{W}_p(t)$ is modified as follows[27]:

$$\begin{aligned}\boldsymbol{W}_p(t+1) = \boldsymbol{W}_p(t) + \mu(t)\Big\{\boldsymbol{W}_p(t) - \frac{1}{2}\sum_{\tau}[&\boldsymbol{B}_R(t-L,\tau)\boldsymbol{y}(t-\tau-L)\boldsymbol{u}^{\mathrm{T}}(t-p) + \\ &\boldsymbol{B}_R^{\mathrm{T}}(t-L,\tau)\boldsymbol{y}(t-L)\boldsymbol{u}^{\mathrm{T}}(t-\tau-p)]\Big\}\end{aligned} \tag{9.3.23}$$

where $\boldsymbol{B}_R(t,\tau) = \boldsymbol{\Lambda}_R^{-1}\boldsymbol{R}_y(t,\tau)$.

We can further develop the optimal algorithm based on the above algorithm. Its basic idea is to propose a new evaluation function based on mass experiments, where the performance will produce better results than evaluation based on theory. The evaluation function is

$$J = \frac{1}{4}\sum_{\tau}\|\boldsymbol{R}(t,\tau)\|^{\circ r} - \frac{1}{2\pi \mathrm{j}}\oint \lg|\det \boldsymbol{W}(z,t)|z^{-1}\mathrm{d}z \tag{9.3.24}$$

where $\|\boldsymbol{A}\|^{\circ r}\sum_{i,j}|a_{ij}|^r$.

when r=2,

$$\begin{aligned}J &= \frac{1}{4}\sum_{\tau}\|\boldsymbol{R}(t,\tau)\|^{\circ 2} - \frac{1}{2\pi \mathrm{j}}\oint lg\,|\,det\,\boldsymbol{W}(z,t)|z^{-1}\mathrm{d}z \\ &= \frac{1}{4}\sum_{\tau}\sum_{i,j}|R_{ij}(t,\tau)|^2 - \frac{1}{2\pi \mathrm{j}}\oint \lg|\det \boldsymbol{W}(z,t)|z^{-1}\mathrm{d}z\end{aligned} \tag{9.3.25}$$

Therefore, the on-line learning iteration of the new algorithm[27,28] is

$$\begin{aligned}&\boldsymbol{W}_p(t+1) = \boldsymbol{W}_p(t) + \mu(t)\{\boldsymbol{W}_p(t) - \\ &\sum_{\tau}\boldsymbol{F}(t-L,\tau)\frac{1}{2}[\langle \boldsymbol{y}(t-\tau-L)\boldsymbol{u}(t-p)^{\mathrm{H}}\rangle +\end{aligned}$$

$$\langle \boldsymbol{y}(t-L)\boldsymbol{u}(t-p-\tau)^{\mathrm{H}}\rangle]\} \tag{9.3.26}$$

where the definition of $\boldsymbol{u}(t)$ is the same as Eq. (9.3.22); however,

$$\boldsymbol{F}(t,\tau) = |\boldsymbol{R}(t,\tau)\circ|^{\circ(r-1)} \circ \mathrm{sgn}[\boldsymbol{R}(t,\tau)] \tag{9.3.27}$$

where $\circ$ is Hadamard product. Table 9.2 shows the function matrix $\boldsymbol{F}(t,\tau)$ with a different r.

Table 9.2 Function matrix $\boldsymbol{F}(t,\tau)$ with different r

r	$\boldsymbol{F}(t,\tau)$ as defined in Eq. (9.2.27)
r=1	$\boldsymbol{F}(t,\tau)=\mathrm{sgn}[\boldsymbol{R}(t,\tau)]$
r=2	$\boldsymbol{F}(t,\tau)=\boldsymbol{R}(t,\tau)$
r=3	$\boldsymbol{F}(t,\tau)=\|\boldsymbol{R}(t,\tau)\circ\|\circ\boldsymbol{R}(t,\tau)$
r=4	$\boldsymbol{F}(t,\tau)=\boldsymbol{R}(t,\tau)^{\circ 2}\circ\boldsymbol{R}(t,\tau)$
r even	$\boldsymbol{F}(t,\tau)=\boldsymbol{R}(t,\tau)^{\circ(r-2)}\circ\boldsymbol{R}(t,\tau)$
r odd	$\boldsymbol{F}(t,\tau)=\boldsymbol{R}(t,\tau)^{\circ(r-2)}\circ\|\boldsymbol{R}(t,\tau)\circ\|$

9.3.4 Frequency-Domain Algorithm

First, define the following nonnegative definite matrix function:

$$\boldsymbol{M}_{\mathrm{S}}(f) = \boldsymbol{S}_y(f)\boldsymbol{S}_y{}^{\mathrm{H}}(f) \tag{9.3.28}$$

If and only if power spectrum $\boldsymbol{S}_y(f)$ of the equalizer output is a diagonal matrix, $\boldsymbol{M}_S(f)$ is also a diagonal matrix. For nonnegative definite matrix $\boldsymbol{S}_y(f)$, we apply the Hadamard inequality to construct the evaluation function:

$$J_{\mathrm{S}} = \frac{1}{L_f}\sum_{k=0}^{L_f-1} J_k(f_k) \tag{9.3.29}$$

where $f_k = \dfrac{k}{L_f}$, k=0, $\cdots$, L_f-1, L_f is the data length in the frequency-domain. In practice, we can choose L_f=L, or the integer power of 2 in order to perform the FFT operation. In addition,

$$J_k(f_k) = \sum_{i=1}^{m} \lg \sum_{j=1}^{m} |S_{y,ij}(f_k)|^2 - \lg|\det\boldsymbol{S}_y(f_k)|^2 \tag{9.3.30}$$

it needs to be pointed out, Eqs. (9.3.29) and (9.3.30) are functions of parameter $\boldsymbol{W}_p(t)$ and Eq. (9.3.30) is the evaluation function at f_k in the frequency-domain.

According to the definition of the correlation function, we have

$$\mathrm{d}\boldsymbol{R}_y(t,\tau) = \langle \mathrm{d}\boldsymbol{y}(t)\boldsymbol{y}^{\mathrm{H}}(t-\tau) + \boldsymbol{y}(t)\mathrm{d}\boldsymbol{y}^{\mathrm{H}}(t-\tau)\rangle$$

$$= \langle d\boldsymbol{X}(z)\boldsymbol{y}(t)\boldsymbol{y}^{H}(t-\tau)+\boldsymbol{y}(t)d\boldsymbol{y}^{H}(t-\tau)\boldsymbol{X}^{H}(z)\rangle$$
$$= d\boldsymbol{X}(z)\boldsymbol{R}_y(t,\tau)+\boldsymbol{R}_y(t,\tau)d\boldsymbol{X}^{H}(z) \tag{9.3.31}$$

At time t, we perform the Fourier transform for the above equation:

$$d\boldsymbol{S}_y(f)=d\boldsymbol{X}(f)\boldsymbol{S}_y(f)+\boldsymbol{S}_y(f)d\boldsymbol{X}^{H}(f) \tag{9.3.32}$$

Compute the differential of $J_k(f_k)$ with respect to $\boldsymbol{X}(f_k)$:

$$\frac{dJ_k(f_k)}{d\boldsymbol{X}(f_k)}=(\boldsymbol{\Lambda}^{-1}-\boldsymbol{M}^{-1})\boldsymbol{S}_y^{*}(f_k)\boldsymbol{S}_y^{T}(f_k)+ \boldsymbol{S}_y^{T}(f_k)(\boldsymbol{\Lambda}^{-1}-\boldsymbol{M}^{-1})\boldsymbol{S}_y^{*}(f_k) \tag{9.3.33}$$

with the modified coefficient differential matrix $d\boldsymbol{X}(f_k)$. Note that $d\boldsymbol{X}(f_k)$ is a linear combination of the coefficient differentials $d\boldsymbol{W}_{ij}(f_k)$ in the matrix polynomial $d\boldsymbol{W}(f_k)$. As long as $\boldsymbol{W}(f_k)$ is nonsingular, $d\boldsymbol{X}(f_k)$ represents a valid search direction to minimize Eq. (9.3.30), because $d\boldsymbol{X}(f_k)$ spans the same tangent space of the matrix as that spaned by $d\boldsymbol{W}(f_k)$[22,28].

Therefore, the random gradient search direction of evaluation function $J_k(f_k)$ at minimal frequency domain f_k is

$$\Delta\boldsymbol{W}(f_k)=-\mu(t)\left[\frac{dJ_k(f_k)}{d\boldsymbol{X}(f_k)}\right]\boldsymbol{W}(f_k) \tag{9.3.34}$$

where $\mu(t)$ is the learning rate. The search direction is the natural gradient search direction and uses the Riemannian metric tensor of all matrix filter space formied by $\boldsymbol{W}(f_k)$[25,26].

Substitute Eq. (9.3.33) into (9.3.34) and we get

$$\Delta\boldsymbol{W}(f_k)=\mu(t)\{(\boldsymbol{M}^{-1}-\boldsymbol{\Lambda}^{-1})\boldsymbol{S}_y^{*}(f_k)\boldsymbol{S}_y^{T}(f_k)+ \boldsymbol{S}_y^{T}(f_k)(\boldsymbol{M}^{-1}-\boldsymbol{\Lambda}^{-1})\boldsymbol{S}_y^{*}(f_k)\}\boldsymbol{W}(f_k) \tag{9.3.35}$$

In order to get updated change directly, perform the inverse Fourier transform on the above formula:

$$\Delta\boldsymbol{W}_p=\frac{1}{L_f}\sum_{k=0}^{L_f-1}\Delta\boldsymbol{W}(f_k)e^{j\frac{2\pi}{L_f}kp} \tag{9.3.36}$$

It should be pointed out that, because we finally get an on-line algorithm at time t, $\Delta\boldsymbol{W}_p$ can be written as

$$\Delta\boldsymbol{W}_p(t)=\boldsymbol{W}_p(t+1)-\boldsymbol{W}_p(t) \tag{9.3.37}$$

Therefore, we get the updated formula of $\boldsymbol{W}_p(t)$:

$$\boldsymbol{W}_p(t+1)=\boldsymbol{W}_p(t)+\frac{\mu(t)}{L_f}\sum_{k=0}^{L_f-1}\boldsymbol{G}(f_k)\boldsymbol{W}(f_k)e^{j\frac{2\pi}{L_f}kp} \tag{9.3.38}$$

where

$$\boldsymbol{G}(f_k)=(\boldsymbol{M}^{-1}-\boldsymbol{\Lambda}^{-1})\boldsymbol{S}_y^{*}(f_k)\boldsymbol{S}_y^{T}(f_k)+ \boldsymbol{S}_y^{T}(f_k)(\boldsymbol{M}^{1}-\boldsymbol{\Lambda}^{1})\boldsymbol{S}_y^{*}(f_k) \tag{9.3.39}$$

Equation (9.3.38) indicates that the algorithm obtained from the natural gradient algorithm has equivariant Property.

In the realization of the above algorithm, it is noted that $\boldsymbol{S}_x(f_k)$ and $\boldsymbol{S}_y(f_k)$ are the Fourier transformation of $\boldsymbol{R}_x(t,\tau)$ and $\boldsymbol{S}_y(t,\tau)$ at time t respectively. Therefore, we evaluate $\boldsymbol{R}_x(t,\tau)$ and $\boldsymbol{R}_y(t,\tau)$ first. The evaluation equations are as follows:

$$\hat{\boldsymbol{R}}_x(t,\tau)=(1-\alpha_x)\hat{\boldsymbol{R}}_x(t-1,\tau)+\alpha_x\boldsymbol{x}(t)\boldsymbol{x}^{\mathrm{T}}(t-\tau) \tag{9.3.40}$$

$$\hat{\boldsymbol{R}}_y(t,\tau)=(1-\alpha_y)\hat{\boldsymbol{R}}_y(t-1,\tau)+\alpha_y\boldsymbol{y}(t)\boldsymbol{y}^{\mathrm{T}}(t-\tau) \tag{9.3.41}$$

where α_x and α_y are influence factors generally selecting, according to experience, $\alpha_x=\alpha_y=0.1\sim0.001$.

Since the initialization of parameter matrix $\boldsymbol{W}_p(t)$ greatly affects the convergence, we can apply the center-tap initialization scheme in practice; i.e., when $m=n$, $\boldsymbol{W}(z)=\boldsymbol{I}z^{-j}$, $0\leqslant j\leqslant L$; when $m<n$, $\boldsymbol{W}(z)=[\boldsymbol{I}\quad \boldsymbol{1}]z^{-j}$, here $\boldsymbol{1}$ is the $m\times(n-m)$ dimension matrix with elements 1.

In practice, there is an error term $\boldsymbol{n}(t)$ that also greatly affects the convergence. Therefore, we present a constraint factor $\beta(f_k)$, which is defined as

$$\beta(f_k)=\sum_{i\neq j}|S_{y,ij}(f_k)|^2 \text{ or } \beta(f_k)=1-\frac{\det\boldsymbol{M}_y(f_k)}{\prod_{i=1}^{m}M_{y,i}(f_k)} \tag{9.3.42}$$

The constraint factor can be used as the condition for iteration termination and used to adaptively adjust the learning rate. When $\beta(f_k)=0$, the power spectrum matrix $\boldsymbol{S}_y(f_k)$ loses its function. The modified updating formula is

$$\boldsymbol{W}_p(t+1)=\boldsymbol{W}_p(t)+\frac{\mu(t)}{L_f}\sum_{k=0}^{L_f-1}\beta(f_k)\boldsymbol{G}(f_k)\boldsymbol{W}(f_k)\mathrm{e}^{j\frac{2\pi}{L_f}kp} \tag{9.3.43}$$

It needs to be pointed out that when $\boldsymbol{S}_y(f_k)$ approaches the diagonal matrix, $\beta(f_k)$ does not strictly approach zero due to the effect of the observation and evaluation errors and the source signal itself. Therefore, we must set a threshold γ, when $|\beta(f_k)|<\gamma$, $\beta(f_k)=0$. The value of γ is determined by experiment, usually, resulting in $\gamma=0.01\sim0.1$.

9.4 Blind Separation for Fewer Sensors than Sources—Underdetermined Model

The blind processing model discussed previously requires the number of observation signals to be equal to or no less than the number of source signals. As the extension of the conventional model the underdetermined model-based blind signal processing can solve the problem of the number of source signals being more than the number of the observed signals. The concept of the underdetermined model was first put forward by Cardoso[29] in 1991. He stated that the underdetermined model could be identified by the fourth-order cumulant. Shamsunder used higher-order statistics to estimate the DOA in the underdetermined model[30]. Cao also made some contributions to the underdetermined model in his early work[4]. In 1999, Taleb and Jutten specifically discussed the underdetermined model[31], but they did not put forward an effective algorithm for the source estimation.

For blind signal separation in the underdetermined model, we usually assume the sparsity of sources as well as the independence among the sources.Although the identifiability of the underdetermined model requires all the sources to be non-Gaussian, it does not mean only sparse signals can be recovered. When some of the sources are not sparse, it is unreasonable to use the sparsity assumption for the source estimation. Applications of the general prior distribution model and its learning mechanism will increase adaptivaty of the algorithm. In this section, the proposed MoG-uICA algorithm will be able to identify different types of sources from the mixures by using the Gaussian mixed model (GMM) for the source priori distribution. The simulation results show that the proposed algorithm will induce increased computation complexity while it accelerates the convergence and removes the necessity of sparsity for all sources, which means that MoG-uICA can recover sparse and non-sparse signals from the mixtures.

Although the proposed MoG-uICA has a good separation performance, it needs to establish some relatively simple assumption on the probability model of source signals, which means that for each source, the observation must be independentally identically distribution each time. For this purpose, we transform the object from the time-domain to the coefficient sequence based on linear transformations. This assumption on coefficients is more reasonable relative to the assumption of signals in the time-domain. Take the coefficient sequence as a new time series, do the underdetermined independent components, analysis, and estimate the transform coefficients and system parameters; then reconstruct the source signals.The usual linear transformation methods include: wavelet transform whose base function is real function, wavelet packet transform, discrete cosine transform (DCT), Hadamard transform, and Hartley transform, etc. For different types of signals, the coefficient sequence may have different probability distributions after linear transformations. Therefore, while we make full use of the superiority of linear transormation, we should also consider that the algorithm has enough flexibility to cope with differences with the aid of general probability modeling, and the algorithm is flexible enough to cope with the differences among the coefficient sequences. The role of GMM is to adapt different distribution, so that no matter what linear transformation is chosen, it can fully and accurately describe the parameter model. Although there is some effect on identification with different initial values for the same linear transformation, this is the intrinsic phenomenon that a local minimum exists. Generally, the application of GMM provides more freedom for the selection of linear transformation. It needs to be pointed out that the idea of sparse signal processing is usually used to process underdetermined model-based problems, and it needs to select a proper transformation method for sparsity. However, there are no strict requirements for sparseness, which is different from sparse signal processing.

For the underdetermined problem in practice, the convolutive model is more suitable for the actual situation, but the procedure for convolution underdetermined problem is more complex. For this problem, we put forward two kinds of methods. One is frequency-domain processing, which applies the nonstationarity of the signal and STFT, using MoG-uICA in the frequency-domain; but the problem to be solved is the permutation and amplitude distortion in the frequency-domain for the same estimated signals. The other method is a processing technique in the time-domain by using a Bayesian filter, which uses the time-domain structure information and state space model of the signal to realize the dual estimation of the source signals and system parameters. This model is suitable not only for a static model, i.e. instantaneous mixing model, but also for a dynamic model, i.e. the convolution mixing model. The Bayesian filter is a kind of non-linear non-Gaussian filter that can be used for parameter estimation (neural network training or system parameter identification), state estimation, and the dual estimation of parameter and state with determined structure information. There is no need to perform an inverse operation of the observation equation in the estimation of hidden state variables. In practice, there is no inverse operation in many cases, which is the basic reason that the Bayesian filter can be used for the underdetermined model problem. Each signal is constructed as a time-varying AR (TVAR) process and then goes into the channel model under the drive of white noise; finally, we get the observation signal, i.e. convolution mixed signal. This forms a two-layer dual estimation of the source signal model and the channel model. The forward processs goes from driven noise to observation signal, while the backward process goes from observation signal to TVAR model. The forward process transfers prediction, and the backward transfers information with each layer estimating the states and parameters of the state space model. Finally then, we obtain a solution to the problem.

9.4.1 Underdetermined Model

Basic model of blind source separation is

$$x = As \tag{9.4.1}$$

or

$$x = As + n \tag{9.4.2}$$

where $A \in R^{m \times n}$. Three main solution differences between the underdetermined model and the basic model include the identifiability of the mixing matrix, the separability of the source signal, and the uniqueness of the solution of underdetermined model.

For two expressions of $\boldsymbol{x}$, $(\boldsymbol{A}, \boldsymbol{s})$ and $(\boldsymbol{B}, \boldsymbol{r})$, Eq. (9.4.1) can be written as

$$x = As = Br \tag{9.4.3}$$

assume the components of $\boldsymbol{s}$ and $\boldsymbol{r}$ are statistically mutually independent and the two expressions are reduced, which means the column vectors of matrix $\boldsymbol{A}$ and $\boldsymbol{B}$ are pairwise independent respectively, so the system identification will be unique either way. In other words, the identifiability is unique, which means each column of $\boldsymbol{A}$ is linearly correlated with one column of $\boldsymbol{B}$ in two reduced expressions of $\boldsymbol{x}$, and *vice versa*. In order to introduce the identifiability of the system, namely the uniqueness of model, here we give the Theorem proved by Darmois[32].

Theorem 9.4 For the two expressions in Eq. (9.4.3), if column vector $\boldsymbol{a}_i$ of $\boldsymbol{A}$ is not linearly correlated with any column vector of $\boldsymbol{B}$, then source signal s_i is a Gaussian signal.

As seen from the Theorem, it is easy to get its converse-negative proposition; if all source signals are non-Gaussian, then the column vectors of the two matrices in both expressions are pairwise linearly correlated. Compared with the identification condition for the determined model with one Gaussian signal, the identification condition for the underdetermined model requires all source signals to be non-Gaussian.

Separability mainly aims at the estimation of the source signals. For a non-underdetermined model, the definition of separability is that there exists a demixing matrix $\boldsymbol{W}$ satisfying $\boldsymbol{\Lambda P s}=\boldsymbol{W x}$, where $\boldsymbol{P}$ and $\boldsymbol{\Lambda}$ are a permutation matrix and a diagonal matrix respectively. For the underdetermined model, only a non-linear algorithm can realize the estimation of the source signals. Therefore, the underdetermined model is only non-linearly separable.

The definition of uniqueness is that if the model is identifiable, $\boldsymbol{s}$ and $\boldsymbol{r}$ have the same distributions except for the indeterminacy of order and amplitude. For the underdetermined situation, we cannot solve with a known mixing matrix, so it is inseparable. But if we can ensure the distribution of the source signals, making the likelihood function of the mixtures unique, then in a statistical sense, we can estimate source signals according to the maximum likelihood method. That is to say, the uniqueness of the underdetermined model lies in the uniqueness of the likelihood function. This algorithm is the idea of an overcomplete ICA[32,33]. The detailed discussion of uniqueness is referred to Ref.[34], which gives upper and lower bounds of the number of observation signals for the given number of source signals. For instance, given the number of observation signals m, there can be at most $m(m+1)/2$ source signals. However, given the number of source signals n, m should satisfy $m(m-1)/2<n<m(m+1)/2$.

It needs to be pointed out that it is difficult to quantitatively describe identification performance for the underdetermined model. Given the number of sensors, it is worth studying whether there is an upper bound or not for the source number. Using higher-order statistics (HOS), Tong did a detailed analysis of MA parameters identification under the underdetermined model[35]; however, in Ref.[36], Common thoroughly analyzed HOS under the underdetermined model. Because of the particularity of the tensor itself, HOS-based tensor algebra can be applied in identifying the underdetermined model, but for the given number of sensors, there is an upper bound of the number of source signals[35,37]. Table 9.3 shows the upper bound applied to fourth-order statistics. Meanwhile, when the order of HOS increases, the upper bounds of the source signals increase accordingly.

Table 9.3 Upper bound of the number of source signals given the number of sensors

Number of sensors	2	3	4	5	6	7	8
Number of sources	3	6	10	15	22	30	42

Taleb used partial differential equation to successfully identify a mixing matrix from two mixtures which contain four noisy signals with noise sources[32]. However, in Table 9.3, the upper bound of the source number is only 3. Therefore, it can be seen that the upper bound of the source numbers change for the different identification algorithms, and there may not be a general upper bound.

9.4.2 Linear and Nonlinear Algorithm

The linear and nonlinear algorithm discussed here mainly aims at estimating the source signals after finding the estimation of the mixing matrix in the instantaneous linear mixing undertermined model, not to estimate the source signals and mixing matrix synchronously. The essential difference between linear or nonlinear mixtures lies in the operation of the observation being linear or nonlinear.

In Eq. (9.4.1), the probability of $\boldsymbol{x}$ is $p(\boldsymbol{x}|\boldsymbol{A},\boldsymbol{s})$, then the posterior probability of the source signal $\boldsymbol{s}$ is

$$p(\boldsymbol{s} \mid \boldsymbol{x},\boldsymbol{A}) = p(\boldsymbol{x} \mid \boldsymbol{s},\boldsymbol{A})p(\boldsymbol{s})/p(\boldsymbol{x}) \tag{9.4.4}$$

Therefore, we can estimate source signal $\boldsymbol{s}$ from the MAP algorithm:

$$\boldsymbol{s} = \arg\max_{\boldsymbol{x}=\boldsymbol{A}\boldsymbol{s}} p(\boldsymbol{s} \mid \boldsymbol{x},\boldsymbol{A}) = \arg\max_{\boldsymbol{x}=\boldsymbol{A}\boldsymbol{s}} p(\boldsymbol{x} \mid \boldsymbol{s},\boldsymbol{A})p(\boldsymbol{s}) \tag{9.4.5}$$

Given mixing matrix $\boldsymbol{A}$ and source signal $\boldsymbol{s}$, $p(\boldsymbol{x}|\boldsymbol{s},\boldsymbol{A})$ is trivial, so Eq. (9.4.5) can be changed into [38~42]

$$\boldsymbol{s} = \arg\max_{\boldsymbol{x}=\boldsymbol{A}\boldsymbol{s}} p(\boldsymbol{s}) \tag{9.4.6}$$

It can be seen that it is important to assume source signals during the estimation of the source signal. This assumption indicates preliminary cognition of the source signal and to supplement information of the system model in some sense.We discuss some characteristics with a different priori hypothesis on source signals below; here we suppose that the mean of all source signals should be zero.

1. Linear algorithm and underdetermined model

Assuming source signals are independent and have Gaussian distribution, Equation (9.4.6) can be written as

$$\boldsymbol{s} = \arg\max_{\boldsymbol{x}=\boldsymbol{A}\boldsymbol{s}} \exp\left(-\sum_i s_i^2\right) \tag{9.4.7}$$

In this case, $\boldsymbol{s}$ are sets which satisfy Eq. (9.4.1) and have minimum Euclidean norm.

$$\boldsymbol{s} = \arg\min_{\boldsymbol{x}=\boldsymbol{A}\boldsymbol{s}} \|\boldsymbol{s}\|_2 \tag{9.4.8}$$

where $\|\boldsymbol{s}\|_2$ indicates the l_2 norm of $\boldsymbol{s}$. To set the Moore-Penrose inverse of $\boldsymbol{A}$, $\boldsymbol{A}^+$, on the observation signal $\boldsymbol{x}$, we can get

$$s = A^{+} x = A^{H}(AA^{H})^{-1} x \tag{9.4.9}$$

But this method cannot get rid of the correlation among signals.

2. Non-linear algorithm and underdetermined model

Suppose source signals are Laplace, the Eq. (9.4.6) can be written as

$$s = \arg\max_{x=As} \exp(-\sum_{i} \| s_i \|) \tag{9.4.10}$$

In this case, s are sets which satisfy Eq. (9.4.1) and have a minimum l_1 norm.

$$s = \arg\min_{x=As} \| s \|_1 \tag{9.4.11}$$

where $\| s \|_1$ indicates the l_1 norm of s.

It can be seen, that in general we can express the distribution of source signals as

$$p(s) \propto \exp(\sum_{i} \| s_i \|_p) \tag{9.4.12}$$

thus Eq. (9.4.11) can be changed into

$$s = \arg\max_{x=As} \exp(-\sum_{i} \| s_i \|_p) \tag{9.4.13}$$

That is to say, source signals satisfy the solution with a minimum norm, the l_p norm of the basic model,

$$s = \arg\min_{x=As} \| s \|_p \tag{9.4.14}$$

where $p \geqslant 0$.

According to the above analysis, only if $p=2$, the recovery algorithm of the source signals is linear, while $p \in [0,1]$, we usually estimate the source signals with an optimization algorithm or approximate calculation method. In addition, when $p \in [0,1]$, it indicates the signals are sparse, while $p \in [0,2]$, it indicates the signals are not sparse. Besides a linear programming algorithm such as an interior point, there are other approximation methods, such as the MCMC[43,44], a mean field[45,46], or a variational algorithm[47,48] to estimate the source signals.

9.4.3 Performance of Algorithms and Evaluation Index

In simulation, because of the particularity of the underdetermined model, the intersymbol interference that is used to quantize the performance of the algorithm is not suitable, because we cannot find a separating matrix and mixing system for the underdetermined model so as to obtain an approximation of the generalized permutation matrix. Therefore, we have to define a new index to estimate the performance of the algorithm from the source signal and mixing system. We define some indexes as follows: One is the Algebraic Matrix Distance Index (AcMDI)[42] or Angle Matrix Distance Index (AeMDI)[49] used to evaluate the identification performance of the source signals. The other is to evaluate the estimation performance of the source signals[50].

1. Comparison of the real mixing matrix and its estimation

This is an index used to compare two matrices with the same dimensions. Column vectors of each matrix constitute a spatial direction set, but each direction in the set does not have a relation with the corresponding column vector sign, nor with the position of the norms in the matrix, and it is only correlated with the relative relation of the components in the vectors. AcMDI[42] is used to describe the difference between the two directional sets so as to characterize the distance between the two matrices and is insensitive to the column vector, sign the norm, and the order.

$$\mathrm{AcMDI}(\boldsymbol{A},\boldsymbol{B}) = 2 - \sum_{\mathrm{rows}} \max(\boldsymbol{A}^{\mathrm{H}}\boldsymbol{B})/n - \sum_{\mathrm{cols}} \max(\boldsymbol{A}^{\mathrm{H}}\boldsymbol{B})/n \tag{9.4.15}$$

where $\boldsymbol{A}$ and $\boldsymbol{B}$ are two matrices with the same dimension, and their column vectors are normalized.

For each column $\boldsymbol{a}_i$ of $\boldsymbol{A}$, we can find a column vector $\boldsymbol{b}_i$ in the group of the column vector of $\boldsymbol{B}$ that is most similar to $\boldsymbol{a}_i$ and this reflects the fact that there is an element $\boldsymbol{a}_i^{\mathrm{H}}\boldsymbol{b}_j$ with a maximum absolute value at the corresponding column j and row i of $\boldsymbol{A}^{\mathrm{H}}\boldsymbol{B}$, while $\sum_{\mathrm{rows}} \max(\boldsymbol{A}^{\mathrm{H}}\boldsymbol{B})$ can be regarded as a classification of the column vectors of $\boldsymbol{B}$ according to the column vectors of $\boldsymbol{A}$. Similarly, for each column of $\boldsymbol{B}$, there must be a similar column vector in the group of the column vector of $\boldsymbol{A}$, and $\sum_{\mathrm{cols}} \max(\boldsymbol{A}^{\mathrm{H}}\boldsymbol{B})$ can also be a classification of the column vectors of $\boldsymbol{A}$ according to the column vectors of $\boldsymbol{B}$. Both of them reflect the similarities between the column vectors of $\boldsymbol{A}$ and $\boldsymbol{B}$ from both directions. The larger the value of AcMDI($\boldsymbol{A}, \boldsymbol{B}$) is, the more difference there is between $\boldsymbol{A}$ and $\boldsymbol{B}$. When $\boldsymbol{A}=\boldsymbol{B}$; AcMDI($\boldsymbol{A}, \boldsymbol{B}$)=0, while the spaces spanned by the column vector of $\boldsymbol{A}$ and $\boldsymbol{B}$ are orthogonal, AcMDI($\boldsymbol{A}, \boldsymbol{B}$)=2 (Now, both $\boldsymbol{A}$ and $\boldsymbol{B}$ must be higher matrices).

This parameter has two properties, namely

$$\mathrm{AcMDI}(\boldsymbol{A},\boldsymbol{B}) = \mathrm{AcMDI}(\boldsymbol{B},\boldsymbol{A}) \tag{9.4.16}$$

$$0 \leqslant \mathrm{AcMDI}(\boldsymbol{A},\boldsymbol{B}) \leqslant 2 \tag{9.4.17}$$

if and only if $\boldsymbol{A}$ and $\boldsymbol{B}$ are orthogonal, AcMDI ($\boldsymbol{A}, \boldsymbol{B}$)=2.

Similarly, we can define another index—the Angle Matrix Distance Index (AeMDI):

$$\mathrm{AeMDI}(\boldsymbol{A},\boldsymbol{B}) = \sum_{(i,j)} \arccos(|\boldsymbol{a}_i^{\mathrm{T}}\boldsymbol{b}_j|) \tag{9.4.18}$$

where $\boldsymbol{A}$ and $\boldsymbol{B}$ have the same dimension and their column vectors are normalized, while (i,j) depends on the similarity between the estimation $\hat{s}_i$ and real source signal.

$$(i,j) = \arg\max_j (\hat{s}_i \otimes s_j / \| s_j \|) \tag{9.4.19}$$

2. Comparison of source signals and their estimation

As the inverse of the underdetermined model cannot be directly found, the estimation performance cannot be evaluated by the estimation of the mixing matrix only; this is much different from the other non-underdetermined model. Therefore, for the underdetermined model, we need three indices to evaluate the performance of the algorithm, which are the SIR (Source to Interference Ratio), the SNR (Source to Noise Ratio), and the SAR (Source to Artifacts Ratio).

For simplicity, the source signal is normalized, $\| s_i \|^2 = \langle s_i, s_i \rangle = 1$. Consider the indeterminacy of the amplitude, and let

$$\hat{s}_i = \alpha s_i + e \tag{9.4.20}$$

where e is the error term.

Define the deviation bias index D_{total}, in order to satisfy when $\alpha=0$, $D_{\text{total}}=\infty$, then we get

$$D_{\text{total}} = \frac{\| \hat{s}_i \|^2 - |\langle \hat{s}_i, s_i \rangle|^2}{|\langle \hat{s}_i, s_i \rangle|^2} \tag{9.4.21}$$

Thus, the estimation $\hat{s}_i$ of the source signal s_i can be divided into two parts; one is the projection along s_i, and the other is the total deviation e_{total}.

$$\hat{s}_i = \langle \hat{s}_i, s_i \rangle s_i + e_{\text{total}}$$

According to the triangular Theorem,

$$\| e_{\text{total}} \|^2 + |\langle \hat{s}_i, s_i \rangle|^2 = \| s_i \|^2$$

assume the source signals are mutually independent, and the additive noise and source signals are also mutually independent; then we can divide the estimation of the source signal $\hat{s}_i$ into four parts. The first one is the projection along s_i, the second one the estimation error e_{interf}, which is caused by other source signals. The third one is error e_{noise} which is caused by observation noise, while the last one is system error e_{artif} caused by the algorithm, which is unavoidable. The four parts above can be expressed as

$$\begin{aligned} \hat{s}_i &= \langle \hat{s}_i, s_i \rangle s_i + e_{\text{interf}} + e_{\text{noise}} + e_{\text{artif}} \\ e_{\text{interf}} &= P_s \hat{s}_i - \langle \hat{s}_i, s_i \rangle s_i \\ e_{\text{noise}} &= P_{s,n} \hat{s}_i - P_s \hat{s}_i \\ e_{\text{artif}} &= \hat{s}_i - P_{s,n} \hat{s}_i \end{aligned} \tag{9.4.22}$$

where

$$P_s \hat{s}_i = \sum_j \langle \hat{s}_i, s_j \rangle s_j$$

$$P_{s,n}\hat{s}_i = P_s\hat{s}_i + \sum_k \langle \hat{s}_i, n_k \rangle n_k / \| n_k \|^2$$

therefore, we can get four indexes: the Source to Distortion Ratio, the Source to Interference Ratio, the Source to Noise Ratio, and the Source to Artifacts Ratio.

$$\mathrm{SDR}(\hat{s}_i) = 10\lg\left(\left| \frac{|\langle \hat{s}_i, s_i \rangle|^2}{\| \hat{s}_i \|^2 - |\langle \hat{s}_i, s_i \rangle|^2} \right| \right) \tag{9.4.23}$$

$$\mathrm{SIR}(\hat{s}_i) = 10\lg \frac{|\langle \hat{s}_i, s_i \rangle|^2}{\| e_{\text{interf}} \|^2} \tag{9.4.24}$$

$$\mathrm{SNR}(\hat{s}_i) = 10\lg \frac{\| \langle \hat{s}_i, s_i \rangle s_i + e_{\text{interf}} \|^2}{\| e_{\text{noise}} \|^2} \tag{9.4.25}$$

$$\mathrm{SAR}(\hat{s}_i) = 10\lg \frac{\| \langle \hat{s}_i, s_i \rangle s_i + e_{\text{interf}} + e_{\text{noise}} \|^2}{\| e_{\text{artif}} \|^2} \tag{9.4.26}$$

As seen above, these four indices are used with known source signals. However, in the real system, they can not be used to quantitatively describe the performance of the algorithm. For simulation, the real source signals and the system are known, and we can directly use the above indices, but in practice, we need to use given training data that realizes channel equalization. It is worth pointing out that, Cichocki[51] put forward a completely different thought, which is to use the sub-band decomposition ICA to realize blind computation of the performance index in a normal model.

9.4.4 Blind Processing of Instantaneous Mixtures Underdetermined Model

1. Bayesian method of statistical model

Compared with other methods, more prior knowledge or a relatively reasonable hypothesis can be added into the estimation of source signals with the statistical reasoning method, while lack of information cannot be estimated by a mathematical method. According to the extra information, the solution can be found from an affine set.The combination of Bayesian reasoning and ICA can give more information about the mixing matrix, observation noise and source signal information or constraints[52] on the basis of an independent assumption of the source signals; thus it fully increases the flexibility of ICA. Although the Bayesian inference method provides a good solution for estimation and learning, it is difficult to realize the exact correct Bayesian reasoning. Thus, it needs some approximation algorithms, such as the Laplace approximation, MCMC approximation and variational approximation, etc[32,47,53~56]. The Laplace approximation method applies the Gaussian function locally, which makes the integral in the inference an approximate analytical expression, but it does not give good results with a small data set, and the computation cost of the Hessian matrix is quite large. The disadvantage of the MCMC approximation is that its convergence rate is very slow. The variational approximation applies a kind of factorized posterior distribution to approximate the real posterior distribution and estimates the parameters of factorized posterior distribution according to the minimization of the KL distance between the two distributions. The disadvantage of this approximation is that it neglects some correlation of the posterior distribution. Højen-Sørensen put forward correction method using a linear response to realize the estimation of the posterior distribution[57,58]. But the parameter of the prior distribution of the source signals is fixed with this method, and it does not provide a parameter learning mechanism to realize the adaptability for sources with a different distribution. That is to say, it needs to preset the parameter of the prior distribution according to experience.

In order to make a better description of the Bayesian ICA, the dependency relationship between the parameters in the whole model can be described by a hierachical-directed acyclic graph, namely the Bayesian network[59], as shown in Fig. 9.2. Each ellipse in the map can be recognized as sub-map, which includes a corresponding detailed assumption; specific details are given in the following.

$$\boldsymbol{x}^t = \boldsymbol{A}\boldsymbol{s}^t + \boldsymbol{n}^t \tag{9.4.27}$$

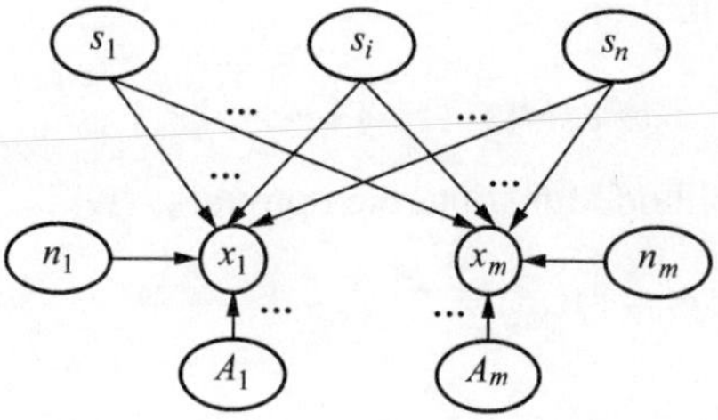

Fig. 9.2 Generative graph of the mixing system

suppose $\boldsymbol{s}^t=[s_1^t,s_2^t,\cdots,s_n^t]^{\mathrm{T}} \in \mathbf{R}^n$ is mutually independent, $\boldsymbol{x}^t=[x_1^t, x_2^t, \cdots, x_m^t]^{\mathrm{T}} \in \boldsymbol{R}^m$ is a weighted linear combination of the source signal under Gaussian white noise. $\boldsymbol{A} \in \mathbf{R}^{m\times n}$ is the mixing matrix, with the noise being a zero-mean isotropic Gaussian noise, and its covariance matrix is diagonalizable. Here we apply the inverse matrix $\boldsymbol{\Lambda}$ of the covariance matrix.

$$p(\boldsymbol{n}^t) = N(\boldsymbol{n}^t \mid 0, \boldsymbol{\Lambda}; \boldsymbol{M}) \tag{9.4.28}$$

$\boldsymbol{M}$ includes assumptions, such as noise, a mixing matrix, source signals, etc. In other words, $\boldsymbol{M}$ contains some specific details of sub-maps in the Bayesian networks.

For the mixtures vector $\boldsymbol{x}^t$, the likelihood function is

$$p(\boldsymbol{x}^t \mid \boldsymbol{A}, \boldsymbol{s}^t, \boldsymbol{\Lambda}) = \left(\det\left(\frac{\boldsymbol{\Lambda}}{2\pi}\right)\right)^{\frac{1}{2}} e^{-\frac{1}{2}(\boldsymbol{x}^t - \boldsymbol{A}\boldsymbol{s}^t)^{\mathrm{T}}\boldsymbol{\Lambda}(\boldsymbol{x}^t - \boldsymbol{A}\boldsymbol{s}^t)} \tag{9.4.29}$$

therefore, the posterior probability distribution of the signal can be expressed as

$$p(\boldsymbol{s}^t \mid \boldsymbol{x}^t, \boldsymbol{A}, \boldsymbol{\Lambda}) = \frac{p(\boldsymbol{x}^t \mid \boldsymbol{s}^t, \boldsymbol{A}, \boldsymbol{\Lambda}) p(\boldsymbol{s}^t)}{p(\boldsymbol{x}^t \mid \boldsymbol{A}, \boldsymbol{\Lambda})} \tag{9.4.30}$$

In addition, the integration with respect to $\boldsymbol{s}^t$ from Eq. (9.4.29) leads to

$$p(\boldsymbol{x}^t \mid \boldsymbol{A}, \boldsymbol{\Lambda}) = \int \mathrm{d}\boldsymbol{s}^t p(\boldsymbol{x}^t \mid \boldsymbol{s}^t, \boldsymbol{A}, \boldsymbol{\Lambda}) p(\boldsymbol{s}^t) \tag{9.4.31}$$

thus, differentiate Eq. (9.4.31) and let the derivative be zero; then we can obtain an estimation of $\boldsymbol{\Lambda}$ and $\boldsymbol{A}$.

$$\widetilde{\boldsymbol{A}} = \boldsymbol{x}^t \langle \boldsymbol{s}^t \rangle \langle \boldsymbol{s}^t \boldsymbol{s}^{t\mathrm{T}} \rangle^{-1} \tag{9.4.32}$$

$$\frac{1}{\widetilde{\boldsymbol{\Lambda}}} = \langle (\boldsymbol{x}^t - \boldsymbol{A}\boldsymbol{s}^t)(\boldsymbol{x}^t - \boldsymbol{A}\boldsymbol{s}^t)^{\mathrm{T}} \rangle \tag{9.4.33}$$

where $\langle \cdot \rangle$ indicates $\langle \cdot \rangle_{s|x,A,\Lambda}$, and both expressions of their estimation are functions of $\langle s \rangle$ and $\langle ss^{\mathrm{T}} \rangle$. It can be seen that the method for estimation of Λ and $\boldsymbol{A}$ is not optimal since it may be a local optimum of the objective function or may lie in a saddle point.

Because we cannot use explicit integration to exactly express the posterior distribution, we have to adopt approximation results[32,53~56]. Here, we adopt variational approximation, which uses a factorized expression to approximate the true posterior distribution:

$$Q(\boldsymbol{s}^t) = \prod_{i=1}^{n} Q(s_i^t) \tag{9.4.34}$$

For a Gaussian likelihood function, the optimal $Q(s_i^t)$ is [60]

$$Q(s_i^t) = \frac{p(s_i^t) \mathrm{e}^{-1/2\lambda_i^t s_i^{t2} + \gamma_i^t s_i^t}}{C} \tag{9.4.35}$$

where $C = \int p(s_i^t) \mathrm{e}^{-1/2\lambda_i^t s_i^{t2} + \gamma_i^t s_i^t} \mathrm{d}s_i^t$, $\lambda_i^t, \gamma_i^t \in \mathbb{R}$. The Kullback-Leibler distance between $Q(\boldsymbol{s}^t)$ and Eq. (9.4.30) is

$$\mathrm{KL} = \int \mathrm{d}\boldsymbol{s}^t Q(\boldsymbol{s}^t) \lg \frac{Q(\boldsymbol{s}^t)}{p(\boldsymbol{s}^t \mid \boldsymbol{x}, \boldsymbol{A}, \boldsymbol{\Lambda})} \tag{9.4.36}$$

define

$$\boldsymbol{\gamma}^t = \boldsymbol{h}^t - (\boldsymbol{J} - \mathrm{diag}(\boldsymbol{J}))\langle \boldsymbol{s}^t \rangle \tag{9.4.37}$$

$$\lambda_i^t = \boldsymbol{J}_{ii} \tag{9.4.38}$$

$$\boldsymbol{J} = \boldsymbol{A}^{\mathrm{T}}\boldsymbol{\Lambda}\boldsymbol{A}$$

$$\boldsymbol{h} = \boldsymbol{A}^{\mathrm{T}}\boldsymbol{\Lambda}\boldsymbol{x}$$

where diag ($\boldsymbol{J}$) indicates the diagonal matrix of the diagonal elements of $\boldsymbol{J}$.

Therefore, the posterior estimation of the source signals can be expressed as

$$\langle s_i^t \rangle = \frac{\partial \lg \int \mathrm{d}s_i{}^t p(s_i{}^t)\,\mathrm{e}^{-1/2\lambda_i^t t s_i^{t2} + \gamma_i^t s_i^t}}{\partial \gamma_i{}^t} \tag{9.4.39}$$

where $\frac{\partial(\cdot)}{\partial \gamma_i^t}$ indicates a partial derivative with respect to γ_i^t.

Consider the posterior correlation problem; to improve the effect of the posterior factorized approximation on the correlation of this model, it needs to modify the variational algorithm. The posterior correlation can be explained by the Markove Blanket (MB). The MB of a node is the parent node of the current node, sub-nodes and other parent nodes of the nodes shown in Fig. 9.3. Each node has a tight link with its MB, while the other links can be neglected. In Fig. 9.3, s_i and s_j have the same node x, given x; no matter whether s_i and s_j are independent or not, they are a posterior correlation.

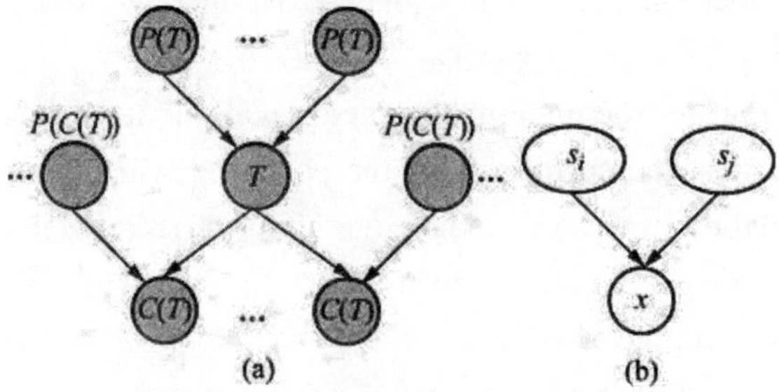

Fig. 9.3
(a) Markove Blanket of T;
(b) posterior probability correlation of the variable in underdetermined model

Generally speaking, whether there exists correlation or not in the posterior distribution depends on the mixing matrix. If you assume prior independence of the source signals, the posterior independence of the source signals depends on Eq. (9.4.29),

$$p(\boldsymbol{x}_t \mid \boldsymbol{A}, \boldsymbol{s}_t, \boldsymbol{\Lambda}) = \left(\det\left(\frac{\boldsymbol{\Lambda}}{2\pi}\right)\right)^{1/2} \mathrm{e}^{-1/2(\boldsymbol{x}_t - \boldsymbol{A}\boldsymbol{s}_t)^{\mathrm{T}}\boldsymbol{\Lambda}(\boldsymbol{x}_t - \boldsymbol{A}\boldsymbol{s}_t)}$$

$$= \left(\det\left(\frac{\boldsymbol{\Lambda}}{2\pi}\right)\right)^{1/2} \mathrm{e}^{-1/2(\boldsymbol{x}_t^{\mathrm{T}}\boldsymbol{\Lambda}\boldsymbol{x}_t - 2\boldsymbol{x}_t^{\mathrm{T}}\boldsymbol{\Lambda}\boldsymbol{A}\boldsymbol{s}^t + \boldsymbol{s}_t^{\mathrm{T}}\boldsymbol{A}^{\mathrm{T}}\boldsymbol{\Lambda}\boldsymbol{A}\boldsymbol{s}^t)}$$

if $\boldsymbol{A}^T\boldsymbol{\Lambda}\boldsymbol{A}$ is a diagonal matrix, the source signals have independent posterior distribution. Generally, it is difficult to satisfy, so a posterior correlation exists, but is not inevitable. However, for an underdetermined model, the existence of the correlation is inevitable.

Because of the positive definiteness of $\boldsymbol{\Lambda}$, non-diagonal matrix $\boldsymbol{\Lambda}$ can be decomposed as $\Lambda = U^T U$ with a Cholesky decomposition.

$$A^T \Lambda A = A^T U^T U A = A'^T A'$$

Therefore, we only consider the simple situation of $\Lambda = I_{m\times m}$.

The mixing matrix $A \in R^{m\times n}$, $m<n$, can be written as $[a_1, a_2, \cdots, a_n]$, $a_i \neq 0$, $i=1,2,\cdots n$; then

$$A^T A = \begin{bmatrix} a_1^T \\ a_2^T \\ \vdots \\ a_n^T \end{bmatrix} [a_1, a_2, \cdots, a_n]$$

Assume $\text{Rank}(A)=m$, $a_i \neq 0$, $i=1,2,\cdots,n$, then there must exist a sub-set $L=\{l_i, l_i \in T, i=1,2,\cdots,m\}$ of $T=\{1,2,\cdots,n\}$, which satisfies

$$\text{Rank}(A_l) = \text{Rank}([a_{l_1}, a_{l_2}, \cdots, a_{l_m}]) = m$$

For a_k, $k \in T/L$, if $A_l^T a_k = 0_{m\times 1}$ then $a_k = 0$, which will contradicte $a_i \neq 0$, $i=1,2,\cdots,n$; therefore, $A_l^T a_k \neq 0_{m\times 1}$, most exist; that is

$$A^T A - \text{diag}(A^T A) \neq 0_{n\times n}$$

This shows that for the underdetermined model, under the assumption that the row is full rank, the correlation of the posterior distribution of the source signals is inevitable. In order to reduce the effect of factorized approximation, the linear response provides a solution that can estimate the posterior correlation of the source signals.

$$\langle s_i^t s_j^t \rangle = \Delta_{ij}^t + \langle s_i^t \rangle \langle s_j^t \rangle \tag{9.4.40}$$

$$\Delta^t = (\Pi_t + J)^{-1} \tag{9.4.41}$$

where

$$\Pi_t = \text{diag}(\Pi_{1t}, \Pi_{2t}, \cdots, \Pi_{nt})$$

$$\Pi_{it} = \left(\frac{\partial \langle s_i^t \rangle}{\partial \gamma_i^t}\right)^{-1} - J_{ii}$$

2. Gaussian mixtures model based Bayesian analysis

In order to find a satisfactory solution from the affine spaces of a solution set, or optimal solution under a preassigned constraint, blind separation for the underdetermined model, in addition to the application of independence, also assumes sparsity of signals[31,53,61~63]. Sparsity assumption means amplitude of the signals is very small or approaching zero and are subjected to Laplace distribution at most times, which is the embodiment of sparsity. If it is processed by a wavelet transform or time-frequency expression, then the coefficients are sparse. Although identification of the underdetermined model requires all source signals are non-Gaussian[34], that does not mean that only sparse signals can be recovered in an underdetermined model. When non-sparse signals exist in source signals, using the sparsity assumption to recover source signals is not reasonable. If all source signals are non-sparse signals, we can use the same prior distribution. If the source signal includes sparse and non-sparse signals, that is the mixture of sparse and non-sparse signals. For example, in the mixture of voice and music, voice is sparse but music is sub-Gaussian, which can be regarded as a non-sparse signal. Sparsity has a close relationship to Gaussian; therefore, prior distribution should be describable, which means it can describe sparsity and non-sparsity with different parameters. Application of the prior distribution model with descriptive ability and a learning mechanism of parameters can, greatly increase the adaptability of the algorithm.

Compared with the normal ICA, the time-frequency analysis has some special advantages. It makes full use of the time-domain and frequency-domain information and can be used in some situations where the independence of source signals is not satisfied. But it needs a disjoint orthogonal hypothesis, which can be considered as a sparsity hypothesis. In most situations, the orthogonal hypothesis cannot be satisfied, which means the signals are not sparse in the time-frequency domain. In order to solve this problem, Luo[64,65] presented a self-splitting competitive learning clustering algorithm, which means that the time-frequency plane is split into several areas, and the number of source signals is no larger than the number of sensors in each local area. The signal recovery in underdetermined model is transformed into a separation of blind signals in a determined or over-determined situation. Sang-Gyun Kim[66] presented a method that generates a hidden observation series related to the observation signals, which make the number of observation signals equal to the number of source signals, so the underdetermined model is transformed into the determined model; and then the source signals can be recovered by the separation algorithm in a determined situation. The common point of the above two methods is to transform the underdetermined model into determined model, while using existing separation methods to recover the non-sparse source signals. We discuss how to use a mean field ICA to recover non-sparse source signals without a model transformation below.

As discussed above, the mean field ICA needs to use a concrete form of prior distribution $p(s_i^t)$ during the computation of the posterior estimation Eq. (9.4.39); therefore, it is better to use a parameteric model rather than a non-parameteric model. The common parameteric models are the Gaussian mixtures, model

(GMM)[56] and the generalized Gaussian model (GGM), etc. For the generalized Gaussian model,

$$p(s)=\frac{\gamma}{2\sigma\Gamma(1/\gamma)}\exp\left(-\frac{1}{\gamma}\left|\frac{s-\mu}{\sigma}\right|^{\gamma}\right) \tag{9.4.42}$$

where $\Gamma(\gamma)=\int_0^{\infty} x^{\gamma-1}\mathrm{e}^{-x}\mathrm{d}x$, $\gamma>0$, μ is the center of distribution, σ is the standard deviation. As γ increases from zero, the sparsity decreases. When $\gamma=1$, it indicates Laplace distribution. When $\gamma=2$, it indicates Gaussian distribution. And when $\gamma=\propto$, it indicates uniform distribution. For the Gaussian mixtures model

$$p(s^t)=\sum_{q=1}^{k}p(q^t=q\mid\boldsymbol{\pi})p(s^t\mid q^t,\mu_q,\beta_q)=\sum_{q=1}^{k}\pi_q N(s^t\mid\mu_q,\beta_q) \tag{9.4.43}$$

where k indicates the number of Gaussian components in the mixing model, variable q $(1\leqslant q\leqslant k)$ indicates the component label of the observation signals in the mixing model, while π_q indicates probability of the observation signals generated by each Gaussian component, namely the mixing coefficient; μ_q and β_q are the mean and variance inverse of the qth component respectively.

Compared with GMM, GGM has fewer parameters, which reduces the computation cost of the parameters learning, but GGM cannot give an explicit integral expression of Eq. (9.4.39); it can only describe single mode distribution. GGM can describe multi-mode distribution, and the composing of mixture' s model may include other distribution types, such as t distribution, Laplace distribution, etc. Therefore, it is more suitable for the prior distribution of source signals. Mark Girolami put forward a variational method that can use the Gaussian function to describe the non-Gaussian distribution, which is similar to GMM; but it is restricted to Gaussian and super-Gaussian[67] situations.

Theoretically, GMM can describe any distribution with enough Gaussian functions. Figure 9.4 illustrates the generating graph of GMM. Here, we use a one-dimension vector to describe the learning process of parameters in the GMM. Suppose the prior distribution of $\boldsymbol{\mu}$, $\boldsymbol{\beta}$, and π[68~70]

$$p(\boldsymbol{\pi})=\mathrm{Dirichlet}(\boldsymbol{\pi}\mid\boldsymbol{\lambda})\qquad\boldsymbol{\lambda}=[\lambda_1,\lambda_2,\cdots,\lambda_k]\in\boldsymbol{R}^k \tag{9.4.44}$$

$$p(\boldsymbol{\mu})=\prod_{q=1}^{k}N(\mu_q\mid m,\tau) \tag{9.4.45}$$

$$p(\boldsymbol{\beta})=\prod_{q=1}^{k}\mathrm{Gamma}(\beta_q\mid b,c) \tag{9.4.46}$$

where m, τ, b, c, and λ are super parameters of $\boldsymbol{\mu}$, $\boldsymbol{\beta}$, and π.

We can apply the variational Bayes-VB to estimate the parameters of GMM, which comes from the mean theory of statistical dynamics. Its advantage is that it can use an analytical expression to approximate the obje-ctive function.

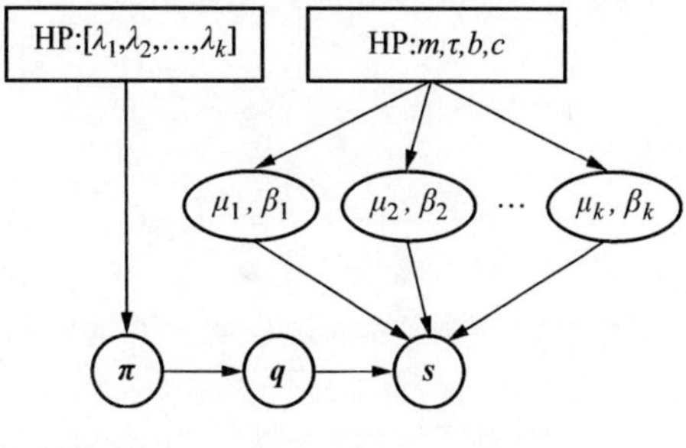

Fig. 9.4 Generating graph of GMM

In order to use GMM to fit the distribution of $\{s^t\}$, suppose $\boldsymbol{W}$ indicates the set of hidden-layer variables and unknown parameters $\boldsymbol{W}=\{\boldsymbol{q}, \pi, \boldsymbol{\mu}, \boldsymbol{\beta}\}, \boldsymbol{s} \in \mathbf{R}^{1\times N}$, and $p'(\boldsymbol{W})$ is the approximation of $p(\boldsymbol{W}|\boldsymbol{s})$,

$$p'(\boldsymbol{W})=p'(\boldsymbol{q})p'(\boldsymbol{\pi})p'(\boldsymbol{\mu})p'(\boldsymbol{\beta})$$

$$p(\boldsymbol{s},\boldsymbol{W})=p(\boldsymbol{s} \mid \boldsymbol{q},\boldsymbol{\mu},\boldsymbol{\beta})p(\boldsymbol{q} \mid \boldsymbol{\pi})p(\boldsymbol{\pi})p(\boldsymbol{\mu})p(\boldsymbol{\beta})$$

$$\lg p(\boldsymbol{s})=\lg \frac{p(\boldsymbol{s},\boldsymbol{W})}{p(\boldsymbol{W} \mid \boldsymbol{s})}$$

$$\begin{aligned}\lg p(\boldsymbol{s}) &= \int \mathrm{d}\boldsymbol{W} p'(\boldsymbol{W}) \lg \frac{p'(\boldsymbol{W})p(\boldsymbol{s},\boldsymbol{W})}{p'(\boldsymbol{W})p(\boldsymbol{W} \mid \boldsymbol{s})} \\ &= \int \mathrm{d}\boldsymbol{W} p'(\boldsymbol{W}) \lg \frac{p'(\boldsymbol{W})}{p(\boldsymbol{W} \mid \boldsymbol{s})} + \\ &\quad \int \mathrm{d}\boldsymbol{W} p'(\boldsymbol{W}) \lg \frac{p(\boldsymbol{s},\boldsymbol{W})}{p'(\boldsymbol{W})} \\ &= F(\boldsymbol{s}) + \mathrm{KL}(p'(\boldsymbol{W}) \mid p(\boldsymbol{W} \mid \boldsymbol{s}))\end{aligned} \tag{9.4.47}$$

$$\begin{aligned}F(\boldsymbol{s}) &= \int \mathrm{d}\boldsymbol{W} p'(\boldsymbol{W}) \lg p(\boldsymbol{s},\boldsymbol{W}) - \\ &\quad \int \mathrm{d}\boldsymbol{W} p'(\boldsymbol{W}) \lg p'(\boldsymbol{W}) \\ &= \langle \lg p(\boldsymbol{s},\boldsymbol{W}) \rangle_{p'(\boldsymbol{W})} + H(\boldsymbol{W})\end{aligned} \tag{9.4.48}$$

$$\mathrm{KL}(p'(\boldsymbol{W}) \mid p(\boldsymbol{W} \mid \boldsymbol{s})) = \int \mathrm{d}\boldsymbol{W} p'(\boldsymbol{W}) \lg \left(\frac{p'(\boldsymbol{W})}{p(\boldsymbol{W} \mid \boldsymbol{s})}\right) \geqslant 0$$

It can be seen from the above

$$\lg p(\boldsymbol{s}) \geqslant F(\boldsymbol{s})$$

if and only if $p'(\boldsymbol{W})=p(\boldsymbol{W}|\boldsymbol{s})$, $\lg p(\boldsymbol{s})=F(\boldsymbol{s})$. Notice $\lg p(\boldsymbol{s})$ is a constant of the unknown set $\boldsymbol{W}$; therefore, the factorized form is properly selected $p'(\boldsymbol{W})$ to maximize $F(\boldsymbol{s})$ while minimizing the KL distance between $p'(\boldsymbol{W})$ and $p(\boldsymbol{W}|\boldsymbol{s})$. According to $F(\boldsymbol{s})$, we can also compare the explanatory capability with different models and different parameters. The larger $F(\boldsymbol{s})$ is, the more accurate the explanation is under condition of the current model and parameters. If

$p'(\mathbf{W})$ is a factorized form, namely

$$p'(\mathbf{W}) = \prod_i p'(w_i)$$

then

$$F(\boldsymbol{s}) = \langle \lg p(\boldsymbol{s},\mathbf{W}) \rangle_{\prod_i p'(w_i)} + \sum_i H(w_i)$$

Optimal $p'(w_i)$ can be computed by $\dfrac{\partial F(\boldsymbol{s})}{\partial p'(w_i)} = 0$:

$$p'(w_i) = \frac{1}{Z_i} \exp(\langle \lg p(\boldsymbol{s},\mathbf{W}) \rangle_{\prod_{j\neq i} p'(w_j)})$$

where Z_i is the normalized constant of $p'(w_i)$.

After obtaining parameterized integral $p(s_i^t)$, we can easily obtain the expression $\langle s_i^t \rangle$, with GMM prior distribution according to Eq. (9.4.39):

$$\langle s_i^t \rangle = \frac{\sum_{q_i=1}^{k} \left(\pi_{i,q_i} \dfrac{\gamma_i^t \beta_{i,q_i}^{-1} + \mu_{i,q_i}}{(\lambda_i^t \beta_{i,q_i}^{-1} + 1)^{\frac{3}{2}}} \mathrm{e}^{\zeta_{q_i}} \right)}{\sum_{q_i=1}^{k} \left(\dfrac{\pi_{i,q_i}}{\sqrt{\lambda_i^t \beta_{i,q_i}^{-1} + 1}} \mathrm{e}^{\zeta_{q_i}} \right)} = \frac{\sum_{q_i=1}^{k} \left(\pi_{i,q_i} \dfrac{\gamma_i^t \beta_{i,q_i}^{-1} + \mu_{i,q_i}}{(\lambda_i^t \beta_{i,q_i}^{-1} + 1)^{\frac{3}{2}}} \mathrm{e}^{\bar{\zeta}_{q_i}} \right)}{\sum_{q_i=1}^{k} \left(\dfrac{\pi_{i,q_i}}{\sqrt{\lambda_i^t \beta_{i,q_i}^{-1} + 1}} \mathrm{e}^{\bar{\zeta}_{q_i}} \right)} \tag{9.4.49}$$

$$\zeta_{q_i} = -\frac{1}{2} \left(\mu_{i,q_i}^2 \beta_{i,q_i}^{-1} - \frac{(\gamma_i^t + \mu_{i,q_i} \beta_{i,q_i})}{\lambda_i^t + \beta_{i,q_i}} \right)$$

in practice, $\bar{\zeta}_{q_i} = \zeta_{q_i} - \max\limits_{q_i=1:k}(\zeta_{q_i})$ is more convenient to realize than ζ_{q_i}.

The implementaton steps of the algorithm are listed in Table 9.4.

Table 9.4 Steps of the algorithm

Step 1	Initialize noise, mixing matrix and source signals.
Step 2	Initialize parameters of the source model.
Step 3	Estimate $\langle s_i^t \rangle$ and posterior correlation $\langle s_i^t s_j^t \rangle$.
Step 4	Use posterior correlation to update the estimation of $\boldsymbol{A}$ and $\boldsymbol{\Lambda}$.
Step 5	Update the source estimation.
Step 6	Update the source model.
Step 7	Return to ③ until the end condition is satisfied.

In order to verify the algorithm, we make simulations for two purposes. One tests the performance of separation and identification; the other tests adaptation, which means the performance of recovering non-sparse signals from source signals. The two simulations apply the system model with a three source signals input and two sensor signals output. The initial value of the source signals can be found by the generalized inverse of the initial mixing matrix or random time series. Some parameters of the top layer of the model also need to be initialized, such as $\lambda_1, \lambda_2, \cdots, \lambda_k, m, \tau, b$ and c, but the parameters of the top layer are insensitive to the estimation of source signals. Set b=1,000, c=0.001, m=0, τ=1,000 and $\lambda_1=\lambda_2=\cdots=\lambda_k=5$.

Identification of the mixing matrix can be characterized by the difference and changes of the estimated value and real value. $\{\theta_i = \angle(\mathbf{A}(:,i), \tilde{\mathbf{A}}(:,i)), i=1,2,3\}$ indicates angles of a corresponding column vector of the mixing matrix between the estimated value and real value. Moreover, the sum of the absolute values of the difference between column vector pairs is defined as the whole performance index of the mixing matrix identification.

$$\theta_{\text{total}} = \sum_i \theta_i$$

in an ideal condition, $\{\theta_i\}$ and θ_{total} will be zero. Figure 9.5 shows waveforms and the amplitude histogram of three source signals. Figure 9.6 shows the parameterized density estimation of the three source signals. Figure 9.7 is the estimation of source signals with MoG-uICA, which coincide basically with Fig. 9.5 in its shape. MoG-uICA iterates approximately 25 times (approximately 500 seconds more or less, depending on hardware), and has good stability.

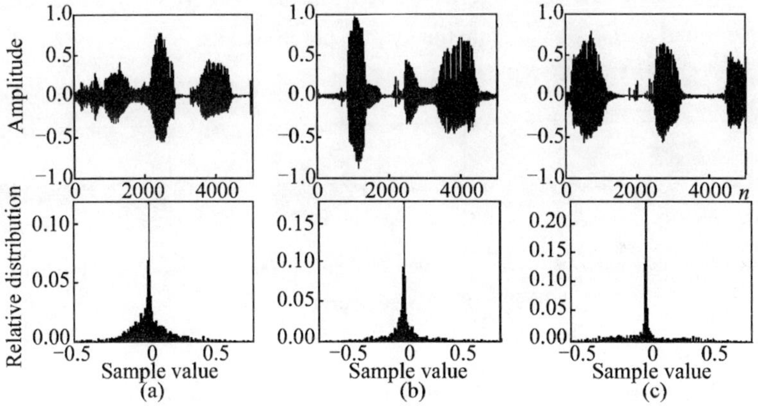

Fig. 9.5 Waveforms of source signals and their amplitude histogram

Besides good identification, this gives good results for the estimation of the source signals. Using the GMM as the prior distribution of the source signals can increase the identification stability, get rid of the sparsity assumption of source signals and restriction of the same distribution as well as for all source signals. In simulation, the source signal includes two segments of voice signals and one a music

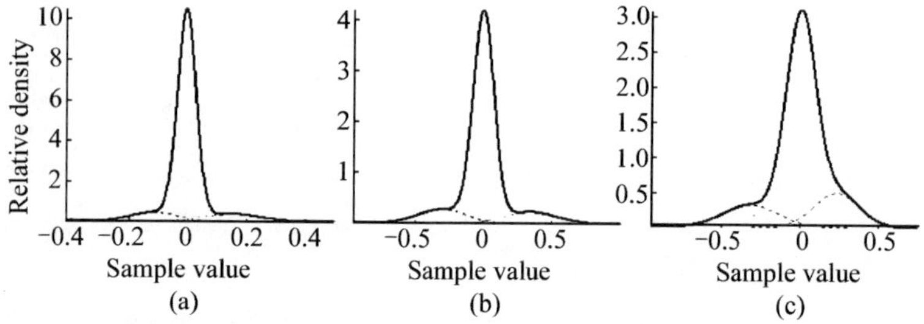

Fig. 9.6 Parameterized density estimation of source signals with MoG-uICA

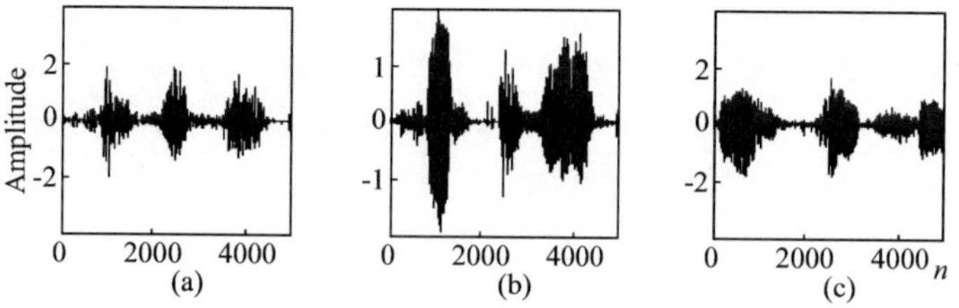

Fig. 9.7 Estimation results of source signals with MoG-uICA

signal, as shown in Fig. 9.8. The three kurtosis values of signals are 10.0630, 4.8352 and −2.8992 respectively. Two parts of the voice signals can be regarded as the sparse super-Gaussian signals of the same type, while the music signal is considered a non-sparse sub-Gaussian signal, the signal-to-noise ratio is 30 dB. Figure 9.9 shows the estimation result of the source signals, and it can be seen. MoG-uICA can recover non-sparse and sparse signals from the mixed source signals.

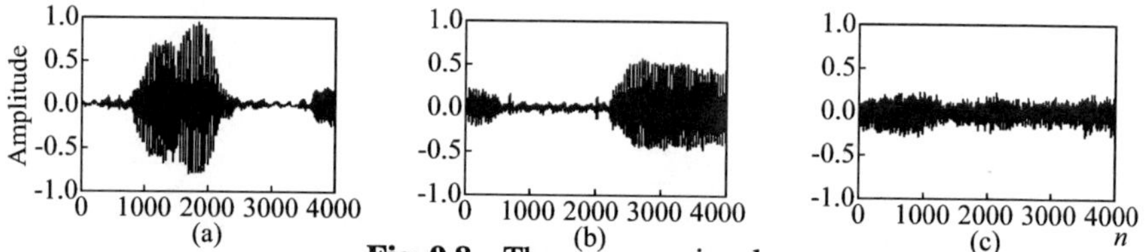

Fig. 9.8 Three source signals

(a, b) voice signals (super-Gaussian, sparse); c) music (sub-Gaussian, non-sparse)

Further research shows that in order to reduce the effect of the hypothesis of the probability model with a practical processing result, we can apply two methods to reduce the difference between a real situation and an hypothesis. One is to use linear transformation, which means a transformation coefficients sequence is processed by underdetermined analysis as a new time series to complete the estimation of the transformation coefficients of the system parameters and source signals. Then source signal is reconstructed. The other is

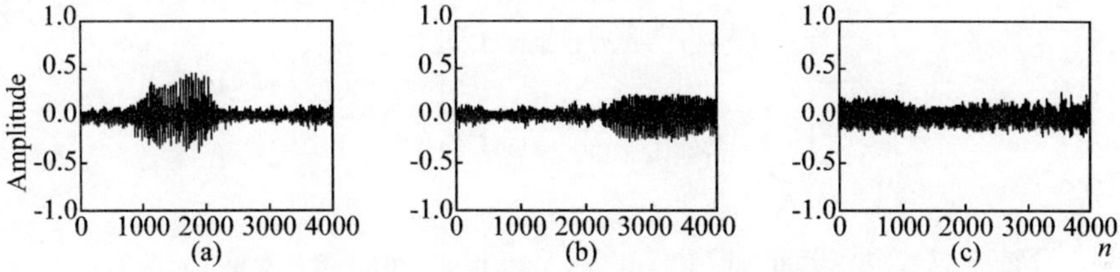

Fig. 9.9 Estimation results of source signals with MoG-uICA

to apply the Markov model to describe the signal[71~73].

Generally, the linear transformation model is

$$c_k = \sum_{n=1}^{N} x[n]\psi_k[n] \tag{9.4.50}$$

where $\psi_k(n)$ is the base function. For a set selected and complete base $\{\psi_k(n)\}$, we can do linear transformation to observation signals $x(n)$. For different kinds of signals, the coefficient's sequences have different probability distributions for different kinds of linear transformation. But, no matter which transformation is adopted, we can obtain the parametrization description of the probability distribution using a general probability model such as the GGM or GMM.

Take wavelet transform for example. Construction of the coefficients vector is shown in Fig. 9.10.

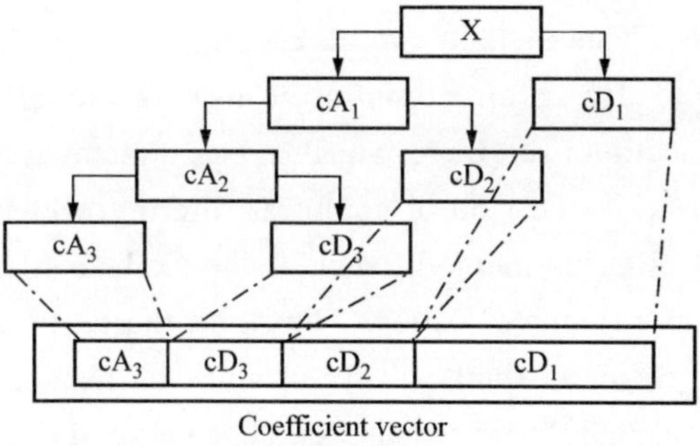

Fig. 9.10 Construction of wavelet transform coefficients vector

The basic flowchart of the whole algorithm is shown in Fig. 9.11.

9.4.5 Underdetermined Model and Bayesian Filter in Convolutive Mixtures

Most documents apply frequency-domain processing on convolutive problems, which transforms the dynamic processes of convolution into an instantaneous mixing model. For an estimation of the convolutive underdetermined model, it almost is instantaneous under determined situation.

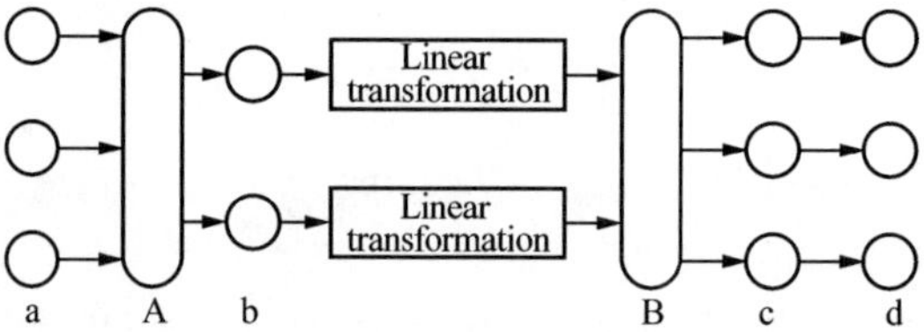

Fig. 9.11 Flowchart of MoG-uICA algorithm with linear transformation
a. Source signal in time domain; b. Observed signal in time domain;
c. Estimated value of transform coefficients of source signals;
d. Reconstructed source signals from the estimated value of coefficients;
A. Mixing system; B. Underdetermined analysis for transform coefficients of observed signals

model, almost always an instantaneous underdetermined situation.

We presented two methods to solve blind processing problems under the convolutive model in this section. One is processing in the frequency-domain, which considers nonstationarity and applies a short time Fourier transform (STFT) and uses the MoG-uICA algorithm in the frequency-domain. The problem to be solved with this method is the permutation and amplitude distortion in the frequency-domain for the same estimated signal. The other is the processing in the time-domain, which applies the time-domain structure and state space model of signals to realize estimation of the source signals and system parameters by using the Bayesian filter-based dual estimation method. This method can be used not only in the instantaneous mixing model but also in the convolutive mixing model. Therefore, the emphasis of this section is on Bayesian filter-based methods.

Bayesian filtering is one kind of nonlinear filtering, which unifies several kinds of adaptive filtering methods, such as the Kalman filter, the extended Kalman filter, the unscented Kalman filter and the particle filter into the framework of the Bayesian filter[74~79]. The significant feature is that it can be used in non-Gaussian non-linear signals, and depends on the application of the Monte Carlo (MCMC) method. Meanwhile, it can predict, filter and smooth according to practical needs. For Bayesian filtering, three points need to be mentioned. One is that it does not need to do any inverse operation about the observation equation during the estimation of hidden variables; this is the main basis for applying Bayesian filtering to the underdetermined model. The second is that the Bayesian filter is a framework of an algorithm and has many selections of realized forms according to practical models. The third is that practical processing methods include recursive on-line algorithms and batch processing algorithms; batch processing algorithms use future and current data to smooth out the current estimation.

Bayesian filter can be described by a state space equation if signals are expressed by a state space model and it is convenient to estimate the source signals. When the mixing system or channel is time-varying, we can use similar methods to form a complex system with hierarchical state equations. The input is observation signals and the system updates the estimated parameters of each layer by the top-down mode.

For a general system, the state space model with additive noise is

$$\begin{aligned} x_{k+1} &= \boldsymbol{F}(x_k, u_k, w) + v_k \\ y_k &= \boldsymbol{H}(x_k, w) + \boldsymbol{n}_k \end{aligned} \tag{9.4.51}$$

With non-additive noise

$$\begin{aligned} x_{k+1} &= \boldsymbol{F}(x_k, u_k, w, v_k) \\ y_k &= \boldsymbol{H}(x_k, w, \boldsymbol{n}_k) \end{aligned} \tag{9.4.52}$$

where $\boldsymbol{x}_k$ is state, $\boldsymbol{w}$ is the parameters of the dynamic system, $\boldsymbol{u}_k$ is the observable external input, $\boldsymbol{y}_k$ is output with noise, $\boldsymbol{v}_k$ and $\boldsymbol{n}_k$ are the process noise and observation noise of the dynamic system respectively. $\boldsymbol{F}(\cdot)$ and $\boldsymbol{H}(\cdot)$ are the state function and observation function of the state space model, and both can be linear or nonlinear.

If the model parameters are known, given the observation sequence and interior state sequence of the estimation model, $\boldsymbol{x}_k$ is actually the conditional mean value of the observation sequence $\boldsymbol{y}_k$ under the meaning of the least mean square error.

$$\hat{\boldsymbol{x}}_k = \mathrm{E}(\boldsymbol{x}_k \mid \boldsymbol{Y}_0^k)$$

where $\boldsymbol{Y}^k_0$ is the observation series at time k. The external input is a determined variable. We need to estimate the posterior distribution $p(\boldsymbol{x}_k|\boldsymbol{Y}^k_0)$ if we want to obtain $E(\boldsymbol{x}_k|\boldsymbol{Y}^k_0)$. In this condition of the posterior distribution, we can obtain optimal estimation in any meaning; The Bayesian method is used to obtain the posterior distribution.

$$p(\boldsymbol{x}_k \mid \boldsymbol{Y}_0^k) = \frac{p(\boldsymbol{x}_k \mid \boldsymbol{Y}_0^{k-1})\, p(\boldsymbol{y}_k \mid \boldsymbol{x}_k)}{p(\boldsymbol{y}_k \mid \boldsymbol{Y}_0^{k-1})} \tag{9.4.53}$$

where

$$p(\boldsymbol{x}_k \mid \boldsymbol{Y}_0^{k-1}) = \int p(\boldsymbol{x}_{k-1} \mid \boldsymbol{Y}_0^{k-1})\, p(\boldsymbol{x}_k \mid \boldsymbol{x}_{k-1})\, \mathrm{d}\boldsymbol{x}_{k-1} \tag{9.4.54}$$

$$p(\boldsymbol{y}_k \mid \boldsymbol{Y}_0^{k-1}) = \int (\boldsymbol{x}_k \mid \boldsymbol{Y}_0^{k-1})\, p(\boldsymbol{y}_k \mid \boldsymbol{x}_k)\, \mathrm{d}\boldsymbol{x}_k \tag{9.4.55}$$

$p(\boldsymbol{x}_k|\boldsymbol{x}_{k-1})$ depends on the state transition equation, while $p(\boldsymbol{y}_k|\boldsymbol{x}_k)$ depends on the observation equation.

Actually, we cannot obtain the analytical expression in most situations, so we need to use some approximate methods. When $\boldsymbol{F}(\cdot)$ and $\boldsymbol{H}(\cdot)$ are differentiable and the noise has Gaussian distribution, the common way is to linearize the non-linear system with the Extended Kalman filter (EKF). However, the EKF is only a first-order approximation and requires $\boldsymbol{F}(\cdot)$ and $\boldsymbol{H}(\cdot)$ differentiable, thereby educe the unscented KF on the basis of deterministic sampling or unscented transform. Here, the deterministic sampling is different from the Monte Carlo sampling. Given a mean value, variance and some necessary parameters, the generation of the sampling data is determined, while the Monte Carlo sampling is one kind of random sampling. Compared with the EKF, UKF is a method of higher-order approximation with the same computational complexity and does not need any linear processing. One of the common points is to approximate real posterior distribution with Gaussian distribution. However, in most situations, especially when $\boldsymbol{v}_k$ and $\boldsymbol{n}_k$ are non-Gaussian, and $\boldsymbol{F}(\cdot)$ and $\boldsymbol{H}(\cdot)$ are nonlinear, it is required to have a more general algorithm, such as particle filtering (PF), sequential importance sampling, sequential Monte Carlo filter, condensation algorithm, or survival of the fittest, which approximate the Bayesian integral with the random sampling. Here we put emphasis on the Unscented PF, which is an integration of UKF and PF. UPF involves the importance of sampling, resampling and a selection of the importance function during sampling. It needs to be pointed out that besides approximate reasoning and the learning algorithms mentioned above, such as EKF/UKF/PF/UPF, there are other approximate algorithms, such as the MCMC[43,44], the mean-value field[45] and variational algorithms[53,54]. Table 9.5 shows the main steps of the PF algorithm.

Table 9.5

Step 1 Initialization: k=0

Get sampling $\boldsymbol{x}_0^i$, i=1, ⋯, N, from prior distribution $p(\boldsymbol{x}_0)$

Step 2 k=1, 2, ⋯

i. Importance sampling i=1, ⋯, N

$$\chi_k^i \sim q(\boldsymbol{x}_k \mid \boldsymbol{x}_{0:k-1}^i, \boldsymbol{Y}_0^k)$$

$$w_k^i = w_{k-1}^i \frac{p(\boldsymbol{y}_k \mid \boldsymbol{x}_k^i) p(\boldsymbol{x}_k^i \mid \boldsymbol{x}_{k-1}^i)}{q(\boldsymbol{x}_k^i \mid \boldsymbol{x}_{0:k-1}^i, \boldsymbol{Y}_0^k)}$$

$$\tilde{w}_k^i = w_{k-1}^i \left(\sum_{j=1}^{N} w_k^j \right)^{-1}$$

ii. Resampling

(continued)

iii. MCMC

iv. Output: sampling data can approximate posterior distribution

$$\hat{p}(\boldsymbol{x}_k \mid \boldsymbol{Y}_0^k) = \frac{1}{N}\sum_{i=1}^{N}\delta(\boldsymbol{x}_k - \boldsymbol{x}_k^i)$$

the optimal MMSE estimation is

$$\hat{\boldsymbol{x}}_k = E(\boldsymbol{x}_k \mid \boldsymbol{Y}_0^k) \approx \frac{1}{N}\sum_{i=1}^{N}\boldsymbol{x}_k^i$$

Step iii is not necessary, because after resampling the larger weighted particles generate many sub-particles, while smaller weighted particles have only a few or even no sub-particles. In order to increase the diversity of particles, we can process each particle with the MCMC. It is important to select the importance functions for PF during sampling, with the support region covering the real posterior distribution with as much overlap as possible. In Fig. 9.12, in order to locate the sampling region at the highest likelihood region, we use UKF (or EKF), so that the particles with smaller weight move to the higher likelihood region at current observation $\boldsymbol{y}_k$.

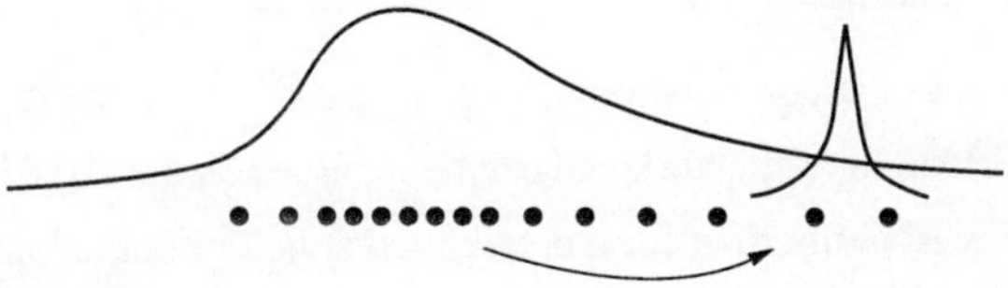

Fig. 9.12 Sketch map of generation of UKF auxiliary sampling function

More specifically, we use UKF to compute the current mean value and variance according to the current $\boldsymbol{y}_k$, mean value and variance of particles as last time, which can be taken as a mean value and variance of the Gaussian importance function. Then we can get recursive computation Table 9.6 the next time, according to the sampling of the Gaussian importance function. Tables 9.5 and 9.6 form the realization framework of the UPF. The UPF can also apply other methods, such as the likelihood PF[77] and applying gradient information, so the generation region of recursive particles can approximate a higher likelihood region. Applying these methods effectively makes up for some problems caused by the sub-optimal sampling function.

Table 9.6 Steps of UKF auxiliary sampling function

$i=1,\cdots,N$, update the distribution of each particle at $k-1$ by using UKF

i. Compute Sigma point

$$\Delta_{k-1}^{(i)a} = \left[\bar{\boldsymbol{x}}_{0:k-1}^{(i)a} \quad \bar{\boldsymbol{x}}_{0:k-1}^{(i)a} + \gamma\sqrt{\boldsymbol{P}_{k-1}^{(i)a}} \quad \bar{\boldsymbol{x}}_{0:k-1}^{(i)a} - \gamma\sqrt{\boldsymbol{P}_{k-1}^{(i)a}}\right]$$

ii. Time update

$$\Delta_{k|k-1}^{(i)x} = \boldsymbol{F}(\Delta_{k-1|k-1}^{(i)x}, \Delta_{k-1|k-1}^{(i)v}) \quad \bar{\boldsymbol{x}}_{k|k-1}^{(i)} = \sum_{j=0}^{2L} \boldsymbol{W}_j^m \Delta_{j,k|k-1}^{(i)x}$$

$$\boldsymbol{P}_{k|k-1}^{(i)} = \sum_{j=0}^{2L} \boldsymbol{W}_j^m (\Delta_{j,k|k-1}^{(i)x} - \bar{\boldsymbol{x}}_{k|k-1}^{(i)})(\Delta_{j,k|k-1}^{(i)x} - \bar{\boldsymbol{x}}_{k|k-1}^{(i)})^{\mathrm{T}}$$

$$\boldsymbol{Y}_{k|k-1}^{(i)} = \boldsymbol{H}(\Delta_{j,k|k-1}^{(i)x}, \Delta_{k-1|k-1}^{(i)n}) \quad \bar{\boldsymbol{y}}_{k|k-1}^{(i)} = \sum_{j=0}^{2L} \boldsymbol{W}_j^m \boldsymbol{Y}_{j,k|k-1}^{(i)}$$

iii. Measurement update

$$\boldsymbol{K}_k = \left(\sum_{j=0}^{2L} \boldsymbol{W}_j^c (\Delta_{j,k|k-1}^{(i)x} - \bar{\boldsymbol{x}}_{k|k-1}^{(i)})(\boldsymbol{Y}_{j,k|k-1}^{(i)} - \bar{\boldsymbol{y}}_{k|k-1}^{(i)})^{\mathrm{T}}\right) \cdot$$
$$\left(\sum_{j=0}^{2L} \boldsymbol{W}_j^c (\boldsymbol{Y}_{j,k|k-1}^{(i)} - \bar{\boldsymbol{y}}_{k|k-1}^{(i)})(\boldsymbol{Y}_{j,k|k-1}^{(i)} - \bar{\boldsymbol{y}}_{k|k-1}^{(i)})^{\mathrm{T}}\right)^{-1}$$

$$\bar{\boldsymbol{x}}_{k|k}^{(i)} = \bar{\boldsymbol{x}}_{k|k-1}^{(i)} + \boldsymbol{K}_k(\boldsymbol{y}_k - \bar{\boldsymbol{y}}_{k|k-1}^{(i)})$$

$$\boldsymbol{P}_{k|k}^{(i)} = \boldsymbol{P}_{k|k-1}^{(i)} - \boldsymbol{K}_k\left(\sum_{j=0}^{2L} \boldsymbol{W}_j^c (\boldsymbol{Y}_{j,k|k-1}^{(i)} - \bar{\boldsymbol{y}}_{k|k-1}^{(i)})(\boldsymbol{Y}_{j,k|k-1}^{(i)} - \bar{\boldsymbol{y}}_{k|k-1}^{(i)})^{\mathrm{T}}\right)\boldsymbol{K}_k^{\mathrm{T}}$$

Sampling $\boldsymbol{x}_k^{(i)} \sim q(\boldsymbol{x}_k^{(i)} \mid \boldsymbol{x}_{0:k-1}^{(i)}, \boldsymbol{Y}_{0:k}) \approx N(\bar{\boldsymbol{x}}_{k|k}^{(i)}, \boldsymbol{P}_{k|k}^{(i)})$

Table 9.6 describes the computation process of UKF[80,81]. Actually, EKF and UKF have the same calculation framework and can be regarded as the same kind of method. On the basis of Bayesian filtering, for a determined structure but unknown parameters and state, we can do dual-estimation of the hidden state (source signals) and system parameters (mixing matrix or channel response). Figure 9.13 shows the sketch map of the dual-estimation algorithm of parameters and state. In the two estimations, the parameters and state form a new state vector, which will lead to a dimension increase of the state space; therefore, we use the dual Bayesian filtering.

Marginal estimation of dual-estimation is

$$p(\boldsymbol{x}_1^N, \boldsymbol{w} \mid \boldsymbol{y}_1^N) = p(\boldsymbol{x}_1^N \mid \boldsymbol{w}, \boldsymbol{y}_1^N)\, p(\boldsymbol{w} \mid \boldsymbol{y}_1^N) \tag{9.4.56}$$

For $\hat{\boldsymbol{x}}_1^N$, Eq.(9.4.56) can be changed into

$$\hat{\boldsymbol{x}}_1^N = \arg\max_{\hat{\boldsymbol{x}}_1^N}(p(\boldsymbol{x}_1^N \mid \boldsymbol{w}, \boldsymbol{y}_1^N)) \quad \text{with } \boldsymbol{w} \text{ fixed} \tag{9.4.57}$$

For $\hat{\boldsymbol{w}}$,

$$\hat{\boldsymbol{w}} = \arg\max_{\boldsymbol{w}}(p(\boldsymbol{w} \mid \boldsymbol{y}_1^N)) \tag{9.4.58}$$

The joint estimation is

$$(\hat{\boldsymbol{x}}_1^N, \hat{\boldsymbol{w}}) = \arg\max_{\hat{\boldsymbol{x}}_1^N, \hat{\boldsymbol{w}}}(p(\boldsymbol{y}_1^N \mid \boldsymbol{x}_1^N, \boldsymbol{w})\, p(\boldsymbol{x}_1^N \mid \boldsymbol{w})\, p(\boldsymbol{w})) \tag{9.4.59}$$

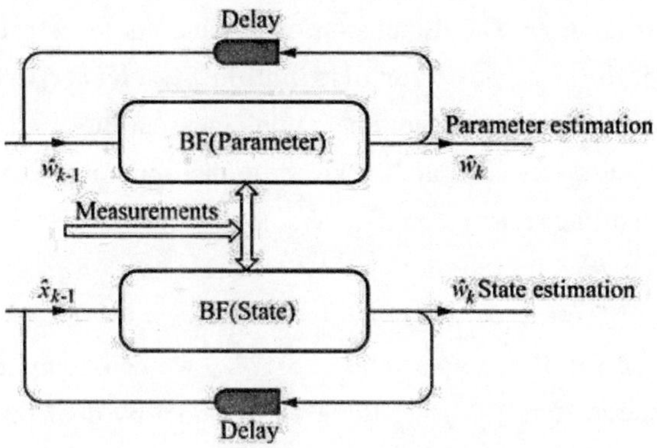

Fig. 9.13 Sketch map of dual-estimation algorithm of parameters and state

Because we can use the state space form to express the process from source signals to observation signals, which forms a hierarchical structural filtering algorithm with a prediction-calibration-prediction mode. To construct each source signal as a time-varying autoregressive model, go into the channel model driven by white noise to finally obtain the observation signals. The process from driven noise to observation is defined as forward, and the process from observation signals to the TVAR model is defined as backward. The forward process transfers prediction, while the backward process transfers new information, and the state and parameters corresponding to the state space model are estimated in each layer. Framework of the algorithm that solves the convolution underdetermined model with dual-estimation of the Bayesian filtering is shown in Fig. 9.14.

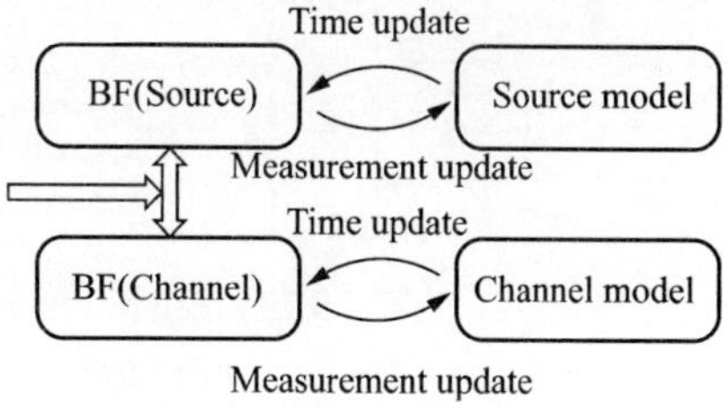

Fig. 9.14 Framework of algorithm which solves the convolution underdetermined model with dual-estimation of Bayesian filter

The model coefficients of source signals are described by the variation of Markov in the framework of algorithms, and the general selection applie to the first-order Gaussian Random Walk.

However, for a complex system it is unavoidable in practice to face many parameters. In order to improve the computational performance and reduce variance of the estimation, we need to select the proper method, such as the Rao-Blackwellisation method of a particle filter[76] to reduce the dimension of parameters. The method is similar to expectation maximization-EM, in which some parameters of analytical expressions in systems are eliminated through integrals. These parameters include the model parameter $\boldsymbol{a}(t)$ of TVAR and channel parameter $\boldsymbol{h}(t)$. The dimension of parameters to be estimated is also reduced to obtain state posterior distribution $p(\gamma^k{}_t|y(1:t))$. This posterior distribution-based estimation is nonlinear non-Gaussian Bayesian filtering.

In order to recursively compute the posterior distribution, it needs to predefine the initial value of the recursive parameters

$$\begin{aligned} & p(\gamma_{t=1:T}, \alpha_{1:T} \mid \boldsymbol{y}(1:T)) \\ & = p(\alpha_{1:T} \mid \gamma_{t=1:T}, \boldsymbol{y}(1:T))p(\gamma_{t=1:T} \mid \boldsymbol{y}(1:T)) \end{aligned} \tag{9.4.60}$$

After we get the set of state particles $\{\gamma_t^k\}_{t=1:T}^{k=1:N_s}$, we can compute channel and time varying parameters $\{\alpha_{1:T}\}$ of the TVAR model by dual-estimation of the Kalman filtering or joint estimation methods.

9.5 FastICA Separation Algorithm of Complex Numbers in Convolutive Mixtures

The convolution mixing model in the frequency-domain has the following same form as the instantaneous linear mixing model in the time-domain:

$$\boldsymbol{x}(z) = \boldsymbol{H}(z)\boldsymbol{s}(z) \tag{9.5.1}$$

$$\boldsymbol{y}(z) = \boldsymbol{W}(z)\boldsymbol{x}(z) = \boldsymbol{C}(z)\boldsymbol{s}(z) \tag{9.5.2}$$

Therefore, we can first transform the mixed signal in the time-domain to a product form in the frequency-domain and then separate them in the frequency domain. The process of blind separation of the convolutive mixtures in the frequency-domain is shown in Fig. 9.15.

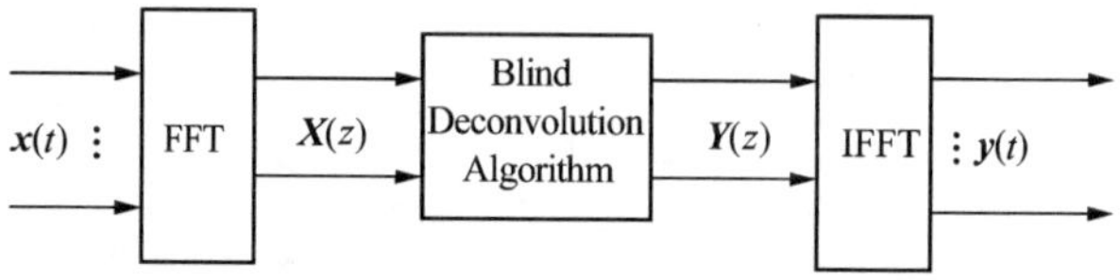

Fig. 9.15 Blind separation sketch of convolutive mixtures in frequency-domain

The convolution mixed signal becomes an instantaneous mixture in the frequency-domain, so the general method is to divide the whole frequency-domain into a B narrow bands, whose center frequency is Z_b, Z_b, $0 \leqslant b \leqslant B-1$. In the Fast Fourier Transform (FFT), B is actually the Fourier transform length. Perform the instantaneous linear blind separation in each band where computation is dependent and calculation is easy. That way we can obtain separation through the instantaneous linear blind separation algorithm first, then perform inverse transformation to obtain source signals in the time-domain. The difference between the frequency-domain and the time-domain separation is that the transformed signal in the frequency-domain is complex, so the instantaneous linear blind separation cannot be used directly on mixed signals in the frequency-domain. In other words, it needs to expand the cost function, constraint function and iterative learning rules of the instantaneous linear blind separation algorithm into the complex number domain.

We have discussed the FastICA algorithm developed by Hyvärinen and Oja in Chapter 3. This algorithm has very fast convergence speed, and unlike gradient-based algorithms, it doesn't need to choose any iterative parameter. Furthermore, it can approximate negentropy using any nonlinear function with high accuracy, so it is a fast and effective method. Thus, effective expansion of the FastICA algorithm to a complex number domain is an important step to speed up the algorithm.

The cost function of the FastICA, as a measure for non-Gaussianity, is maximal negentropy. In other words, it is equivalent to minimizing mutual information between output vectors. Hyvärinen uses the following equations as approximate negentropy in the FastICA and suggests that this approximation is better than the traditional higher-order cumulant method in accuracy:

$$J(y_i) \approx c[E(G(y_i)) - E(G(v))]^2 \tag{9.5.3}$$

where G is a non-quadratic function and v is a Gaussian random variable with zero-mean and unit-variance. For $\boldsymbol{y}=\boldsymbol{W}\boldsymbol{x}$, the cost function of Eq. (9.5.3) can be written as

$$J_G(\boldsymbol{W}) = [E(G(\boldsymbol{W}_i^{\mathrm{T}}\boldsymbol{x})) - E(G(v))]^2 \tag{9.5.4}$$

Thinking about the speed and stability of the algorithm, Hyvärinen gives three kinds of nonlinear functions in both theoretical and practical in application, and their corresponding first derivatives are:

$$G_1(u) = \frac{1}{a_1}\lg\cosh(a_1 u), \quad g_1(u) = G_1'(u) = \tanh(a_1 u) \tag{9.5.5}$$

$$G_2(u) = -\frac{1}{a_2}\exp(-a_2 u^2/2),$$

$$g_2(u) = G_2'(u) = u\exp(-a_2 u^2/2) \tag{9.5.6}$$

$$G_3(u) = -\frac{1}{4}u^4, \quad g_3(u) = G_3'(u) = u^3 \tag{9.5.7}$$

where a_1, a_2 are constants, and $1 \leqslant a_1 \leqslant 2, a_2 \approx 1$.

For the above three cost functions,

(1) G_1 is a general cost function with better performance;

(2) G_2 is used when the source signal is super-Gaussian distribution or has a separation system requiring strong stability;

(3) G_3, or a kurtosis-based cost function is suitable for separation of sub-Gaussian signals without singular points.

Thinking of the constraint condition that output vectors should be mutually unrelated, this blind separation problem can be considered as a maximization problem of negentropy with decorrelation as a constraint condition:

$$\begin{cases} \text{Maximization: } \sum_{i=1}^{n} J_G(\boldsymbol{W}_i), \text{where } \boldsymbol{W}_i, i = 1, \cdots, n; & (9.5.8) \\ \text{Constricant: } E((\boldsymbol{W}_k^{\mathrm{T}}\boldsymbol{x})(\boldsymbol{W}_j^{\mathrm{T}}\boldsymbol{x})) = \delta_{jk}, \\ \quad \text{when } j = k, \delta_{jk} = 1, \text{else } \delta_{jk} = 0 & (9.5.9) \end{cases}$$

Based on the Kuhn-Tucker condition[82], under constraint condition $E\{(\boldsymbol{W}^{\mathrm{T}}\boldsymbol{x})^2\} = \|\boldsymbol{W}\|^2 = 1$, maximizing cost function $E\{G(\boldsymbol{W}^{\mathrm{T}}\boldsymbol{x})\}$ can result in the final evaluation function of the FastICA in the real number field:

$$E\{\boldsymbol{x}g(\boldsymbol{W}^{\mathrm{T}}\boldsymbol{x})\} - \beta\boldsymbol{W} = 0 \tag{9.5.10}$$

where $\beta = E\{\boldsymbol{W}_0^{\mathrm{T}}\boldsymbol{x}g(\boldsymbol{W}_0^{\mathrm{T}}\boldsymbol{x})\}$, $\boldsymbol{W}_0$ is the optimal solution and is updated by the iterative result each time. The cost function and constraint condition of the FastICA in the real number field can be expanded into the complex number field[83,84].

Assume source signal $\boldsymbol{s}=\boldsymbol{s}_r+\mathrm{j}\boldsymbol{s}_i$ is a whitened signal, i.e. $\boldsymbol{s}$ has a zero-mean and unit variance, and its real and imaginary parts are mutually uncorrelated. Therefore, $E(\boldsymbol{s}\boldsymbol{s}^{\mathrm{H}})=\boldsymbol{I}$; here $\boldsymbol{s}^{\mathrm{H}}$ indicates the conjugate transpose of $\boldsymbol{s}$. Suppose the mixed signal and the unmixing matrix are the complex number form $\boldsymbol{x}=\boldsymbol{x}_r+\mathrm{j}\boldsymbol{x}_i$, $\boldsymbol{W}=\boldsymbol{W}_r+\mathrm{j}\boldsymbol{W}_i$. With $\boldsymbol{s}$ as a whitened signal, there is $E(\boldsymbol{x}\boldsymbol{x}^{\mathrm{H}})=\boldsymbol{I}$.

Similarlly, based on the Kuhn-Tucker condition, under constraint condition $E((\boldsymbol{W}^{\mathrm{H}}\boldsymbol{x})^2) = \|\boldsymbol{W}\|^2 = 1$, the maximum cost function $E(G|\boldsymbol{W}^{\mathrm{H}}\boldsymbol{x}|^2)$ can be obtained from the first derivative of the following equation:

$$\nabla E(G(|\boldsymbol{W}^{\mathrm{H}}\boldsymbol{x}|^2)) - \beta\nabla E(|\boldsymbol{W}^{\mathrm{H}}\boldsymbol{x}|^2) = 0 \tag{9.5.11}$$

With differentiation of both real and imaginary parts of $\boldsymbol{W}$ in Eq. (9.5.11), the first item on the left side of the equation can be written as

$$\nabla E(G\{|\boldsymbol{W}^{\mathrm{H}}\boldsymbol{x}|^2)\} = \begin{pmatrix} \dfrac{\partial}{\partial \boldsymbol{W}_{1r}} \\ \dfrac{\partial}{\partial \boldsymbol{W}_{1i}} \\ \vdots \\ \dfrac{\partial}{\partial \boldsymbol{W}_{nr}} \\ \dfrac{\partial}{\partial \boldsymbol{W}_{ni}} \end{pmatrix} E\{G(|\boldsymbol{W}^{\mathrm{H}}\boldsymbol{x}|^2)\}$$

$$=2\begin{pmatrix} E(\mathrm{Re}(x_1(\boldsymbol{W}^{\mathrm{H}}\boldsymbol{x})^*)g(|\boldsymbol{W}^{\mathrm{H}}\boldsymbol{x}|^2)) \\ E(\mathrm{Im}(x_1(\boldsymbol{W}^{\mathrm{H}}\boldsymbol{x})^*)g(|\boldsymbol{W}^{\mathrm{H}}\boldsymbol{x}|^2)) \\ \vdots \\ E(\mathrm{Re}(x_n(\boldsymbol{W}^{\mathrm{H}}\boldsymbol{x})^*)g(|\boldsymbol{W}^{\mathrm{H}}\boldsymbol{x}|^2)) \\ E(\mathrm{Im}(x_n(\boldsymbol{W}^{\mathrm{H}}\boldsymbol{x})^*)g(|\boldsymbol{W}^{\mathrm{H}}\boldsymbol{x}|^2)) \end{pmatrix} \tag{9.5.12}$$

where $*$ indicates a complex conjugate. Rewrite it in its vector form:

$$\nabla E(G(|\boldsymbol{W}^{\mathrm{H}}\boldsymbol{x}|^2)) = 2E(\boldsymbol{x}(\boldsymbol{W}^{\mathrm{H}}\boldsymbol{x})^* g(|\boldsymbol{W}^{\mathrm{H}}\boldsymbol{x}|^2)) \tag{9.5.13}$$

From the second item on the left side of the equation, we obtain

$$\nabla E(|\boldsymbol{W}^{\mathrm{H}}\boldsymbol{x}|^2) = 2\begin{pmatrix} \mathrm{Re}(\boldsymbol{W}_1) \\ \mathrm{Im}(\boldsymbol{W}_1) \\ \vdots \\ \mathrm{Re}(\boldsymbol{W}_n) \\ \mathrm{Im}(\boldsymbol{W}_n) \end{pmatrix} \tag{9.5.14}$$

In the solution of Eq. (9.5.11) by the Newton method, we first use the following equation to approximate the Jacobian matrix of $\nabla E(G(|\boldsymbol{W}^{\mathrm{H}}\boldsymbol{x}|^2))$:

$$\nabla^2 E(G(|\boldsymbol{W}^{\mathrm{H}}\boldsymbol{x}|^2)) \approx 2E(g(|\boldsymbol{W}^{\mathrm{H}}\boldsymbol{x}|^2) + |\boldsymbol{W}^{\mathrm{H}}\boldsymbol{x}|^2 g'(|\boldsymbol{W}^{\mathrm{H}}\boldsymbol{x}|^2))\boldsymbol{I} \tag{9.5.15}$$

Similarly, the Jacobian matrix of $\beta\,\nabla E(|\boldsymbol{W}^{\mathrm{H}}\boldsymbol{x}|^2)$ is

$$\beta\nabla^2 E(|\boldsymbol{W}^{\mathrm{H}}\boldsymbol{x}|^2) = 2\beta\boldsymbol{I} \tag{9.5.16}$$

According to Eqs. (9.5.15) and (9.5.16), the Jacobian matrix of the cost function can be

$$J = 2(E(g(|\boldsymbol{W}^{\mathrm{H}}\boldsymbol{x}|^2) + |\boldsymbol{W}^{\mathrm{H}}\boldsymbol{x}|^2 g'(|\boldsymbol{W}^{\mathrm{H}}\boldsymbol{x}|^2)) - \beta)\boldsymbol{I} \tag{9.5.17}$$

From the learning rules of the Newton iterative method, we obtain

$$\begin{aligned} \boldsymbol{W}^{+} &= \boldsymbol{W} - \frac{\nabla E(G(|\boldsymbol{W}^{\mathrm{H}}\boldsymbol{x}|^2)) - \beta\nabla E(|\boldsymbol{W}^{\mathrm{H}}\boldsymbol{x}|^2)}{\nabla^2 E(G(|\boldsymbol{W}^{\mathrm{H}}\boldsymbol{x}|^2)) - \beta\nabla^2 E(|\boldsymbol{W}^{\mathrm{H}}\boldsymbol{x}|^2)} \\ &= \boldsymbol{W} - \frac{E(x(\boldsymbol{W}^{\mathrm{H}}\boldsymbol{x})^* g(|\boldsymbol{W}^{\mathrm{H}}\boldsymbol{x}|^2)) - \beta\boldsymbol{W}}{E(g(|\boldsymbol{W}^{\mathrm{H}}\boldsymbol{x}|^2) + |\boldsymbol{W}^{\mathrm{H}}\boldsymbol{x}|^2 g'(|\boldsymbol{W}^{\mathrm{H}}\boldsymbol{x}|^2)) - \beta} \end{aligned} \tag{9.5.18}$$

Multiply both sides of Eqs. (9.5.18) by $[-E(g(|\boldsymbol{W}^{\mathrm{H}}\boldsymbol{x}|^2) + |\boldsymbol{W}^{\mathrm{H}}\boldsymbol{x}|^2 \cdot g'(|\boldsymbol{W}^{\mathrm{H}}\boldsymbol{x}|^2)) + \beta]$, and the learning rules of $\boldsymbol{W}$ is

$$\boldsymbol{W}^{+} = E(\boldsymbol{x}(\boldsymbol{W}^{\mathrm{H}}\boldsymbol{x})^* g(|\boldsymbol{W}^{\mathrm{H}}\boldsymbol{x}|^2)) - E(g(|\boldsymbol{W}^{\mathrm{H}}\boldsymbol{x}|^2) + |\boldsymbol{W}^{\mathrm{H}}\boldsymbol{x}|^2 g'(|\boldsymbol{W}^{\mathrm{H}}\boldsymbol{x}|^2))\boldsymbol{W} \tag{9.5.19}$$

Meanwhile, in order to avoid the indeterminacy of the amplitude of the output vectors, perform normalization to $\boldsymbol{W}$ in each iteration step:

$$W_{\mathrm{new}} = \frac{\boldsymbol{W}^{+}}{\|\boldsymbol{W}^{+}\|} \tag{9.5.20}$$

In this way, by combining Eq. (9.5.19) and Eq. (9.5.20), we obtain the iterative rules of the instantaneous linear blind separation of the FastICA in a complex number field.

Now using the complex number FastICA algorithm, perform the blind separation of the real convolutive mixtures in the time-domain. The FastICA algorithm can extract the independent component in the mixed signals one by one and also separate each independent component at one time. Taking the two source signals and the two observed signals as an example to illuminate the specific steps in the application of FastICA in separating each independent component results in the following sketch:

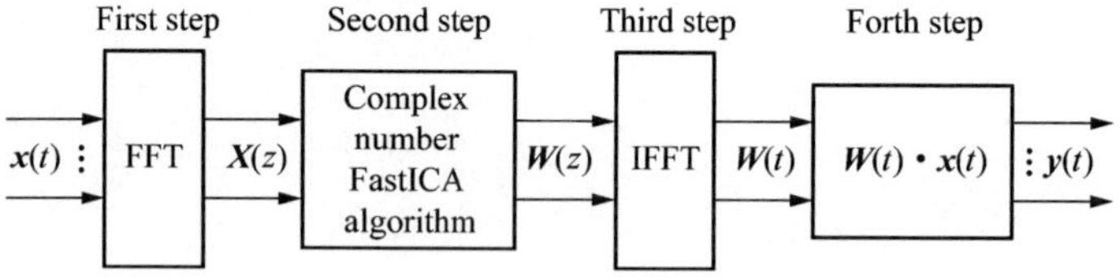

Fig. 9.16 Sketch of blind separation of FastICA algorithm in complex number field

(1) First, pre-whiten the observed convolution mixed signal to eliminate the second-order correlation and make its covariance matrix into a unit matrix. Divide the convolution mixed signal into l segments, then perform the Fourier transform for each segment. The observed signal is transformed into a complex form after the Fourier transform: $x_j(z) = (x_{j1r} + jx_{j1i}, x_{j2r} + jx_{j2i}, \cdots, x_{jnr} + jx_{jni})$, j=1,2.

(2) Use FastICA to do the instantaneous linear blind separation for each segment of the complex signals.

① Initialize the deconvolution matrix: $\boldsymbol{W}$=randn (2, 2)+j · randn (2,2) and set the iterative number.

② According to Eq. (9.5.19), compute

$$\boldsymbol{W}_1 = E(\boldsymbol{x}(\boldsymbol{W}^{\mathrm{H}}\boldsymbol{x})^* g(|\boldsymbol{W}^{\mathrm{H}}\boldsymbol{x}|^2)) \tag{9.5.21}$$

then

$$\boldsymbol{W}_2 = E\{g(|\boldsymbol{W}^{\mathrm{H}}\boldsymbol{x}|^2) + |\boldsymbol{W}^{\mathrm{H}}\boldsymbol{x}|^2 g'(|\boldsymbol{W}^{\mathrm{H}}\boldsymbol{x}|^2)\}\boldsymbol{W} \tag{9.5.22}$$

where the nonlinear function in the cost function equation (9.5.11) is $G(y) = \lg(a+y)$, a=0.1, then obtain $g(y)=G'(y)=1/(a+y)$, $g'(y)=-1/(a+y)^2$;

③ Iteratively update $\boldsymbol{W}$

$$\boldsymbol{W} = \boldsymbol{W}_1 - \boldsymbol{W}_2 \tag{9.5.23}$$

④ According to Eq. (9.5.20), normalize $\boldsymbol{W}$:

$$\boldsymbol{W}_{\text{new}} = \boldsymbol{W} / \|\boldsymbol{W}\| \tag{9.5.24}$$

⑤ After every iteration, the iteration number is reduced by 1. If the iteration number becomes zero, end the process; otherwise go on to step 2;

(3) When the iteration process is over, take the inverse transformation of the demixing matrix in the frequency-domain back to the time-domain:

$$\boldsymbol{W}_t \leftarrow \mathrm{Ifft}(\boldsymbol{W}_f) \tag{9.5.25}$$

(4) Calculate the solution of this segment in the time domain:

$$\boldsymbol{y}_l = \boldsymbol{W}_t \boldsymbol{x} \tag{9.5.26}$$

Perform blind separation to l segments of the observed signal one by one. Calculate the correlation coefficient c_{ij} between the output of this segment and the previous one according to the following equation:

$$c_{ij} = \frac{|E(x_i y_j)|}{|E(x_i)||E(y_j)|} \tag{9.5.27}$$

then adjust the order of the output vectors to keep signals in different channels to be consistent, and reorder it to obtain the output signal in the time-domain. For two source signals and two mixed signals, calculate the correlation coefficient between every two adjacent segments of the four output signals: c_{11}, c_{12}, c_{21} and c_{22}. If the outputs of the first channel in both of the first two segments are the first signal and the outputs of the second channel are the second signal, then $c_{11} > c_{12}$, $c_{22} > c_{21}$. If the output of the first channel in the previous segment is the first signal and the output of the second channel is the second signal, but in the next segment, the output of the first channel is the second signal and the output of the second channel is the first signal, then $c_{12} > c_{11}$, $c_{21} > c_{22}$. Assume that we keep the first condition for both segments, where the output of the first channel is the first signal and the output of the second channel is the second signal, then we can compare c_{11}/c_{12} and c_{21}/c_{22} and can decide whether we need to exchange the two rows of output $y_t(t)$. For example, if $c_{11}/c_{12} > c_{21}/c_{22}$, keep the order of the second segment; otherwise, if $c_{11}/c_{12} < c_{21}/c_{22}$, exchange the output vectors of the second segment.

To compare the iterative speed and the separation effect with the blind separation in the time-domain, use the same recorded vocal signal to perform the experiments. The two signals recorded by two microphones in an office room use the English pronunciation of numbers one to ten and the section of a music signal. The vocal signal source, music source and two microphones are located in the four vertexes of a square with a length of 60 cm respectively. The sampling rate is 16 kHz, for a total of 110, 000 samples. The recorded signals are shown in Fig. 9.17 (a) and (b); the separation results of the convolutive mixtures by using the complex FastICA algorithm are shown in Fig. 9.17 (c) and (d); for comparison, the separation results using the time-domain separation algorithm are shown as Fig. 9.17 (e) and (f).

Initiate the unmixing matrix $\boldsymbol{W}$ to a random complex 2×2 matrix. We can see

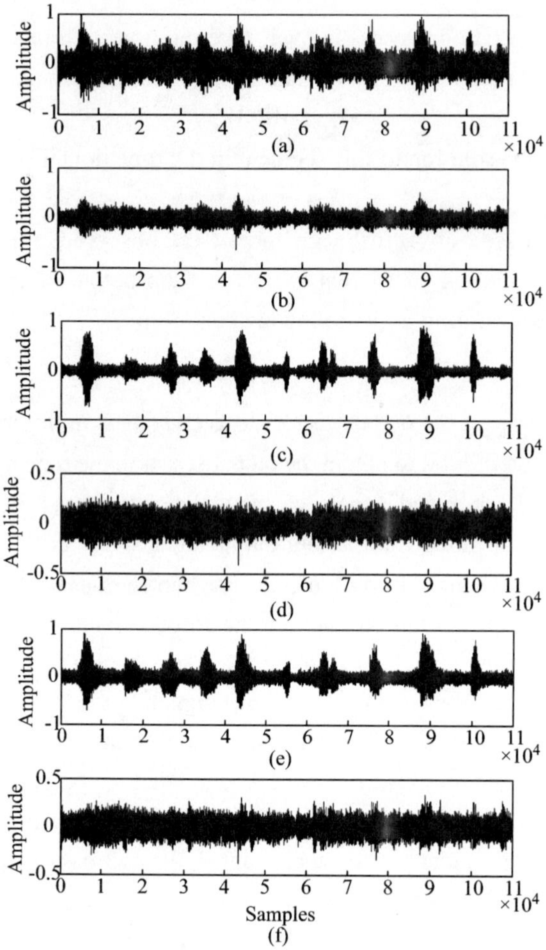

Fig. 9.17 Comparision of blind separation results of recorded voice and music signal in time-and frequency-domain

from the simulation that the complex FastICA convolutive mixtures', separation method can separate the vocal signal and the background music signal better. By comparison of Fig. 9.17 (c) and (e), (d) and (f), the separated vocal signal based on the complex FastICA method contains less of the music signal and is much clearer than the result of the time-domain method; meanwhile, the music signal separated by the frequency-domain method is better in restraining the vocal signal. Another significant advantage of the frequency-domain method is the higher iterative speed. In this experiment, the algorithm converges after sixty iteration steps within 41 seconds, which is much faster than the 15 minutes it takes in the time-domain method.

9.6 On-line Complex ICA Algorithm Based on Uncorrelated Characteristics of Complex Vectors

In recent years, general practical applications promote the research of a complex ICA (CICA). For example, for the deconvolution in the frequency domain, there is computation of the complex signal[85]. In the functional magnetic resonance imaging (fMRI), the amplitude and phase of the complex fMRI image are helpful for revealing the physical mechanism of magnetic resonance[86,87]. In radar signal processing, the Doppler frequency of foil strip interference and echo signal show large differences, and it is possible to extract the objective echo signal from the foil strip interference by using the CICA algorithm. In communication fields, there is the separation problem of the circular distribution source[88]. And in the acoustic field, especially under water, the vector sensors require complex processing[89] etc.

The existing CICA algorithms can be classified into two kinds. One is expanded from a real signal to a complex signal. For example, the expansion of the JADE algorithm is an early CICA algorithm, which just changes the transposed symbol of the separating matrix to the conjugate transposed symbol, but without the help of the characteristics of complex numbers. Other algorithms mainly use a real nonlinear function and complex nonlinear function to generate an independent component. In the processing method of the complex numbers by a nonlinear function, the greatest challenge is the conflict between boundedness and analyticity in the complex domain, which is expatiated in the Liouville Theorem. To solve this problem, there generally are three methods. The first method is defined as the Complex Split Nonlinear Function—CSNF[90,91], and it calculates the real and imaginary parts of a complex number separately, because the real and imaginary parts of complex numbers are real numbers, so a real nonlinear function is still used. In this way, the boundedness can be guaranteed, showever, analyticity cannot be satisfied. The second method is to calculate the amplitude of the variables, also using the real nonlinear function[92,93], which requires statistical characteristics of the signals. For example, Ref.[92] took independent component analysis through the expanded Hebbian learning rules, but it requires the complex signal to be a circular distribution, and at most contains one Gaussian signal; and Ref.[94] requires that the real and imaginary parts of a complex signal are uncorrelated. These two methods mentioned above can solve some special problems successfully, but in the aspect of a higher-order statistics calculation, they are not so effective as the complex nonlinear function[95]. The last method is to extract the complex independent component in a complex number field using higher-order statistics generated by a complex nonlinear function[96~98].

1. Basic explanation of complex random vector[99,100]

Similar to a real random vector, the co-covariance matrix between two complex random vectors $\boldsymbol{x}=\boldsymbol{x}_R+j\boldsymbol{x}_I$ and $\boldsymbol{y}=\boldsymbol{y}_R+j\boldsymbol{y}_I$ can be defined as $\boldsymbol{C}_{xy}=E[(\boldsymbol{x}-\boldsymbol{m}_x)(\boldsymbol{y}-\boldsymbol{m}_y)^H]$, where $\boldsymbol{m}_x=E(\boldsymbol{x}),\boldsymbol{m}_y=E(\boldsymbol{y})$, and H indicates a transposed complex conjugate. Meanwhile, the pseudo covariance matrix between $\boldsymbol{x}$ and $\boldsymbol{y}$ is defined as $\widetilde{\boldsymbol{C}}_{xy}=E[(\boldsymbol{x}-\boldsymbol{m}_x)(\boldsymbol{y}-\boldsymbol{m}_y)^T$; then the covariance matrix between the real and imaginary part of $\boldsymbol{X}$ and $\boldsymbol{Y}$ is

$$\mathrm{cov}[\boldsymbol{x}_R,\boldsymbol{y}_R]=\frac{1}{2}\mathrm{Re}\{\boldsymbol{C}_{xy}+\widetilde{\boldsymbol{C}}_{xy}\};\mathrm{cov}[\boldsymbol{x}_I,\boldsymbol{y}_I]=\frac{1}{2}\mathrm{Re}\{\boldsymbol{C}_{xy}-\widetilde{\boldsymbol{C}}_{xy}\};$$

$$\mathrm{cov}[\boldsymbol{x}_I,\boldsymbol{y}_R]=\frac{1}{2}\mathrm{Im}\{\boldsymbol{C}_{xy}+\widetilde{\boldsymbol{C}}_{xy}\};\mathrm{cov}[\boldsymbol{x}_R,\boldsymbol{y}_I]=-\frac{1}{2}\mathrm{Im}\{\boldsymbol{C}_{xy}-\widetilde{\boldsymbol{C}}_{xy}\}$$

If $\boldsymbol{x}$ and $\boldsymbol{y}$ are uncorrelated, then the above covariance matrices all equal zero.

Lemma 1 If and only if $\boldsymbol{C}_{xy}=\boldsymbol{0}$, and $\widetilde{\boldsymbol{C}}_{xy}=\boldsymbol{0}$ at the same time, then the two complex random vectors $\boldsymbol{x}$ and $\boldsymbol{y}$ are uncorrelated.

Definition For complex random vector $\boldsymbol{x}=\boldsymbol{x}_R+j\boldsymbol{x}_I$, if its correlation matrix is equal to zero, then $\boldsymbol{x}$ is called proper. Obviously, the sub-vector of a proper complex random vector must be proper. The composite vector of two proper complex random vectors $\boldsymbol{x}_1$ and $\boldsymbol{x}_2$ is $\boldsymbol{x}$; if $\boldsymbol{x}$ is proper, then $\boldsymbol{x}_1$ and $\boldsymbol{x}_2$ is combining proper.

2. On-line enhancing CICA algorithms

For a CICA problem, the mixing model is as follows:

$$\boldsymbol{x}(k)=\boldsymbol{A}\boldsymbol{s}(k) \tag{9.6.1}$$

here all elements are complex numbers.

For convenience, assume the source signal and observed signal have the same number. To ensure successful extraction of the independent component, the following five assumptions are made:

① The mixing system is time-invariant and mixing matrix $\boldsymbol{A}$ is full rank;

② All source signals are uncorrelated;

③ For each signal, different sample points should be mutually uncorrelated;

④ At most, one Gaussian signal can exist;

⑤ Source signals are not proper.

Generally, the following equation is used to estimate independent components:

$$y(k) = Wx(k) \tag{9.6.2}$$

and the learning rule

$$W(k+1) = W(k) + \mu F[x(k), W(k)] \tag{9.6.3}$$

where the estimate matrix W equals A^{-1}, μ is the learning rate, and the estimate function $F[x(k), W(k)]$ is a special matrix function which satisfies:

$$E\{F[x(k), W(k)]\} = 0 \tag{9.6.4}$$

where $E(\cdot)$ is the expectation operation.

Ref. [95] presents a CICA algorithm based on a natural gradient method,

$$\Delta W = \mu[I - \Phi(u(k))u(k)^{\mathrm{H}}]W \tag{9.6.5}$$

where $u(k)=Wx(k)$, $\Phi(\cdot)$ is an analytical complex nonlinear function.

Refer to Eq. (9.6.3), where the estimate function in Eq. (9.6.5) is

$$F(x(k), W(k)) = [I - \Phi(u(k))u(k)^{\mathrm{H}}]W(k) \tag{9.6.6}$$

as long as the result in the right brackets equals zero, this equation can satisfy the requirement in Eq. (9.6.4). Based on this point, a new CICA algorithm is developed.

Actually, uncorrelation between complex vectors $\Phi(u(k))$ and $u(k)^{\mathrm{H}}$ in Eq. (9.6.6) can effectively generate the CICA. Therefore, besides using the natural gradient method to enhance the learning rate, the uncorrelation between two vectors is still a problem to be solved. According to Lemma 1, add a new item in Eq. (9.6.6) to form a new evaluation function:

$$\begin{aligned} F(x(k), W(k)) &= \{[I - \Phi(u(k))u(k)^{\mathrm{H}}] + [C - \Phi(u(k))u(k)^{\mathrm{T}}]\}W(k) \\ &= [C - \Phi(u(k))u(k)^{\mathrm{H}} - \Phi(u(k))u(k)^{\mathrm{T}}]W(k) \end{aligned} \tag{9.6.7}$$

then the new CICA algorithm is

$$\begin{aligned} W(k+1) = W(k) + \mu[C - \Phi(u(k))u(k)^{\mathrm{H}} \\ - \Phi(u(k))u(k)^{\mathrm{T}}]W(k) \end{aligned} \tag{9.6.8}$$

where matrix $W(k)$ estimates the kth sample. We call this algorithm the strengthened CICA (SCICA) algorithm.

The selection of a complex nonlinear function in Eqs. (9.6.5) and (9.6.6) can be referred to in chapter of this book.

3. Simulation

The complex FastICA and JADA are good, but they can only apply in batch processing. Therefore, in this section, we will compare both the on-line and the batch methods.

Use the following evaluation index as the evaluation criteria of the performance of the algorithm:

$$\mathrm{PI}=\sum_{i=1}^{n}\left(\sum_{h=1}^{n}\frac{|c_{ih}|}{\max\limits_{p=1,\cdots,n}(c_{ip})}-1\right)+\sum_{h=1}^{n}\left(\sum_{i=1}^{n}\frac{|c_{ih}|}{\max\limits_{p=1,\cdots,n}(c_{ph})}-1\right) \tag{9.6.9}$$

where $(c_{th})=\boldsymbol{AW}$. The smaller the PI is, the better the performance of the CICA.

Now there are two source signals $\boldsymbol{s}(k)=\boldsymbol{r}(k)[\cos(\boldsymbol{f}(k))+j\sin(\boldsymbol{f}(k)-\pi/4)]$ and $j=\sqrt{-1}$, real vector $\boldsymbol{r}(k)$ and $\boldsymbol{f}(k)$ are radius and phase respectively, $\boldsymbol{r}(k)=[r_1(k), r_2(k)]^{\mathrm{T}}$, $\boldsymbol{f}(k)=[f_1(k), f_2(k)]^{\mathrm{T}}$; where $r_1(k)$ and $r_2(k)$ have different probability distributions and $f_1(k)$ and $f_2(k)$ have uniform distributions. The length of the signal is 5,000 points, and $\Phi=2\tanh(\cdot)$ is chosen for its complex nonlinear function.

Simulation-1 Signal $r_1(k)$ and $r_2(k)$ are non-Gaussian with kurtosis 6.6 and an exponential distribution, and kurtosis 5.4 and a geometric distribution respectively. The mixing matrix is randomly generated. The learning rate of the strengthened CICA and CICA are 0.06 and 0.0003.

Simulation-2 Signals $r_1(k)$ and $r_2(k)$ have exponential distribution and normal distribution, one signal being super-Gaussian and the other being closed Gaussian, with kurtosis 6.3 and 0.6 respectively. The mixing matrix is randomly generated. The learning rate of the strengthened CICA and CICA are 0.07 and 0.00013.

Figures 9.18, 9.19 and 9.20 show that the new algorithm fully utilizes the fact of the mutual independence between source signals when extracting complex independent components, because it combines the Property of a pseudo-covariance matrix with the evaluation function based on the natural gradient method. Whether in on-line or batch mode, the speed of the convergence and separation effects of the independent component of the enhanced CICA are much better than that of the CICA. And in the on-line mode, the stability of the strengthened CICA is better. Table 9.7 shows that compared to the complex FastICA, the estimate result of the complex independent component algorithm is better due to the analyticity of the complex nonlinear function. This on-line algorithm has good prospects for practical applications.

Table 9.7 Comparison of four kinds of methods

Algorithm	Strengthened CICA	CICA	Complex FastICA	JADA
PI	0.08	0.15	0.34	0.37

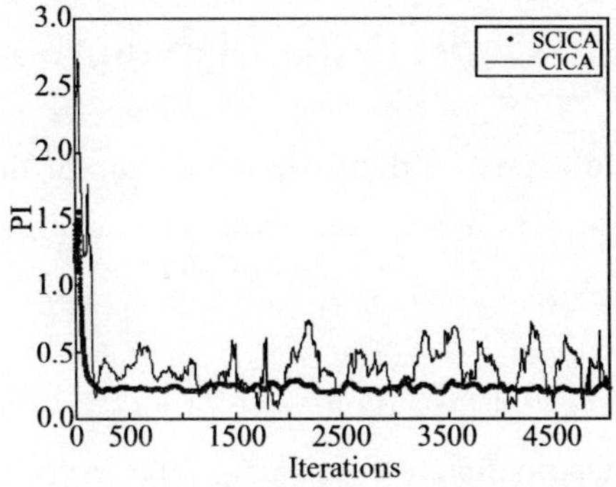

Fig. 9.18 On-line separation effect of SCICA and CICA to a non-Gaussian distribution source signal

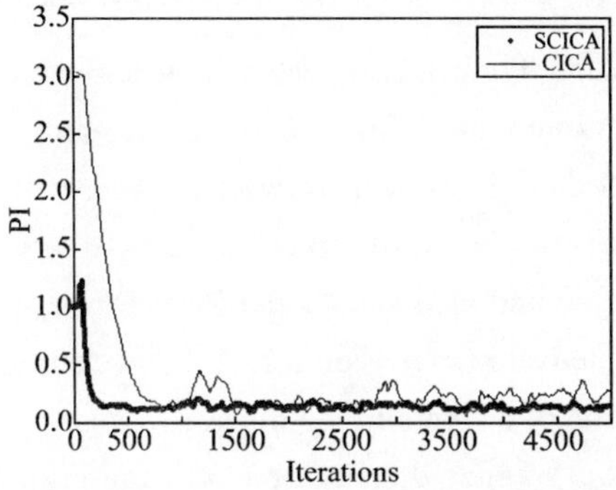

Fig. 9.19 On-line separation effect of SCICA and CICA to a non-Gaussian and a close-Gaussian source signals

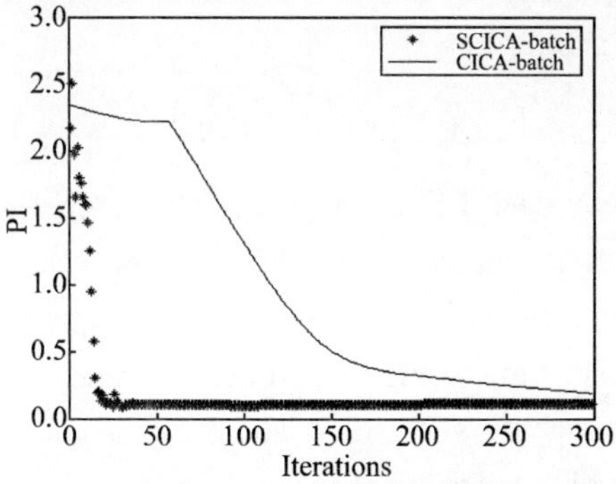

Fig. 9.20 Batch separation effect of SCICA and CICA to a non-Gaussian and a close-Gaussian source signals

9.7 ICA-Based Wigner-Ville Distribution

The Wigner distribution is a two-dimensions function of time and frequency, with a bilinear transform. The Wigner distribution of signal $x(t)$ is defined as[101]

$$\begin{aligned} W_x(t,\omega) &= \int_{-\infty}^{+\infty} x\left(t+\frac{\tau}{2}\right) x^*\left(t-\frac{\tau}{2}\right) e^{-j\tau\omega}\, d\tau \\ &= \int_{-\infty}^{+\infty} X\left(\omega+\frac{\theta}{2}\right) X^*\left(\omega-\frac{\theta}{2}\right) e^{jt\theta}\, d\theta \end{aligned} \tag{9.7.1}$$

and the cross-Wigner distribution of signal $x(t)$, $g(t)$ is defined as

$$\begin{aligned} W_{x,g}(t,\omega) &= \int_{-\infty}^{+\infty} x\left(t+\frac{\tau}{2}\right) g^*\left(t-\frac{\tau}{2}\right) e^{-j\tau\omega}\, d\tau \\ &= \int_{-\infty}^{+\infty} X\left(\omega+\frac{\theta}{2}\right) G^*\left(\omega-\frac{\theta}{2}\right) e^{jt\theta}\, d\theta \end{aligned} \tag{9.7.2}$$

i.e. the definition in the time-domain is equivalent to that in the frequency-domain.

The Wigner distribution not only has lots of characteristics that we expect for joint density but also other simple characteristics. These characteristics satisfy the macroscopic Property required by time-frequency distribution (such as real value, total energy, etc.), and also satisfy the local Property (such as marginal characteristics and limited support). Therefore, it has a definite physical meaning of the energy distribution of a signal in the time or frequency domain.

Ville introduced the Wigner distribution into the signal processing field in 1948, which is called the Wigner-Ville distribution. Replace signal $x(t)$ in the Wigner distribution with its analytical signal $A[x(t)]$, and using the framework of the operator theory of quantum mechanics, the Wigner-Ville distribution (WVD) should be

$$W(t,\omega) = \frac{1}{2\pi}\int_{-\infty}^{\infty} A\left[x\left(t-\frac{\tau}{2}\right)\right]^* A\left[x\left(t+\frac{\tau}{2}\right)\right] e^{-j\tau\omega}\, d\tau \tag{9.7.3}$$

The WVD has many useful time-frequency characteristics, but it has a key problem with multi-component signal interferences and a distribution cross-term. Cross interference term means that the WVD of the signal will generate some meaningless oscillatory components at the time-frequency center coordinates of every two signal components when the signal contains multiple components. In a mathematical sense, the existence of the cross-term is caused by the time and frequency disturbances of the nonlinear transformation.

Assume $x(t)$ is the sum of n components $x_k(t)$:

$$x(t) = \sum_{k=1}^{n} x_k(t)$$

According to the definition of the cross-Wigner spectrum, we have

$$W_x(t,\omega) = \sum_{k=1}^{n} W_{x_k}(t,\omega) + \sum_{k=1}^{n}\sum_{l=1}^{n} 2\mathrm{Re}[W_{x_k,x_l}(t,\omega)] \tag{9.7.4}$$

for two signals, if $x(t)=x_1(t)+x_2(t)$,

$$\begin{aligned} W(t,\omega) &= \int x^*\left(t-\frac{\tau}{2}\right)x\left(t+\frac{\tau}{2}\right)\mathrm{e}^{-\mathrm{j}\tau\omega}\,\mathrm{d}\tau \\ &= \int\left[x_1^*\left(t-\frac{\tau}{2}\right)+x_2^*\left(t-\frac{\tau}{2}\right)\right] \\ &\quad \left[x_1\left(t+\frac{\tau}{2}\right)+x_2\left(t+\frac{\tau}{2}\right)\right]\mathrm{e}^{-\mathrm{j}\tau\omega}\,\mathrm{d}\tau \\ &= W_{11}(t,\omega)+W_{22}(t,\omega)+W_{12}(t,\omega)+W_{21}(t,\omega) \end{aligned}$$

where

$$W_{11}(t,\omega) = \int x_1^*\left(t-\frac{\tau}{2}\right)x_1\left(t+\frac{\tau}{2}\right)\mathrm{e}^{-\mathrm{j}\tau\omega}\,\mathrm{d}\tau$$

$$W_{22}(t,\omega) = \int x_2^*\left(t-\frac{\tau}{2}\right)x_2\left(t+\frac{\tau}{2}\right)\mathrm{e}^{-\mathrm{j}\tau\omega}\,\mathrm{d}\tau$$

$$W_{12}(t,\omega) = \int x_1^*\left(t-\frac{\tau}{2}\right)x_2\left(t+\frac{\tau}{2}\right)\mathrm{e}^{-\mathrm{j}\tau\omega}\,\mathrm{d}\tau$$

$$W_{22}(t,\omega) = \int x_2^*\left(t-\frac{\tau}{2}\right)x_1\left(t+\frac{\tau}{2}\right)\mathrm{e}^{-\mathrm{j}\tau\omega}\,\mathrm{d}\tau$$

there is $W_{12}(t,\omega)=W_{21}^*(t,\omega)$, so:

$$W(t,\omega)=W_{11}(t,\omega)+W_{22}(t,\omega)+2\mathrm{Re}\{W_{12}(t,\omega)\}$$

where $2\mathrm{Re}\{W_{12}(t,\omega)\}$ is the cross-term. Assume W_{11} and W_{22} exist in the neighborhood of point (t_1,ω_1) and point (t_2,ω_2) in the plane domain (t,ω), then the interference term exists in the neighborhood of the center point $(t_m,\omega_m)=\left(\frac{t_1+t_2}{2},\frac{\omega_1+\omega_2}{2}\right)$ and is called outside interference. The cross-spectrum interference term provides a false spectrum distribution, so that disturbs the physical explanation of the Wigner spectrum.

Aiming to solve the problem of the cross intereference term, different kernel functions are used to reject the cross interference in the time-frequency distribution. For example, the Choi-Williams spectrum, which is equivalent to the Wigner spectrum, uses an exponential function as a kernel function which can reduce cross interference by a chosen parameter, but it also reduces aggregation of the time-frequency of the signal to a certain extent. Ref. [102] presented an ICA-based Wigner-Ville distribution that preserves the aggregation of the time-frequency as well as reduces cross interference. Unlike the Wigner-Ville distribution that directly calculates multiple component signals, this presented algorithm first constructs a Wigner-Ville distribution matrix of observed signals whose diagonal elements are independent components extracted by the time-frequency blind separation algorithm, and other non-diagonal elements are cross-term between independent components. Using the indeterminacy of blind separation, we can eliminate the non-diagonal elements by the approximate joint diagonalization method. Then we can reconstruct a Wigner-Ville distribution without cross-term through the amplitude recovery and sum of the auto-term, which improves the resolution of the time-frequency distribution.

Unlike directly summing up the component signal in cross-term analysis, assume that M observed signals are $x_i(t)(i=1, \cdots, M)$ and perform the linear weight sum of each component signal $s_i(t)(i=1, \cdots, N)$ to form observed signals $x_i(t) = \sum_{j=1}^{N} a_{ij} s_j(t)$. WVD of the ith observed signal is as follows:

$$\mathrm{WVD}_{x_i}(t,f) = \sum_{k=1}^{N} a_{ik}^2 \mathrm{WVD}_{kk}(t,f) + \sum_{m=1}^{N} \sum_{\substack{n=1 \\ n \neq m}}^{N} a_{im} a_{in} \mathrm{WVD}_{mn}(t,f)$$

$$i = 1, \cdots, M \tag{9.7.5}$$

where $\mathrm{WVD}_{mn}(t, f)$ is the cross-Wigner-Ville distribution of the mth and nth component signal. Assume the number of observed signals is not less than the number of component signals ($M \geqslant N$) to establish the Wigner-Ville distribution matrix of the observed signal vector $\boldsymbol{x}(t)=[x_1(t), \cdots, x_M(t)]^{\mathrm{H}}$:

$$\mathrm{WVD}_x(t,f) = \int_{-\infty}^{+\infty} \boldsymbol{x}\left(t+\frac{\tau}{2}\right) x^{\mathrm{H}}\left(t+\frac{\tau}{2}\right) \mathrm{e}^{-\mathrm{j}2\pi f \tau} \mathrm{d}\tau$$

$$= \boldsymbol{A} \begin{bmatrix} \mathrm{WVD}_{11}(t,f) & \mathrm{WVD}_{12}(t,f) & \cdots & \mathrm{WVD}_{1N}(t,f) \\ \mathrm{WVD}_{21}(t,f) & \mathrm{WVD}_{22}(t,f) & \cdots & \mathrm{WVD}_{2N}(t,f) \\ \vdots & \vdots & \vdots & \vdots \\ \mathrm{WVD}_{M1}(t,f) & \mathrm{WVD}_{M2}(t,f) & \cdots & \mathrm{WVD}_{MN}(t,f) \end{bmatrix} \boldsymbol{A}^{\mathrm{H}} \tag{9.7.6}$$

where $\boldsymbol{A}=(a_{ij})_{M \times N}$, and the diagonal elements of the WVD matrix are auto terms of the independent component signals, while other non-diagonal elements are cross-terms between components.

Similar to the time-frequency blind separation process, first perform prewhitening of the observed signal and then obtain the whitened signal through the whitening matrix $\boldsymbol{W}$. The WVD matrix of the whitened signal $\boldsymbol{y}(t)$ is as follows:

$$\mathrm{WVD}_y(t,f) = \boldsymbol{W}[\mathrm{WVD}_x(t,f)]\boldsymbol{W}^{\mathrm{H}} = \boldsymbol{U}[\mathrm{WVD}_s(t,f)]\boldsymbol{U}^{\mathrm{H}} \tag{9.7.7}$$

here $\boldsymbol{U}=\boldsymbol{W}\boldsymbol{A}$ is a unitary matrix, so the following formula is given:

$$\boldsymbol{U}^{\mathrm{H}}[\mathrm{WVD}_y(t,f)]\boldsymbol{U} = \mathrm{WVD}_s(t,f) \tag{9.7.8}$$

the above equation in the auto-term time-frequency point (t, f) is a diagonal matrix; that is, the WVD matrix of whitened signal can be transformed into a diagonal matrix by a unitary matrix. Considering calculation error and disturbance noises, the perfect diagonalization of the WVD matrix is impossible. Therefore, the approximate joint diagonalization method is used so that the non-diagonal elements are relatively small compared to the diagonal elements. The WVD diagonalization contrast function is

$$C(\boldsymbol{U}) = \sum_{i=1}^{K} \mathrm{off}\{\boldsymbol{U}^{\mathrm{H}}[\mathrm{WVD}_y(t_i, f_i)]\boldsymbol{U}\} \tag{9.7.9}$$

Search the set of the unitary matrix to optimize the WVD diagonalization contrast function. The unitary matrix $\hat{\boldsymbol{U}}$ that minimizes the WVD diagonalization contrast function is the mixing unitary matrix to be identified. After we obtain the mixing unitary matrix, the WVD of the component signals can be obtained by the following equation:

$$\mathrm{WVD}_{ii}(t, f) = \hat{\boldsymbol{u}}_i^{\mathrm{H}}[\mathrm{WVD}_y(t, f)]\hat{\boldsymbol{u}}_i \qquad i = 1, \cdots, N \tag{9.7.10}$$

where $\boldsymbol{u}_i$ is the ith column vector of the unitary matrix $\hat{\boldsymbol{U}}$.

Due to the indeterminacy of the blind separation results, the WVD_{ii} (t, f) obtained has also the indeterminacy of both amplitude and permutation. The indeterminacy of the permutation will cause the permutation of added items in the sum formula of the auto-terms, but the resulting WVD does not change, and the indeterminacy of the permutation does not have any effect on the reconstruction. Therefore, only the amplitude of the component signals at the observation point is required. Using the inverse method of the block matrix to recover the component signals at the observation point, the whitened observation vector is decomposed as follows:

$$\boldsymbol{y}(t) = \boldsymbol{v}_1(t) + \cdots + \boldsymbol{v}_N(t) \tag{9.7.11}$$

where $\boldsymbol{v}_i(t)$ is the signal vector of the ith component in the observation point. $\hat{\boldsymbol{U}}^{\mathrm{H}}$ is the separating matrix after the mixing unitary matrix $\hat{\boldsymbol{U}}$ is identified through the approximate joint diagonalization method. The WVD of the ith component signal in the jth observed signal $\boldsymbol{v}_i^j(t)$ can be calculated as follows:

$$\mathrm{WVD}_{v_i^j}(t, f) = \hat{u}_i^j \hat{\boldsymbol{u}}_i^* \mathrm{WVD}_y(t, f) \hat{\boldsymbol{u}}_i (\hat{u}_i^j)^* \tag{9.7.12}$$

$$i = 1, \cdots, N \quad j = 1, \cdots, N$$

where $\hat{u}_i^j$ is the jth element of the column vector $\hat{\boldsymbol{u}}_i$.

If the WVD of each component in the observed signal is obtained, the WVD of reconstructed signal can be calculated by the sum of the auto-terms:

$$\mathrm{WVD}_{x_i}(t, f) = \mathrm{WVD}_{v_1^i}(t, f) + \cdots + \mathrm{WVD}_{v_N^i}(t, f)$$

$$i = 1, \cdots, M \tag{9.7.13}$$

The WVD of the above equation doesn't contain cross-terms, so it provides good time-frequency characteristics.

Here is a simulation example: Taking two linear frequency modulation signals $\sin(75\pi t^2)$ and $\sin(75\pi t^2+100\pi t)$ as original signal components, with the sampling rate as 500 Hz, the observed signal is formed by the original signal components passing through the mixing matrix generated randomly by a computer. Therefore, the observed signals contain two component signals whose WVD distributions are shown in Fig. 9.21. It is shown that there is cross interference term between two auto-terms in addition to that of the original signal, and the amplitude of the cross interference term is even larger than the amplitude of the original signal, which severely disturbs the time-frequency energy analysis. The reconstructed WVD of the observed signal by the ICA-based method is shown in Fig. 9.22, and the cross-term is completely eliminated.

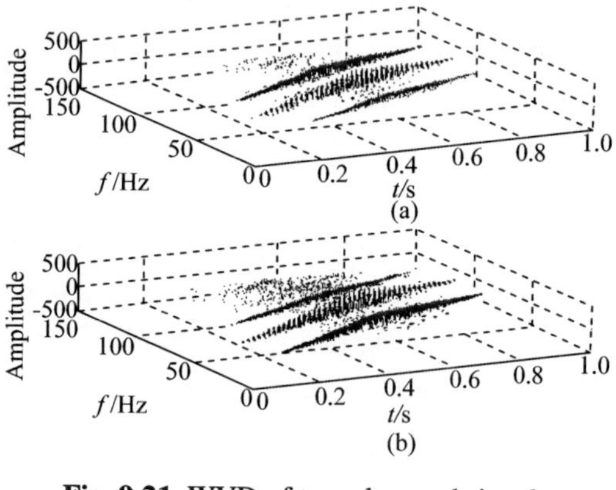

Fig. 9.21 WVD of two observed signals

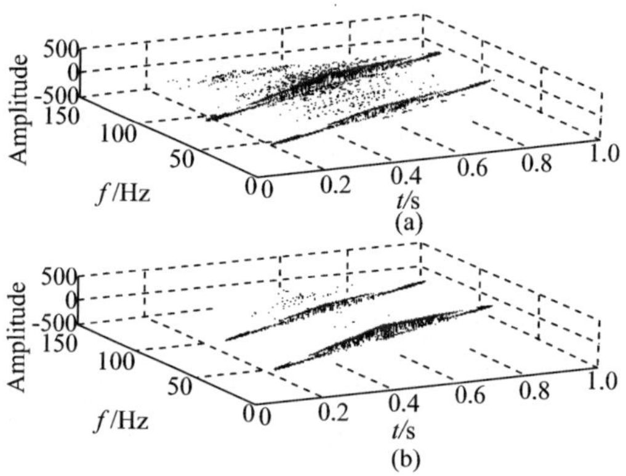

Fig. 9.22 WVD of reconstructed signals

9.8 ICA Feature Extraction

As one of the main important methods of blind separation, the ICA attracts general attention. However, people did not begin to use ICA for feature extraction in pattern recognition until the end of the 1990s. The reason is that the traditional method supposes the features to be extracted come from Gaussian distribution, while the ICA is unable to handle signals that contain two or more Gaussian distributed components. Actually, for many recognition process, only clear features are needed and do not need to be recover the real source signals. In feature extraction, the ICA focuses on maximizing the nongaussianity of features that can reduce the dimension of the feature space and simplify the computation. Although the ICA feature extraction method has many applications in speech recognition[103~105], facial recognition, and colored image recognition, this section discusses only the basic concept and examples of acoustics because of space limitations.

9.8.1 Basic Concept of ICA Feature Extraction

For the basic mathematical model of ICA $\boldsymbol{x}(t)=\boldsymbol{A}\boldsymbol{s}(t)$, the higher-order statistics feature is used to characterize the output signals in the extraction process of independent components, so the iterative process of the ICA algorithm is also the extraction process of the higher-order statistics feature. If $\boldsymbol{x}(t)$ in the model no longer expresses a mixture of several source signals but a source signal itself, then the mutually statistically independent $\boldsymbol{s}(t)$ and real number matrix $\boldsymbol{A}$ extracted through the ICA are the signal features. Assumming the signal to be linearly combined by a series of a mutually independent vectors $\boldsymbol{s}$ and every vector corresponds to a base function, then the signal can be written as: $\boldsymbol{x}=\boldsymbol{A}\boldsymbol{s}$. Every column of the mixing matrix $\boldsymbol{A}$ is called "base function" of that signal, i.e. the higher-order statistics feature vector of the signal. At this time, the purpose of the ICA is to decompose $\boldsymbol{x}$ into a superposition of base functions through the independent component decomposition of the signal.

Similar to the mathematical model of blind separation, assume observed signal $\boldsymbol{x}$ be composed of base function $\boldsymbol{A}$ and an independent coefficient $\boldsymbol{s}$ in feature extraction. For observed variable $\boldsymbol{x}$, it needs to transform the one-demension form into the matrix form:

$$\boldsymbol{x} = \boldsymbol{A}\boldsymbol{s} = \sum_{i=1}^{N} a_i s_i \tag{9.8.1}$$

ICA feature extraction is also used to search an unmixing matrix $\boldsymbol{W}$ to make coefficient $\boldsymbol{s}$ as mutually independent as possible.

$$\boldsymbol{u} = \boldsymbol{W}\boldsymbol{x} \tag{9.8.2}$$

where $\boldsymbol{u}$ is the estimation of $\boldsymbol{s}$, and the signal is zero-mean and unit-variance. For a single source signal, the purpose of the ICA is to decompose the signal into a superposition of base functions through independent component decomposition. Compared to the blind source separation problem, the extraction of the base function is similar to blind separation. To obtain demixing matrix $\boldsymbol{W}$ through the ICA algorithm make $\boldsymbol{WA}=\boldsymbol{DP}$, where $\boldsymbol{DP}$ indicates the scalar transform of unit matrix. The base function $\boldsymbol{A}$ can be obtained by solving the unmixing matrix $\boldsymbol{W}$, i. e., $\boldsymbol{A}=\boldsymbol{W}^{-1}$. Therefore, any one of the ICA algorithms for the blind separation of the mixed signal is suitable for ICA feature extraction.

Take the infomax algorithm developed by Bell and Sejnowski as an example. It chooses a logistic function as a nonlinear function $g(y)=(1+\mathrm{e}^{-y})^{-1}$,whose derivative $g'(y)=\mathrm{e}^{-y}/(1+\mathrm{e}^{-y})^2$ has a super-Gaussian distribution Property; this function conforms to the probability distribution of speech and music signals $p(u_i)=g'_i(u_i)$ and does well in the blind separation of speech and music signals. How accurately the probability function fits with the independent component $\boldsymbol{s}$ is not only important in blind signal separation but also a key for feature extraction. According to experiment results, when using the ICA in feature extraction of a male speech signal with generalized Gaussian model $p(x) \propto \exp(-|x|^q)$, to fit the probability function of the mutually independent coefficient, the estimated q is between 1/3 and 3/4.

Therefore, in order to better fit the probability distribution of the independent components, the ICA based on the Generalized Gaussian Model (GGM) is used for feature extraction of acoustic signals. The updated iterative learning rule is

$$\Delta \boldsymbol{W} \propto \frac{\partial \lg p(\boldsymbol{s})}{\partial \boldsymbol{W}} \boldsymbol{W}^{\mathrm{T}} \boldsymbol{W} = \eta[\boldsymbol{I} - \varphi(\boldsymbol{s})\boldsymbol{s}^{\mathrm{T}}]\boldsymbol{W} \tag{9.8.3}$$

where $\lg p(\boldsymbol{s})$ is the log-likelihood function; $p(\boldsymbol{s})$ is the generalized Gaussian probability density function; η is the learning rate; and vector $\varphi(\boldsymbol{s})$ is defined as $\varphi(\boldsymbol{s}) = \frac{\partial \lg p(\boldsymbol{s})}{\partial \boldsymbol{s}}$. Thus,

$$\varphi_i(s_i) = -q c \sigma_i^{-q} \mid s_i - \mu_i \mid^{q-1} \operatorname{sgn}(s_i - \mu_i) \tag{9.8.4}$$

where $\mu_i=E(s_i)$, and $\sigma_i = \sqrt{E[(s_i - \mu_i)^2]}$ are the mean and variance of the signal respectively. The exponential parameter q is the parameter defined by the generalized Gaussian probability density function from the signal data.

9.8.2 ICA Feature Extraction and Characteristics Analysis of Speech and Music Signals

According to the basic principle of the ICA feature extraction, we must extract the higher-order statistics features from different speech and music signals by using the ICA algorithm based on the GGM, then analyse and discuss characteristics of the ICA base function and projection coefficient[105].

Four kinds of speech and music signals are chosen for the ICA feature extraction: male speech, female speech, jazz, and harmonica music signals. As shown in Fig 9.23, each signal length of 5 s, with a sampling rate of 8 kHz and a sample number of 40,000.

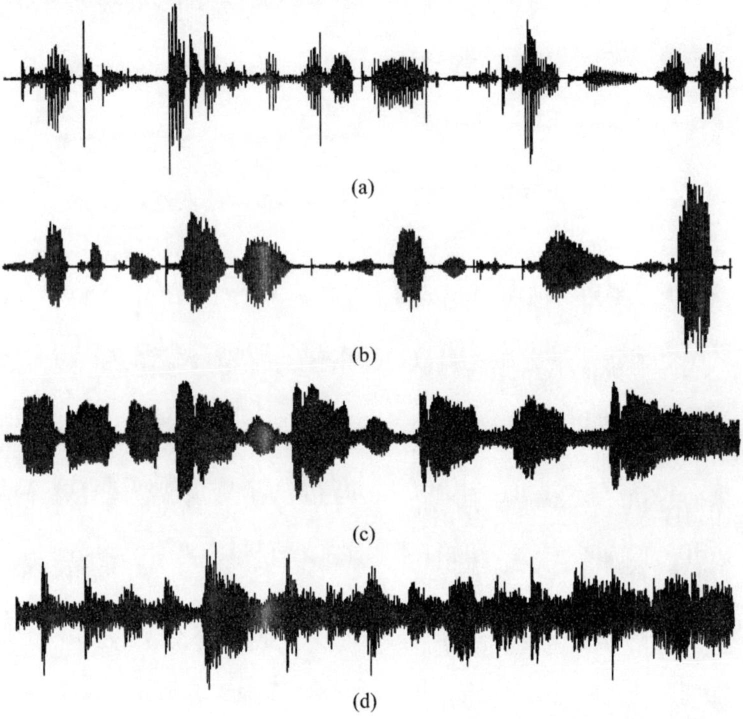

Fig. 9.23 Four kinds of acoustic signals

For these four groups of signals, we decompose them into a matrix form first. The speech and music signals with the 8 kHz sampling rate could be treated as a stationary signal in an interval of 5 ms, so taking every 40 data (i.e. in an interval of 5 ms) as a column, divide the original signal into $40 \times 1,000$ observed signal matrices. To speed up the convergence, the observed signal matrix should be a zero-mean and covariance matrix with a unit-variance through centering and pre-whitening. The iterative process of the ICA algorithm initiates the demixing matrix $\boldsymbol{W}$ as a 40×40 dimensional unit matrix, with the learning rate η of 0.2. As the iterative steps increase, the learning rate η reduces to 0.05 gradually. Figure 9.24 shows 40 base functions extracted from these four acoustic signals.

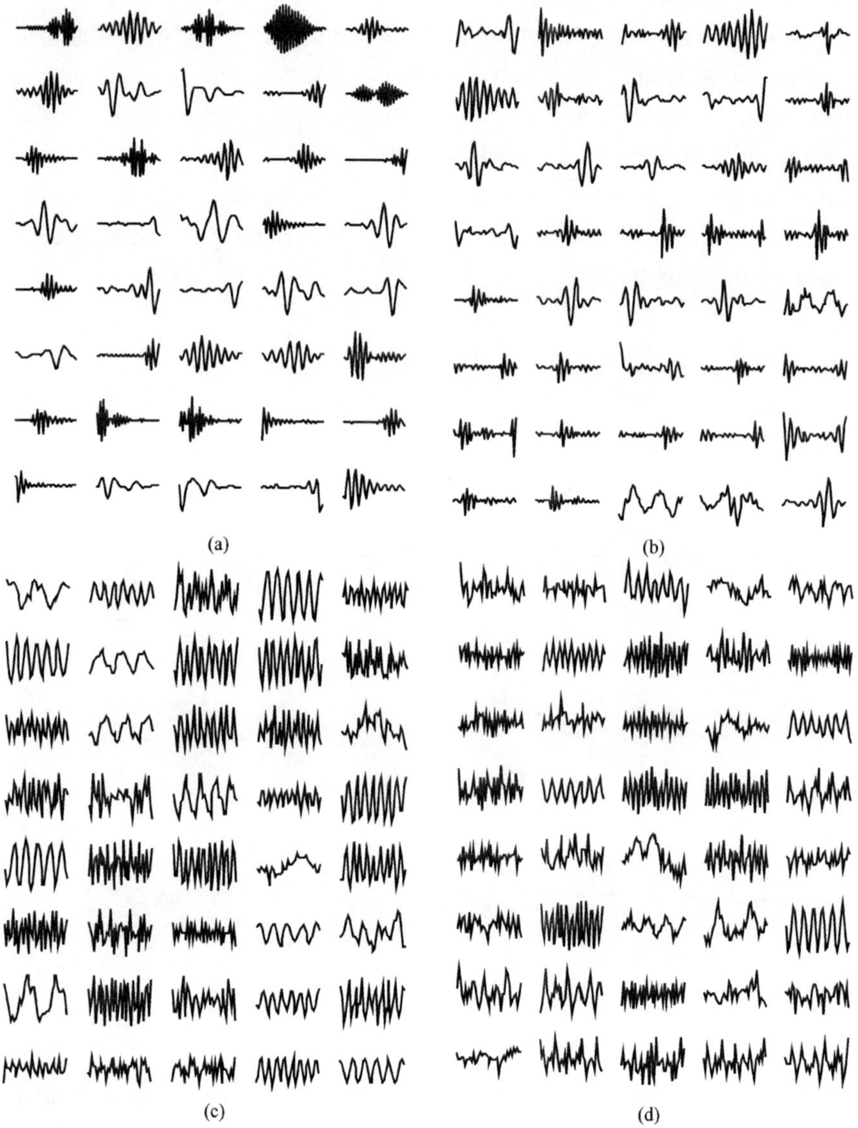

Fig. 9.24 Base function of four acoustic signals

It can be seen from the time-domain figures of the base functions of speech and music signals that the male speech and female speech signals (shown in Fig. 9.24 (a) and (b)) have a stronger local Property in the time-domain and its base functions are concentrated in a shorter time interval. However, the base functions of the jazz and harmonica signals don't have this local Property in the time-domain. Compared with the base functions of the traditional Fourier transform, no matter whether it's for a speech signal or a music signal, there is no local Property in the time-domain.

To verify the Property of the base function of the ICA, a comparison is made between the base functions extracted by the ICA and traditional methods, including the DFT, DCT and PCA the methods.

Equation (9.8.5) shows the DFT transform of signal, which assumes $x(n)$ is a periodic function with the period of N. Based on this assumption, the base functions have a similar Property for a different frequency sometimes, and a large amount of information is lost during the reconstruction of the signal, which is decomposed by these base functions if the signal doesn't have periodicity, as shown in Fig. 9.25(a). Similarly, the DCT in Eq. (9.8.6) also assumes that the signal is a periodic function with a period of $2N$, as shown in Fig. 9.25(b).

$$X_{\mathrm{DFT}}(k) = \sum_{n=1}^{N} x(n)\cos\frac{2\pi(k-1)}{N}(n-1) \tag{9.8.5}$$

$$X_{\mathrm{DCT}}(k) = \sum_{n=1}^{N} x(n)\cos\frac{\pi(k-1)}{2N}(2n-1) \tag{9.8.6}$$

the PCA base function decomposition of the signal is a process of removing the correlation. The PCA base function is the eigenvector of the covariance matrix $\langle \boldsymbol{x}\boldsymbol{x}^{\mathrm{T}} \rangle$, which is only suitable to second-order feature extraction but incompetent for higher-order statistics properties. As shown in Fig. 9.25(c), the PCA base function also has global Property in the time-domain.

It can be seen from Fig. 9.25 that the base functions of the DFT and DCT are similar in a global Property in the time-domain, and the PCA base functions also have global Property in the time-domain. Moreover, for any speech or music signal, the base functions of the DFT and DCT are constant. In contrast, the ICA base functions for different signals are different, especially because they have a stronger local Property for speech signals.

Furthermore, making a further analysis of the frequency feature of the ICA basis function. Figure 9.26 (a)~(d) show the corresponding frequency spectrum of the base functions in Fig. 9.25, (a)~(d) respectively.

It can be seen that the base functions of different signals all have a local Property in the frequency-domain. That is to say, the feature of the base function can characterize the signal without overlap, so the ICA base function can effectively be used feature in the pattern recognition and classification of acoustics.

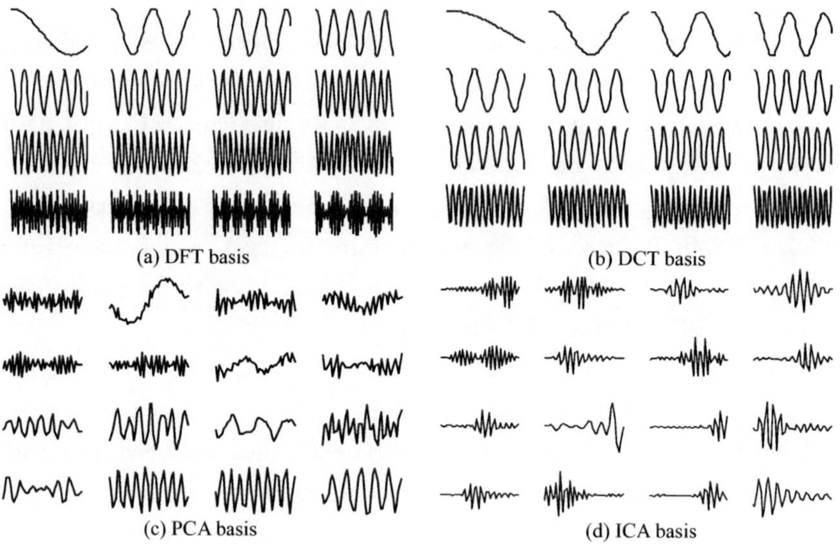

Fig. 9.25 Base functions of DFT, DCT, PCA and ICA

In the feature extraction of the ICA algorithm, the voice signal is decomposed into a base function vector and mutually independent coefficients. This process can be regarded as linear transformation in the sense of a base function, because the result of the algorithm is the original signal matrix multiplied by the demixing matrix $\boldsymbol{W}$, which transforms the original signal into a mutually independent vector matrix. It can be regarded as the projection of the base function in every direction. Take the male speech signal as an example and discuss the Property after ICA transformation, which is the Property of the independent coefficients $\boldsymbol{s}$. The projection coefficients of the male speech signal can be obtained through traditional methods and ICA transformation. Figure 9.27 shows the log histogram of DFT, DCT, PCA and ICA.

Figure 9.27 shows that the signals transformed by PCA and ICA have more super-Gaussianity than DFT and DCT, in which the super-Gaussianity is most obvious for the ICA (solid line in the figure). The sparsity of such distribution is very useful for speech signal coding. This also gives a demonstration of the efficiency of higher order feature extraction and decomposition by ICA; meanwhile, it also shows that the structural characteristics of higher-order statistics can realize the attribute of speech itself.

Similarly, this feature extraction method can also be adopted for the radiated noise from ships and the ocean environment noise.

Figure 9.28 shows the radiated noise from some ships over a calm sea surface

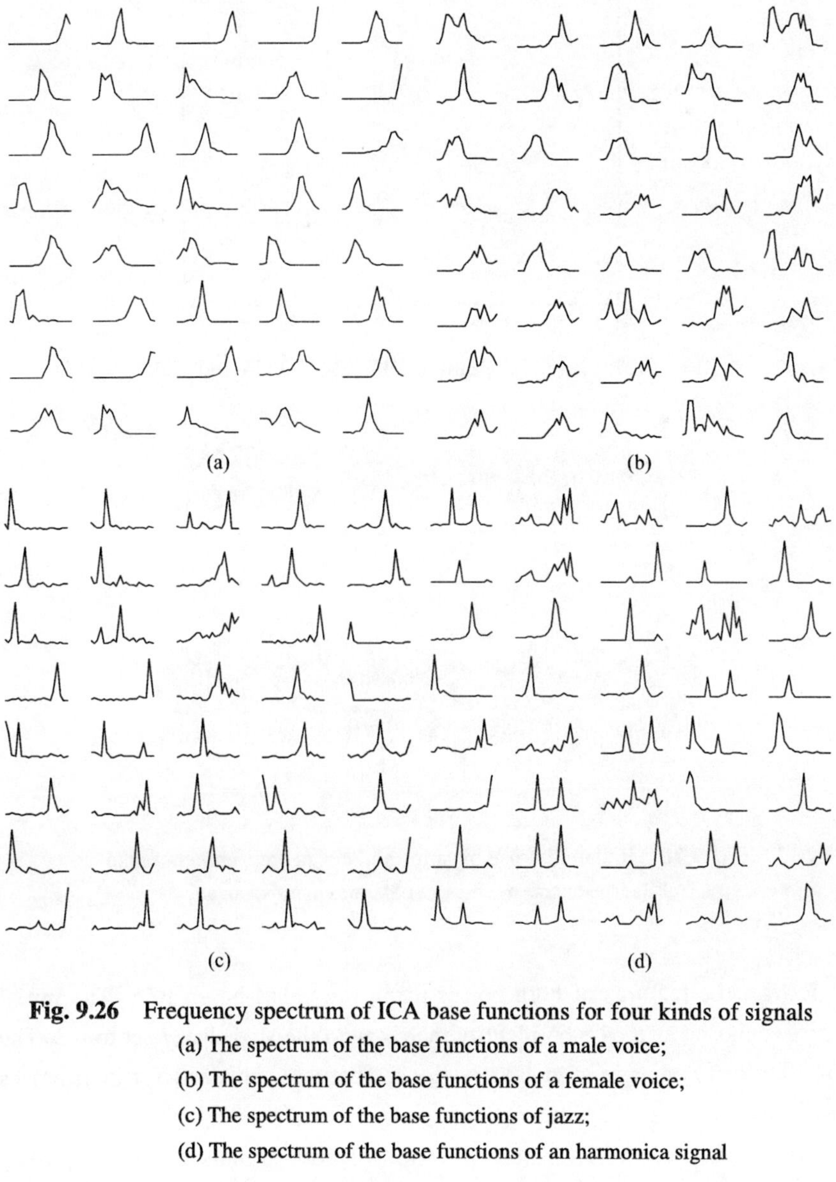

Fig. 9.26 Frequency spectrum of ICA base functions for four kinds of signals
(a) The spectrum of the base functions of a male voice;
(b) The spectrum of the base functions of a female voice;
(c) The spectrum of the base functions of jazz;
(d) The spectrum of the base functions of an harmonica signal

and the ocean environment noise that has not been disturbed by the ship.The sampling rate is 1 kHz and the duration is 20 seconds for each signal. Assume that the signal is stationary within a shorter duration of time. Transform each signal into a $64 \times 2{,}000$ observation matrix; overlae the segments of sampling signal by 10 samples each; then whiten and centralize the observation matrix.

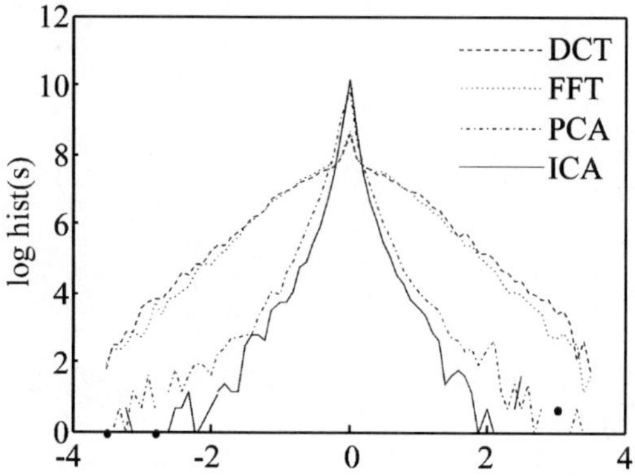

Fig. 9.27 Log-histogram of DFT, DCT, PCA and ICA

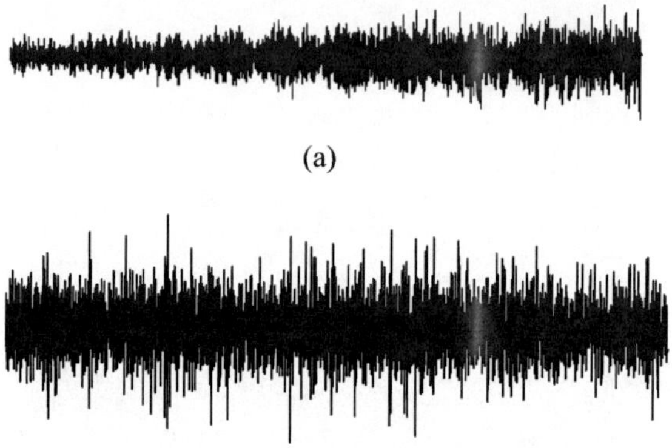

(a)

(b)

Fig. 9.28 Radaited noise by ships and ocean environment noise
(a) Radaited noise by ships; (b) Marine environment noise

Extract the features of both noises using the GGM-based ICA. The initial matrix is chosen as a 64×64 identity matrix with the step size η set to 0.3. The algorithm can converge after 60 iterations. The base functions for both noises are illustrated in Fig. 9.29 and Fig. 9.30.

The features of the ICA base function of both signals in the frequency domain are shown in Fig. 9.31 and Fig. 9.32.

It can be seen from Fig. 9.31 that for the radiated noise by ships, the base functions have good local properties and low frequency features in the frequency-domain, which is consistent with the fact that most identifiable information about radiated noise from ships is located in a low frequency. For the ocean environment noise as shown in Fig. 9.32, the base functions are distributed in a wider frequency

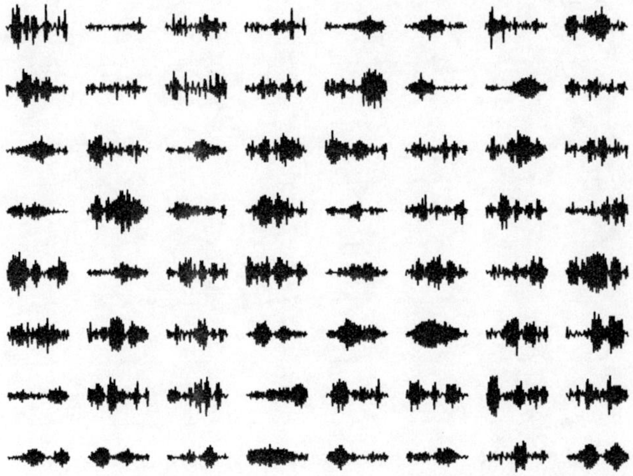

Fig. 9.29 Base functions for radiated noise by ship

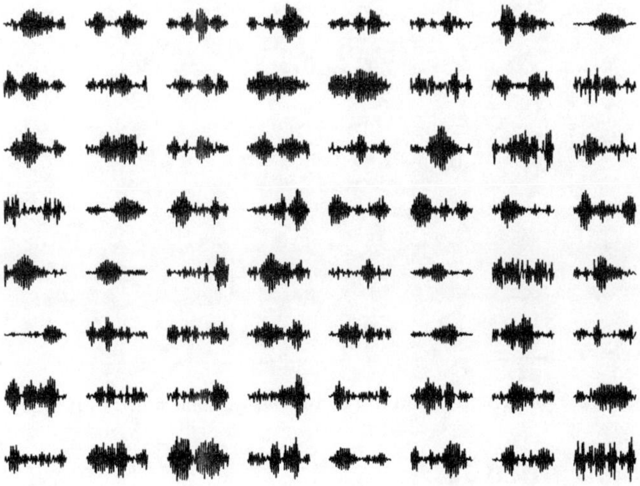

Fig. 9.30 Base functions for ocean environment noise

range, which is the main difference between the two noises. The coeffieients of the radiated noise have better sparsity after ICA transformation. This shows that the higher order feature extraction and decomposition can reflect the intrinsic attribute of signals and is helpful to make a distinction between both noises. The local Property of the ICA base functions and the sparsity of the coefficients are very helpful for denoising and signal coding.

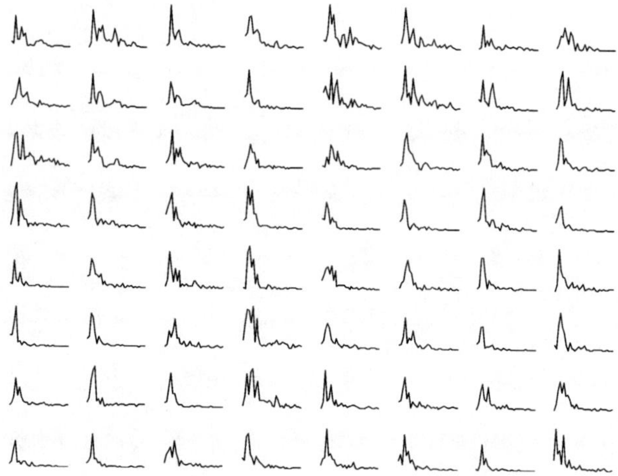

Fig. 9.31 Feature of base function of radiated space ship in frequency domain.

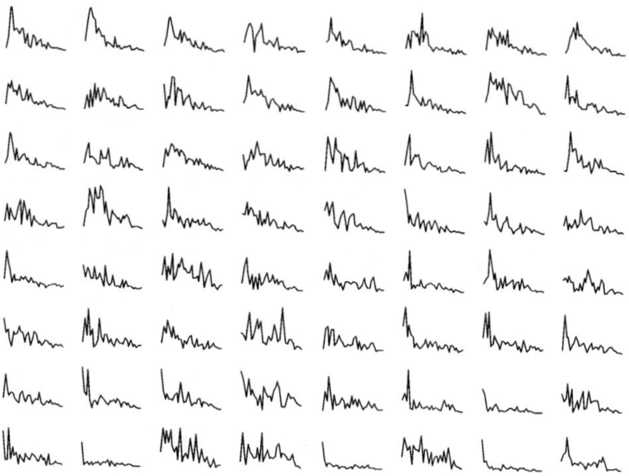

Fig. 9.32 Feature of base function of ocean environment noise in frequency domain

9.9 Constrained ICA

9.9.1 Introduction

As described in previous chapters, the ICA has an intrinsic order indeterminacy, which makes it hard to perform an inter-subject analysis, since it's tricky to pick up the same interested component from different subjects. A straightforward solution is to sort the ICA output using prior information. like the experiment design. Alternatively, prior information could be incorporated into the blind source separation process as a constraint to select the interested components, as was done in the so-called constrained ICA (cICA). cICA was first proposed by Lu and Rajapakse[106~108] and explored by many other researchers[109~115]. In this section, we will focus on the algorithm proposed by Lu and Rajapakse[106~108].

9.9.2 Algorithm

Assume the observed signal $\boldsymbol{x}=(x_1, x_2,\cdots, x_n)^{\mathrm{T}}$ to be a linear mixture of several unknown mutually independent sources $\boldsymbol{s}=(s_1, s_2,\cdots, s_m)^{\mathrm{T}}$: $\boldsymbol{x}=\boldsymbol{A}\boldsymbol{s}$, where $\boldsymbol{A}$ is an $n\times m$ matrix ($m\leqslant n$); a standard independent component analysis (ICA) seeks an $m\times n$ unmixing matrix $\boldsymbol{W}$, such that $\boldsymbol{s}$ can be approximately recovered by $\boldsymbol{y}=\boldsymbol{W}\boldsymbol{x}$, with $\boldsymbol{y}=(y_1, y_2,\cdots, y_m)^{\mathrm{T}}$. With an appropriate measure of output dependence as the object function $O(\cdot)$, the ICA can be formed as an optimization problem to minimize this object function. A widely used object function is the sum of the negentropy of the output components:

$$O(\boldsymbol{y}) = \sum_i J(y_i) \tag{9.9.1}$$

A reliable and flexible approximation of negentropy is as follows:

$$J(y_i) \approx \rho[E\{G(y_i)\} - E\{G(v)\}]^2 \tag{9.9.2}$$

where ρ is a positive constant, $G(\cdot)$ is a nonquadratic function, and v is a Gaussian variable with a zero mean and unit variance.

Without any prior information about $\boldsymbol{s}$ and $\boldsymbol{A}$, the ICA solution is up to the scale indeterminacy and order indeterminacy. Since a permutation matrix $\boldsymbol{P}$ and a nonzero scalar λ and their inverses can be inserted into the mixing model without disturbing the observed signal, $\boldsymbol{x}=\boldsymbol{A}(\lambda\boldsymbol{P})^{-1}\lambda\boldsymbol{P}\boldsymbol{s}$, the independent sources can be recovered as $\lambda\boldsymbol{P}\boldsymbol{s}$, *i. e.*, with an arbitrary scale and an arbitrary order. The idea of the cICA[106~108] is to incorporate prior information into the ICA learning process to automatically pick up the interested components. Using $\boldsymbol{q}(\boldsymbol{y})=(q_1(\boldsymbol{y}),\cdots, q_M(\boldsymbol{y}))^{\mathrm{T}}$ ($M\leqslant m$) to represent the constraint function, the cICA learning process can be expressed as below:

$$\begin{aligned} &\text{maximize } O(\boldsymbol{y}) \\ &\text{subject to } \boldsymbol{q}(\boldsymbol{y}) \leqslant 0 \text{ and } \boldsymbol{h}(\boldsymbol{y}) = 0 \end{aligned} \tag{9.9.3}$$

where $\boldsymbol{h}(\boldsymbol{y})=(h_1(\boldsymbol{y}),\cdots,h_N(\boldsymbol{y}))^{\mathrm{T}}$ defines the N equality constraints. To facilitate an easy ICA algorithm and to bound the learning step, the separating matrix $\boldsymbol{W}$ is usually required to be a unit, and the equality constraint is $\boldsymbol{h}(\boldsymbol{y})=\boldsymbol{W}\boldsymbol{W}^{\mathrm{T}}-\boldsymbol{I}$.

By converting the inequality constraints into equality constraints, $q_i(\boldsymbol{y})+$

$z_i^2=0(i=1,\cdots,M$, and z_i is a slack variable), the inequality constrained optimization problem described by Eq. (9.9.3) can be solved by the Lagrange multipliers method using the following augmented Lagrangian function[116]:

$$L(\boldsymbol{W},\boldsymbol{\mu},z,\boldsymbol{\lambda}) = O(\boldsymbol{y}) + \boldsymbol{\mu}^{\mathrm{T}}(\boldsymbol{q}(\boldsymbol{y}) + z^2) + \frac{1}{2}\gamma \| \boldsymbol{q}(\boldsymbol{y}) + z^2 \|^2 + \boldsymbol{\lambda}^{\mathrm{T}}\boldsymbol{h}(\boldsymbol{y}) + \frac{1}{2}\gamma \| \boldsymbol{h}(\boldsymbol{y}) \|^2 \tag{9.9.4}$$

where $\boldsymbol{\mu}=(\mu_1,\cdots,\mu_M)^{\mathrm{T}}$ and $\boldsymbol{\lambda}=(\lambda_1,\cdots,\lambda_N)^{\mathrm{T}}$ are two sets of Lagrange multipliers, γ is the scalar penalty parameter, $z=(z_1,\cdots,z_i,\cdots,z_M)^{\mathrm{T}}$, and $\|\cdot\|$ means the Euclidean norm. The last item $\frac{1}{2}\gamma\|\cdot\|^2$ is the penalty term to ensure that the local convexity assumption holds at a solution of the optimization problem.

With Eq. (9.9.4), the optimal value of z can be obtained by carrying out a maximization of $L(\boldsymbol{W},\boldsymbol{\mu},z,\boldsymbol{\lambda})$ with respect to z[116] as follows:

$$\begin{aligned}\max_{z} L(\boldsymbol{W},\boldsymbol{\mu},z,\boldsymbol{\lambda}) &= \max_{z^2} L(\boldsymbol{W},\boldsymbol{\mu},z,\boldsymbol{\lambda}) \\ &= \max_{z^2}\{\boldsymbol{\mu}^{\mathrm{T}}(\boldsymbol{q}(\boldsymbol{y}) + z^2) + \frac{1}{2}\gamma \| \boldsymbol{q}(\boldsymbol{y}) + z^2 \|^2\} \\ &= \max_{\psi\geqslant 0}\{\boldsymbol{\mu}^{\mathrm{T}}(\boldsymbol{q}(\boldsymbol{y}) + \boldsymbol{\psi}) + \frac{1}{2}\gamma \| \boldsymbol{q}(\boldsymbol{y}) + \boldsymbol{\psi} \|^2\}\end{aligned} \tag{9.9.5}$$

where $\boldsymbol{\psi}=z^2$, and $\boldsymbol{O}$ means a zero vector with a dimension of M. Since the function in braces above is quadratic in $\boldsymbol{\psi}$, it will reach its global minimum when its derivative with respect to ψ is zero. Correspondingly we can obtain

$$\boldsymbol{\mu} + \gamma[\boldsymbol{q}(\boldsymbol{y}) + \boldsymbol{\psi}] = 0 \tag{9.9.6}$$

and

$$\boldsymbol{\psi} = z^2 = -\frac{1}{\gamma}[\boldsymbol{\mu} + \gamma \boldsymbol{q}(\boldsymbol{y})] \tag{9.9.7}$$

As z_i^2 is always no less than 0, the final solution of Eq. (9.9.5) is

$$(z^*)^2 = \max\left\{0, -\frac{1}{\gamma}[\boldsymbol{\mu} + \gamma \boldsymbol{q}(\boldsymbol{y})]\right\} \tag{9.9.8}$$

After substituting $(z^*)^2$ into Eq. (9.9.4), the augmented Lagrangian function can be redescribed as

$$\begin{aligned}L_{z^*}(\boldsymbol{W},\boldsymbol{\mu},\boldsymbol{\lambda}) &= J(\boldsymbol{y}) - \sum_i \frac{1}{2\gamma}[\max^2\{\mu_i + \gamma q_i(\boldsymbol{y}),0\} - \mu_i^2] + \boldsymbol{\lambda}^{\mathrm{T}}\boldsymbol{h}(\boldsymbol{y}) + \frac{1}{2}\gamma \| \boldsymbol{h}(\boldsymbol{y}) \|^2 \\ &= J(\boldsymbol{y}) - Q + \boldsymbol{\lambda}^{\mathrm{T}}\boldsymbol{h}(\boldsymbol{y}) + \frac{1}{2}\gamma \| \boldsymbol{h}(\boldsymbol{y}) \|^2\end{aligned} \tag{9.9.9}$$

with $Q = \sum_i \frac{1}{2\gamma}[\max^2\{\mu_i + \gamma q_i(y), 0\} - \mu_i^2]$.

A Newton-like learning method can then be used to update $\boldsymbol{W}$ in order to maximize L_{z^*}:

$$\boldsymbol{W}_{k+1} = \boldsymbol{W}_k - \eta(L''_{z^*\boldsymbol{W}_k^2})^{-1} L'_{z^*\boldsymbol{W}_k} \tag{9.9.10}$$

where $L'_{z^*\boldsymbol{W}_k}$ is the first derivative of L_{z^*} with respect to $\boldsymbol{W}$ at the kth iteration:

$$L'_{z^*\boldsymbol{W}_k} = \bar{\rho}E\{xG'(\boldsymbol{y})\} - \boldsymbol{\mu}E\{xq'(\boldsymbol{y})\} - \boldsymbol{\lambda}E\{\boldsymbol{xy}\}$$

here $\bar{\rho} = \pm\rho$, with the sign coincident with $E\{G(\boldsymbol{y})\} - E\{G(\boldsymbol{v})\}$, and $G'_y(\boldsymbol{y})$ and $\boldsymbol{q}'(\boldsymbol{y})$ are the first derivatives of $G_y(\boldsymbol{y})$ and $\boldsymbol{q}(\boldsymbol{y})$ with respect to $\boldsymbol{y}$.

$L''_{z^*\boldsymbol{W}_k^2}$ is the Hessian matrix and can be approximated by

$$L''_{z^*\boldsymbol{W}_k^2} = \zeta(\boldsymbol{W}_k)\boldsymbol{R}_{xx} \tag{9.9.11}$$

where $\zeta(\boldsymbol{W}_k) = \bar{\rho}E\{G''_{y^2}(\boldsymbol{y})\} - \boldsymbol{\mu}E\{\boldsymbol{q}''_{y^2}(\boldsymbol{y})\} - \boldsymbol{\lambda}$, $\boldsymbol{R}_{xx} = E\{\boldsymbol{xx}^{\mathrm{T}}\}$, $G''_{y^2}(\boldsymbol{y})$ and $\boldsymbol{q}''_{y^2}(\boldsymbol{y})$ are the second derivatives. With the above approximation, the cICA learning rule can then be given by

$$\boldsymbol{W}_{k+1} = \boldsymbol{W}_k - \eta(\boldsymbol{R}_{xx})^{-1} L'_{z^*\boldsymbol{W}_k} / \zeta(\boldsymbol{W}_k) \tag{9.9.12}$$

With the prewhitening process, $\boldsymbol{R}_{xx} = E\{\boldsymbol{xx}^{\mathrm{T}}\} = \boldsymbol{I}$, the above iteration can be further simplified as

$$\boldsymbol{W}_{k+1} = \boldsymbol{W}_k - \eta L'_{z^*\boldsymbol{W}_k} / \zeta(\boldsymbol{W}_k) \tag{9.9.13}$$

The learning process for $\boldsymbol{\lambda}$ can be avoided by adding the following normalization step:

$$\boldsymbol{W}_{k+1} = \boldsymbol{W}_{k+1} / \| \boldsymbol{W}_{k+1} \| \tag{9.9.14}$$

the learning rule for another Lagrange multiplier $\boldsymbol{\mu}$ can be obtained as below by setting the derivative of $L_z^*(\boldsymbol{W}, \boldsymbol{\mu}, \boldsymbol{\lambda})$ with respect to $\boldsymbol{\mu}$ to zero and then taking some trivial mathematical operations:

$$\mu_{k+1} = \max\{0, \mu_k + \gamma q(\boldsymbol{W}_k)\} \tag{9.9.15}$$

The cICA learning algorithm can be described as

① Zero mean and prewhitening the input data;

② Choose a proper penalty parameter γ;

③ Initiate $\boldsymbol{W}$ and $\boldsymbol{\mu}$;

④ Determine the sign of $\bar{\rho}$;

⑤ Update $\boldsymbol{\mu}$ with Eq. (9.9.15);

⑥ Update $\boldsymbol{W}$ with Eq. (9.9.13);

⑦ Normalize $\boldsymbol{W}$ with Eq. (9.9.14);

⑧ Go back to step 4 until $|\boldsymbol{W}_k^{\mathrm{T}} \boldsymbol{W}_{k+1}|$ approaches 1 with some error.

9.9.3 Applications and Discussions

cICA has shown great promise in its applications to fetal ECG extractions[111], and fMRI data analysis[108,109,114]. A caveat is that by incorporating prior information into the blind separation process, we are actually sacrificing the data-driven Property of the ICA, which can consequently give bias to the blind source separation. Additionally, the appropriate choice and adjustment of the threshold for the constraints to be as close to the reference signals as possible is crucial for a successful cICA-based component extraction.

9.10 Particle Filtering Based Nonlinear and Noisy ICA

9.10.1 Introduction

With the development of independent component analysis (ICA), noisy or nonlinear ICA has become a more and more important area[117,118]. Noisy ICA solutions are mainly based on the local linear projective noise reduction[118,119]and Bayesian approach or the maximum-likelihood method with generative parametric density models for independent distributions [120~122]. These methods all propose to extract the clean sources from the noisy mixtures directly. As for the nonlinear ICA, it has been a popular topic since Taleb and Jutten studied the post nonlinear ICA [123].The Informax technique and Bayesian methods are two main tools used for nonlinear ICA[124,125]. Meanwhile, nonlinear and noisy ICA progresses slowly because of the difficult nature of the problem, and few studies have been published. Lappalainen and Honkela's method required that the source distribution is Gaussian. Attias assumed the source distribution to be a mixture of Gaussians. Maeda *et al.* proposed a nonlinear and noisy extension of the ICA [126] where the distributions of the sources were fixed. As yet, no very robust methods have been presented.

To enhance noisy speech, particle filtering (PF) was used without extensive a priori knowledge[127]. In Ref.[127], the relation between noisy speech and estimated clean speech is linear. In this study, we extend that relation to be nonlinear. In the state-space problems, if the models are nonlinear or noise is non-Gaussian or the system is dynamic, the traditional Kalman filter (KF) is no longer effective as in the linear and Gaussian cases, and PF is particularly useful in dealing with those problems[128,129]. Along these lines, we adopt PF to resolve the nonlinear and noisy ICA in this study[130]. The special assumptions are that the post nonlinear function and the distribution of background noise are already known or can be evaluated.

The key step is to estimate the noise free mixtures from the noisy mixtures and the linear mixtures from the nonlinear mixtures dimension by dimension through PF. In this study noise means free that the noise is essentially removed from the data. Within this framework, both the nonlinear and noisy determined ICA, as well as the nonlinear and noisy underdetermined ICA problems are perfectly treated.

9.10.2 Problem Statement

Generally, the nonlinear and noisy ICA can be described as

$$z_i(t) = h\left[\sum_{j=1}^{n} a_{ij} s_j(t)\right] + v_i(t) \tag{9.10.1}$$

where, $h(\cdot)$ is a post nonlinear function that is differentiable, $z_i(t)$ is the observed signal, $s_j(t)$ is the independent source, $v_i(t)$ is the background noise, $\boldsymbol{A}=\{a_{ij}\}$ is the mixing matrix, and $i=1,\cdots,m$, and $j=1,\cdots,n$. The simpler model is the classical ICA that may be stated as

$$x_i(t) = \sum_{j=1}^{n} a_{ij} s_j(t) \tag{9.10.2}$$

Currently available noisy ICA or nonlinear ICA methods directly extract sources from the mixtures. However, combining Eqs. (9.10.1) and (9.10.2), we find that the nonlinear and noisy ICA will not become so difficult as long as the linear and clean mixtures $x_i(t)$ are estimated. To reach this goal, the next five assumptions are made in our framework.

Assumption 1: Sources can be nonstationary;

Assumption 2: The post nonlinear function is known a priori or can be estimated;

Assumption 3: Distributions of background noise are known a priori or can be estimated;

Assumption 4: Sources should be mutually independent;

Assumption 5: The mixing matrix is of full rank.

With assumption 1, the time-varying autoregressive (TVAR)[131] model can be used to describe the evolution of the clean linear mixtures as follows:

$$x_i(t) = \boldsymbol{\alpha}_i(t-1) \cdot \boldsymbol{x}_i(t-1) + e_i(t-1) \tag{9.10.3}$$

where, $\boldsymbol{\alpha}_i(t-1)=[\alpha_i(t-1,0),\cdots,\alpha_i(t-1,p-1),\cdots,\alpha_i(t-1,p-1)]$ is a row vector, P is the TVAR order, $\alpha_i(t-1,p)$ is the TVAR coefficient, $\boldsymbol{x}_i(t-1)=[x_i(t-1),\cdots,x_i(t-p-2),\cdots,x_i(t-p-2)]^{\mathrm{T}}$ is the column vector, and $\boldsymbol{e}_i(t-1)$ is the excitation noise. Different categories of clean mixtures should correspond to different excitation noises. Obviously, the transition function is the time-varying row vector:

$$f_i(t-q)=\boldsymbol{\alpha}_i(t-q) \tag{9.10.4}$$

Assumption 2 and assumption 3 help to establish the relationship between the clean linear mixtures $x_i(t)$ and noisy nonlinear mixtures $z_i(t)$. Eventually, we construct the dynamic state-space model as

$$\begin{aligned}\boldsymbol{\theta}_i(t)&=\{\boldsymbol{f}_i(t-q)\boldsymbol{x}_i(t-q)+\boldsymbol{e}_i(t-q)\}_{q=1}^{Q}\\ z_i(t)&=h[\boldsymbol{x}_i(t)]+\boldsymbol{v}_i(t)\end{aligned} \tag{9.10.5}$$

where $\boldsymbol{x}_i(t)=\boldsymbol{f}_i(t-1)\ \boldsymbol{x}_i(t-1)+\boldsymbol{e}_i(t-1)$, $Q=P$, q being the index for the state.

Assumption 4 and assumption 5 guarantee that sources can be extracted by ICA after the clean linear mixtures is estimated.

Equation (9.10.5) defines a nonlinear, non-Gaussian and dynamic system where neither the state nor the transition function is known. So, the problem belongs to the dual estimation [132].

9.10.3 Particle Filtering

PF is a sequential Monte Carlo methodology where the basic idea is the recursive computation of relevant probability distributions using the concepts of importance sampling and approximation of the probability distributions with discrete random measures[128].

Consider a system with a state-space representation,

$$\begin{aligned}\boldsymbol{\alpha}_t&=f(\boldsymbol{\alpha}_{t-1})+\boldsymbol{u}_t\\ \boldsymbol{\beta}_t&=h(\boldsymbol{\alpha}_{t-1})+\boldsymbol{v}_t\end{aligned} \tag{9.10.6}$$

where $\boldsymbol{\alpha}_t$ are unobserved states of the system, $\boldsymbol{\beta}_t$ are the observations at time t, $f(\cdot)$ and $h(\cdot)$ are the prespecified state evolution and observation function, and $\boldsymbol{u}_t$ and $\boldsymbol{v}_t$ are process and observation noises respectively.

The main task of sequential signal processing is the estimation of the state $\boldsymbol{\alpha}_t$ recursively from the observations $\boldsymbol{\beta}_t$. In general, there are three probability distribution functions of interest, and they are the filtering distribution $p(\boldsymbol{\alpha}_t|\boldsymbol{\beta}_{0:t})$, the predictive distribution $p(\boldsymbol{\alpha}_{t+l}|\boldsymbol{\beta}_{0:t}), l\geqslant 1$, and the smoothing distribution $p(\boldsymbol{\alpha}_t|\boldsymbol{\beta}_{0:T}), T>t$. All the information about $\boldsymbol{\alpha}_t$ regarding filtering, prediction, or smoothing is captured by these distributions respectively, and so the main goal is their tracking, which is obtaining $p(\boldsymbol{\alpha}_t|\boldsymbol{\beta}_{0:t})$ from $p(\boldsymbol{\alpha}_{t-1}|\boldsymbol{\beta}_{0:t-1})$, $p(\boldsymbol{\alpha}_{t+l}|\boldsymbol{\beta}_{0:t})$ from $p(\boldsymbol{\alpha}_{t+l-1}|\boldsymbol{\beta}_{0:t})$, or $p(\boldsymbol{\alpha}_t|\boldsymbol{\beta}_{0:T})$ from $p(\boldsymbol{\alpha}_{t+1}|\boldsymbol{\beta}_{0:T})$.

Assume that the required pdf $p(\boldsymbol{\alpha}_{t-1}|\boldsymbol{\beta}_{0:t-1})$ at time $t-1$ is available. Using the system model in Eq. (9.10.6), we may obtain the prior pdf of the state at time t via the Chapman-Kolmogorov equation

$$p(\boldsymbol{\alpha}_t \mid \boldsymbol{\beta}_{0:t-1}) = \int p(\boldsymbol{\alpha}_t \mid \boldsymbol{\alpha}_{t-1})\, p(\boldsymbol{\alpha}_{t-1} \mid \boldsymbol{\beta}_{0:t-1})\, \mathrm{d}\boldsymbol{\alpha}_{t-1} \tag{9.10.7}$$

at time step t, a measurement $\boldsymbol{\beta}_t$ becomes available, and this may be used to update the prior via Bayes' rule

$$p(\boldsymbol{\alpha}_t \mid \boldsymbol{\beta}_{0:t}) = \frac{p(\boldsymbol{\beta}_t \mid \boldsymbol{\alpha}_t)\, p(\boldsymbol{\alpha}_t \mid \boldsymbol{\beta}_{0:t-1})}{p(\boldsymbol{\beta}_t \mid \boldsymbol{\beta}_{0:t-1})} \tag{9.10.8}$$

where the normalizing constant

$$p(\boldsymbol{\beta}_t \mid \boldsymbol{\beta}_{0:t-1}) = \int p(\boldsymbol{\beta}_t \mid \boldsymbol{\alpha}_t)\, p(\boldsymbol{\alpha}_t \mid \boldsymbol{\beta}_{0:t-1})\, \mathrm{d}\boldsymbol{\alpha}_t \tag{9.10.9}$$

depends on the likelihood functions $p(\boldsymbol{\beta}_t|\boldsymbol{\alpha}_t)$ defined by the measurement model in Eq. (9.10.6) and the known statistics of observation noises. In the updated stage Eq. (9.10.8), the measurement $\boldsymbol{\beta}_t$ is used to modify the prior density to obtain the required posterior density of the current state.

The recurrence relations Eqs. (9.10.7) and (9.10.8) form the basis for the optimal Bayesian solution that solves the problem of recursively calculating the exact posterior density. This recursive propagation of the posterior density is only a conceptual solution in that, in general, it cannot be determined analytically; but particle filtering can approximate the optimal Bayesian solution. The key idea is to represent the required posterior density function. In particle filtering, the distributions are approximated by discrete random measures defined by particles and weights assigned to the particles. Assume $\boldsymbol{\alpha}_t^{(k)}$ are the particles of $\boldsymbol{\alpha}_t$, $w_t^{(k)}$ are their corresponding weights, and K is the number of particles used in the approximation; let $\boldsymbol{\alpha}_t^{(k)} \sim q(\alpha_t), k=1,\cdots,K$ be samples that are easily generated from a proposed $q(\cdot)$ called an importance density. The filtering distribution Eq. (9.10.8) can then be approximated as

$$p(\boldsymbol{\alpha}_t \mid \boldsymbol{\beta}_{0:t}) \approx \sum_{k=1}^{K} w_t^{(k)}\, \delta(\boldsymbol{\alpha}_t - \boldsymbol{\alpha}_t^{(k)}) \tag{9.10.10}$$

where $\delta(\cdot)$ is the Dirac delta function, and

$$w_t^{(k)} \propto \frac{p(\boldsymbol{\alpha}_{0:t}^{(k)} \mid \boldsymbol{\beta}_{0:t})}{q(\boldsymbol{\alpha}_{0:t}^{(k)} \mid \boldsymbol{\beta}_{0:t})}. \tag{9.10.11}$$

Let the importance density be chosen to factorize such that

$$q(\boldsymbol{\alpha}_{0:t} \mid \boldsymbol{\beta}_{0:t}) = q(\boldsymbol{\alpha}_t \mid \boldsymbol{\alpha}_{0:t-1}, \boldsymbol{\beta}_{1:t})\, q(\boldsymbol{\alpha}_{0:k-1} \mid \boldsymbol{\beta}_{0:k-1}) \tag{9.10.12}$$

and then, easily associate with Eqs. (9.10.8), (9.10.11) and (9.10.12), so an update weight can be obtained according to

$$w_t^{(k)} \propto w_{t-1}^{(k)} \frac{p(\boldsymbol{\beta}_t \mid \boldsymbol{\alpha}_t^{(k)}) p(\boldsymbol{\alpha}_t^{(k)} \mid \boldsymbol{\alpha}_{t-1}^{(k)})}{q(\boldsymbol{\alpha}_t^{(k)} \mid \boldsymbol{\alpha}_{t-1}^{(k)}, \boldsymbol{\beta}_t)} \tag{9.10.13}$$

Assume now that the posterior distribution $p(\boldsymbol{\alpha}_{0:t-1}|\boldsymbol{\beta}_{0:t-1})$ is approximated by the discrete random measurement. Then assuming that we can generate $f(\boldsymbol{\alpha}_t|\boldsymbol{\alpha}_{t-1})$ and $h(\boldsymbol{\beta}_t|\boldsymbol{\alpha}_t)$ point-wise, we generate for each state trajectory $\boldsymbol{\alpha}_{0:t-1}^{(k)}$, a random sample from a proposed distribution $\boldsymbol{\alpha}_t^{(k)} \sim q(\boldsymbol{\alpha}_t|\boldsymbol{\alpha}_{1:t-1}^{(k)}, \boldsymbol{\beta}_{1:t})$. Then, the updated weight should be

$$w_t^{(k)} \propto \frac{h(\boldsymbol{\beta}_t \mid \boldsymbol{\alpha}_t^{(k)}) f(\boldsymbol{\alpha}_t^{(k)} \mid \boldsymbol{\alpha}_{t-1}^{(k)})}{q(\boldsymbol{\alpha}_t^{(k)} \mid \boldsymbol{\alpha}_{1:t-1}^{(k)}, \boldsymbol{\beta}_{1:t})} \tag{9.10.14}$$

For example, in some simple scenarios, the proposed distribution can be chosen as

$$q(\boldsymbol{\alpha}_t \mid \boldsymbol{\alpha}_{0:t-1}^{(k)}, \boldsymbol{\beta}_{0:t}) = f(\boldsymbol{\alpha}_t \mid \boldsymbol{\alpha}_{0:t-1}^{(k)}) \tag{9.10.15}$$

so that the importance weight simplifies to

$$w_t^{(k)} \propto h(\boldsymbol{\beta}_t \mid \boldsymbol{\alpha}_t^{(k)}) \tag{9.10.16}$$

Usually, a resampling step[19] is performed, such that the probability that $\boldsymbol{\alpha}_t^{(k)}$ is selected is proportional to $w_t^{(k)}$. Then the resulting samples are an approximate set of unweighted random samples drawn from the filtering distribution $p(\boldsymbol{\alpha}_t|\boldsymbol{\beta}_{0:t})$.

9.10.4 Experiment and Analysis

As we mentioned above, available nonlinear and noisy ICA solutions are currently few. Thus, we try to validate our method through the linear and noisy ICA in comparison with DSS [133] first; then, we perform simulations about the nonlinear and noisy ICA. Our proposed procedure is briefly outlined below.

Step1: Let $x_i(t)=z_i(t)$ and then estimate the TVAR coefficients from $x_i(t)$;

Step2: Substitute the TVAR coefficients into Eq. (9.10.5) and apply PF to estimate the linear and clean mixtures;

Step3: Substitute the estimated linear and clean mixtures into Eq. (9.10.5) and apply PF to estimate the TVAR coefficients;

Step4: Iterate between step 2 and step 3 until the TAVR coefficients vary slowly between adjacent estimations;

Step5: Perform ICA on the estimated linear and clean mixtures to separate sources.

Hereinafter, we denote our method as the PF-ICA.

To meet the nonstationary assumption, we choose different speech phonemes of different people as sources. The excitation noise in Eq. (9.10.3) is pink noise. The background noises have Gaussian distribution and extreme value distribution respectively. Extreme value distribution is often used to model the smallest or largest value among a large set of independent, identically distributed random values representing measurements or observations. It is a super-Gaussian distribution. Here, we adopt it to show that PF has advantages in dealing with non-Gaussian noise problems.

Next, we define different signal-to-noise ratios (SNR) and the performance index (PI) of our approach. The input SNR is defined as

$$\mathrm{SNR1} = \frac{1}{m}\sum_{i=1}^{m} 10\lg 10\left(\frac{\sum_{t=1}^{T}\{h[x_i(t)]\}^2}{\sum_{t=1}^{T} v_i^2(t)}\right) \tag{9.10.17}$$

Then we select a parameter to reflect how PF functions behave. It is the SNR related to the linear and clean mixtures $x_i(t)$ and the estimated linear and clean mixtures $\hat{x}_i(t)$, defined as

$$\mathrm{SNR2} = \frac{1}{m}\sum_{i=1}^{m} 10\lg 10\left(\frac{\sum_{t=1}^{T}\{h[x_i(t)]\}^2}{\sum_{t=1}^{T}\{h[x_i(t)]-h[\hat{x}_i(t)]\}^2}\right) - \mathrm{SNR1} \tag{9.10.18}$$

For the determined problem, we may estimate the independent phonemes from the estimated linear and clean mixtures by the FastICA. The performance index is chosen as in Ref.[134],

$$\mathrm{PI1} = \sum_{t=1}^{n}\left(\sum_{h=1}^{n}\frac{|c_{th}|}{\max\limits_{p=1,\cdots,n}(c_{tp})}-1\right)+\sum_{h=1}^{n}\left(\sum_{t=1}^{n}\frac{|c_{th}|}{\max\limits_{p=1,\cdots,n}(c_{ph})}-1\right) \tag{9.10.19}$$

Where, $\boldsymbol{C}=(c_{th})=\boldsymbol{WA}$, and $\boldsymbol{W}$ is the demixing matrix. It is obvious that the smaller the PI is, the better the performance.

For the underdetermined problem, the independent phonemes may be estimated by MoG-uICA[135]. The corresponding performance index is

$$\mathrm{PI2} = \frac{1}{n}\sum_{i=1}^{n} 10\lg 10\left(\frac{\sum_{t=1}^{T} s_i^2(t)}{\sum_{t=1}^{T}(s_i(t)-\hat{s}_i(t))^2}\right) \tag{9.10.20}$$

All sources are sampled at 8,000 Hz. The phonemes are 187.5 ms and 375 ms duration respectively. During the entire simulation, the TVAR model order is 20. With 2,000 particles generated, a Rao-Blackwellized PF and smoother[127] are used in every simulation. For each simulation, sources are different, and the mixing matrix is randomly generated. The variance of the sources and mixtures remain equal to one. By adjusting the variances of the noises, the input SNR varies from 0 dB to 10 dB respectively. At each input SNR, 100 simulations are performed, and the averaged SNR2, PI1 and PI2 are shown below.

1. Linear and noisy determined ICA

Two noisy mixtures are generated according to Eq. (9.10.1). Here, $h(\cdot)=1(\cdot)$ in Eq. (9.10.5). The background noise is Gaussian and the length of each phoneme is 187.5 ms. As the system is linear and the noise is Gaussian, we adopt the KF to estimate the clean mixtures. By this, we show that PF outperforms KF in the dynamic system even if the system is linear and Gaussian. With DSS, we choose the nonlinear function $\tanh(\cdot)$ as the denoising function, since the sources are speech. With PF, the average benefit is SNR2 ≈ 7.3 dB, and with KF, the average SNR2 is about 1.5 dB. The difference in performance results from the dynamic system defined by Eq. (9.10.5). As described above, the TVAR model order is 20; therefore, the state length is 20 samples. Such a short length cannot guarantee KF to converge. However, PF has the advantage of using many more particles to overcome this shortcoming. Figure 9.33 illustrates the performances of different procedures. It is apparent that PF-ICA behaves the best. The lower the input SNR, the more advantage PF-ICA shows. This simulation proves our proposed method is effective. Next we use our method in the nonlinear and non-Gaussian noisy cases.

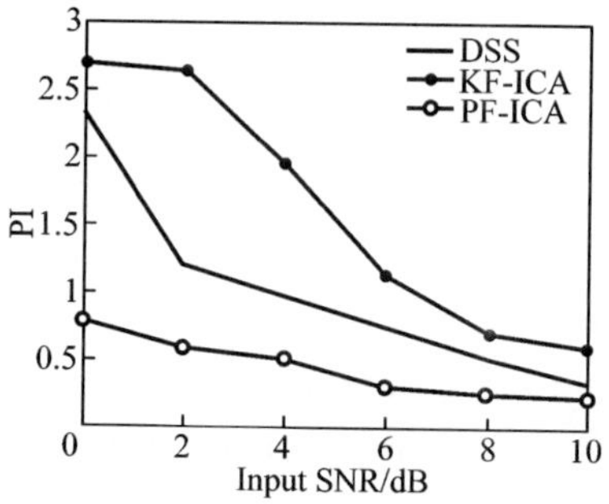

Fig. 9.33 Performance index of DSS, KF-ICA and PF-ICA with linear and noisy determined ICA

2. Nonlinear and noisy determined ICA

Two noisy mixtures are considered, and $h(\cdot)=\tanh(\cdot)$ is chosen. The background noise is of an extreme value distribution and the length of each phoneme is 187.5 ms. The average benefit of PF is SNR2=6.7 dB. Figure 9.34 shows how PF-ICA behaves. Compared with Fig. 9.33, PF-ICA achieves almost identical performance in both nonlinear and linear cases. This reflects that the PF-ICA has particular advantage in a nonlinear and nongaussian noisy ICA.

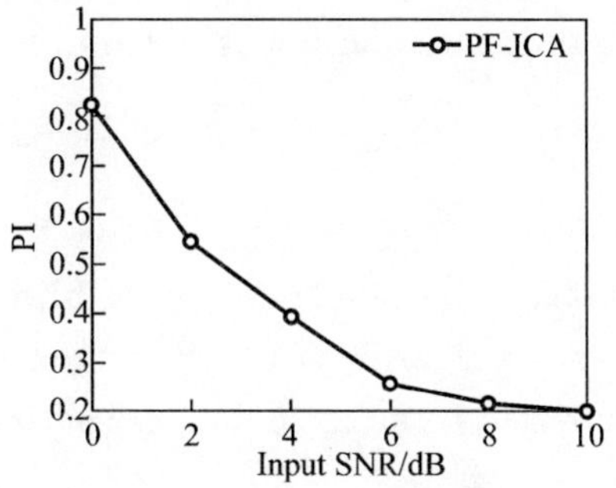

Fig. 9.34 Performance index of PF-ICA with nonlinear and noisy determined ICA

3. Nonlinear and noisy underdetermined ICA

Three sources are used, and $h(\cdot)=\tanh(\cdot)$ is selected also. Two noisy mixtures are simulated. The background noise has an extreme value distribution and the length of each phoneme is 375 ms. PF obtains an average benefit of SNR2=6.4 dB. Figure 9.35 shows how the PF-ICA behaves. Obviously, this difficult problem is solved under our framework.

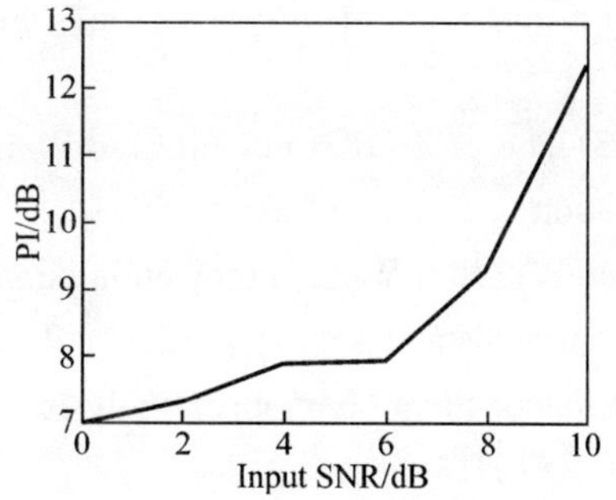

Fig. 9.35 Performance index of PF-ICA with nonlinear and noisy underdetermined ICA

9.10.5 Conclusion

With the nonstationary characteristics of sources, the dynamic state space model can be set up for single channel mixtures; thus the PF can convert the noisy mixtures to clean mixtures and the nonlinear model to a linear model dimension by dimension, provided that the noise distribution and the post nonlinear function are already known. Then, linear ICA algorithms can extract sources from the linear mixtures evaluated by the PF. Under this framework, the nonlinear and noisy determined ICA and the nonlinear and noisy underdetermined ICA can be well resolved. Moreover, our framework is promising in a hierarchical ICA as long as the nonlinear function is accurately estimated between different layers.

References

[1] Cichocki A, Amari S I (2002) Adaptive blind signal and image processing—Learning algorithms and applications.Wiley, New York

[2] Hyvärinen A, Karhunen J, Oja E (2001) Independent component analysis. Wiley, New York

[3] Li Y, Wang J (2002) Sequential blind extraction of instantaneously mixed sources. IEEE Transactions on Signal Processing 50(5): 997–1006

[4] Cao X R, Liu R W (1996) General approach to blind source separation. IEEE Transactions on Signal Processing 44(3): 562–571

[5] Friedman J, Tukey J W (1974) A projection pursuit algorithm for exploratory data analysis. IEEE Transactions on Computers C-23(9): 881–889

[6] Huber P J (1985) Projection pursuit (with discussion). Annals of Statistics 13(2): 435–525

[7] Jones M C (1983) The projection pursuit algorithm for exploratory data analysis. Dissertation, University Bath

[8] Jones M C, Sibson R (1987) What is projection pursuit? (with discussion). Journal of the Royal Statistical Society A 150(1): 1–37

[9] Comon P (1994) Independent component analysis, a new concept? Signal Processing 36(3): 287–314

[10] Silverman B W (1985) Density estimation for statistics and data analysis. Chapman and Hall, New York

[11] Boscolo R, Pan H, Roychowdhury V (2001) Non-parametric ICA. In: Lee T W, Jung T W, Makeig S and Sejnowski T (eds) Proceedings Third International Conference Independent Component Analysis and Blind Signal Separation. San Diego, 2001, pp 13–18

[12] Boscolo R H, Pan H, Roychowdhury V (2004) Independent component analysis based on non-parametric density estimation. IEEE Transactions

on Neural Networks 15(1): 55–65

[13] Cardoso J F (1998) Blind signal separation: Statistical principles. Proceedings of IEEE Special Issue on Blind Identification and Estimation 9: 2009–2025

[14] Karvanen J, Eriksson J, Koivunen V (2000) Pearson system based method for blind separation. In: P. Pajunen and J. Karhunen (eds) Proceedings of Second International. Workshop on Independent Component Analysis and Blind Signal Separation, Helsinki, 2000, pp 585–590

[15] Vlassis N, Motomura Y (2001) Efficient source adaptivity in independent component analysis. IEEE Transactions on Neural Networks 12(5): 559–566

[16] Jia P, Zhang H Y, Shi X Z (2003) Blind source separation based on nonparametric density estimation. Circuits, System and Signal Processing 22(1): 57–67

[17] Tong L, Liu R W, Soon V C, et al (1991) Indeterminacy and identifiability of blind identification. IEEE Transactions on Circuits and Systems 38(5): 499–509

[18] Choi S, Cichocki A, Amari S (2002) Equivariant nonstationary sources eparation. Neural Networks 15(1): 121–130

[19] Choi S, Cichocki A, Amari S (2002) Natural gradient learning for second-order nonstationary source separation. In: Neural Networks, 2002. IJCNN '02. Proceedings of the 2002 International Joint Conference on. Hawaii, 2002, 1: 654–658

[20] Hachem W, Desbouvries F, Loubaton P (1999) On the identification of certain noisy FIR convolutive mixtures. In: Proceedings of First International Workshop on Independent Component Analysis and Blind Signal Separation. Aussois, 1999, pp 401–405

[21] Gorokhov A, Loubaton P (1997) Subspace-based techniques for blind separation of convolutive mixturess with temporally correlated sources.Circuits and Systems I: Fundamental Theory and Applications, IEEE Transactions on Circuist and Systems 44(9): 813–820

[22] Amari S I, Cardoso J F (1997) Blind source separation—Semiparametric statistical approach. IEEE Transaction on Signal Processing 45(11): 2692–2700

[23] Kawamoto M, Inouye Y (2003) Blind deconvolution of MIMO-FIR

channels driven by colored inputs using second-order statistics. In: ISCAS '03. Proceedings of the 2003 International Symposium, Bangkok, 2003, pp 48–51

[24] Horn R A, Johnson C R (1985) Matrix analysis. Cambridge University Press, Cambridge

[25] Amari S I, Douglas S C, Cichocki A et al (1997) Novel on-line adaptive learning algorithms for blind deconvolution using natural gradient approach. In: Proceedings of 11th IFAC Symp. System Identification, Japan, 1997, pp 1007–1012

[26] Amari S I (1998) Natural gradient works efficiently in learning. Neural Computation 10(2): 251–276

[27] Sheng X Z, Shi X Z (2005) On-line SOS-based multichannel blind equalization algorithm with noise. Signal Processing 85: 1602–1610

[28] Sheng X Z, Shi X Z (2005) On-line blind equalization algorithmof an FIR MIMO channel system for non-stationary signals. IEE Proc.-Vis. Image Signal Process 152(5): 575–581

[29] Cardoso J F (1991) Super symmetric decomposition of the fourth order cumulant tensor. Blind identification of more sources than sensors. In: ICASSP '91, Toronto, 1991, pp 3109–3112

[30] Shamsunder S, Giannakis G (1993) Modeling of nongaussian array data using cumulants: DOA estimation of more sources than sensors. Signal Processing 30(3): 279–297

[31] Taleb A, Jutten C (1999) On underdetermined source separation. In: ICASSP '99, Phoenix, 1999, 3: 1445–1448

[32] Taleb A (2001) An algorithm for the blind identification of n independent signals with 2 sensors. In: Sixth Inter- national Symposium on Signal Processing and its Applications (ISSPA 2001), Kuala Lumpur, 2001, 1: 5–8

[33] Lewicki M S, Sejnowski T J (2000) Learning overcomplete representation. Neural Computation 12(2): 337–365

[34] Eriksson J, Koivunen V (2004) Identifiability and separability and uniqueness of linear ICA models. IEEE Signal Processing Letters 11(7): 601–604

[35] Tong L (1996) Identification of multichannel MA parameters using high-order statistics. Signal Processing 53: 195–209

[36] Comon P (1996) Decompositions of quantics in sums of powers of linear

forms. Signal Processing 53: 96–107

[37] Chen B N (2001) Frequency domain blind multiple input multiple output system identification. Dissertation, Drexel University

[38] Li Y Q, Cichocki A, Amari S I (2003) Sparse component analysis for blind source separation with less sensors than sources. In: Fourth International Symposium on Independent Component Analysis and Blind Signal Separation, Nara, 2003

[39] Theis F J, Jung A, Puntonet C G et al (2003) Linear geometric ICA: Fundamentals and algorithms. Neural Computation 15(2): 419–440

[40] Takigawa I, Kudo M, Nakamura A et al (2004) On the minimum l1-norm signal recovery in underdetermined source separation. In: Independent Component Analysis and Blind Signal Separation: Fifth International Conference, ICA 2004, Granada, 2004; Proceedings Carlos G. Puntonet(ed) Springer, New York, 2005, pp 193–201

[41] Theis F J, Lang E W (2002) Formalization of the two-step approach to overcomplete BSS. In: N. Younan (ed) SIP 2002. Proceedings of 4th Intern. Conf. on Signal and Image Processing (SIP'02), Hawaii, 2002, pp 207–212

[42] Waheed K, Fathi M, Salem F M (2003) Algebraic overcomplete independent component analysis. In: 4th International Symposium on Independent Component Analysis and Blind Source Separation, Nara, 2003, pp 1077–1082

[43] Kendall W, Liang F M, Wang J S (2010) Markov chain Monte Carlo: innovations and applications in statistics, physics, and bioinformatics [DB/OL]. http://www.ims.nus.edu. Sg/Programs/mcmc/. Cited 23 Jan 2010

[44] Doucet A,Godsill S J, Andrieu C (2000) On sequential Monte Carlo sampling methods for Bayesian filtering. Statist. Comput. 10(3): 197–208

[45] Tanaka T (1998) A theory of mean field approximation. In: NIPS. Breckenridge, Colorado, 1998, pp 351–360

[46] Haykin S (1999) Neural networks: A comprehensive foundation, 2nd edn. Prentice Hall, Upper Saddle River

[47] Beal M J (2003) Variational algorithm for approximate Bayesian inference. Dissertation, University of London

[48] Jaakkola T (2001) Tutorial on variational approximation methods. In: Opper M and Saad D (eds) Advanced Mean Field Methods: Theory and

Practice. The MIT Press, Cambridge, MA, pp 129–159

[49] Zhang Y Y, Shi X Z, Chen C H (2006) Gaussian mixtures model based bayesian analysis for underdetermined blind source separation. Journal of Circuits System and Signal Processing 25(1): 81–94

[50] Grinoval R, Benaroya L,Vincent E et al (2003) Proposals for performance measurement in source separation. In: Proceedings of 4th International Symposium on Independent Component Analysis and Blind Signal Separation (ICA2003), Nara, 2003, pp 763–768

[51] Cichocki A, Zurada J M (2004) Blind signal separation and extraction: Recent trends, future perspectives, and applications. In: ICAISC, Zekopane, 2004, pp 30–37

[52] Cichocki A, Georgiev P (2003) Blind source separation with matrix constraints. IEICE Transactions Fundamentals, E86-A(1): 522–531

[53] Girolami M (2001) A variational method for learning sparse and over-complete representations. Neural Computation 13(12): 2517–2532

[54] Ghahramani Z, Beal M J (2001) Graphic models and variational methods. In: Opper M and Saad D (eds) Advanced Mean Field Methods-Theory and Practice. The MIT Press, Cambridge, MA

[55] Davies M, Mitianoudis N (2004) Simple mixtures model for sparse over-complete ICA. IEE proceedings in Vision, Image and Signal Processing 151(1): 35–43

[56] Valpola H (Coordinator) (2003) BLISS IST-1999-14190, Blind source separation and applications: Technical report on Bayesian method. Deliverable D17 Report Version: Final Report Preparation Date: June 27, 2003

[57] Højen-Sørensen, Winther O, Hansen L K (2002) Mean field approaches to independent component analysis. Neural Computation 14(4): 889–918

[58] Pedro A.d.F.R, Højen-Sørensen, Winther O et al (2000) Ensemble learning and linear response theory for ICA. In: NIPS, Vanconver, 2000, pp 542–548

[59] Csató L, Fokoué E, Opper M et al (2000) Efficient approaches to Gaussian process classification. In: Solla S A, Leen T K and Müller K R (eds) Advances in Neural Information Processing Systems. The MIT Press, Cambridge, MA

[60] Kreutz-Delgado K, Murray J F, Rao B D (2003) Dictionary learning

algorithm for sparse representation. Neural Computation 15(2): 349–396

[61] Zibulevsky M, Pearlmutter B A, Bofill P, et al (2000) Blind source separation by sparse decomposition. In: Szu H H, Vetterli M, Campbell W J and Buss J R (eds) Proceedings of SPIE, Vol. 4056, Wavelet Applications VII, April 2000, pp 165–174

[62] Bofill P, Zibulevsky M (2002) Blind separation of more sources than mixturess using sparsity of their short time Fourier transform. In: Proceedings of the International Workshop on Independent Component Analysis and Blind Signal Separation, Helsinki, 2000, pp 87-92

[63] Nguyen L T, Adel Belouchrani A, Boashash B (2001) Separating more sources than sensors using time frequency distributions. In: Proceedings of the International Symposium on Signal Processing and Its Applications (ISSPA), Kuala-Lumpur, 2001

[64] Luo Y, Lambotharan S, Chambers J A (2004) A new block based time frequency approach for underdetermined blind source separation. In: Proceedings of IEEE International Conference on Acoustics, Speech, and Signal Processing, 2004. (ICASSP '04), 2004, 5:V-537–540

[65] Luo Y, Chambers J A (2003) A modified underdetermined blind source separation using competitive learning. In: Proceedings of the 3rd International Symposium on Image and Signal Processing and Analysis, 2003, 2: 966–969

[66] Kim S G, Yoo C D (2004) Underdetermined independent component analysis by data generation. In: Fifth International Conference, ICA 2004. Granada, 2004, pp 22–24

[67] Attias H (1999) Independent factor analysis. Neural Computation 11(4):803–851

[68] Palmer J A, Wipf D P, Rao B D (2005) Variational representation of non-Gaussian priors. UCSD, San Diego

[69] Chan K, Lee T W, Sejnowski T J (2003) Variational Bayesian learning of ICA with missing data. Neural Computation 15(8): 1991–2011

[70] Ghahramani Z, Beal M J (2000) Variational inference for Bayesian mixtures of factor analyses. In: Solla S A, Leen T K and Müller K (eds) Advances in Neural Information Processing System. The MIT Press, 2000, 12: 449–455

[71] Choudrey R, Penny W D, Roberts S J (2001) An ensemble learning

approach to independent component analysis. In: Neural Networks for Signal Processing X, 2000. Proceedings of the 2000 IEEE Signal Processing Society Workshop, Sydney, 2001, 1: 435–444

[72] Choudrey R A (2002) Variational methods for Bayesian independent component analysis. Dissertation, University of Oxford

[73] Choudrey R A, Roberts S (2003) Bayesian ICA with hidden Markov model sources. In: Proceedings of ICA-2003, Nara, 2003, pp 809–814

[74] Chen Z (2003) Bayesian filtering: From Kalman filters to particle filters, and beyond. Adaptive Systems Lab, McMaster University, McMaster

[75] Haykin S, Huber K, Chen Z (2004) Bayesian sequential state estimation for MIMO wireless communications. Proceedings of the IEEE 92(3): 439–454

[76] Doucet A, Godsill S J, Andrieu C (2000) On sequential Monte Carlo sampling methods for Bayesian filtering. Statist. Comput. 10(3): 197–208

[77] Arulampalam M S, Maskell S, Gordon N et al (2002) A tutorial on particle filters for online nonlinear/non-Gaussian Bayesian tracking. IEEE Transactions Signal Processing 50(2): 174–188

[78] Djuric P M, Kotecha J H, Zhang J Q et al (2003) Particle filtering. Signal Processing Magazine 20(5): 19–38

[79] Doucet A, de Freitas J F G, Gordon N J (2001) Sequential Monte Carlo methods in practice. Springer, New York

[80] Julier S J, Uhlmann J K (2004) Unscented filtering and nonlinear estimation. Proceedings of the IEEE Aerospace and Electronic Systems 92(3): 401–422

[81] Haykin S (2001) Kalman filtering and neural networks. Wiley, New York

[82] Gordon N, Salmond D, Smith A F M (1993) Novel approach to nonlinear and non-Gaussian Bayesian state estimation. Proc. Inst. Elect. Eng., F 140(1): 107–113

[83] Bingham E, Hyvärinen A (2000) ICA of complex-valued signals: A fast and robust deflationary algorithm. In: Proceedings of International Joint Conference on Neural Networks (IJCNN2000), Como, 2000, pp 357–362

[84] Bingham E, Hyvärinen A (2000) Fast and robust deflationary separation of complex-valued signals. In: Proceedings of European

Conference on Signal Processing (EUSIPCO2000), Tampere, 2000, I: 23-26

[85] Sawada H, Mukai R, Makino S (2005) Frequency-domain blind source separation. In: Speech Enhancement, Benesty J, Makino S and Chen J (eds) Speech Enhancement. Springer, New York

[86] Calhoun V, Adali T (2003) Complex ICA for FMRI analysis: Performance of several approaches. In: Proceedings of ICASSP '03, Hong Kong, 2003, pp II -717–720

[87] Calhoun V, Adali T, Pearlson G et al (2002) On complex informax applied to functional MRI data. In: Proceedings of ICASSP '02, Orlando, 2002, pp I-1009~ I-1012

[88] Fiori S (2002) Blind separation of circularly-distributed sources by neural extended APEX algorithm. Neurocomputing 34(2): 239–252

[89] Chen H W, Zhao J W (2005) Coherent signal subspace wideband optimal beamforming for acoustic vector-sensor array. Acta Acustica 30(1): 76–82

[90] Sawada H, Mukai R, Makino S (2002) A polar-coordinate based activation function for frequency domain blind source separation. In: Proceedings of ICASSP, Orlando, 2002, pp 1001–1004

[91] Smaragdis P (1998) Blind separation of convolved mixtures in the frequency domain. Neurocomputing 22(1): 21–34

[92] Fiori S (2003) Extended Hebbian learning for blind separation of complex-valued sources. IEEE Transactions on Circuits and Systems II: Analog and Digital Signal Processing 50(4): 195–202

[93] Annemüller J, Sejnowski T J, Makeig S (2003) Complex spectral domain independent component analysis of electroencephalographic data. In: Proceedings of ICA Workshop, Nara, 2003

[94] Bingham E, Hyvärinen A (2000) A fast fixed-point algorithm for independent component analysis of complex-valued signals. Int. J. Neural Systems 10(1): 1–8

[95] Adali T, Kim T, Calhoun V (2004) Independent component analysis by complex nonlinearities. In: Proceedings of ICASSP '04, Montreal, 2004, 5: V-525–528

[96] Kim T, Adali T (2003) Approximation by fully-complex multilayer perceptrons. Neural Computation 15(7): 1641–1666

[97] Calhoun V, Adali T (2002) Complex Informax: Convergence and

approximation of Informax with complex nonlinearities. In: Proceedings of the 12th IEEE Workshop on Neural Networks for Signal Processing, Baltimore, MD, 2002, pp 307–316

[98] Kim T, Adali T (2002) Fully complex multi-layer perceptron network for nonlinear signal processing. Journal of VLSI Signal Processing 32 (1): 29–43

[99] Eriksson J, Koivunen V (2004) Complex-valued ICA using second order statistics. In: Proceedings of the IEEE Signal Processing Society Workshop on Machine Learning for Signal Processing, Sao Luis, 2004, pp 183–191

[100] Neeser F D, Massey J L (1993) Proper complex random processes with applications to information theory. IEEE Transactions on Information Theory 39(4):1293–1302

[101] Shi X Z (2003) Signal processing and soft-computing. Higher Education Press, Beijing

[102] Wu J B, Chen J, Zhong P (2003) Time frequency-based blind source separation for elimination of cross-terms in Wigner distribution. Electronics Letters 39(5):475–477

[103] Lee J H, Jung H Y (2000) Speech feature extraction using independent component analysis. In: Proceedings of ICASP, Istanbul, 2000, 3: 1631–1634

[104] Bell A J, Sejnowski T J (1996) Learning the higher-order structure of a nature sound. Network:Computation in Neural System 7(2): 261–266

[105] Kong Wei (2005) Research on blind source separation algorithm and its application in acoustic signals. Dissertation, Shanghai Jiao Tong University (Chinese)

[106] Lu W, Rajapakse J C (2000) Constrained independent component analysis. In: Leen T K, Dietterich T G, and Tresp V (eds) Advances in Neural Information Processing Systems 13 (NIPS2000). The MIT Press, Cambridge, MA, 2000,16: 570–576

[107] Lu W, Rajapakse J C (2003) Eliminating indeterminacy in ICA. Neurocomputing 50(2): 271–290

[108] Lu W, Rajapakse J C (2005) Approach and applications of constrained ICA. IEEE Transactions Neural Netw 16(1): 203–212

[109] Ille N, Beucker R, Scherg M (2001) Spatially constrained independent

component analysis for artifact correction in EEG and MEG. Neuroimage 13(6): S159–S159

[110] Balsi M, Filosa G Valente G (2005) Constrained ICA for functional magnetic resonance imaging. In: Proceedings of the 2005 European Conference on Circuit Theory and Design, Cork, 2005, pp II-67–II I-70

[111] Lin Q H, Zheng Y R, Yin F L (2007) A fast algorithm for one-unit ICA-R. Information Sciences 177(5): 1265–1275

[112] Liao X, Carin L (2002) A new algorithm for independent component analysis with or without constraints. In: Proceedings of the Second IEEE Sensor Array and Multichannel (SAM) Signal Processing Workshop, Rosslyn, 2002, pp 413–417

[113] Zhang H J, Guo C, Shi Z (2008) A new constrained fixed-point algorithm for ordering independent components. Journal of Computational and Applied Mathematics 220(1–2): 548–558

[114] Calhoun V D, Adali T, Stevens M C (2005) Semi-blind ICA of fMRI: A method for utilizing hypothesis-derived time courses in a spatial ICA analysis. Neuroimage 25(2): 527–538

[115] Huang D S, Mi J X (2007) A new constrained independent component analysis method. IEEE Transactions on Neural Networks 18(5): 1532–1535

[116] Bertsekas D P (1996) Constrained optimization and Lagrange multiplier methods. Athena Scientific, Belmont, MA

[117] Jutten C,Karhunen J (2003) Advances in nonlinear blind source separation. In: Proceedings of ICA 2003, Nara, 2003, pp 245–256

[118] Särelä J, Valpola H (2005) Denoising source separation. Journal of Machine Learning Research 6(3): 233–272

[119] Grubera P, Stadlthannera K, Matthias B (2006) Denoising using local projective subspace methods. Neurocomputing 69(13–15): 1485–1501

[120] Hyvarinen A (1998) Independent component analysis in the presence of Gaussian noise by maximizing joint likelihood. Neurocomputing 22(1): 49–67

[121] Moulines E, Cardoso J F, Gassiat E (1997) Maximum likelihood for blind signal separation and deconvolution of noisy signals using mixtures models. In: Proceedings of ICASSP, Munich, 1997

[122] Chan K, Lee T W, Sejnowski T J (2003) Variational Bayesian learning of ICA with missing data. Neural Computation 15(10): 1991–2011

[123] Taleb A, Jutten C (1997) Nonlinear source separation: The postlinear mixturess. In: Proc. Europ. Symp. on Artificial Neural Networks (ESANN'97), Bruges, 1997, pp 279–284

[124] Almeida L B (2004) Linear and nonlinear ICA based on mutual information—the MISEP method. Signal Processing 84(2): 231–245

[125] Valpola H, Karhunen J (2002) An unsupervised ensemble learning method for nonlinear dynamic state-space models. Neural Computation 14(11): 2647–2692

[126] Maeda S I, Song W J, Ishii S (2005) Nonlinear and noisy extension of independent component analysis: Theory and its application to a pitch sensation model. Neural Computation 17(1): 115–144

[127] Fong W, Godsill S J, Doucet A et al (2002) Monte Carlo smoothing with application to audio signal enhancement. IEEE Transactions on Signal Processing 50(2): 438–449

[128] Djuric P M, Kotecha J H, Zhang J Q et al (2003) Particle filtering. IEEE Signal Processing Magazine 20(1): 19–38

[129] Arulampalam M S, Maskell S, Gordon N et al (2002) A tutorial on particle filters for online nonlinear/non-Gaussian Bayesian tracking. IEEE Transactions on Signal Processing 50(2):174–188

[130] Cong F Y, Xu S, Zhou S H et al (2007) Particle filtering based noisy blind source separation. Electronics Letters 43(9): 547–549

[131] Hall M G, Oppenheim A, Willsky V et al (1983) Time varying parametric modeling of speech. Signal Processing 5: 267–285

[132] Kitagawa G (1998) A self-organizing state-space model. J. Am. Stat. Assoc. 93(433): 1203–1215

[133] Särelä J, Volpola H (2005) Denoising source separation. J. Mach. Learn. Res. 6(3): 233–272

[134] Amari S, Cichocki A, Howard Y H (1996) A new learning algorithm for blind signal separation. In: Advances in Neural Information Processing Systems 8. MIT Press, Massachusettes, pp 757–763

[135] Zhang Y Y, Shi X Z, Chen C H (2006) A Gaussian mixtures model for underdetermined independent component analysis. Signal Processing 86(7): 1538–1546

Chapter 10 Data Analysis and Application Study

10.1 Target Enhancement in Active Sonar Detection

10.1.1 Introduction

The active sonar detection usually includes the following: beamforming, a matched filter and normalization. They explore the geometrical information of the array, correlation between the target echo and the transmitted echo, and numerical smoothing respectively. Although these methods are versatile, researchers try to find better detection solutions under heavy reverberation, for example, to use longer and larger arrays, to cancel more reverberation, to design better matched filters and so on. Currently, reverberation models and specialized processing methods are two attractive techniques[1]. As for the first method, receivers with differing range resolutions may encounter different statistics for a given waveform. Furthermore, a given receiver may encounter different statistics at different ranges[2]. The Weibull, log-normal, Rician, multi-modal Rayleigh and non-Rayleigh distributions have been used to describe sonar reverberation[3,4]. The matched filter is the optimal detector with the background of Gaussian noise; therefore, these studies imply that it is necessary to reduce the reverberation before the signal is fed into the matched filter. Hence, the second method aims to cancel reverberation. Reverberation can be assumed to be a sum of returns issued from the transmitted signal, and a data matrix may be generated under the segmentation of the data received by active sonar[5,6]. Then the principal component inverse (PCI)[7~13] can be applied to separate reverberation and target echo with the data matrix. However, some important prior knowledge, such as the target power, should be provided[11]. The PCI and other methods are shown to behave very well in Doppler cases[11~13]. In case prior knowledge is hard to obtain and the Doppler effect does not exist, it would be highly desired to cancel reverberation and enhance the target echo with easily obtainable but minimal prior knowledge even in some complicated undersea

X. Shi, *Blind Signal Processing*

situations.

The essence of the PCI implicitly assumes that the target echo and reverberation are not correlated. Along this line, to enhance the desired target, this study aims to fully exploit this implicit information and to add the blind source separation (BSS)[14] technique between the beamforming and the matched filter. BSS is an important statistical signal processing method. The advantage of BSS is that it does not need much prior knowledge and makes full use of the simple and apparent statistical properties, such as non-Gaussianity, nonstationarity, colored characteristics, uncorrelation, independence, and so on. We studied BSS on cancelling reverberation in Ref.[15]. From the perspective of BSS, the data received by an active sonar is convolutive[16]. Consequently, we perform a convolutive BSS to null oceanic reverberation. In this section, illustrations are shown on matched filter outputs under low signal-to-reverberation ratios (SRR), and some examples of cancelling real sea reverberation are given.

10.1.2 Problem Formulation

1. Analysis of active sonar data

We assume that reverberation is a sum of returns generated from the transmitted signal. The active sonar data was regarded as the convolution of the path impulse response and the transmitted signal[16]. The active sonar data can be expressed as

$$d(t) = h(t) * e(t) + n(t) = \sum_{k=1}^{K} h_k(t-\tau_k) \cdot e(t) + n(t) \tag{10.1.1}$$

where, $*$ denotes the convolution operation, τ_k is the propagation delay, $h_k(t)$ is the path impulse response, and $e(t)$ is the transmitted signal. In the reverberation dominant circumstance, what we pursue is to decompose the active sonar data into the target echo component $\tilde{d}(t)$ and reverberation component $r(t)$, given by

$$\tilde{d}(t) = \sum_{k=1}^{K_1} \underbrace{h_k(t-\tau_k)}_{t\arg et} \cdot e(t) + \underbrace{n(t)}_{noise} \tag{10.1.2}$$

$$r(t) = \sum_{k=K_1+1}^{K} \underbrace{h_k(t-\tau_k)}_{reverberation} \cdot e(t) + \underbrace{n(t)}_{noise} \tag{10.1.3}$$

where we assume $K_1 \ll (K-K_1)$, i.e. the reverberation is very heavy, and the number of target echo returns is much less than the number of reverberation returns. Consequently, it is apparent that $\tilde{d}(t)$ and $r(t)$ are of different non-

Gaussianity. Here, we define three correlation coefficients.

$$C_1 = mean\{corr|h_i(t-\tau_i)e(t), h_j(t-\tau_j)e(t)|\}, \quad i,j \in [1, K_1]$$
$$C_2 = mean\{corr|h_i(t-\tau_i)e(t), h_j(t-\tau_j)e(t)|\}, \quad i,j \in [K_1+1, K]$$
$$C_3 = mean\{corr|h_i(t-\tau_i)e(t), h_j(t-\tau_j)e(t)|\}, \quad i \in [1, K_1], j \in [K_1+1, K]$$

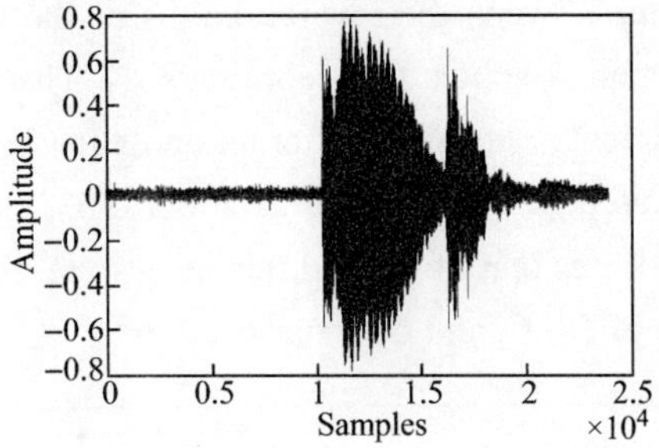

Fig. 10.1 Target echo waveform

where C_1, C_2 and C_3 are three average correlation coefficients between target echo returns, between reverberation returns, and between the target echo returns and reverberation returns respectively. We assume that C_1 and C_2 are nearly comparable, and $C_1 \gg C_3$, $C_2 \gg C_3$. To explain the correlation coefficients, an example of real sea reverberation is shown next. Consider a sinusoidal transmitted signal with a frequency of 1,750 Hz, with a duration of 200 ms. The array has 36 sensors with a distance between sensors of 0.5 m, and the received signal is sampled at 6,000 Hz.

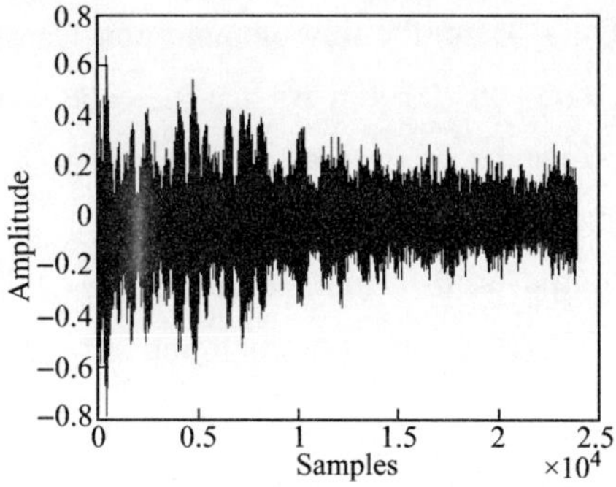

Fig. 10.2 Reverberation waveform

Figure 10.1 contains only the waveform of the target echo and background noise. The horizontal ordinate is the sample number, and the vertical ordinate is the normalized amplitude. This waveform describes $\tilde{d}(t)$ in Eq. (10.1.2). Heavy reverberation generated by a sine wave is covered in Fig. 10.2 where no target echo exists and the background noise is imbedded. In Fig. 10.3, the upper and lower plots are the values of C_2 and C_3 respectively. A sinusoid target echo of 1,750 Hz is simulated with the length of 200 ms. A block with the same length of the target is performed on the reverberation in Fig. 10.2. There is no overlap between the adjacent blocks. Thus, the upper plot shows the correlation coefficients between the reverberations of the adjacent blocks, and the average value, which is almost 0.3, represents C_2. The lower plot describes the correlation coefficients between the target and each blocked reverberation one by one, and the average value, which is about 0.03, symbolizes C_3 in Fig. 10.3. C_2 is almost ten times that of C_3, and this means that the above assumption about the relationship of C_1, C_2 and C_3 is reasonable.

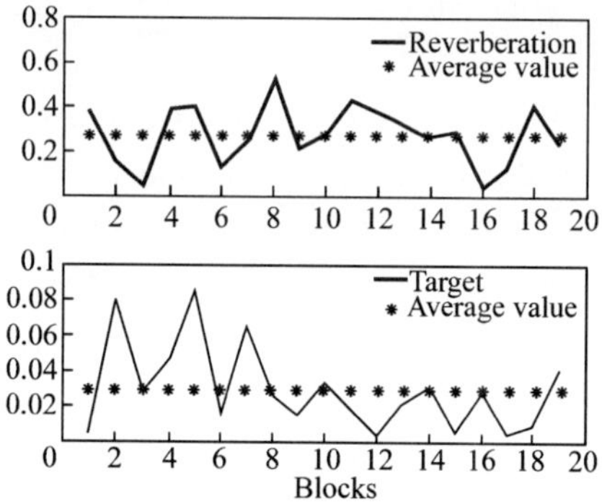

Fig. 10.3 Values of C_2 and C_3

The two waveforms in Fig. 10.1 and Fig. 10.2 have a standard kurtosis of 8.1 and 1.6[17] respectively. Here, the time-domain non-Gaussianity is enough to discriminate between $\tilde{d}(t)$ and $r(t)$. Experience indicates that detection for $\tilde{d}(t)$ must be very evident. However, in the real world problem, it is not an easy job to cancel the reverberation to the degree shown in Fig. 10.1. So we use the kurtosis of the matched filter output to discriminate between $\tilde{d}(t)$ and $r(t)$. We assume that the echo is located in $r(t)$ between the time index t_1 and t_2; the definition of SRR follows,

$$\text{SRR} = 10 \lg 10\left[\sum_{t=0}^{t_2-t_1} e(t)^2 \Big/ \sum_{t=t_1}^{t_2} r(t)^2\right] \tag{10.1.4}$$

it is apparent that the smaller the SRR is, the heavier the reverberation.

2. Target enhancement procedure by BSS

Currently, convolutive BSS is mainly based on higher-order or second-order statistics. The higher-order statistics-based methods require sources to be independent of each other, but the reverberation and the target echo do not meet that requirement in an active sonar system. Correspondingly, the second-order methods only need sources to be uncorrelated, and we have demonstrated above that the reverberation and target echo may have a slight correlation. Hereinafter we assume the target echo returns and the reverberation returns are the sources. The assumption for uncorrelation between them has already been made in the last subsection 10.1.2. If the convolutive BSS is directly applied to the data received by the array, the convolutive BSS may destroy the order of the signals received by different sensors and may result in beamforming failure. To avoid this shortcoming, we use data after beamforming.

Beamforming maps array signals to beam bin outputs that correspond to the main lobe and the adjacent beam bins. A data matrix is constructed to approximate the multiple-input/ multiple-output (MIMO) problem. The first row of the data matrix is the output corresponding to the main lobe, and the second is the most adjacent beam bin that is output to the main lobe. Other beam bins are not adopted in this study. Actually, this is reasonable, because here, beamforming functions as a pre-process to the convolutive BSS. Conventional beamforming is a linear transform, and the geometric information of the array is applied to make the desired signal focused on the main lobe. The output of the main lobe usually covers most of the desired signal information, and the adjacent beam bin output to the main lobe helps the two-inputs/two-outputs problem. Through beamforming, when we map the array signals onto two outputs, most of the energy of the desired signals is kept.

The sources are assumed to be the target echo returns and reverberation returns, and the two selected outputs of beamforming are the received signals. Therefore, the two outputs of the BSS should be the target echo and reverberation. After matched filtering is performed on the outputs, the one with the larger kurtosis is regarded as the target and is substituted the for original main lobe output. In this way, the target in the main lobe can be enhanced.

10.1.3 Convolutive BSS Algorithm

1. Separation algorithm

Generally speaking, for a given source-receiver pair, the waveform arrives at a different travel time, owing to varying path lengths[18]. All of these make reverberation non-stationary. The transmitted signal is the sinusoid or hyperbolic frequency modulated (HFM) signal here, so the reverberation must retain the time-structured feature; reverberation is nonstationary and colored. In this section, we will deduce the convolutive BSS algorithm based on the characteristics of reverberation.

Consider an acoustic scenario, where J sensors receive multiple filtered copies of statistically uncorrelated signals. Mathematically, the received signals can be expressed as a convolution, i.e.

$$x_j(t) = \sum_{i=1}^{I}\sum_{p=0}^{P-1} h_{ji}(p)s_i(t-p), j = 1,\cdots,J, t = 1,\cdots,L \quad (10.1.5)$$

where $h_{ji}(p)$ models the P-point impulse response from sources to sensors, and L is the length of the received signal. Equation (10.1.5) can be stated in a more compact matrix-vector form:

$$\boldsymbol{x}(t) = \sum_{p=0}^{P-1} \boldsymbol{H}(p)\boldsymbol{s}(t-p) = \boldsymbol{H}(t) * \boldsymbol{s}(t) \quad (10.1.6)$$

where $\boldsymbol{x}(t)=[x_1(t),\cdots,x_j(t)]^{\mathrm{T}}$ is the received signal vector, $\boldsymbol{s}(t)=[s_1(t),\cdots,s_I(t)]^{\mathrm{T}}$ is the source vector, and $\boldsymbol{H}(p)$ is the mixing filter matrix. The objective of the convolutive BSS is to find the deconvolutive filter matrix $\boldsymbol{W}(q)$ of length $\boldsymbol{Q}$ for deconvolution, that is

$$\boldsymbol{y}(t) = \boldsymbol{W}(t) * \boldsymbol{x}(t) \quad (10.1.7)$$

where $\boldsymbol{y}(t)$ denotes the deconvoluted sources.

For non-stationary sources, a moving block is usually performed on the received signals, and it is usually assumed that the windowed signal in each block is nearly stationary. Let $\hat{s}(n,m)$ and $\hat{x}(n,m)$ denote the blocked sources and received signals respeetively in the mth block. Thus, Eq. (10.1.6) approximately becomes

$$\hat{\boldsymbol{x}}(n,m) = \boldsymbol{H}(n) * \hat{\boldsymbol{s}}(n,m), \quad n = 1,\cdots,N, m = 1,\cdots,M \quad (10.1.8)$$

where m is the index for blocks, M is the total number of blocks, N is the length of each block, and $N \geqslant 2P$; thus, a linear convolution can be approximated by a circular convolution. A hamming window is adopted to block the signals, and the overlap between the adjacent blocks is 0.5. After the Fast Fourier Transform (FFT) is applied on the three components in Eq. (10.1.8), we obtain

$$\hat{\boldsymbol{X}}(w,m) = \boldsymbol{H}(w)\hat{\boldsymbol{S}}(w,m), w = 1,\cdots,N \quad (10.1.9)$$

where $\boldsymbol{w}$ denotes the frequency bin. Thus, the convolutive mixtures in the time-domain turn into the instantaneous mixtures in the frequency-domain. This is the basic idea of the frequency method for the convolutive BSS[19~21].Then, the separation is done in each frequency bin,

$$\hat{\boldsymbol{Y}}(w,m) = \hat{\boldsymbol{W}}(w)\hat{\boldsymbol{X}}(w,m) \quad (10.1.10)$$

where $\hat{W}(w)$ and $\hat{Y}(w,m)$ are the separating matrix and the separated signals in the frequency bin ω respectively.

Since the reverberation is colored, $\hat{S}_j(\omega,m)$ must be colored, too; then the autocorrelation matrix of $\hat{S}(w,m)$ is

$$R_{\hat{S}}(w,\tau) = E[\hat{S}(w,m)\hat{S}^{H}(w,m-\tau)] \tag{10.1.11}$$

where τ is the time delay index related to the degree of the colored characteristic. Under the uncorrelation of the sources, $R_{\hat{S}}(w,\tau)$ must be a diagonal matrix, so must $R_{\hat{Y}}(w,\tau)$.

$$\begin{aligned} R_{\hat{Y}}(w,\tau) &= E[\hat{Y}(w,m)\hat{Y}^{H}(w,m-\tau)] \\ &= E[(\hat{W}(w)X(w,m))\cdot(\hat{W}(w)X(w,m-\tau))^{H}] \\ &= \hat{W}(w)R_{\hat{X}}(w,\tau)\hat{W}^{H}(w) \end{aligned} \tag{10.1.12}$$

If we assume the sources have an unit variance, i.e. $R_{\hat{S}}(w,\tau)=I$, then we extend the cost function with the colored characteristics,

$$\Phi = \frac{1}{4}\sum_{\tau=0}^{T-1} \| \hat{W}(w)R_{\hat{X}}(w,\tau)\hat{W}^{H}(w) - I \|_{F}^{2} \tag{10.1.13}$$

where $\|\cdot\|_F$ is the Frobenius norm and T is the total number of delayed samples.

For a $l_{row}\cdot l_{col}$ full column rank matrix D, according to Ref.[26],

$$\| D \|_{F}^{2} = \sum_{i=1}^{l_{row}}\sum_{j=1}^{l_{col}} d_{i,j}^{2} = Trace(DD^{T}) \tag{10.1.14}$$

where $Trace[\cdot]$ is the trace of a matrix. Substituting Eq. (10.1.14) into Eq. (10.1.13), we obtain

$$\begin{aligned} \Phi = \frac{1}{4}\sum_{\tau=0}^{T-1} Trace\{&[\hat{W}(w)R_{\hat{X}}(w,\tau)\hat{W}^{H}(w) - I]\cdot \\ &[\hat{W}(w)R_{\hat{X}}(w,\tau)\hat{W}^{H}(w) - I]^{H}\} \end{aligned} \tag{10.1.15}$$

Through the gradient decent algorithm, we differentiate Φ with respect to $\hat{W}(w)$ and obtain

$$\frac{\partial\Phi}{\partial\hat{W}(w)} = \sum_{\tau=0}^{T-1}[\hat{W}(w)R_{\hat{X}}(w,\tau)\hat{W}^{H}(w) - I]\hat{W}(w)R_{\hat{X}}(w,\tau) \tag{10.1.16}$$

By using an unsupervised neural network, the learning algorithm for the separated matrix follows,

$$\begin{aligned} \hat{W}(w,r+1) = \hat{W}(w,r) + \mu(w)\sum_{\tau=0}^{T-1}&[\hat{W}(w,r)R_{\hat{X}}(w,\tau)\hat{W}^{H}(w,r) - I] \\ &\hat{W}(w,r)R_{\hat{X}}(w,\tau) \end{aligned} \tag{10.1.17}$$

where $\mu(w)$ is the learning rate and r is the iteration number.

2. Correct permutation

After $\hat{W}(w)$ is completely computed in each frequency bin, the deconvolutive filter matrix may be obtained by the inverse fast Fourier transform (IFFT), that is

$$[\boldsymbol{W}(1),\cdots,\boldsymbol{W}(q),\cdots,\boldsymbol{W}(Q)] = IFFT[\hat{\boldsymbol{W}}(1),\cdots,\hat{\boldsymbol{W}}(w),\cdots,\hat{\boldsymbol{W}}(N)] \tag{10.1.18}$$

However, the BSS may introduce the permutation problem. If the permutation is not corrected, the same column of a different $\hat{W}(w)$ possibly corresponds to the different sources; so does $\hat{W}(q)$. This may lead to failure in deconvolution. For speech signals, the envelope of the spectrum of the adjacent bins is correlated, and different sources are assumed to be uncorrelated. With this characteristic, the permutation is corrected in each frequency bin. In Ref. [19], the example with speech was given. After applying the short-time FFT on the signal, we obtain a data matrix. If the row is the number of the frequency bins and the column is the number of blocks, the correlation is between the adjacent two rows. It is worth mentioning that this correlation is not between the adjacent bins of the spectrogram, i.e. the FFT of data blocks.

To $\hat{s}(n,m)$ in Eq. (10. 1. 8), the DFT is

$$\begin{aligned}\hat{S}(w,m) &= \sum_{n=0}^{N-1}\hat{s}(n,m)\cdot\exp(-\mathrm{j}2\pi nw/N)\\ &= \sum_{n=0}^{N-1}\hat{s}(n,m)\cdot\cos(2\pi nw/N) - \mathrm{j}\cdot\sum_{n=0}^{N-1}\hat{s}(n,m)\sin(2\pi nw/N)\end{aligned} \tag{10.1.19}$$

$$\hat{S}(w+1,m) = \sum_{n=0}^{N-1}\hat{s}(n,m)\cos[2\pi n(w+1)/N] - \mathrm{j}\cdot\sum_{n=0}^{N-1}\hat{s}(n,m)\sin[2\pi n(w+1)/N] \tag{10.1.20}$$

It is apparent that $\hat{S}(w,m)$ and $\hat{S}(w+1,m)$ must be somewhat correlated because the phase difference is quite small. Equations (10.1.19) and (10.1.20) imply that the correlation between the adjacent blocks of unmixed signals is not limited to the speech signals. This correlation does not result from the physical mechanism but from the computation method of the DFT. Hence, we correct the permutation through the correlation between the two vectors $\hat{S}(w)$ and $\hat{S}(w+1)$ where $\hat{S}(w)=[\hat{S}(w,1),\cdots,\hat{S}(w,m),\cdots,\hat{S}(w,M)]$. Three correlation coefficients are defined here. For simplicity, we take two sources only.

$$\begin{aligned}\mathrm{Corr1}(w) &= \mathrm{corr}[\hat{\boldsymbol{S}}_1(w),\hat{\boldsymbol{S}}_1(w+1)],\\ \mathrm{Corr2}(w) &= \mathrm{corr}[\hat{\boldsymbol{S}}_2(w),\hat{\boldsymbol{S}}_2(w+1)],\\ \mathrm{CorrCross}(w) &= \mathrm{corr}[\hat{\boldsymbol{S}}_1(w),\hat{\boldsymbol{S}}_2(w+1)]\end{aligned}$$

We randomly generate two white Gaussian noises in the example to prove the statement made above. The sampling frequency is 8,000 Hz and the size of each

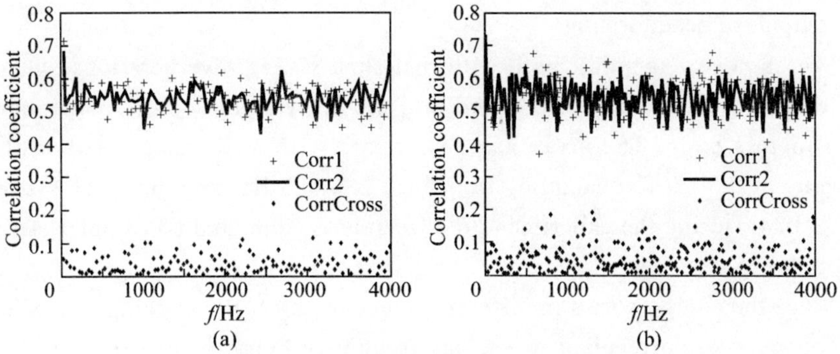

Fig. 10.4 Corr1, Corr2, and CorrCross
(a) The size frame is 25ms; (b) The size frame is 50ms

frame is 25 and 50 ms respectively. The hamming window is adopted to block the signals, and the overlap between the adjacent blocks is 0.5. From the two plots in Fig. 10.4, it is apparent that the correlation that we use in the study does really exist.

We take the example of the two inputs two outputs system to simplify the statement on how to correct the permutation. For $\hat{Y}(w,m)$, we first define the order of the first frequency bin for the reference order and then compute the correlation coefficient Corr1: $\hat{Y}_1(1,m)$ to $\hat{Y}_2(1,m)$ and Corr2: $\hat{Y}_1(1,m)$ to $\hat{Y}_1(2,m)$. If Corr1 is larger than Corr2, it means that $\hat{Y}_1(1,m)$ and $\hat{Y}_2(1,m)$ correspond to the same source and no permutation existed. In the case that Corr1 is smaller than Corr2, it implies that $\hat{Y}_1(1,m)$ and $\hat{Y}_2(1,m)$ belong to the same source, and the first column and the second column of $\hat{W}(2)$ should be exchanged to correct the permutation. In the next frequency bins, permutation corrections are also along this line to guarantee that the same column of $\hat{W}(w)$ corresponds to the same source. Generally, for scale indeterminacy, normalizing the permuted separating matrix is a fast and effective method, and this is helpful for maintaining a stable convergence,also[22].After the uncertainties are resolved, the proper deconvolutive filter matrix may be obtained by Eq. (10.1.18).

10.1.4 Experiments and analysis

In active sonar, the target echo and oceanic reverberation are considered as sources. In Section 10.1.3, the convolutive BSS algorithm requires that sources are uncorrelated, and we have shown that the target echo return and reverberation returns may have a very slight correlation, which means the approximation to the convolutive BSS is reasonable with the data matrix from the outputs of beamforming.

We use two examples, both with real shallow sea reverberations, where all targets are simulated and added into the array signals. The sea is about 170 meters deep. The array has 36 sensors with a distance between the sensors of 0.5 m. The sampling frequency is 6,000 Hz, and the block size is N=1, 024. During the experiment, the weather is calm, and no Doppler effect is produced.

Since the energy varies in different frequency bins, the learning rate $\mu(w)$ is supposed to correspond to various frequency bins, so then it is defined as[23]

$$\mu(w) = \frac{0.6}{\sqrt{\sum_{\tau=0}^{T-1} |\boldsymbol{R}_X(w,\tau)|^2}} \tag{10.1.21}$$

Example 1 This example shows that our approach is as effective as the PCI. The transmitted signal is HFM with the center frequency of 2,250 Hz, a bandwidth of 500 Hz, and a duration of 2 seconds. An HFM target echo is simulated and added to the reverberation with the SRR=−24 dB. The target echo appears 1.3 seconds after the transmitted signal. From the upper plot to the lower plot, Figure 10.5 shows the matched filter outputs corresponding to the main lobe after beamforming, the deconvoluted desired target, the deconvoluted reverberation, and the output by the PCI respectively. This figure implies that the convolutive BSS can separate the desired target echo and the reverberation from the data matrix that is formulated by outputs of the main lobe and the adjacent bin of beamforming. It is apparent that both the convolutive BSS and the PCI are effective. However, the convolutive BSS does not require the prior knowledge of the target energy, and this is a necessary condition for the PCI. This is the advantage of our approach in the fact that the target energy is expensive knowledge and is hard to obtain in active sonar target detection.

Example 2 The transmitted signal is the HFM with the center frequency of 1,750 Hz, a bandwidth of 500 Hz, and a duration of 4 seconds. During the experiment, we sent the transmitted signal hundreds of times. At last, 100 periods of reverberation were selected according to a hundred emissions. An HFM target echo is added at a different time with the SRR=−7 dB to simulate the situation in which the targets have different distances from the platform of the active sonar. We make the variance of the desired deconvolutive signal equal to the original main lobe, then the original main lobe is substituted by this

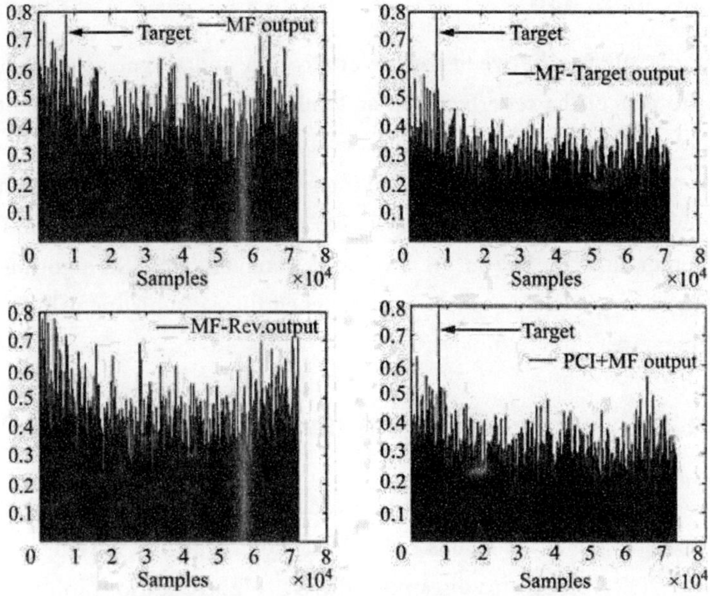

Fig. 10.5 Matched filter outputs with an HFM simulated target at SRR= − 24 dB

desired signal for the next detection steps.

Figures 10.6 (a) and (b) are the main lobe after beamforming and the deconvoluted desired target respectively. With a comparison of both the right and left plots in Fig. 10.6, it is obvious that the SRR has been improved about 2 dB, and the time-frequency feature of the target is more apparent; i.e. the target is enhanced.

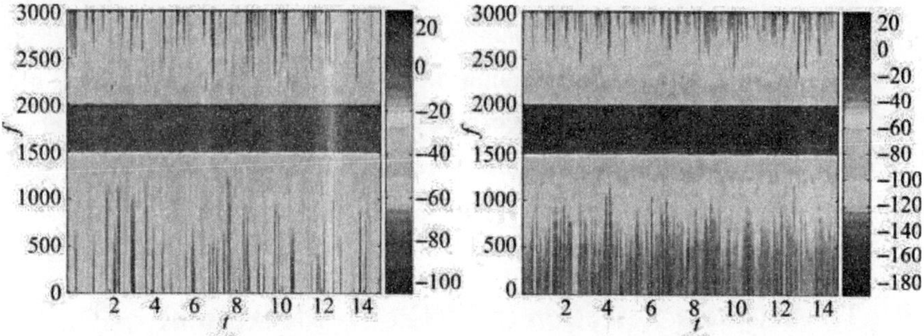

(a) (b)

Fig. 10.6 Spectrogram with an HFM simulated target at SRR=−7 dB
(a) The main lobe after beamforming; (b) The deconvoluted desired target

10.1.5 Conclusion

Spatial uncorrelation between the reverberation and target echo proved to be useful to cancel the reverberation and enhance the target at the main lobe simultaneously. Based on this extra information, a new active sonar detection procedure-beamforming+BSS+matched filter+normalization can outperform the classical method with 2 dB larger output. The nonstationarity and colored feature of the reverberation are helpful to develop the convolutive BSS algorithm, and the non-Gussianity of the outputs of the matched filter can distinguish the target echo from reverberation returns.

10.2 ECG Artifacts Rejection in EEG with ICA

The electroencephalogram (EEG) is the research subject in neural electrophysiology experiment, and it contains rich information on neural activities. During experiments, the electromyography (EMG), electrooculography (EOG), electrocardiogram (ECG), line frequency, and movement of the participants may generate artifacts. It is necessary to reduce these interferences before the EEG analysis, especially in experiments of small animals and brain-damaged subjects. The heart rate of human beings is about 72 beat/min (1.2 Hz), and belongs to the δ frequency band. The body of a small animal is very small, and its heart rate is faster. For example, the heart rate of a big rat is about four times that of a human being's, and its main frequency and key harmonic wave overlap with the θ and α wave; but the frequency band of these two waves is most sensitive and is an active region among the brain is activities. Moreover, in the case of a brain-damaged subject, the EEG is already weak, while the interferences become much more severe. If the artifacts are not removed, the analysis will not be correct. Next, two examples are presented. One experiment is on a small animal and the other is on a brain-damaged participant. All data are from the hospital at the School of Medicine, Johns Hopkins University.

10.2.1 Artifacts Rejection in EEG of Little Animal

When the small animal's heart beat stops, the EEG, ECG and blood pressure are all recorded. The experiment paradigm is as follows: In the first ten minutes, Wistar rats are under anaesthesia, and the baseline of the EEG and ECG are recorded. To reduce the effect of the Halothane on the EEG, five minutes of pure oxygen washing is implemented. To acquire the anoxia and ischemic damages under different levels, the oxygen supply to rats is stopped for three or five minutes. After this, oxygen is supplied immediately and cardiopulmonary resuscitation (CPR) is carried out to bring the blood pressure back to normal. Under such an experiment, after damaging the brain and normalizing the blood pressure, the ECG artifact is very apparent and is related to the location of the electrodes. Figure 10.7 shows such a period of the EEG signal.

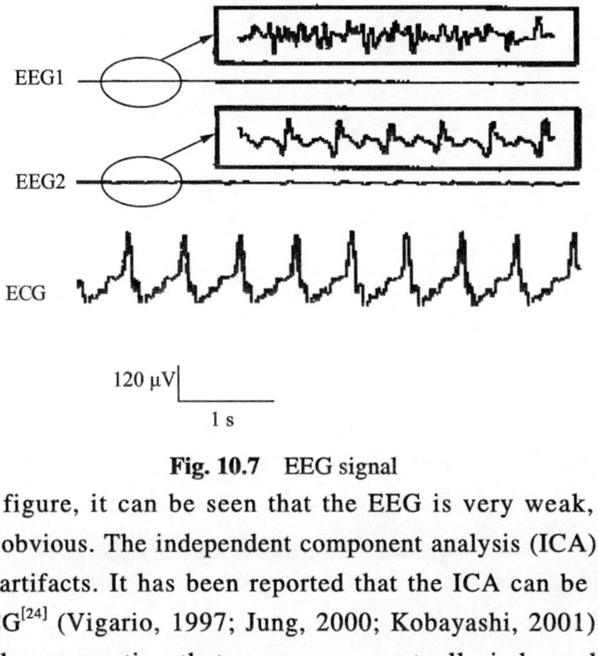

Fig. 10.7 EEG signal

From this figure, it can be seen that the EEG is very weak, while the ECG is quite obvious. The independent component analysis (ICA) is applied to reject the artifacts. It has been reported that the ICA can be applied to reject the ECG[24] (Vigario, 1997; Jung, 2000; Kobayashi, 2001). The ICA is based on the assumption that sources are mutually independent, and a demixing matrix is deduced to minimize the cost function. If the independent components corresponding to the ECG are set to zero and the inverse of the demixing matrix is multiplied, the ECG artifact can be rejected. Figure 10.8 illustrates the algorithm procedure.

Figure 10.9 demonstrates the power spectrum of the EEG before and after the ECG artifact rejection. It is apparent that the ICA rejects the ECG.

Meanwhile, the effect of the artifact rejection is also studied in the monitoring of the anoxia and ischemic damage through the EEG. Figure 10.10 shows that before the ECG is removed, the recovery of the EEG is faster, though this is a feint due to the artifacts of the ECG. The normalized EEG power of the θ band contains the characteristics of the spike after ischemic recovery.

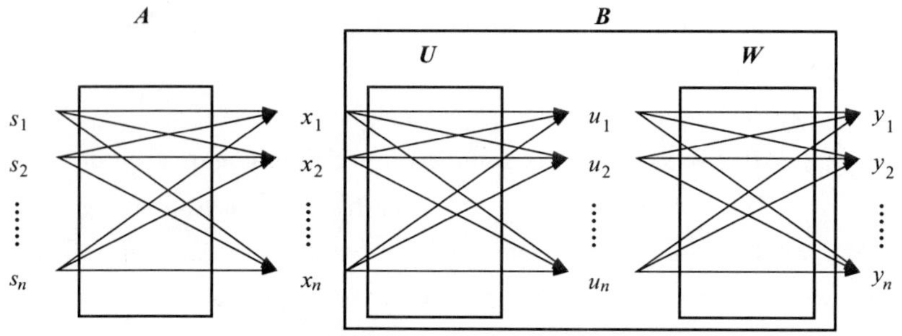

Fig. 10.8 ICA algorithm procedure

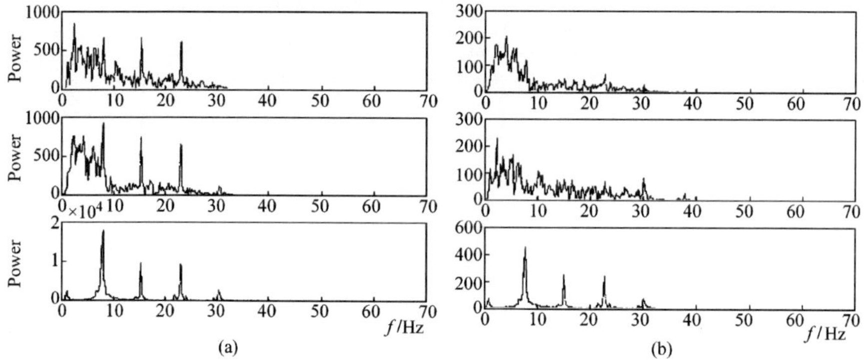

Fig. 10.9 Comparison of the spectrum
(a) original EEG spectrum; (b) EEG spectrum after artifact rejection by ICA

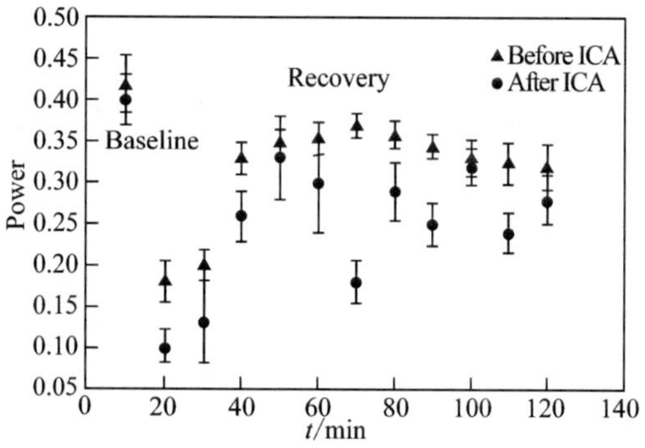

Fig. 10.10 Change of normalized EEG power of θ band [25] (courtesy of Tong SB, Bezerianos A, Paul J.Used with permission)

10.2.2 ICA-Based ECG Artifact Rejection in Clinical EEG[26]

This section presents an example of the EEG of clinical brain damage to demonstrate the application of the ICA in clinical data analysis. The CEEGRAPH of the Bio-Logic Systems is adopted to record the EEG. The single electrode record and the different electrode display are used to meet doctor requirements and to allow researchers to conveniently acquire raw data for further signal processing.

The frequency interval of recording the EEG by the CEEGRAPH is 128 Hz. In fact, only 0~30 Hz is the desired frequency band; therefore, some pre-processing should be necessary. Except for the noise reduction and filtering, the ICA is performed to reject the artifacts. Figure 10.11 shows the results: (a) The EEG data has a distinct ECG artifact at the dual electrodes of T3-C3, C3-Cz, Cz-C4, C4-T4, O1-O2 and the ECG signal, with the length of the data of 4s, 1,024 points; (b) Among six estimated independent components by the ICA, the second component corresponds to the inverse phase of the ECG artifact; (c) by setting the IC2 component to zero and multiplying the inverse matrix of the demixing matrix, the 2nd~6th channels correspond to the EEG after rejecting the a rtifacts.

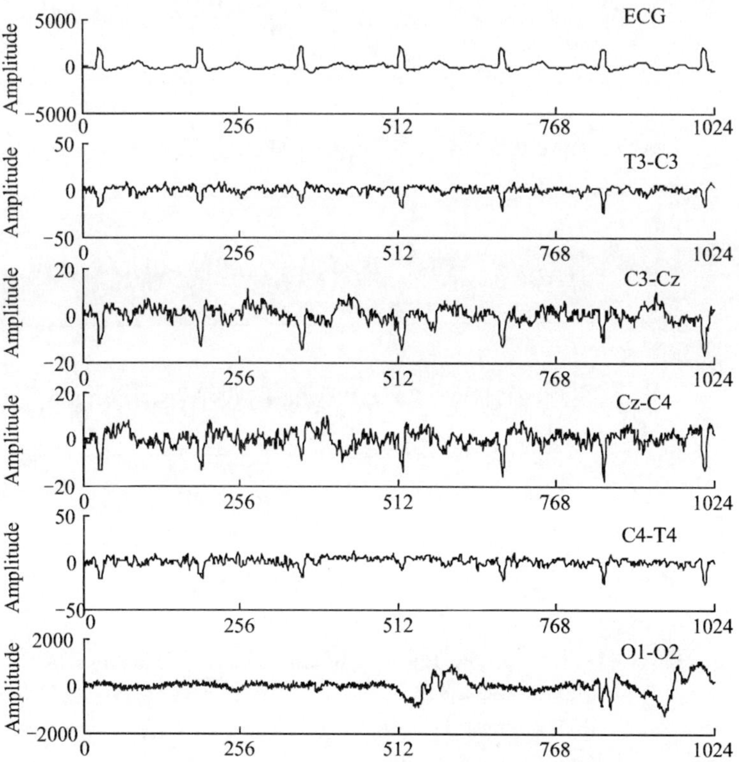

(a)

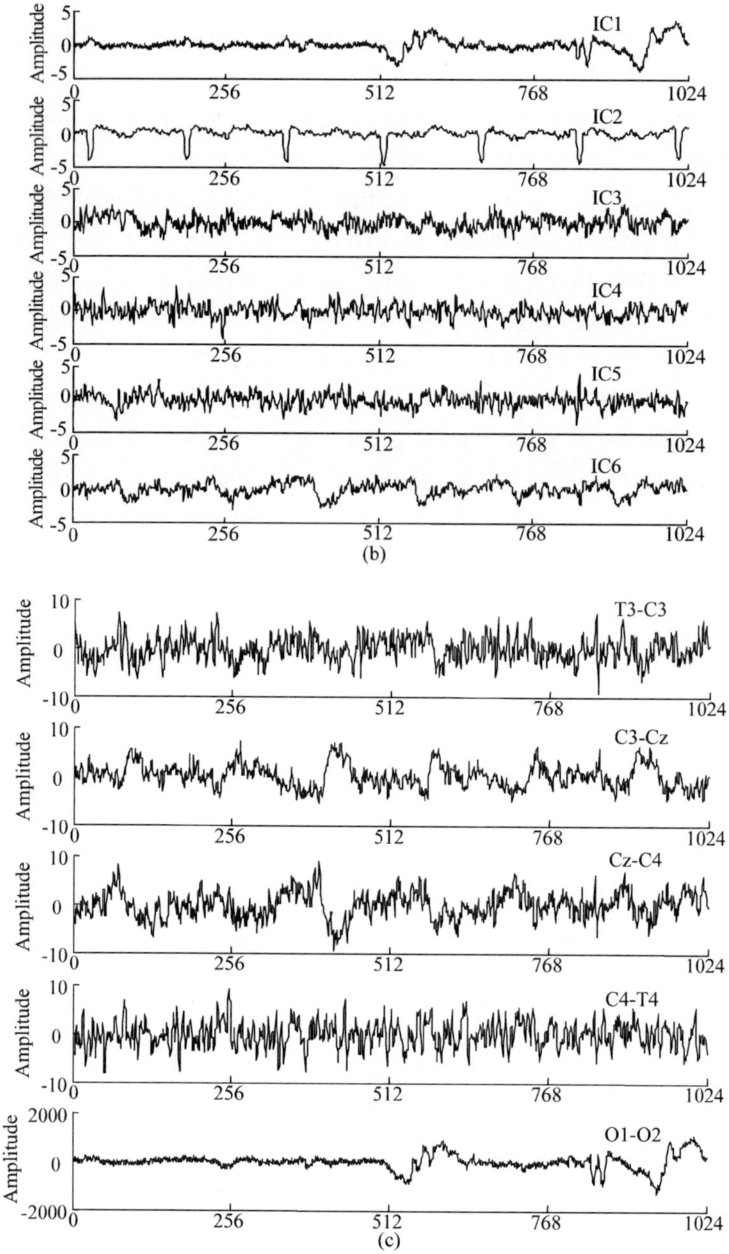

Fig. 10.11 ECG artifact rejection of clinical anoxia EEG by ICA

(a) Raw EEG and ECG;

(b) Six estimated independent components by ICA;

(c) EEG after ECG artifact rejection

10.3 Experiment on Underdetermined Blind Separation of A Speech Signal

To exploit the effectiveness of the underdetermined blind separation algorithm for real-world data, an experimental study on this issue has been performed[27] in a semi-anechoic chamber environment at Shanghai Jiao Tong University.

10.3.1 Experiment Setup

The semi-anechoic chamber is a room with its walls and ceiling covered by acoustically absorbent fiber board. The board is shaped to the wedge and sharp and is shown in Fig. 10.12. Due to acoustical absorption, the reverberation is reduced greatly inside the room. Therefore, the mixing model is simplified and it approaches a free-field acoustical environment.

Fig. 10.12 Acoustically absorbent fiber wall of semi-anechoic chamber

During the experiment, the main apparatus used are a computer, signal analyzer, multichannel data recorder, precise full directivity microphone, and acoustic signal sources. The sampling frequency is 10 kHz. Figure 10.13 illustrates the configuration of sources and sensors. The paradigm is as follows:

(1) According to Fig. 10.13, connect the apparatus, calibrate the microphone by using a sound level calibrator, and adjust the gain of the amplifiers to match the calibrated sensitivity of the microphone;

(2) Play each source independently and record the corresponding signal;

(3) Play all sources simultaneously and record data from each microphone.

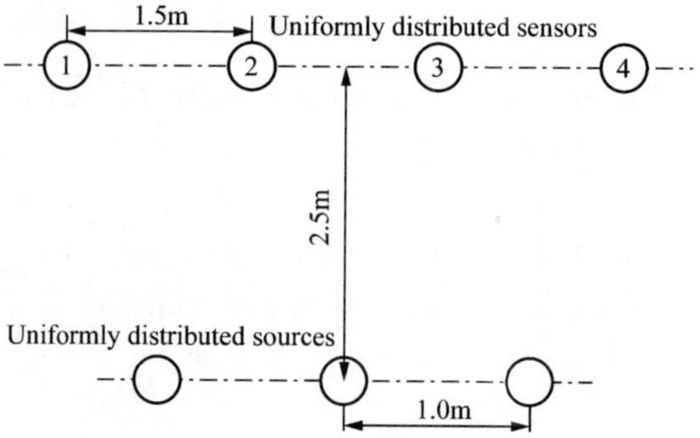

Fig. 10.13 Configuration of sources and sensors

10.3.2 Data Preprocessing

In many applications of the BSS, the key issue is to recover the source waveform. Mean value removal, normalization and whitening do not cause a negative effect on the recovery of the waveform; moreover, appropriate preprocessing could greatly improve the quality of the recovery waveform.

For observation sequence $\boldsymbol{x}_{m\times 1}(t), t=1,2,\cdots,T$, remove the mean value:

$$\boldsymbol{x}_0(t)=\boldsymbol{x}(t)-E(\boldsymbol{x}(t))$$

Whiten the $\boldsymbol{x}(t)$ to eliminate the correlation among observed signals, the whitening matrix $\boldsymbol{T}$ meets

$$z(t)=\boldsymbol{Tx}(t)=\boldsymbol{TAs}(t)$$

$$E(zz^{\mathrm{T}})=\boldsymbol{T}E(\boldsymbol{xx}^{\mathrm{T}})\boldsymbol{T}^{\mathrm{T}}=\begin{bmatrix}\lambda_1 & 0\\ 0 & \lambda_2\end{bmatrix}$$

where $z(t)$ is the whitened $\boldsymbol{x}(t)$.

10.3.3 Data Analysis

In the experiment, sources consist of two segments of speech and one segment of light music, as shown in Fig. 10.14, with the observed mixtures at Channel 2 and 3 shown in Fig. 10.15. After the necessary preprocessing, the MoG-uICA algorithm is used to estimate sources as demonstrated in Fig. 10.16. In practice, it is very difficult to completely recover sources. By comparison with true sources in Fig. 10.14 and estimated sources in Fig. 10.16, it is apparent that the separation is successful despite some differences in details. Figure 10.17 shows the estimation from channels 1 and 4. Since the real observation is a convolutive process, especially for channels 1 and 4, the results are acceptable, although they are worse than the outcomes in Fig. 10.16.

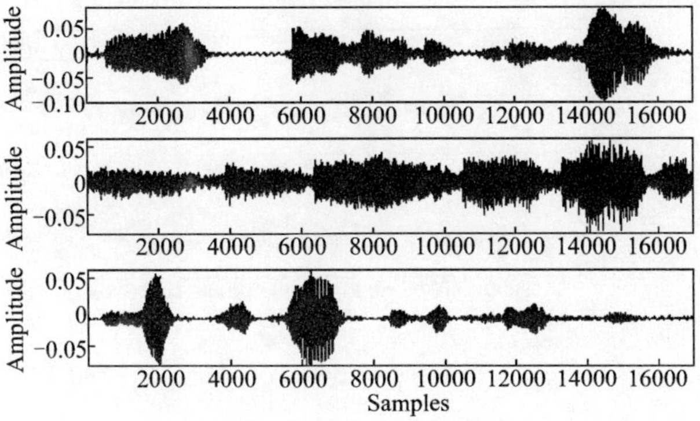

Fig. 10.14 Source signals
(Two segments of speech and one segment of music)

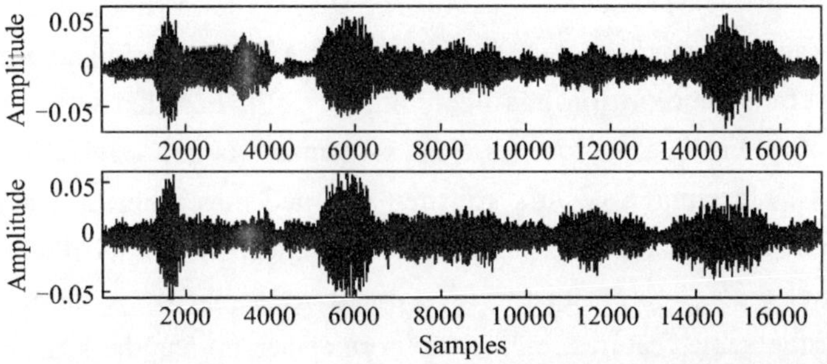

Fig. 10.15 Observed mixtures (channels 2 and 3)

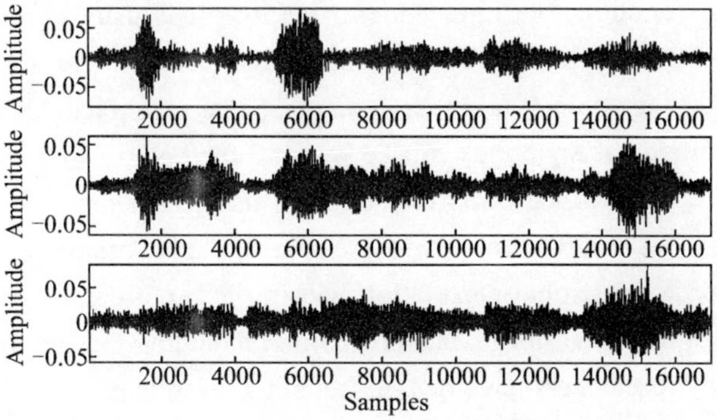

Fig. 10.16 Estimation based on channels 2 and 3

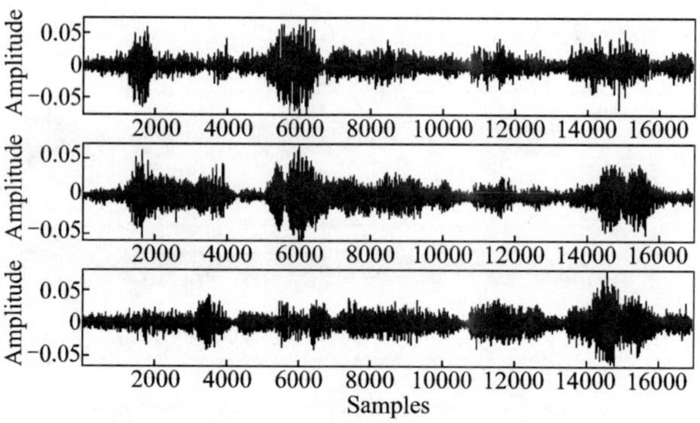

Fig. 10.17 Estimation based on channels 1 and 4

10.4 ICA in Human Face Recognition[28]

Face recognition is an important technique for automatic face identification, which has proven particularly useful for suspect identification in the crime detection and in human-machine interactions[29]. As a standard pattern recognition problem, face recognition has been successfully handled with PCA-based methods[30~32]. In these approaches, each face image is first stacked into a single vector. All face image vectors are then grouped into a big face data matrix. The eigenfaces[31] of the face patterns are subsequently extracted as the mutual orthogonal eigen-vectors of the face image data matrix. A human face has basically the same features, e.g. every normal person has the same number of facial organs and nearly the same spatial topological relations between these organs. Therefore, human facial images are highly redundant, resulting in an efficient eigenface-based facial image data representation and face recognition. While this is enough for subject identification, it is usually not enough for considering the multiple features embedded in the facial images, such as different illuminations, viewpoints, and face reactions. The eigenfaces that account for most of the data variance may not be optimal for coding such aspects; much of the important information is contained in the higher-order statistics of the facial images. As described in previous chapters, the ICA is designed to explore the higher-order statistical correlation within the input data; it is therefore, a natural choice for facial image data processing, including face recognition. Since 1997[33], many researchers have demonstrated the usefulness of ICA in face data analysis[34~37]. Here we only provide a brief introduction to the ICA-based face recognition process; more details can be found in Refs.[36,37].

In face recognition, facial images are usually sampled using the same spatial grid (in an $M_x \times M_y$ pixel matrix, M_x is the number of rows, and M_y is the number of columns) and intensity is normalized. The $M_x \times M_y$ image matrix is generally of a zero mean and stacked columnwise into a single vector, and an $N \times M_x M_y$ image matrix is subsequently formed for N face images.

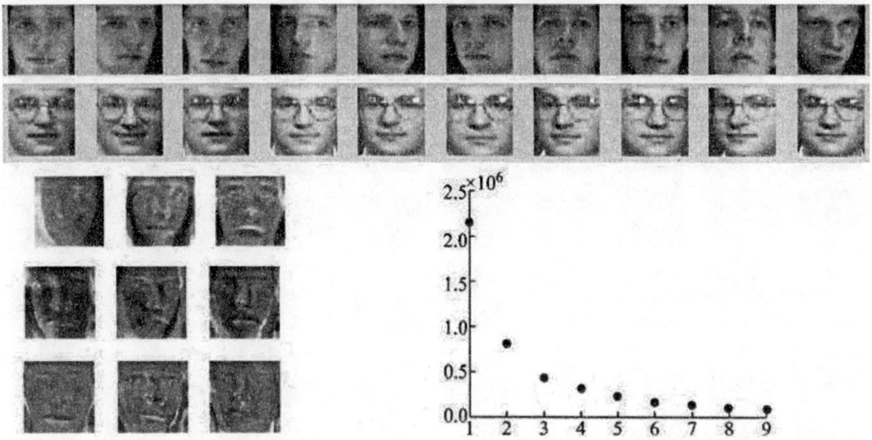

Fig. 10.18 An example of PCA-based face image processing. The top row shows image samples (64 × 64) from the ORL-face database (Samaria and Harter, 1994) The 9 eigenfaces are displayed in the bottom left panel with the associated eigenvalues shown in the bottom right plot.

Denote the big image matrix with $\boldsymbol{X}=\{\boldsymbol{x}_i\}_{i=1,\cdots,N}$, and the correlation matrix can be calculated as

$$\boldsymbol{R}_x = \frac{1}{N}\sum_{i=1}^{N} \boldsymbol{x}_i \boldsymbol{x}_i^{\mathrm{T}} = \frac{1}{N}\boldsymbol{X}\boldsymbol{X}^{\mathrm{T}} \tag{10.4.1}$$

As $M_x M_y$ is usually very large (4,096 in the example shown in Fig. 10.18); it is nontrivial or even unmanageable for the computer to directly calculate the correlation matrix $\boldsymbol{R}_x$ and perform the corresponding PCA decomposition. Actually, the number of face images (N) is generally far fewer than the number of pixels within each facial image, resulting in an efficient alternative to the PCA decomposition approach[31]. Instead of calculating and decomposing $\boldsymbol{R}_x$, the correlation matrix of $\boldsymbol{X}^{\mathrm{T}}$ is estimated as $\boldsymbol{R}_{x^{\mathrm{T}}}=\boldsymbol{X}^{\mathrm{T}}\boldsymbol{X}$. Since $\boldsymbol{R}_{x^{\mathrm{T}}}$ is a manageable $N \times N$ matrix, its eigenvectors $\boldsymbol{u}_i$ and the associated eigenvalues λ_i can be directly calculated using the standard PCA decomposition (usually through SVD), and the eigenvectors of $\boldsymbol{R}_x$ can then be calculated as

$$\boldsymbol{v}_i = \frac{1}{\sqrt{\lambda_i}}\boldsymbol{X}\boldsymbol{u}_i, \qquad i = 1,\cdots,N \tag{10.4.2}$$

The eigenfaces are actually the eigenvectors unwrapped into the original image format. In the example shown in Fig. 10.18, the 9 eigenfaces (eigenvectors) in the bottom left panel were calculated from the 10 facial images shown in the top row with the associated eigenvalues shown in the bottom right plot.

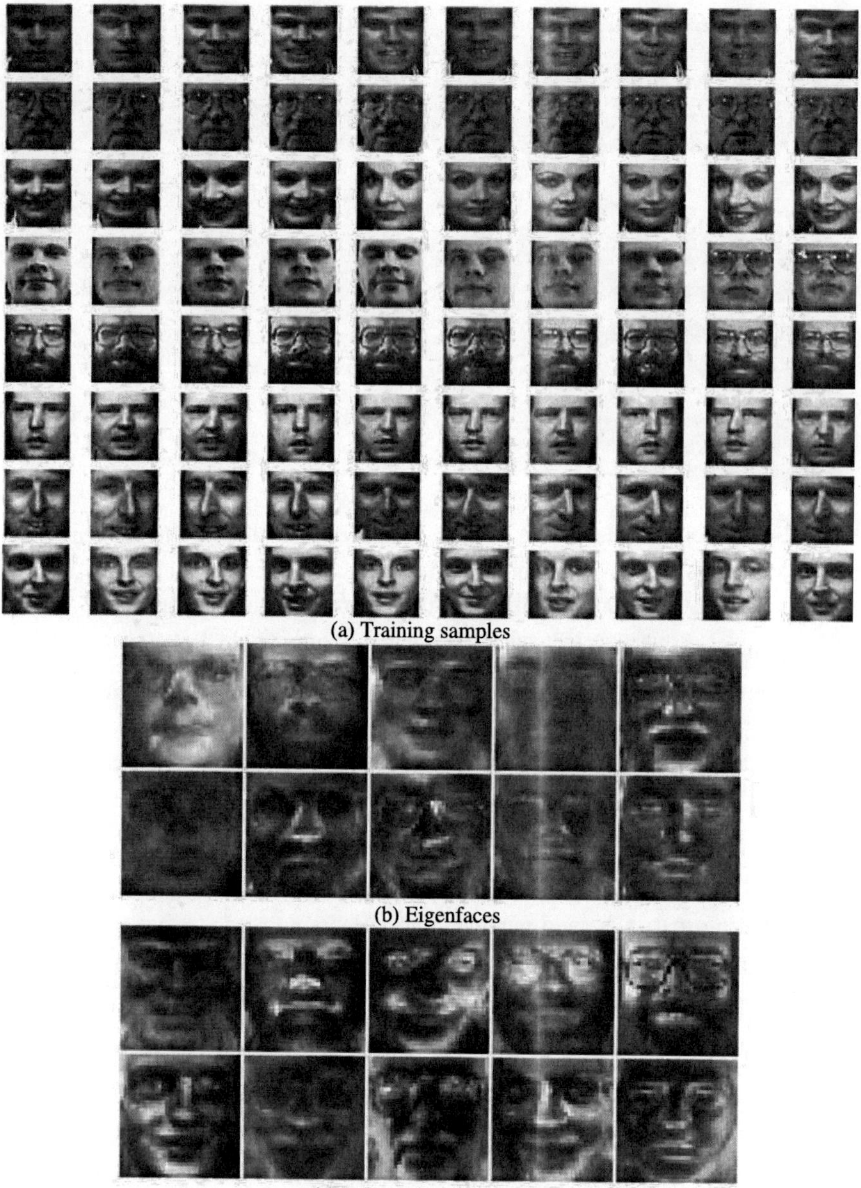

(a) Training samples

(b) Eigenfaces

(c) Independent feature faces

Fig. 10.19 Multi-subject face feature extraction using PCA and ICA

In the PCA-based face recognition, each facial image is represented by the calculated eigenfaces. In other words, the facial image data can be efficiently projected into a lower dimensional subspace spanned by the eigenfaces. Face recognition can then be conveniently conducted in this dimension-reduced space. As stated above, considering only the correlation within the data is not enough to capture the informative higher-order statistical information. To further pursue the remaining higher-order information, the ICA can be applied on the PCA extracted eigenfaces to search for a set of mutually independent (approximate) feature vectors s_i. Consequently, each original facial image is projected into the subspace spanned by these mutually independent feature bases.

Figure 10.19 shows an example of an ICA-based face feature recognition result. Ten images from each of the 8 subjects are selected from the ORL-face database. Each image is clipped to 32×32 pixels. The source images are displayed in Fig. 10.19(a). Figure 10.19(b) shows the 10 eigenfaces extracted using PCA. Figure 10.19(c) shows the corresponding 10 independent feature faces estimated with a FastICA. Apparent differences between the eigenfaces and the independent feature faces extracted with the ICA can be easily found in Fig. 10.19, as well as in the individual subject case shown in Fig. 10.20, where the eigenfaces show sequentially decreased image energy, while the independent

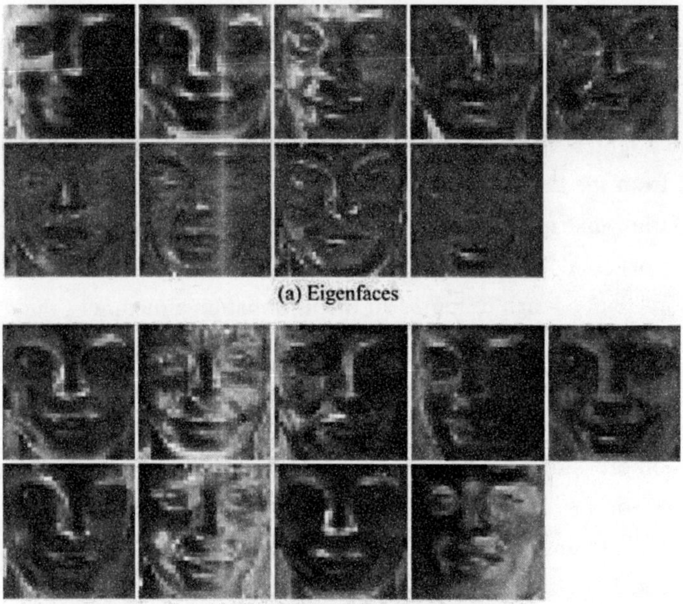

(a) Eigenfaces

(b) Independent feature faces

Fig. 10.20 Single subject face feature extraction using PCA and ICA feature faces do not show such a pattern.

The feature distribution in the feature space is shown in Fig. 10.21.

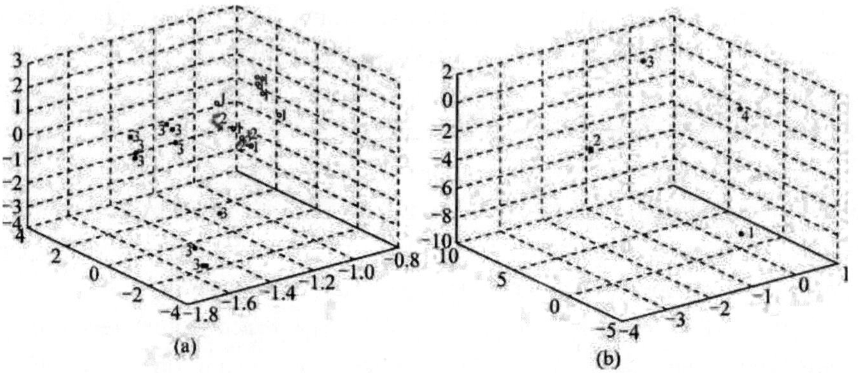

Fig. 10.21 Image feature distribution in feature space
(a) Distribution in PCA feature space; (b) Distribution in ICA featured space

A comprehensive comparison of the PCA and ICA in face recognition can be found in a recent publication[37], where the authors showed that the ICA is generally better than the PCA, though the latter also performs well.

10.5 ICA in Data Compression

Data compression saves data size by removing redundant information of the original data. A powerful tool for multidimensional data compression is the PCA, which represents the original data with the mutual orthogonal eigenbases extracted from the correlation matrix. As mentioned in the previous sections, the correlation matrix captures up-to-the-second-order statistical information. When the original data space is non-Gaussian, second-order statistical information is not enough to represent the information structure of the raw data; higher-order statistical information needs to be considered. Since the ICA is designed to remove the higher-order correlation of the raw data, it could be a potentially useful tool for data compression. Current applications of the ICA for data compression mainly focuses on image compression[38~40]. Here we introduce a working example of the ICA-based plenoptic-illumination function compression[41]. In this section, the concept of the plenoptic-illumination function is introduced first; the independent base image (IBI) is introduced second; the third part relates to compression on the IBIs and the IBI-based raw image representation coefficients; results and discussions are given at the end.

10.5.1 Plenoptic-Illumination Function and Image-Based Relighting

Fig. 10.22 An example of an IAI dataset

The plenoptic-illumination function (PIF)[42,43] is an extension of the original plenoptic function (Adelson and Bergen, 1991), which provides an explicit illumination component for illumination control in image-based modeling and rendering (IBMR). The basic idea of the PIF is to interpolate the scene image under the desired lighting condition from the prior sampled scene images under various illumination conditions. This interpolation process is also called image-based relighting (IBL). Those prior sampled scene images under various lighting conditions are called illumination adjustable images (IAIs). As an example, a few IAIs from an IBL raw database are displayed in Fig. 10.22. While the PIF-based IBL provides an easy solution for illumination control in the IBMR, its real applications suffer from the huge data storage problem caused by the numerous prior sampled images. Data compression is then essential for a practical use of IBL.

For simplicity, we only consider the directional light source, a fixed viewpoint, and a prospective projection in this subsection.

Denote IAI by $P(x, y, \boldsymbol{L})$, where (x, y) is the coordinate in the projection plane, and $\boldsymbol{L}$ is the sampled light direction, for an arbitrary light direction $\boldsymbol{D}$; a new scene image can be generated by

$$P(x,y,\boldsymbol{D}) = \sum_{\boldsymbol{L}_i \in S_D} \boldsymbol{W}_{L_i,D} P(x,y,\boldsymbol{L}_i) \tag{10.5.1}$$

where $\boldsymbol{S_D}$ means the collection of the nearest neighbors of direction $\boldsymbol{D}$, varying due to the different interpolation methods used. Here we use a bilinear interpolation algorithm, where $\boldsymbol{S_D}$ consists of the four nearest neighbors of $\boldsymbol{D}$, and the interpolation weight $\boldsymbol{W}_{L_i,D}$ is calculated as

$$W_{L_i,D} = \frac{(\text{dist}(\boldsymbol{L}_i,\boldsymbol{D}))^{-1}}{\sum_{L_i \in S_D} (\text{dist}(\boldsymbol{L}_j,\boldsymbol{D}))^{-1}} \tag{10.5.2}$$

where dist(·) is the geodesic distance in the unit sphere between two directions.

10.5.2 Independent Base Image-Based Relighting

1. Independent Base Image

It is a natural and essential requirement for any image compression algorithm to consider how the human vision system (HVS) perceives and transmits vision signals as the HVS will give the final and most important justification regarding the quality of any kind of compressed image. According to Barlow's efficient coding theory[44], the vision signal processing of the HVS is a maximum information transfer and redundancy reduction process. Two coding methods have been proposed to implement this model: compact coding and sparse coding[45~47]. Equivalent to the PCA, compact coding was demonstrated to be unable to capture all the vision encoding features of the mammalian vision system. In contrast, sparse coding is more efficient for modeling the perception process of the mammalian vision system. Very similar to the receptive field of cortical simple cells in the primary visual cortex, the image bases extracted by sparse coding are spatially localized, oriented, and bandpassed[48,49]. As the sparseness maximum is also a maximum non-Gaussian process, it is exactly an ICA solution mentioned in previous chapters. Therefore, it is reasonable to use ICA to search for an efficient representation of images. Research has shown that the independent image bases extracted by the ICA have nearly the same features as those extracted by sparse coding [50]. To sum up, the PCA is inefficient to explore the statistical information structure for non-Gaussian signals (such as the IAIs), resulting in an inefficient eigen base-based image representation. The ICA provides a potentially effective tool for extracting image bases for an efficient image representation. In the following we will give an ICA-based IAI data compression method. The ICA will be used first to extract the independent image base; regular image compression will be applied in these independent bases (and will be called an independent base image (IBI)). Although there may not exist physically meaningful independent components in the IAI data, it is still interesting and potentially useful to assess how the apparent IBI works for representing IAI data.

2. ICA decomposition for IAI dataset

Concatenating each column of the IAI matrix into a vector $\boldsymbol{p}_i(i=1,2,\cdots, M)$, all the IAI vectors can then be grouped into an IAI data matrix $\boldsymbol{P}$, which is then whitened and dimension-reduced using the PCA as in the regular the ICA preprocessing:

$$\hat{P} = CG \tag{10.5.3}$$

$$C = PG^{T} \tag{10.5.4}$$

where $\hat{P}$ means the approximation to P, G is the eigen base matrix, each row of G is an eigen base vector (formed by sequentially concatenating all columns) g_j, j=1,⋯, M'. M' is the number of eigen bases remaining after dimension reduction, depending on approximate precision According to Refs.[50,51], M' for the IAI can be a number between 5 and 8. C is the coefficient matrix of the eigen base the representation.

With ICA on G, we can get M' IBIs, u_j, j=1,⋯,M', and the associated representation coefficient matrix B:

$$U = BG \tag{10.5.5}$$

$$G = B^{-1}U \tag{10.5.6}$$

Insert Eq. (10.5.6) into Eq.(10.5.3), and can get

$$\hat{P} = CB^{-1}U = RU \tag{10.5.7}$$

with R=CB^{-1}=$\{r_{ij}\}$, indicating the IBI-based IAI representation coefficient. Similar to the eigen base-based IAI representation, Equation (10.5.7) also gives an optimal IAI reconstruction method in terms of the minimal root of the mean square of errors (MSE). An example of IAI decomposition is illustrated in Fig. 10.23.

Fig. 10.23 Illustration of ICA decomposition of IAI dataset

3. IBI-based relighting

Insert Eq. (10.5.7) into Eq. (10.5.1), we can get the independent image base or the IBI-based relighting:

$$P(x,y,\mathbf{D}) \approx \sum_{j=1}^{M'} \sum_{i' \in S_D} W_{i',D} r_{i',j} u_j(x,y) \tag{10.5.8}$$

with

$$W_{j,D}^{*} = \sum_{i' \in S_D} W_{i',D} r_{i',j} \tag{10.5.9}$$

Equation (10.5.8) can be simplified as

$$P(x,y,\mathbf{D}) \approx \sum_{j=1}^{M'} W_{j,D}^{*} u_j(x,y) \tag{10.5.10}$$

giving an arbitrary directional light source with direction defined by D, the corresponding new scene image can then be obtained through Eq. (10.5.10), and the interpolation coefficients can alternatively be obtained from the IBI-based IAI representation coefficients as described by Eq. (10.5.9). Therefore, using the IBI-based relighting, the original required huge IAI dataset can be reduced into a few IBIs and the corresponding representation coefficients of each original IAI. With a different number of IBI, we can get a different compression ratio.

4. IBI compression

An IBI is an energy relocated "image." As shown in Fig. 10.23, spatial correlations still exist between different pixels, which can be reduced using a regular image compression method like the wavelets-based compression method.

10.5.3 Algorithm and Implementation

1. Algorithm

The ICA plus wavelets-based IAI data compression contains the following major steps:

Step1. IAI images RGB to YIQ conversion;

Step2. ICA decomposition on the YIQ images;

Step3. IBI compression with wavelets transform.

This first step is based on the fact that human eyes have different sensitivity to different channels of the YIQ format image. This conversion then gives an option to pursue a different compression ratio for these 3 channels.

2. Implementation

(1) Preprocessing

Running an ICA on IAIs directly might be highly computationally extensive since each IAI could consist of a large number of pixels. To reduce the computation burden, each IAI is divided into a series of nonoverlapped blocks in the same size. The blocks from all IAI at the same block location are then grouped into a sub-IAI set, and the original IAI data compression now becomes a problem of compressing all the sub-IAI sets. All sub-IAI sets are zero-meaned before the ICA processing.

(2) Extracting IBI and the IBI representation coefficients

Using the algorithm mentioned above, M' independent sub-images (named as sub-IBIs in the following text) will be extracted from each sub-IAI dataset. As stated before, M' is a number between 5 and 8. To give high quality image rendering, we choose M' to be 10 for the Y channel, and 5 for the I and Q channels, respectively. With the M' sub-IBIs, the representation coefficients can be obtained by $\boldsymbol{R}=\hat{\boldsymbol{P}}\boldsymbol{U}^{-1}$, and the presampled sub-IAI under lighting condition $\boldsymbol{L}_i$ can be reconstructed as

$$P(x,y,\boldsymbol{L}_i) \approx \text{mean sub-image} + \sum_{j=1}^{M'} r_{ij} u_j(x,y) \tag{10.5.11}$$

note that $P(x,y,\boldsymbol{L}_i)$ in the above equation is a sub-IAI; u_j means the jth independent base of the current sub-IAI dataset; r_{ij} is the associated representing coefficient of $P(x,y,\boldsymbol{L}_i)$, i.e., the length of the projection of $P(x,y,\boldsymbol{L}_i)$ on u_j.

Correspondingly, giving the light direction $\boldsymbol{D}$, a new sub-scene image can be generated from the sub-IBIs using the following equation:

$$P(x,y,\boldsymbol{D}) \approx \text{mean sub-image} + \sum_{j=1}^{M'} W^*_{j,\boldsymbol{D}} u_j(x,y) \tag{10.5.12}$$

where $\boldsymbol{W}^*$ is the same as in Eq. (10.5.9).

(3) Wavelet transform-based compression for the mean image and the sub-IBIs

Three-level wavelet transform is performed on the mean image using the 9-7 biorthogonal wavelets, and then each channel is compressed by bit allocation, quantization and coding. The target bit rate is 3 bits/pixel to secure a high quality reconstruction for the mean image. To save the data storage of the quantization tables required for compression of all the sub-IBI sets, all sub-IBI sets are put together into M' combined IBIs (CIBIs) according to the spatial locations of each sub-IBI. These CIBIs are then assigned a target bit rate using the bit allocation method reported in Ref.[52], and then compressed with the same wavelet transform based image compression method as used for the mean image compression.

As with saving bit rates for storing side information, the m'th ($m'=1, 2, \cdots, M'$) independent base image representation coefficients associated with all sub-IAIs in all sub-IAI sets are grouped into a coefficient source. Bit allocation, quantization, and bit encoding are then used to compress these M' coefficient sources.

10.5.4 Results and Discussions

One-thousand-and-two-hundred ding images with a fixed viewpoint (note: all data were acquired at the Chinese University of Hong Kong) are used to evaluate the above mentioned ICA-based IAI compression methods. The light direction is equally changed, along with the longitude and the altitude. A subset of the whole dataset is displayed in Fig. 10.22. The original image size is 512×512, and each IAI is split into a series of nonoverlapped 16×16 image blocks.

1. Determining the number of IBIs to be retained

The number of IBIs M' to be retained is critical for both the compression ratio and the reconstruction quality. Previously, we mentioned an approximate value between 5 and 8. To determine an optimal value of M' for the dataset to be compressed, a different number of IBIs are used to reconstruct the original data, and the mean square error (MSE) is calculated between the original images and the reconstructed images to measure reconstruction quality. The results are shown in Fig. 10.24, where we can see when M' is 8; further increasing M' yields only a small reconstruction quality improvement. This improvement is even minor when M' is greater than 10. Based on this evaluation, M' is chosen to be 10, 5 and 5 for the Y, I, and Q channel in the following experiment.

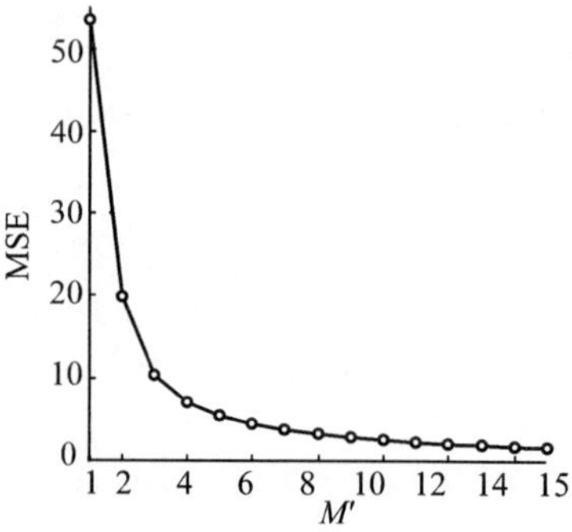

Fig. 10.24 IAI reconstruction errors using different numbers of independent image bases

2. Overall compression results

The ding IAI dataset is compressed with different target bit rates using the proposed ICA-based method, the PCA-based method, the JPEG2000 (Taubman and Marcellin 2002), and the MPEG (Watkinson 2001) algorithms. For the ICA-and PCA-based compressions, the raw data are first split into sub-IAI blocks. Each sub-IAI block is then reduced into a number of IBIs (10, 5, 5 for Y, I, and Q channels) and the associated representing coefficients. The IBIs and representing coefficients are then recombined into CIBIs and large representing coefficient sources using the recombination procedure as stated before. Different target bit rates are then used to compress the mean image, CIBIs, and representing coefficient sources. Since the mean image will be required in scene rendering for any light direction, a relatively high bit rate, 3.0 bps, is used for each channel to ensure a high reconstruction quality, which is 52.84 dB in terms of a peak signal-to-noise ratio (PSNR). The target bit rates are set to 2~6 bps for CIBIs and 4~7 bps for the combined representing coefficient sources. The size of the final compressed file is from 6.08 MB to 10.83 MB, and the compression ratio is 83:1 to 148:1. Scene images are generated from these compressed data with 100 arbitrary light directions, and the PSNRs between them and their counterparts generated from the raw IAI data set are calculated to provide a measure for evaluating the proposed method. Figure 10.25 shows the mean PSNR curves of the four used methods at different real compression ratios. From this figure, we can see that the ICA-based IAI compression method is superior to the JPEG2000 and MPEG in terms of a better compression ratio/reconstruction error performance. This outperformance is mainly due to the fact that the JPEG2000-based IAI compression only considers the redundant information within each IAI separately. The MPEG takes only the redundant information between adjacent IAIs into account, though it can reduce both intra-IAI and inter-IAI information redundance. Interestingly but not surprisingly, the PCA-based IAI compression method outperformed the ICA-based method, as shown in Fig. 10.25. This may be caused by two possible reasons. First, the bit allocation approach used in both methods[50] is based on the signal energy (covariance). As stated before, covariance can only describe the second statistical information; therefore, it is far enough away to describe the all-order statistic independence required in the ICA. Second, for a reconstruction error measure, the PSNR is essential for a second-order statistics measure, which is also not enough for the independent image bases. Up to this point, a more appropriate bit allocation method and a more suitable reconstruction error measure are required to compare the ICA-based data compression to the PCA-based compression method.

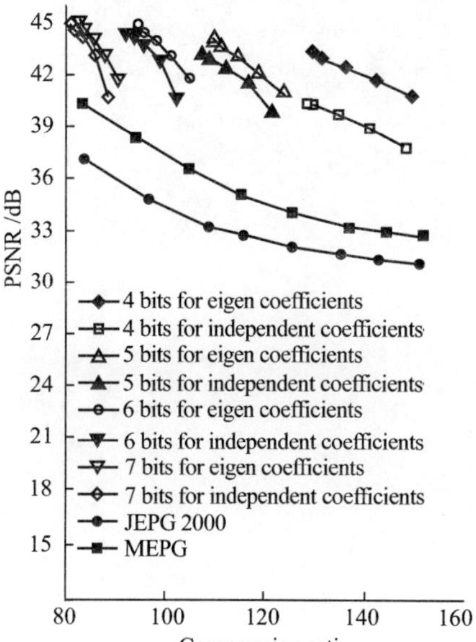

Fig. 10.25 Reconstruction vs real compression ratio curves of different compression methods

Figure 10.26 shows the zoomed versions of the scene images under the same directional light generated from the compressed data using different compression methods. The compressed data size of the PCA and ICA methods are 7.59 MB and 8.03 MB, with a compression ratio of 118:1 and 112:1, with the mean

PSNR of 42.11 dB and 41.6 dB, respectively. The compression ratio using the JPEG2000 and MPEG are 94:1 and 92:1, with a mean PSNR of 34.8 dB and 38.1 dB for each method respectively. Obviously, we can see that the ICA-based method yields a better rendered image than the JPEG2000 and MPEG.

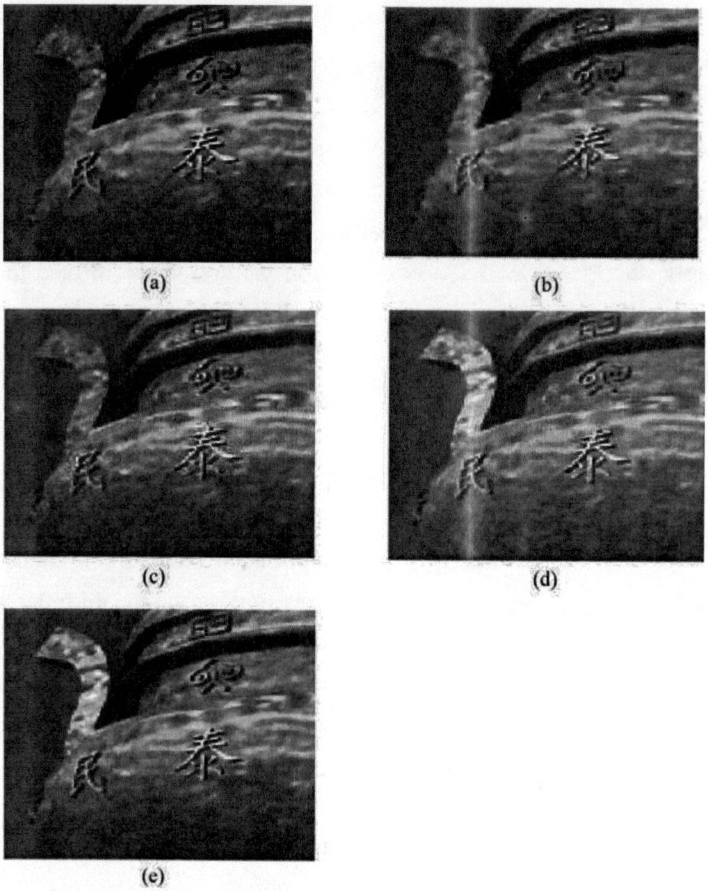

Fig. 10.26 Scene images under the same directional light source generated from (a) the original data; (b) JPEG2000 compressed data; (c) MPEG compressed data (d) the data compressed using ICA-based method; (e) the data compressed using PCA-based method

In summary, the ICA can be used to extract the independent components of multidimensional data, which can then be used to compress the huge original data set. Future work is still needed to explore an appropriate bit allocation method and an appropriate quality measure.

10.6 Independent Component Analysis for Functional MRI Data Analysis

Functional magnetic resonance imaging (fMRI)[53] is one of the most important biomedical technologies that has emerged in the last decades. Due to its noninvasiveness and high spatial temporal resolution, it has been widely used in normal brain function studies, brain disease studies, neuropsychological disorder studies, and presurgical plans, etc. Data analysis is an important component of fMRI study. To examine brain functions from the 4 dimensional fMRI data introduced within-subject variability and across-subject variability, various methods have been developed to process fMRI data, including the ICA-based approach. This section gives an introduction of the ICA in fMRI data analysis. We will briefly review the fMRI technique, the data manipulation problem of the ICA-based fMRI data analysis, and an example of the ICA based fMRI data analysis.

10.6.1 FMRI and Its Data Analysis

FMRI is based on blood flow changes in local vasculatures that accompany neural activities in the brain. These fluctuations of local blood flow induce a corresponding local alternation of deoxyhemoglobin (dHb) concentration, because the change in blood flow occurs without an equivalent change in oxygen concentration[54~57]. Since dHb is paramagnetic, its alternation changes the local T2* value and consequently alters the T2* weighted magnetic resonance imaging (MRI) signal[58~62]. Measuring local T2* MRI signal changes then provides a way to assess neural activity in the brain. Bearing this in mind, we see that certain brain function-targeted stimuli can be exposed to the subjects while simultaneously acquiring brain MR images. The functional activity spectrum is then obtained using various data analysis methods to locate the stimuli-invoked brain regions.

An example of an fMRI experiment is shown in Fig. 10.27, where the stimuli are a checkerboard with black and white colors reversing at 8 Hz and displayed intermittently in the right and left visual fields for 30 seconds at a time during scanning. The baseline is 30 seconds of black screen with a white crosshair in the center of the screen. The whole experiment consists of 3 cycles.

Standard fMRI data analysis includes: head motion correction (called realignment), functional images and structural image coregistration, spatial

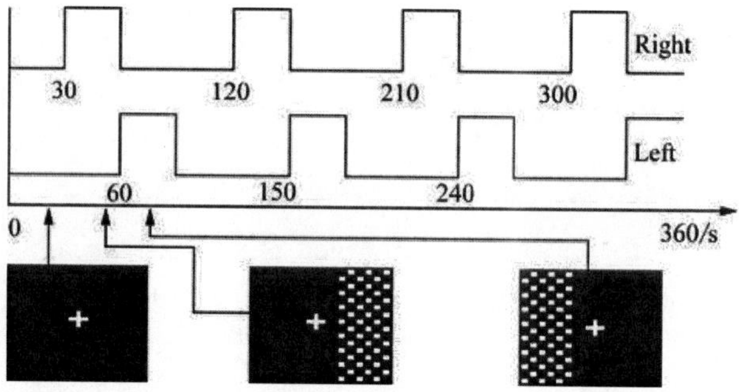

Fig. 10.27 An fMRI paradigm with separate right and left visual stimuli

smoothing, and statistical analysis. Spatial normalization is another necessary step for assessing brain activations which is usually performed using a random effect analysis.

Suppose an fMRI image, $y_i(i=1,\cdots,t)$, t is the number of acquisitions, has k_x, k_y, k_z voxels along the x,y,z direction, respectively. The 3D image data can be reformed into a one-dimensional vector $\boldsymbol{x}_i=[x_{1i}, x_{2i}, \cdots, x_{ki}]^{\mathrm{T}}$ by sequentially concatenating all columns within each slice and then sequentially concatenating all slices; then all the fMRI images can be stacked into a $k\times t$ matrix $\boldsymbol{X}=[\boldsymbol{x}_1,\boldsymbol{x}_2, \cdots,\boldsymbol{x}_t]$, where $k=k_xk_yk_z$. After preprocessing (motion correction, smoothing and coregistration), the general linear model (GLM) is used to perform the statistical analysis of the fMRI data[63]. To model the neural hemodynamic response of the brain, the design paradigm is usually convolved with a canonical hemodynamic response function (HRF) to be used as the reference function, $\boldsymbol{R}$, and the GLM procedure can be described as

$$\boldsymbol{X} = \boldsymbol{R}\boldsymbol{\beta} + \boldsymbol{\varepsilon} \tag{10.6.1}$$

where $\boldsymbol{\beta}$ represents the fitting parameter and $\boldsymbol{\varepsilon}$ is the residual noise, which is usually assumed to be white noise $\boldsymbol{\varepsilon}\sim N(\mu,\sigma^2)$.

With the least squared error method, $\boldsymbol{\beta}$ can be estimated as

$$\hat{\boldsymbol{\beta}} = (\boldsymbol{R}^{\mathrm{T}}\boldsymbol{R})^{-1}\boldsymbol{R}^{\mathrm{T}}\boldsymbol{X} \tag{10.6.2}$$

the same GLM analysis is conducted for every voxel acquired in the experiment. And the significance of the estimated brain activation magnitude (reflected by Eq. (10.6.2)) can be assessed using a null hypothesis that there is no task-invoked brain activation at all. Under this null hypothesis, $\boldsymbol{\beta}$ in Eq. (10.6.1) is zero, and a T statistic can be formulated as

$$\frac{\boldsymbol{c}^{\mathrm{T}}\hat{\boldsymbol{\beta}} - \boldsymbol{c}^{\mathrm{T}}\boldsymbol{\beta}}{\sqrt{\hat{\sigma}^2 \boldsymbol{c}^{\mathrm{T}}(\boldsymbol{R}^{\mathrm{T}}\boldsymbol{R})^{-1}\boldsymbol{c}}} \sim t_{k-p} \tag{10.6.3}$$

where $\boldsymbol{c}$ is the contrast vector with the same number of columns as $\boldsymbol{R}$. In the experiment shown in Fig. 10.27, $\boldsymbol{R}=(\boldsymbol{r}_1,\boldsymbol{r}_2,\boldsymbol{r}_3)$. A contrast vector of $\boldsymbol{c}=(1,0,0)$ means assessing the brain activation invoked by the first task (right field visual stimuli) as compared to the baseline. The rank of reference function matrix $\boldsymbol{R}$ is p.

With the same contrast vector, a statistical parametric map (SPM) can be obtained as a collection of all voxels' t values using Eq. (10.6.3). Brain activations invoked by the tasks can then be revealed by thresholding the SPM with a given significance threshold (T or P values). To provide a decent spatial location in the brain, the thresholded SPM usually is overlapped on the high resolution structural image.

Figure 10.28(a) shows a series of fMRI data from a representative participant using the experiment paradigm as shown in Fig. 10.27. Only one cycle is shown. Figure 10.28(b) and (c) are the time courses of the left and the right side task conditions from the voxel indicated in Fig. 10.28. The combined reference function (without the HRF convolution here) is shown in Fig. 10.28(d). Note that the reference function shown in Fig. 10.28(d) is with an arbitrary scale, and its value is only for indicating different conditions. After a GLM analysis, a contrast between the left (or right) side task and the resting state is defined and the corresponding SPMs thresholded at $P<0.00001$ are displayed in Fig. 10.29. As a sensory motor task is a well characterized fMRI task, it is not a surprise to see that the activations are exactly within the expected brain regions, i.e. the left

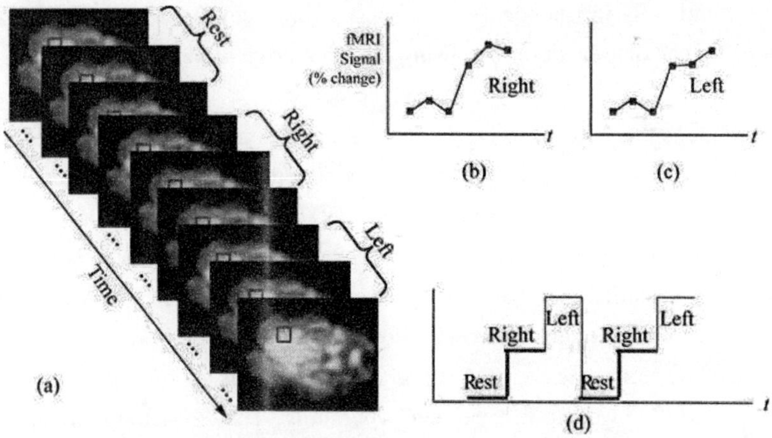

Fig. 10.28 fMRI signal of one experimental cycle of the paradigm specified in Fig. 10.27

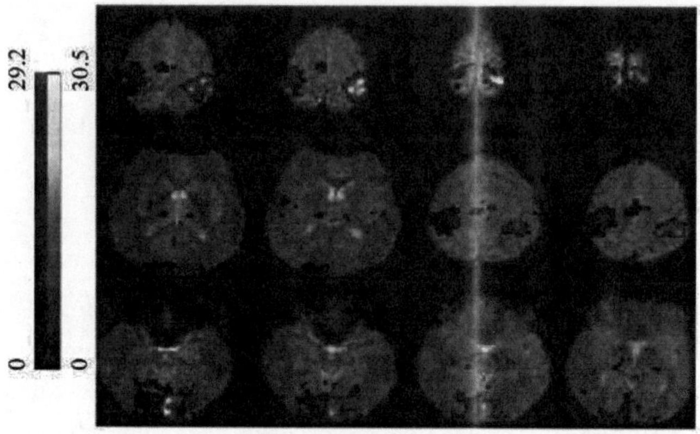

Fig. 10.29 GLM-based statistical analysis results of the sensorymotor fMRI data

(right) side sensory motor task induces brain activations in the right (left) side visual cortex and motor cortex.

10.6.2 ICA-Based fMRI Data Analysis

ICA-based fMRI data analysis can be roughly divided into three classes: spatial ICA (sICA)[64], temporal ICA (tICA)[65], and spatial temporal ICA (stICA)[66], depending on which dimension (or dimensions) of ICA is applied on the $k \times t$ matrix $\boldsymbol{X}=[\boldsymbol{x}_1, \boldsymbol{x}_2, \cdots, \boldsymbol{x}_t]$. We will focus on sICA here, although it is not hard to extend to the tICA and stICA cases.

Given the fMRI data $\boldsymbol{X}$, sICA is to seek a linear unmixing matrix $\boldsymbol{W}$ in order to get the mutually independent components (row vectors) through $\boldsymbol{S}=\boldsymbol{W}\boldsymbol{X}$. This relationship can also be described using the generative model:

$$\boldsymbol{X} = \boldsymbol{W}^{-1}\boldsymbol{S} = \boldsymbol{A}\boldsymbol{S} \tag{10.6.4}$$

This formula is then similar to Eq. (10.6.1), $\boldsymbol{S}$ corresponds to $\boldsymbol{\beta}$ and $\boldsymbol{A}$ corresponds to $\boldsymbol{R}$. What differs is that $\boldsymbol{R}$ in GLM is predefined based on prior information, while $\boldsymbol{A}$ in sICA is purely estimated from data without any prior information. Moreover, GLM is performed separately voxel by voxel, but sICA is designed to estimate the whole brain data-based components. As a result, sICA provides a way to assess the functional connectivities between different brain regions.

We see this in the regular ICA algorithm, sICA consists of a data dimension reduction, whitening, and independency pursuing. The extracted component is usually converted into a Z map ((x-mean)/std), so that the spatial brain activation patterns can be located by thresholding the Z map with a certain threshold. The following is a brief description of the whole procedure.

The first step is to estimate the covariance matrix of $\boldsymbol{X}$ through cov($\boldsymbol{X}$) followed by an eigenvector decomposition (EVD) or singular value decomposition (SVD):

$$\mathrm{cov}(\boldsymbol{X}) = \boldsymbol{Q}\boldsymbol{D}\boldsymbol{Q}^{\mathrm{T}} \tag{10.6.5}$$

where $\boldsymbol{Q}=[\boldsymbol{q}_1,\boldsymbol{q}_2,\cdots,\boldsymbol{q}_k]$ is the matrix consisting of eigenvectors of cov($\boldsymbol{X}$), and $\boldsymbol{D}=\mathrm{diag}(\boldsymbol{d}_1,\cdots, \boldsymbol{d}_k)$ is a diagonal matrix with non-zero elements only on the diagonal, i.e. the eigenvalues of cov($\boldsymbol{X}$). The third step is to project the original data into the subspace spanned by a few principal eigenvectors of cov($\boldsymbol{X}$):$\widetilde{\boldsymbol{Q}} = [\boldsymbol{q}_1, \boldsymbol{q}_2,\cdots, \boldsymbol{q}_k]$, and then whiten them as

$$\widetilde{\boldsymbol{X}} = \widetilde{\boldsymbol{Q}}\boldsymbol{D}^{1/2}\widetilde{\boldsymbol{Q}}^{\mathrm{T}}\boldsymbol{X} \tag{10.6.6}$$

where $\widetilde{\boldsymbol{D}}^{1/2}=\mathrm{diag}(d_1^{1/2},\cdots,d_m^{1/2})$.

The last step is to perform ICA on the whitened data $\widetilde{\boldsymbol{X}}$ to get

$$\boldsymbol{S} = \boldsymbol{U}\widetilde{\boldsymbol{X}} \tag{10.6.7}$$

and the demixing matrix $\boldsymbol{W}=\boldsymbol{U}\widetilde{\boldsymbol{Q}}\widetilde{\boldsymbol{D}}^{1/2}\widetilde{\boldsymbol{Q}}^{\mathrm{T}}$. The inverse matrix $\boldsymbol{W}^{-1}$ is the time course associated with the independent components.

10.6.3 Huge Data Compression Issue

The interested number of independent components is usually limited, but the acquired number of images in the fMRI is generally huge. Data compression then is a necessary step for running an ICA for fMRI data analysis. As stated in the previous chapters, data compression in the ICA is often done by PCA. Standard PCA needs to calculate and decompose the covariance matrix of the original data. However, It is very difficult to do this for a large fMRI dataset, because of the huge memory space required and the enormous computation burdens involved. This could be even worse for the tICA since the spatial dimension k of the fMRI data is usually much greater than temporal dimension t, and the corresponding $k \times k$ covariance matrix cannot be manipulated in any of the current available high-end computers. For example, the fMRI images to be used in this section have $64 \times 64 \times 30$ voxels in each image; for the tICA, the covariance matrix will be $122{,}880 \times 122{,}880$. Even removing the background voxels, the matrix will still be around $110{,}000 \times 110{,}000$!

For a modest fMRI dataset, which is manageable with a modern computer, data compression along the temporal dimension can be alternatively performed by applying PCA on the spatial dimension of the fMRI data. Theoretically, $\boldsymbol{X}^{\mathrm{T}}\boldsymbol{X}$ has the same eigenvalues as $\boldsymbol{X}\boldsymbol{X}^{\mathrm{T}}$. Let λ_j denote the j eigenvalue, $\boldsymbol{v}_j$ and $\boldsymbol{q}_j$ denote the associated eigenvectors of X^TX and XX^T, respectively, we can have

$$X^TXv_j = \lambda_j v_j \quad (10.6.8)$$

$$XX^Tq_j = \lambda_i q_j \quad (10.6.9)$$

left multiplying Eq. (10.6.9) with X^T, we get

$$X^TXX^Tq_j = \lambda_j X^T q_j \quad (10.6.10)$$

comparing Eq. (10.6.9) and Eq. (10.6.10), we can get $v_j = X^T q_j$, which means the PCA decomposition on X^TX can be performed through PCA decomposition on XX^T.

However, this solution becomes problematic when the temporal dimension is too big, so that it is impossible to manipulate XX^T in the computer. Even when the temporal dimension is modest, it is also impossible to calculate and decompose XX^T directly when the data size is too big to be loaded into the memory. A practical solution is through the neural network-based PCA algorithm. Here, we use a cascade recursive least square-based PCA (CRLS-PCA)[67~69].

CRLS-PCA is a state-of-the-art sequential neural PCA. As shown in Fig 10.30, it uses a cascade recurrent neural network to extract the PCs sequentially with the same learning procedures. The eigenvectors are extracted as the link weights $q_j(t')$, and the eigenvalues are the norm of the extracted PCs, $d_j = y_j$. The first PC is extracted from the original data, while others are estimated from the residual errors, which are updated by subtracting the obtained reconstructed data using only the new estimated eigenvector after each PC extraction. The learning procedures for the jth $j(1 \leqslant j \leqslant k$, k is the number of input vectors) eigenvector $q_j(t')$ $(t'=1, 2, \cdots, t)$ are

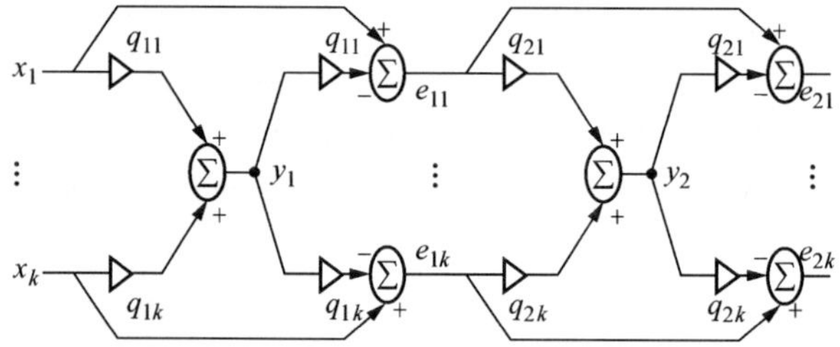

Fig. 10.30 Illustration of CRLS-PCA algorithm

$$q_j(t') = q_j(t'-1) + \frac{y_j(t')}{\eta_j(t')}(e_{j-1}(t') - q_j(t'-1)y_j(t')) \quad (10.6.11)$$

where $y_j(t') = q_j^T(t'-1)e_{j-1}(t')$ is the previous output of the jth neuron before updating. Then $e_{j-1}(t')=[e_{j-1,1}(t'),e_{j-1,2}(t'),\cdots,e_{j-1,k}(t')]$ is the reconstructed error. For the first neuron, $e_0=X$.

The learning rate η_j in the above equation is adapted according to the current signal energy through the RLS learning adaptation:

$$\eta_j(0) = \frac{\sum_{i=1}^{k}\sum_{l=1}^{t} e_{j(il)}^2}{t}, \eta_j(t') = \eta_j(t'-1) + y_j^2(t') \qquad (10.6.12)$$

And the input for the next neuron (i.e., the reconstruction error of the current neuron) is calculated by

$$e_j(t') = e_{j-1}(t) - y_j(t)q_j(t) \qquad (10.6.13)$$

with an updated output

$$y_j(t') = q_j^{\mathrm{T}}(t')e_j(t') \qquad (10.6.14)$$

CRLS-PCA is fast to converge, due to iterative least square-based learning. Further improvement of its speed and accuracy can be achieved with an additional orthogonality deflation step[70]:

$$q_j(0) = p - [q_1, q_2, \cdots, q_{j-1}][q_1, q_2, \cdots, q_{j-1}]^{\mathrm{T}} p \qquad (10.6.15)$$

here p is a random vector and $q_1, q_2, \cdots, q_{j-1}$ are the first $j-1$ principal components that have been estimated.

10.6.4 Results of sICA and tICA

The FMRI data were acquired using the experimental paradigm shown in Fig. 10.27. Both the sICA and the tICA were applied after motion correction and spatial smoothing. For the sICA, we used both the standard PCA and CRLS-PCA for data reduction and then used Infomax to perform the ICA decomposition. For the tICA, we used the indirect PCA method stated above in this section and the CRLS-PCA for data reduction; the ICA decomposition was also conducted using Infomax.

Using different data reduction methods as stated above showed the task-related components extracted with the sICA and tICA to be quite similar. Figure 10.31 shows the two task-related independent components extracted by the sICA, which are very similar to those detected by the GLM as shown in Fig. 10.29. A unique Property of the ICA-based fMRI data analysis is that we can assess not only the spatial activation patterns but also the temporal behavior of these patterns. As shown in Fig. 10.31, the temporal profile of the bilateral activations are not completely independent, which is consistent with the real physical and psychological response mechanism that the unilateral task could evoke from both the contralateral and ipsilateral brain responses.

Similar spatial activation patterns are also found in the tICA decomposition results as displayed in Fig. 10.32. However, the time courses estimated by the tICA are different from those by the sICA. These differences are due to the fact that the tICA is applied along the temporal dimension of the fMRI data, and the extracted time courses are required to be mutually independent, which is not true for the sICA. Moreover, the acquisition time difference between the different MR image slices can also introduce a different appearance to the extracted time courses. A more accurate tICA decomposition should then consider this acquisition time difference through precompensation or by using a specific ICA algorithm that explicitly models the time delays in the observed signal.

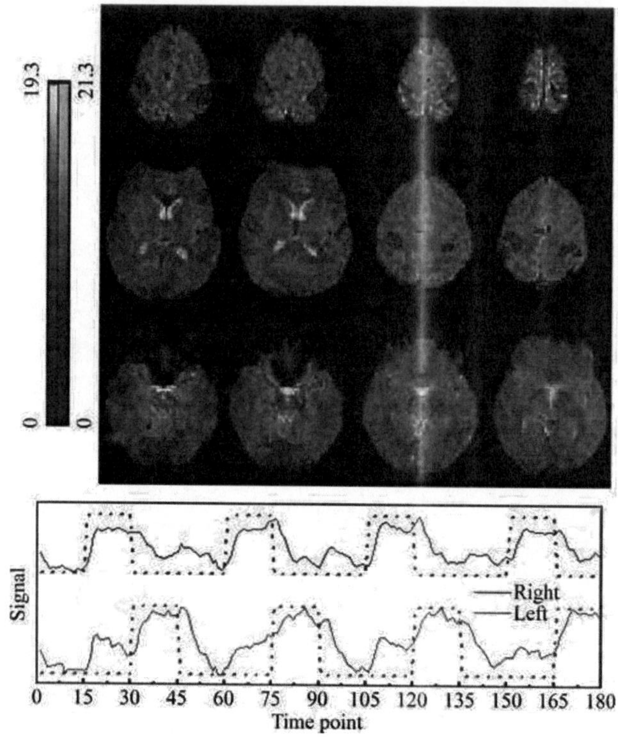

Fig. 10.31 The two task-related spatial components and their time courses revealed by sICA. The spatial components were converted into Z maps and were thresholded with a significance level of $P < 0.0001$

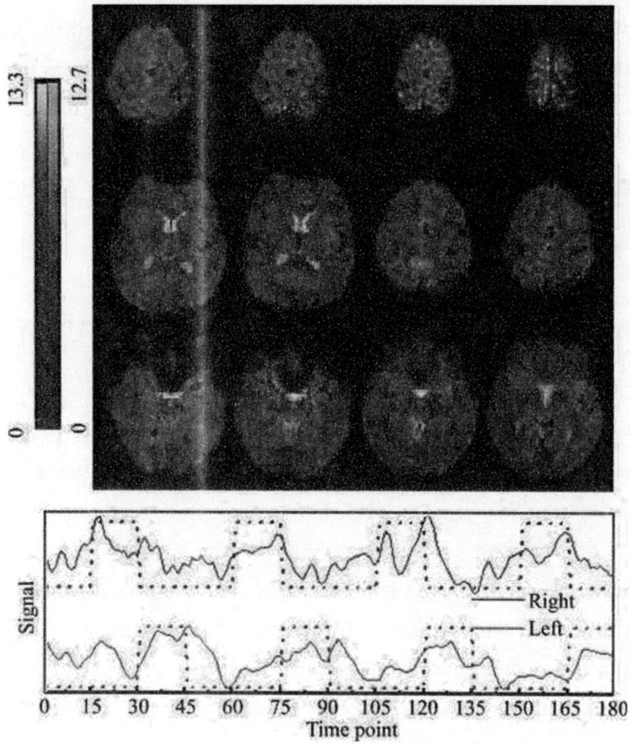

Fig. 10.32 The two task-related time courses and the associated spatial maps extracted by tICA. The significance level for thresholding the spatial maps (converted into Z maps) is $P < 0.00001$. The dot lines in the bottom panel indicate the experimental paradigm

10.6.5 Summary

The ICA provides a way to assess both the spatially connected activation patterns and their temporal behaviors, which are equally important for brain function studies. While the standard GLM is only applicable to the fMRI data with an explicitly known reference function, the ICA is purely data driven, and is applicable to any fMRI data, even those without a specific experimental paradigm, such as the resting fMRI scan. We only mentioned the task-related components in the experiment described in this section to give a feeling of how well the ICA performs in comparison to the standard GLM. However, more information about brain response can also be obtained by exploring those irrelevant to the task. Since this is out of the scope of this section, interested readers can find detailed information in Refs.[64~66]. From a technical point of view, there is still much room for development in methodology, such as how to perform group analysis in the ICA-based fMRI data analysis, how to determine an appropriate number of independent components in the data, and how to retain the principal components or minor components during the data reduction.

10.7 Speech Separation for Automatic Speech Recognition System

10.7.1 Problems of Using Conventional ICA for Speech Signal Separation

Speech recognition is an important component of artificial intelligence and has a wide application in robot and human-machine interactions. While human ears can be well adapted to the real environment to recognize any speech, machine-based speech recognition suffers from the various complicated noise in real practice. Recent research in the practice of speech separation has made great progress due to the advances made in blind source separation.

Empirically, a speech signal is characterized as colored and nonstationary and can be described with an autoregress (AR) model[71]:

$$s(t) = \sum_{k=1}^{P} a(k)s(t-k) + v(t) \tag{10.7.1}$$

where $v(t)$ is a Gaussian signal with a zero mean and variance of σ^2; $a(k)$ is the kth coefficient; and P is the order of the AR model. This model simply says the speech signal at any time is equal to a linear summation of the signal at the previous p time points.

With M microphones and N speech sources, the speech signal can be described as

$$x_i(t) = \sum_{k=0}^{K} h_{ij}(k)s_j(t-k) \qquad i = 1,\cdots,M, j = 1,\cdots,N \tag{10.7.2}$$

where $x_i(t)$ is the received signal in the ith microphone, $s_j(t)$ is the jth source signal, and $h_{ij}(k)$ is the impulse response function between the jth source and the ith microphone.

Since a speech signal is not i.i.d.[72], which is required in a multichannel signal deconvolution model, a standard multichannel signal deconvolution method cannot be directly applied. Based on the features of a speech signal, however, speech signal separation can be solved by a second-order or a higher-order blind deconvolution method. Although several second-order methods have been proposed based on the nonstationary features of a speech signal either in the frequency-domain[73] or in the time-domain[74], or have been based on the nonstationarity and color of a speech signal as in Ref.[75], the most popular solution is the higher-order method through independent component analysis. Consequently, we will introduce only the ICA-based multichannel speech signal separation in this section.

Using the Infomax algorithm, the learning process in the time-domain is described by the following equations:

$$\boldsymbol{W}_l(k+1) = \boldsymbol{W}_l(k) + \mu[\boldsymbol{W}_l(k) - f(\boldsymbol{y}(k-L))\boldsymbol{u}^{\mathrm{T}}(k-l)] \tag{10.7.3}$$

$$\boldsymbol{y}(k) = \sum_{l=0}^{L} \boldsymbol{W}_l(k)\boldsymbol{x}(k-l) \tag{10.7.4}$$

$$\boldsymbol{u}(k) = \sum_{q=0}^{L} \boldsymbol{W}_{L-q}(k)\boldsymbol{y}(k-q) \tag{10.7.5}$$

In the frequency-domain, the corresponding learning process is

$$\widetilde{\boldsymbol{W}}(k+1) = \widetilde{\boldsymbol{W}}(k) + \mu[\boldsymbol{I} - f(\boldsymbol{u})\boldsymbol{u}^{\mathrm{H}}]\widetilde{\boldsymbol{W}}(k) \tag{10.7.6}$$

where μ is the learning rate, $f(\cdot)$ is the nonlinear activation function, and $\boldsymbol{W}_l(k)$ and $\widetilde{\boldsymbol{W}}(k)$ are the weights in the time-domain and frequency-domain, respectively.

Assuming the temporal learning process converges after N steps, that is

$$\boldsymbol{W}_l(k) - f(\boldsymbol{y}(k-L))\boldsymbol{u}^{\mathrm{T}}(k-l) = \boldsymbol{0} \tag{10.7.7}$$

then we can get

$$\boldsymbol{I}\delta_l - E(f(\boldsymbol{y}(k-L))\boldsymbol{y}^{\mathrm{T}}(k-L-l)) = \boldsymbol{0} \tag{10.7.8}$$

and

$$E\{f_i(y_i(k-L))y_j(k-l)\} = \delta_{ij}\delta_l \tag{10.7.9}$$

From Eq. (10.7.9) we can see that after the learning process converges, the

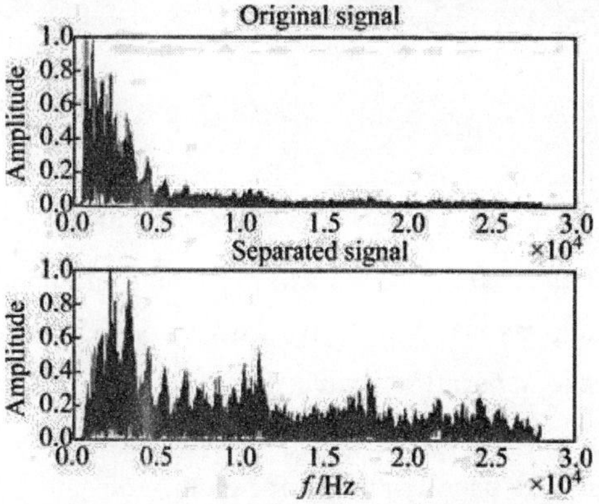

Fig. 10.33 Spectrum of a real speech signal and one separated from a mixed speech source using regular ICA

output signal is uncorrelated in both the temporal and spatial domains; in other words, the speech signal is whitened. From Fig. 10.33, we can see that after signal separation, the signal spectrum was obviously smoothed, and the energy was spread markedly to the whole spectrum.

While this phenomenon is a natural result of the ICA decomposition, what we really want is some separated audio signal natural to human ears, rather than the whitened signal. Therefore, modifications should be added to both the parameter initialization and the ICA learning process.

10.7.2 Null Beamforming-Based Initialization

Saruwatari *et al.* proposed an initialization method based on the null beamforming theory[76]. In array signal processing, the directivity pattern of the array system is defined by

$$F_l(f,\theta) = \sum_{k=1}^{2} W_{lk}(f) \cdot \exp[j2\pi f d_k \sin(\theta_k)/c] \tag{10.7.10}$$

where d_k is the coordinator of the kth microphone and θ is the arrival direction of one of the multiple sound sources.

As illustrated in Fig. 10.34, given two sources and two microphones, if the arrival directions of the two sources are known, the contributions of different sources can then be suppressed using the directivity pattern of the microphones. Masking the signal from direction θ_2, we can get

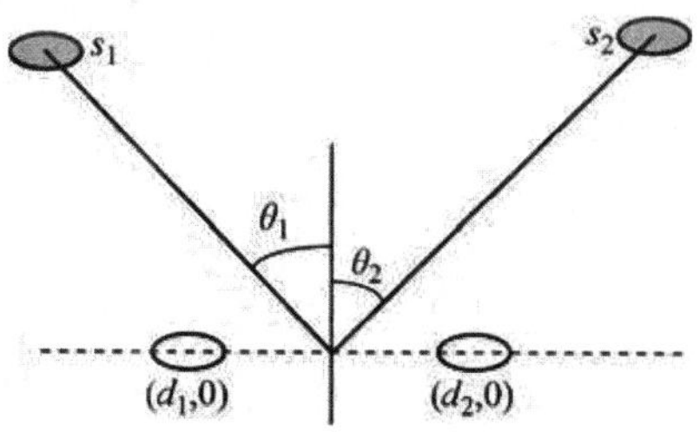

Fig. 10.34 Null beamforming paradigm for a two-source two-microphone speech separation case

$$F_1(f,\theta_1) = \sum_{k=1}^{2} W_{1k}(f) \cdot \exp[j2\pi f d_k \sin(\theta_1)/c] = 1 \tag{10.7.11}$$

$$F_1(f,\theta_2) = \sum_{k=1}^{2} W_{1k}(f) \cdot \exp[j2\pi f d_k \sin(\theta_2)/c] = 0 \tag{10.7.12}$$

from Eqs. (10.7.11) and (10.7.12), we can determine the parameters in the first row of the separating matrix. Similarly, we can determine the parameters in the second row of the separating matrix by masking the signal arriving from the direction of θ_1. As the directions of sources are generally unknown in BSS, arbitrary initial values are usually chosen for them.

The above method put constraints on both the target direction and the null directions through Eqs. (10.7.11) and (10.7.12), and can be named as a full constrained null beamforming initialization. A semi-constrained method can be obtained by nulling only sources from certain directions and leaving the target signal alone. In other words, only Eq. (10.7.12) is required.

10.7.3 Minimal Distortion Principle for Speech Signal ICA Separation

As we can see from previous chapters, the PCA is generally used to reduce data dimensionality and to remove the second-order correlations in the ICA for the instantaneous mixtures' problem. However, this process will also cause unnecessary whitening in the case of the convolutive mixtures, as described in previous paragraphs in this section. To solve this problem, Matsuoka[77] proposed a Minimal Distortion Principle (MDP) rule to control the ICA learning process for separating speech signals. As shown in Fig. 10.35, for each source, the MDP is designed to estimate the signal received by the microphones after it goes through the information channels. In other words, it is to estimate the contributions of each source to the received signal $y_1(t)=h_{11}(t)*s_1(t)$ rather than the source itself $s_1(t)$. This modification can be done by adding a new constraint in the learning process namely diag $(E[(\boldsymbol{y}(t)-\boldsymbol{x}(t))\boldsymbol{y}^{\mathrm{T}}(t)])=0$.

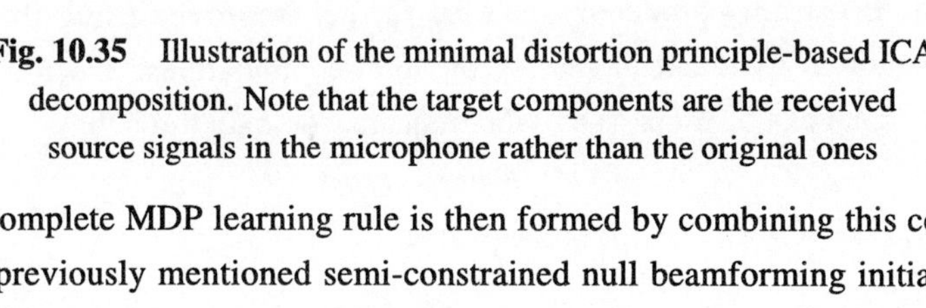

Fig. 10.35 Illustration of the minimal distortion principle-based ICA decomposition. Note that the target components are the received source signals in the microphone rather than the original ones

The complete MDP learning rule is then formed by combining this constraint and the previously mentioned semi-constrained null beamforming initialization. With this learning rule, the separated speech signal is least subjected to the separator, i.e. not whitened during the ICA decomposition. The learning algorithm, after incorporating this new constraint, is

$$\Delta W_l(k) = -\mu \sum_{r=0}^{P} \{\text{off-diag} f(\boldsymbol{y}(k-L))\boldsymbol{y}^{\mathrm{T}}(k-L-l+r) + \beta \,\text{diag}(\boldsymbol{y}(k-L)-\boldsymbol{x}(k-L))\boldsymbol{y}^{\mathrm{T}}(k-L-l+r)\} \quad (10.7.13)$$

with the null beamforming method, the output is also least subjected to the interference of other sources. Initializing with an alternating arrival source angle and repeating the above separation process, we can get the contribution to each microphone from the source. As shown in Fig. 10.36, the contributions of source $s_1(t)$ to the two microphones are extracted after two steps with alternated initial source angles. This separation is similar to the case of 1 source 2 microphones, and is a typical single input/multiple output (SIMO) problem. An advantage of using an SIMO model is that many established algorithms can be adopted in the SIMO (or multiple input/multiple output, MIMO) to reduce the signal reflection due to the non-minimal phase system impulse response.

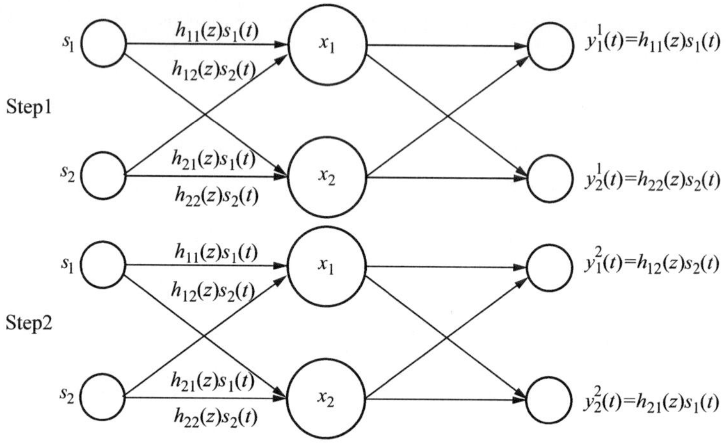

Fig. 10.36 Extracting the contributions of the same source to each microphone using the MDP learning rule after two steps

The MDP learning procedures can be further improved using the binary masking method[78] without increasing the total computations. Specifically, the separated signal can be processed in the frequency domain through

$$\hat{y}_1(f,t) = m_1(f,t)y_1(f,t) \tag{10.7.14}$$

where $m_1(f,t)=1$ if $y_1(f,t)>y_2(f,t)$ and $m_1(f,t)=0$ if $y_1(f,t)<y_2(f,t)$.

Similarly

$$\hat{y}_2(f,t) = m_2(f,t)y_2(f,t) \tag{10.7.15}$$

where $m_2(f,t)=1$ if $y_2(f,t)>y_1(f,t)$ and $m_2(f,t)=0$ if $y_2(f,t)<y_1(f,t)$.This simple masking works well in real practice. But it is worth noting that it is based on the assumption that there is no overlap between the spectrums of the different sources, which is rarely true for speech signals. More complicated mask functions can be used to suppress the interference from other sources, as in this

case where there is spectrum overlapping; however, these functions can also introduce some distortions to the extracted source signal contribution.

Since the MDP is designed to suppress the interference of other sources and to prevent the output signal from whitening, it doesn't help the separation performance of the ICA much. The separation performance can be improved through considering the temporal characteristics of a speech signal. For example, a temporal autocorrelation exists in a regular speech signal within a time window of 2 ms. With a sampling rate of 8 kHz, this is equivalent to 16 sampling points. Affected by the impulse response function, a temporal autocorrelation is only present after 50 timepoints pass through a system. Therefore, a separation performance improvement can be achieved by removing the temporal correlation before the ICA separation using a method like the linear predictive coding (LPC). More details can be found in Ref.[79].

Figure 10.37 shows a paradigm which summarizes the main steps involved in the ICA-based speech separation using the MDP learning rule.

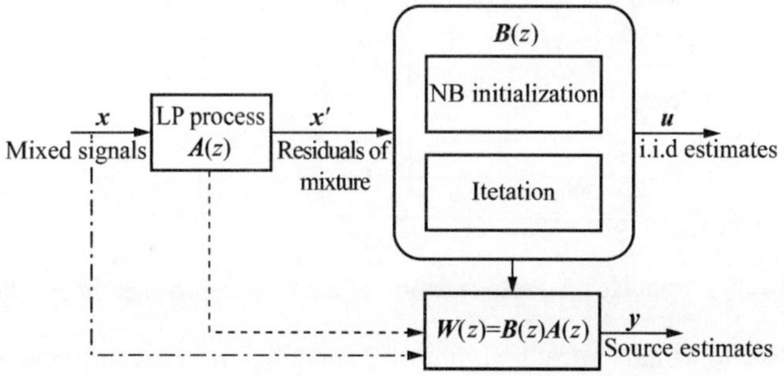

Fig. 10.37 A complete speech signal separation system

10.7.4 Evaluations

In the following, simulations are conducted to evaluate the speech signal separation system described above (also shown in Fig. 10.37).The impulse response function is obtained using the method described in Ref.[83]. Fig. 10.38 shows the experiment settings. The room for mixing the speech sources is 3m × 4m × 2.7m. Two microphones are used with 4 cm distance in between; both are 1.15 m away from the sources; and the receiving angles of the microphones are $-45°$ and $45°$, respectively. The impulse response functions are shown in Fig. 10.39. The two figures in the top row, from the left to the right, are those of source 1 - micro 1 and source 1 - micro 2; the two figures in the bottom row from the left to the right are those of source 2 - micro 1 and source 2 - micro 2.

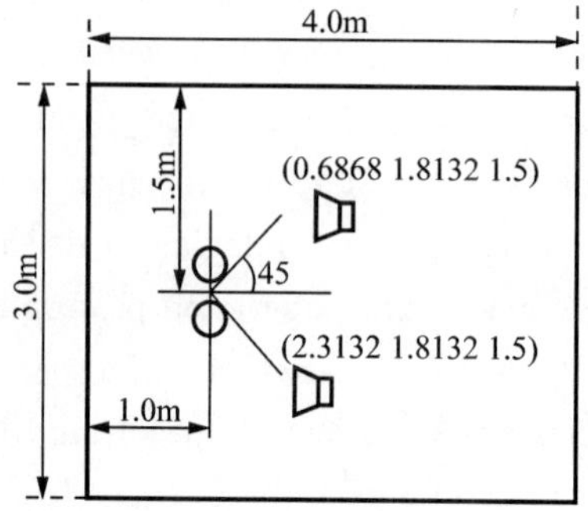

Fig. 10.38 Speech source mixtures room settings

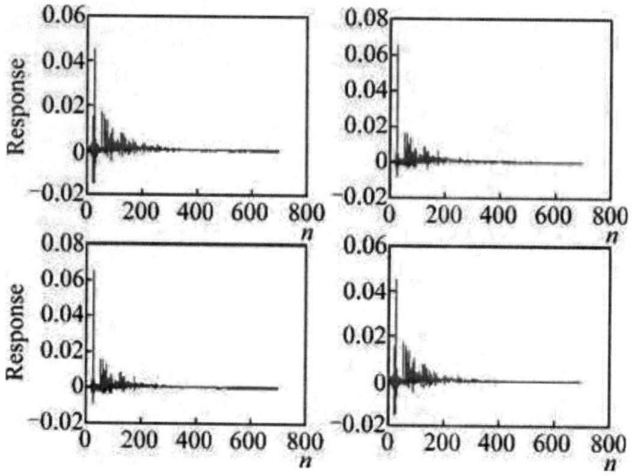

Fig. 10.39 Impulse response function of each source to each microphone

A female and a male speech signal of 7 seconds are used in the simulations. The sampling rate is 8 kHz. Two learning parameters in Eq. (10.7.13) are chosen to be μ=0.002. β=0.00002. Assuming the source directions are −60° and 60°, with a 64 point deconvolution filter (LPC filter), two sets of initial separating matrices are obtained using the fully constrained and incomplete constrained initialization methods as stated above. The matrices are shown in Fig. 10.40. Binary masking is also used in this experiment.

The same two initializations are also conducted without applying the LPC, and the system functions using the same fully constrained initialization and incomplete constrained initialization methods are shown in Fig. 10.41. Figure 10.42 shows the spectrums of the first source signal and its separated version (received by the microphone) from the mixed signal using the two kinds of initialization methods but without using the LPC prewhitening.

The performance of the separation is measured using the signal-to-noise ratio

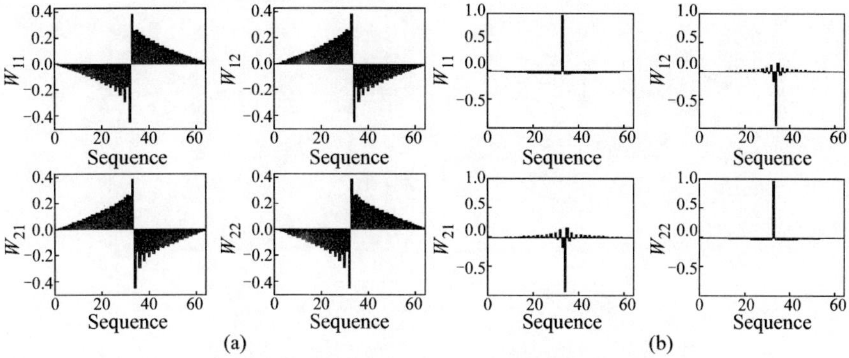

Fig. 10.40 Initial separating matrices using

(a) fully constrained initialization method;

(b) incomplete constrained initialization method

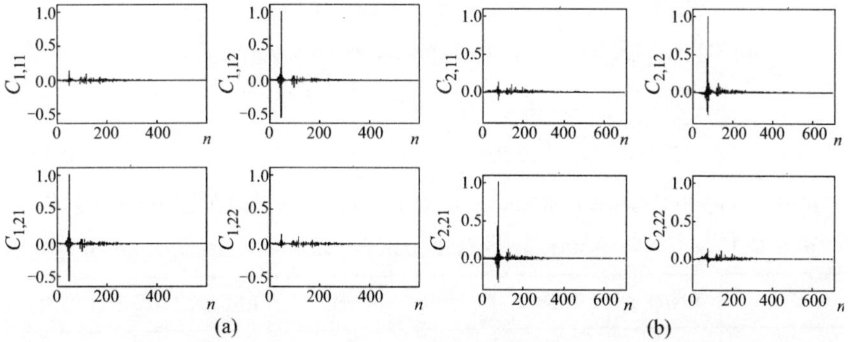

Fig. 10.41 System function of the separating system using

(a) fully constrained initialization method;

(b) incomplete constrained initialization method

improvement (SNRI), i.e. the difference of the output SNR and the input SNR:

$$\mathrm{SNRI}-1 = 10\lg 10 \frac{\sum\limits_{t} |H_{ll}(z)s_l(t)|^2}{\sum\limits_{t} |H_{ln}(z)s_n(t)|^2} \tag{10.7.16}$$

Since the MDP-based speech signal separation is to estimate the contribution of the source signal to the receiver (microphone), sound fidelity (SF) is also used to quantify the separation results. For a two-source two-microphone system, the SF of source 1 to microphone 1 is defined as

$$\mathrm{SF11} = 10\lg 10 \frac{\sum\limits_{t} |G_{11}(z)s_1(t)|^2}{\sum\limits_{t} |G_{11}(z)s_1(t) - H_{11}(z)s_1(t)|^2} \tag{10.7.17}$$

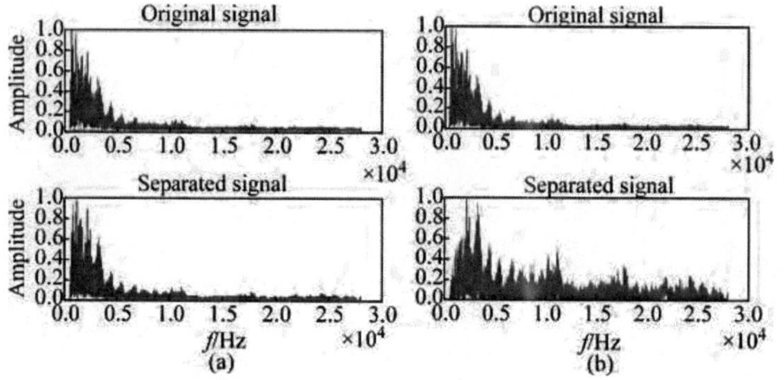

Fig. 10.42 Spectrums of the 1st source and its separated version using (a) fully constrained initialization method; (b) incomplete constrained initialization method

where $G(z)=W(z)H(z)$ is the system function. SF12, SF21, and SF22 can be defined similarly, so the SF for the two speech sources can be defined as

$$\text{SF1} = (\text{SF 11} + \text{SF 21})/2 \tag{10.7.18}$$

$$\text{SF2} = (\text{SF 12} + \text{SF 22})/2 \tag{10.7.19}$$

Table 10.1 lists the results of all these measures for both initialization methods.

Table 10.1 Performance measures of different initialization methods without LPC

	SNRI-1	SNRI-2	SF1	SF2	BM SNRI-1	BM SNRI-2
Incomplete constrained	13.27	14.17	1.49	1.44	18.94	20.17
Fully constrained	12.61	13.61	11.59	11.09	16.6	17.51

From the results shown above, we can see that different initializations without the LPC yield a different separation performance even with the same MDP learning rule. The incomplete constrained initialization method yields a better SNRI but does not preserve the frequency features of the speech signal. The spectrum shown in Fig. 10.42 which shows a clear whitened pattern by using the incomplete constrained initialization method. However, the separation performance is markedly improved with binary masking. By contrast, the fully constrained initialization method gives a worse separation but better speech quality and frequency feature are preserved, which will benefit some post-processing procedures such as dereverbration.

With the LPC, the SNRI-1, and SNRI-2 of the separation with an incomplete constrained initialization are 14.59 and 15.46 respectively. Compared to the results listed in Table 10.1, separation with the LPC prewhitening yields 1.32 dB and 1.29 dB SNRI gain.

10.8 Independent Component Analysis of Microarray Gene Expression Data in the Study of Alzheimer's Disease (AD)

With advances in the DNA microarray technology, it is now possible to quantify expression levels of thousands of genes. Implicit in the work of gene expression analysis today is the assumption that every gene in the human genome does not express itself independently of another gene. Therefore, to understand the coordinated effects of multiple genes, researchers extract the underlying features from any multivariable microarray data set to reduce dimensionality and redundancy. Moreover, due to the inherent noise in most array systems, single control and experiment designs are not always reliable, and it is difficult to understand the functionality of the complicated correlations between the large numbers of genes present in the genome.

Two kinds of unsupervised analysis methods for microarray data, principal component analysis (PCA) and independent component analysis (ICA), have been developed recently to accomplish this task. Since they use correlation-based learning rules, and since only covariance among the observed variables is used in the estimation, the PCA features are only sensitive to second-order statistics. The failure of correlation-based learning algorithms is that they reflect only the amplitude spectrum of the signal and ignore the phase spectrum. Extracting and characterizing the most informative features of the signals, however, require higher-order statistics.

The ICA has been used successfully in processing electroencephalograms (EEG), magnetoencephalograms (MEG), and functional magnetic resonance images[80] (fMRI). In the first phase of the ICA biomedical applications, Hori and his colleagues proposed an ICA-based gene classification method to effectively classify yeast gene expressions into biologically meaningful groups. Liebermeister proposed an ICA method for microarray data analysis to extract expression modes whereby each mode represents a linear influence of a hidden cellular variable. Following this discovery, The ICA has been widely used in microarray data analysis for feature extraction, gene clustering, and classification[81].

10.8.1 Gene Expression in ICA Model

The gene expression data provided by microarray technology is considered a linear combination of some independent components having specific biological interpretations.

Let the $n \times m$ matrix $\boldsymbol{X}$ denote the microarray gene expression data with m genes under n samples; x_{ij} in $\boldsymbol{X}$ is the expression level of the jth gene in the ith sample. Generally speaking, the number of genes m is much larger than that of the samples n, $m \gg n$. Suppose that the data have been preprocessed and normalized, i.e. each sample has a zero mean and standard deviation, then the ICA model for the gene expression data is $\boldsymbol{X}=\boldsymbol{AS}$. Then the transpose, $\boldsymbol{X}^{\mathrm{T}}$, is used in the ICA model: $\boldsymbol{X}^{\mathrm{T}}=\boldsymbol{AS}$, where $\boldsymbol{X}^{\mathrm{T}}$ denotes the same $n \times m$ matrix in the ICA model.

In the ICA model of microarray data, the columns of $\boldsymbol{A}=[a_1,a_2,\cdots,a_n]$ are the $n \times n$ latent variable vectors of the gene microarray data; $\boldsymbol{S}$ denotes the $n \times m$ gene signature matrix or expression mode, in which the rows of $\boldsymbol{S}$ are statistically independent to each other; and the gene profiles in $\boldsymbol{X}$ are considered to be linear mixtures of statistically independent components $\boldsymbol{S}$ combined by an unknown mixing matrix $\boldsymbol{A}$. Figure 10.43 presents the vector framework of the ICA model. Refs.[82] and [83] gave detailed analysis for matrix $\boldsymbol{S}$ and $\boldsymbol{A}$. The mth row matrix $\boldsymbol{A}$ contains the weights with which the expression levels of the n genes contribute to the mth observed expression profile. Hence the assignment for the observed expression profiles with different classes is valid for the rows of $\boldsymbol{A}$, and each column of $\boldsymbol{A}$ can be associated with one specific expression mode. Since the mth column of $\boldsymbol{A}$ contains the weights with which s_m contributes to all observations, this column should show large or small entries according to the class labels.

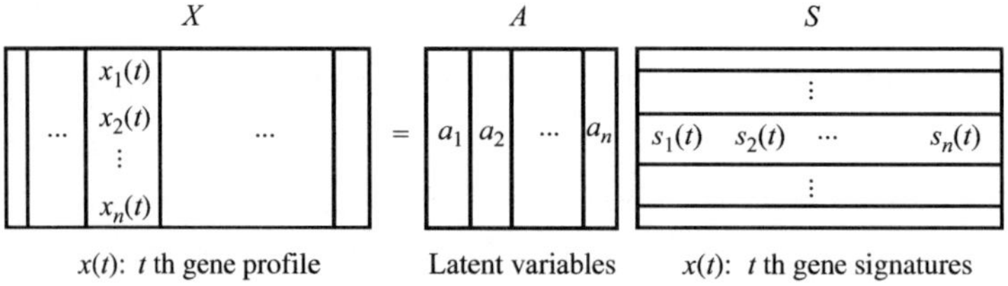

Fig. 10.43 ICA vector model of microarray gene expression data

To obtain $\boldsymbol{S}$ and $\boldsymbol{A}$, the demixing model can be expressed as $\boldsymbol{Y}=\boldsymbol{WX}$, where $\boldsymbol{W}$ is an $n \times n$ demixing matrix. Figure 10.44 shows the processing of the ICA algorithms on microarray gene expression data.

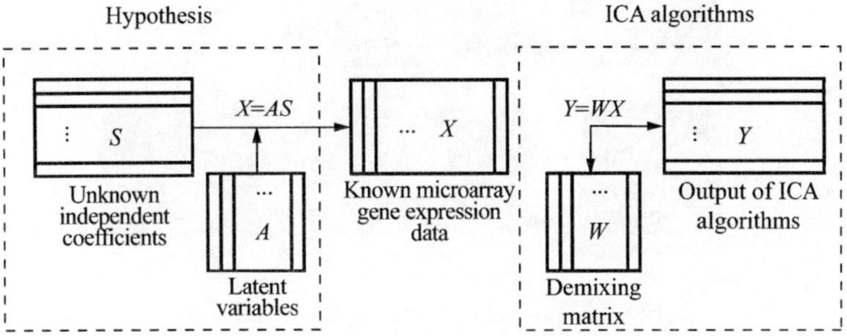

Fig. 10.44 Theoretical framework of ICA algorithms on microarray gene expression data

10.8.2 Application of ICA in Alzheimer's Disease Microarray Data

1. ICA decomposed AD microarray data into biological processes

We perform the FastICA method on 13 (8 control and 5 severe AD) samples with 3,617 genes. The main modeling hypothesis underlying the application of the ICA to the gene expression data is that the expression level of a gene is determined by a linear combination of biological processes, some of which try to express that, while other contending processes try to suppress it.

Figure 10.45 gives the ICA decomposition results of the 13×3617 AD microarray data matrix $\boldsymbol{X}$. Each observed sample is considered a linear combination of gene signatures captured by the ICA under the weights of the rows of matrix $\boldsymbol{A}$. It decomposes matrix $\boldsymbol{X}$ into the latent variable matrix $\boldsymbol{A}$ (13×13) and gene signature matrix $\boldsymbol{S}$ (13×3617). In the FastICA algorithm, the nonlinear function $g(u)=\tanh(au)$ was used as the probability density distribution of the outputs u during the iteration, where a is a constant. As the FastICA algorithm relies on random initializations for its maximization and faces the problem of convergence to local optima, Chiappetta *et al.*[84] presented consensus components by rerunning the FastICA algorithm with the random initializations, while Himberg *et al.*[85] proposed resampling the ICA components and used estimated centrotypes as the representatives of the ICA components. In our study, we iterated the FastICA 50 times to alleviate the instability of the slightly different results in each iteration, and the final components were estimated as the centrotypes of the iterated estimates of each component.

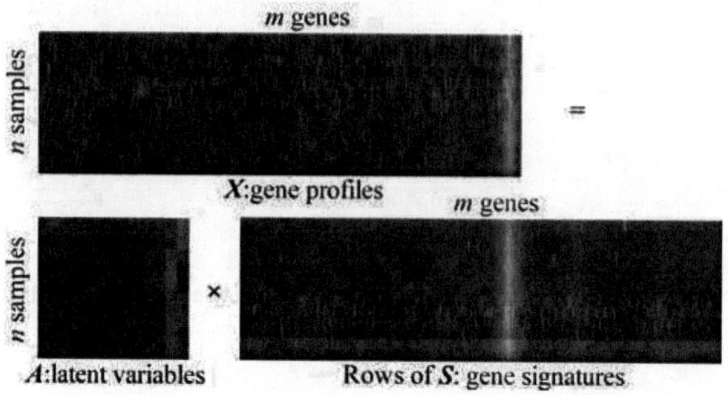

Fig. 10.45 ICA decomposition results of AD microarray gene expression data

2. ICA improved clustering results of AD samples

(1) Sample clustering by matrix ***A***

The ICA essentially seeks a new representation of the observed expression profile matrix ***X*** with the columns of matrix ***A*** representing the new basis vectors (latent variables). Each row of ***A*** contains the weight with which the gene signatures contribute to the observed expression profiles. Unsupervised hierarchical clustering is applied to the rows of ***A*** to validate the efficiency of the ICA outputs (Fig. 10.46(a)). The clustering method is performed on the rows of matrix ***A***, and the last two latent variables which are considered as of noise or with no biological relevance to AD, since their values are similar across all samples are removed. See the graphical representation of matrix ***A*** in Fig. 10.45. As can be seen from the clustering dendrogram, the first eleven ICA latent variables capture sufficient biologically significant information from sample data. The control and severe AD samples can be clearly discriminated. Similar hierarchical clustering is done on the principal components (PCs) outputted by the PCA on the same original data (where the gene expression measurements are the variables and the samples are the observations) in Fig. 10.46(b). In the PCA method, the observed samples are represented as a linear combination of the PCs with associated gene scores. Here the first ten PCs that captured 95.5% variance, are selected to represent the original data, and the remaining PCs with lower variance containing noise are removed for the clustering. Figure 10.46(b) shows that it can cluster some of the control samples and severe AD samples; however, the severe AD sample 'AD-2' cannot be discriminated from the control samples.

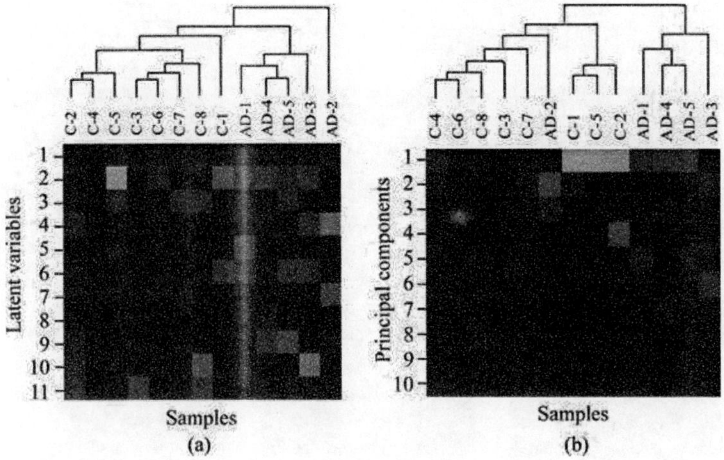

Fig. 10.46

(a) Hierarchical clustering of the ICA outputs with the last two 'common' components of matrix A removed;

(b) Hierarchical clustering of the principal components, the number of the principal components $k=10$

(2) Sample clustering by reconstructed data sequence

ICA is an adaptive process in which the independent components are as sparse as possible. Lee *et al.*[82] demonstrated that the underlying biological processes are more super-Gaussian than the original inputs microarray gene data. For this Property, one may assume that the activities of components with small absolute values are noise or redundancy information; set them to zero, retaining just a few components with large activities, $\boldsymbol{S} \rightarrow \boldsymbol{S}_{new}$. Then the newly reconstructed data is obtained by $\boldsymbol{X}_{new}=\boldsymbol{A}\boldsymbol{S}_{new}$. Figure 10.47 shows that the unsupervised hierarchical clustering is applied to the normalized raw data and the data reconstructed by the PCA and the FastICA respectively.

For the original data (Fig. 10.47(a)), some of the control samples and severe AD samples have been clustered together, but the highest hierarchical split does not separate the two classes as expected. For the data reconstructed by both PCA and ICA, the clustering results are greatly improved (Fig. 10.47(b) and 10.47(c)). In the PCA method, the first ten components associated with a larger variance are selected to reconstruct data that captures most of the information (the cumulated contribution of their eigenvalues exceeded 95.5%) of the original data, whereas the remaining components with a lower variance containing noise are removed. For the ICA method, it extracted m (13) gene signatures (rows

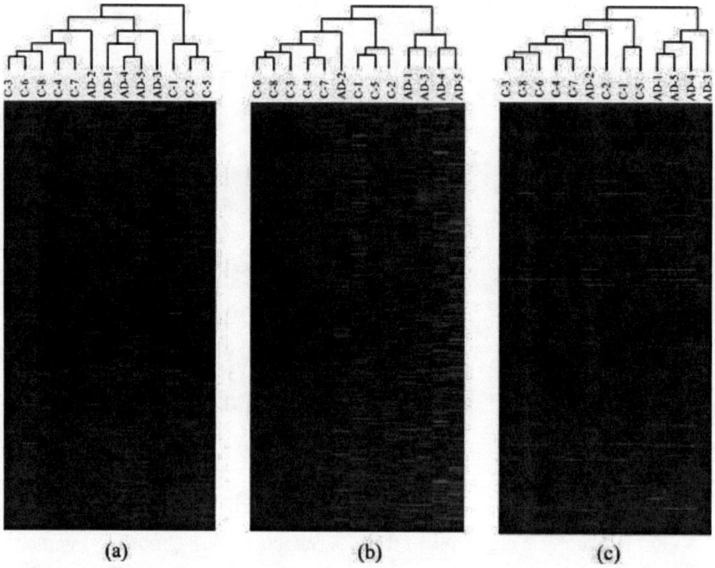

Fig. 10.47 Unsupervised hierarchical clustering of the normalized raw data
(a) The data reconstructed by PCA; (b) The data reconstructed by FastICA;
(c) C1-8: control samples, AD1-5: severe AD samples

of matrix $\boldsymbol{S}$) that are mutual statistically independent as underlying biological processes. Each independent component is as sparse as possible; only a few relevant genes are significantly affected, leaving the majority of genes relatively unaffected. The filtering capacity of the ICA is achieved by setting the entries in each gene signature whose values are less than the threshold to zero. Then the reconstructed data gives a clearer clustering result for discrimination of severe AD samples compared to the original data.

3. ICA identified significant genes for AD

In the ICA outputs matrix, each row of matrix $\boldsymbol{A}$ contains the weights with which the expression levels of the m genes contribute to the corresponding observed expression profile (row of matrix $\boldsymbol{X}$). Therefore, the profile order of rows of $\boldsymbol{A}$ is the same as that of the observed expression profiles, and each column of $\boldsymbol{A}$ is associated with the corresponding gene signature (IC). Figure 10.48 shows the Hinton diagram representation of matrix $\boldsymbol{A}$ derived by the FastICA.

The original data set consists of five severe AD microarray gene expressions (first five rows) and eight control samples (last eight rows), so this assignment is also valid for the rows of $\boldsymbol{A}$. From Fig. 10.48, it can be seen that the 4th and 5th

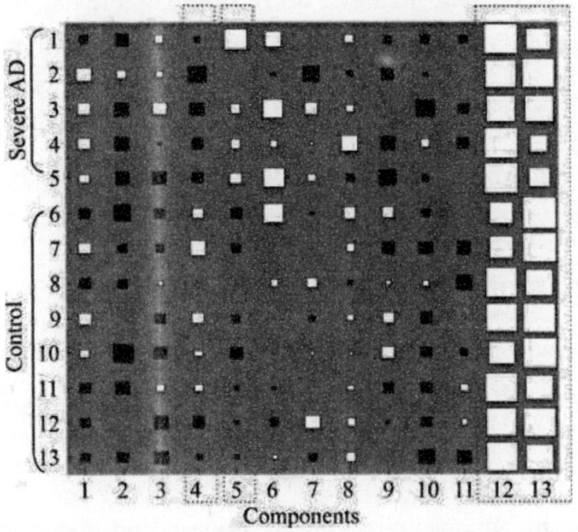

Fig. 10.48 Hinton diagram representation of latent variable matrix ***A***
The size of each square corresponds to the amount a_{nm} of component m in sample n. White and black represent positive and negative values, respectively

columns of ***A*** can discriminate between severe AD and control samples, since the sign of the values is distinctly different. For example, the values of the 4th latent variable (4th columns of ***A***) are all negative for severe AD samples (first five rows) and positive for most of the control samples (last eight rows); and the values of the 5th latent variable (5th columns of ***A***) are all positive for severe AD samples (first five rows) and positive for all the control samples (last eight rows). So the identified latent variables can be related to their corresponding gene signatures (the 4th and 5th row of matrix ***S***) to find those individual genes that strongly contribute to that component. In addition, the last two

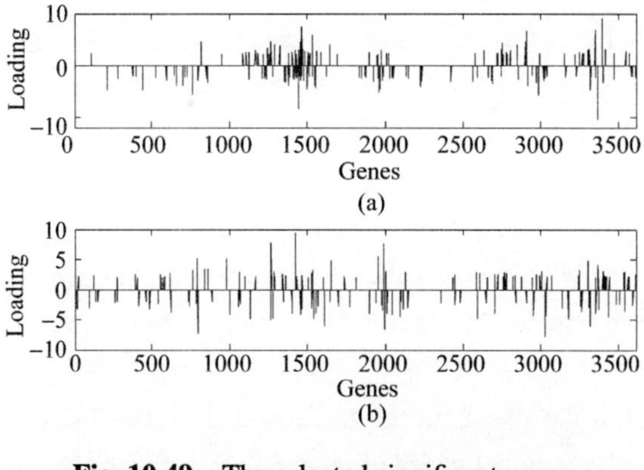

Fig. 10.49 The selected significant genes

columns of matrix A seem to have no biological relevance to AD because their values are similar across all the samples. These 'common' components can be ignored in later investigations. Figure 10.49 shows the corresponding gene signatures in matrix S (4th and 5th row of S). By setting a suitable threshold (here the threshold = 2), significant genes can be selected from the gene signatures.

10.8.3 Discussions

The ICA identified many genes and related pathways that play a prominent role in the AD and relate the activation patterns of these to the AD phenotypes. More than fifty upregulated significant genes in the severe AD were extracted in immunoreactions, metal protein, ribosomal protein, membrane protein, cytoskeleton protein, lipoprotein, binding protein and membrane protein, etc. The ICA also found 37 downregulated significant genes in the above categories, and some oncogenes and phosphoricproteins were expressed at a low level.

In immunoreactions, genes such as AMIGO2, BTG1, CD24, CD44, CDC42EP4, IFITM1, IFITM2, IRF7, IFI44L, IL4R, IRAK1 and NFKBIA were found upregulated in the severe AD; CD22 / MAG were downregulated. Importantly, with the molecular biological experiments, we found novel associations, such as inflammation and the expression of APP.

In metal protein, the ICA found many high expressed genes such as CAMK2B, CALM1, CAPZA2, CHGB, LOC728320/LTF, MPPE1, MT1F, MT1M, MBP, SCGN, SLC24A3, SLC7A11, ZIC1, ZBTB20, ZNF500, ZNF580, ZNF652 and ZNF710; some genes in severe AD such as CABP1, CACNG3, CAMK2B, CAMK1G, CAPZB, MET, ZNF365 and TFRC, were expressed at a low level.

The ICA found genes in the neurofilament family, such as APLP2, CYP26B1, NEFH, NPY, NTRK2, SERPINI1, OLIG2 and NRSN2, which were expressed at a low level in severe AD clinical symptoms.

The ICA also found some significant genes of ribosomal protein, cytoskeleton protein, lipoprotein, binding protein and membrane protein, etc. Specially, the high expression of APOE and its alleles in the severe AD, which were regarded as the best known genetic risk factor for late-onset sporadic AD, were selected as significant genes.

Our results indicate that the ICA outperforms the PCA and SVM-RFE methods in identifying novel genes in related pathways in AD. The ICA can find significant genes in many kinds of categories to understand microarray, in that this approach exploits higher-order statistics to identify gene expression profiles. That means the ICA can help elucidate the molecular taxonomy of AD and design molecular biological experiments to find gene biomarkers for AD.

References

[1] Barnard T J, Khan F (2004) Statistical normalization of spherically invariant non-Gaussian clutter. IEEE Journal of Oceanic Engineering 29(2): 303–309

[2] Fialkowski J M, Gauss R C, Drumheller D M (2004) Measurements and modelling of low-frequency near-surface scattering statistics. IEEE Journal of Oceanic Engineering 29(2): 197–214

[3] LePage K D (2004) Statistics of broad-band bottom reverberation predictions in shallow-water waveguides. IEEE Journal of Ocean Engineering 29(2): 330–346

[4] Preston J R, Abraham D A (2004) Non-Rayleigh reverberation characteristics near 400 Hz observed on the New Jersey shelf. IEEE Journal of Oceanic Engineering 29(2): 215–235

[5] Kirsteins I P, Tufts D W (1994) Adaptive detection using low rank approximation to a data matrix. IEEE Transactions on Aerospace and Electronic Systems 30(1): 55–67

[6] Kirsteins I P, Tufts D W (1989) Rapidly adaptive nulling of interference. In: IEEE International Conference on Systems Engineering Fairborn, 1989, pp 269–272

[7] Freburger B E, Tufts D W (1998) Case study of principal component inverse and cross spectral metric for low rank interference adaptation. In: Proceedings of ICASSP '98, 1998, 4: 1977–1980

[8] Freburger B E, Tufts D W (1998) Adaptive detection performance of principal components inerse, cross spectral metric and the partially adaptive multistage Wiener filter. In: Signals, Systems & Computers, 1998. Conference Record of the Thirty-Second Asilomar Conference. Pacific Grove, 1998, 2: 1522–1526

[9] Freburger B E, Tufts D W (1997) Rapidly adaptive signal detection using the principal component inverse (PCI) method. In: Conference Record of the Thirty-First Asilomar Conference on Signals, Systems & Computers, Pacific Grove, 1997, 1:765–769

[10] Palka T A, Tufts D W (1998) Reverberation characterization and suppression by means of principal components. In: OCEANS '98 Conference Proceedings, Nice, 1998, 3: 1501–1506

[11] Ginolhac G, Jourdain G (2002) Principal component inverse algorithm for detection in the presence of reverberation. IEEE Journal of Oceanic Engineering 27(2) : 310–321

[12] Ginolhac G, Jourdain G (2000) Detection in presence of reverberation. In: OCEANS 2000 MTS/IEEE Conference and Exhibition, Providence, 2000, 2: 1043–1046

[13] Carmillet V, Amblard P O, Jourdain G (1999) Detection of phase-or frequency-modulated signals in reverberation noise. JASA 105(6):357–389

[14] Bell A J, Sejnowski T J (1995) An information-maximization approach to blind separation and blind deconvolution. Neural Computation 7(6):1129–1159

[15] Cong F Y (2005) Blind signal separation and reverberation cancelling with active sonar data. In: Proceedings of ISSPA 2005, Sydney, 2005, pp 523–526

[16] Edelson G S, Kirsteins I P (1994) Modeling and suppression of reverberation components. In: IEEE Seventh SP Workshop on Statistical Signal and Array Processing, Quebec City, 1994, pp 437–440

[17] Hyvärinen A (1999) Fast and robust fixed-point algorithms for independent component analysis. IEEE Transactions Neural Networks 10(3): 626–634

[18] Abraham D A, Lyons A P (2004) Simulation of non-Rayleigh reverberation and clutter. IEEE Journal of Oceanic Engineering 29(2): 347–362

[19] Cong F Y, Chen C H, Ji S (2004) Approach based on colored character to blind deconvolution for speech signals. In: Proceedings of ICIMA 2004, Chengdu, 2004, pp 396–399

[20] Sawada H, MukaiR, Araki S et al (2004) A robust and precise method for solving the permutation problem of frequency-domain blind source separation. IEEE Transactions on Speech and Audio Processing 12(5): 530–538

[21] Ikram M Z, Morgan D R (2005) Permutation inconsistency in blind speech separation: Investigation and solutions. IEEE Transactions on Speech and Audio Processing 13(1): 1–13

[22] Lu W, Rajapakse J C (2003) Eliminating indeterminacy in ICA. Neurocomputing 50(1): 271–290

[23] Ikram M Z (2001) Multichannel blind separation of speech signals in a reverberant environment. Dissertation, Georgia Institute of Technology

[24] Vigario R N (2002) Extraction of ocular artefacts from EEG using independent component analysis. Electroencephal. Clin. Neurophysiol 103: 395–404

[25] Tong S B, Bezerianos A, Paul J (2001) Removal of ECG intererence from the EEG recordings in small animals using independent component analysis. Journal of Neuroscience Methods 108: 11–17

[26] Tong S B (2002) Nonextensive entropy analysis of electroencephalogram and its application study in monitoring brain injury from cardiac arrest. Dissertation, Shanghai Jiao Tong University (Chinese)

[27] Zhang Y Y (2006) Algorithm research on blind signal processing over underdetermined model. Dissertation, Shanghai Jiao Tong University (Chinese)

[28] Hu Y (2003) Research on blind source separation and its applications in image processing. Dissertation, Shanghai Jiao Tong University (Chinese)

[29] Samal A, Iyengar P (1992) Automatic recognition and analysis of human faces and facial expression:A survey. Pattern Recognition 25(1): 65–77

[30] Cottrell G, Metcalfe J (1991) Face, gender and emotion recognition using holons. In: Advances in Neural Information Processing Systems.D. Touretzky. Morgan Kaufmann, San Mateo, CA, 1991, 3: 564–571

[31] Turk M, Pentland A (1991) Eigenfaces for recognition. Journal of Cognitive Neuroscience 13(1): 71–86

[32] Penev P S, Atick J J (1996) Local feature analysis: A general statistical theory for object Representation. Network: Comput. Neural Syst. 7(3): 477–500

[33] Bartlett M S, Sejnowski T J (1997) Viewpoint invariant face recognition using independent component analysis and attractor networks. In: M. Mozer, M. Jordan and T. Petsche (eds) Advances in Neural Information Processing Systems. The MIT Press, Cambridge, MA, 1997, 9: 817–823

[34] Liu C, Wechsler H (1999) Comparative assessment of independent

component analysis (ICA) for face recognition. In: International Conference on Audio and Video Based Biometric Person Authentication, Washington, 1999

[35] Yuen P C, Lai J H (2000) Independent component analysis of face images. In: IEEE Workshop on Biologically Motivated Computer Vision, Seoul, 2000

[36] Bartlett M S, Movellan J R, Sejnowski T J (2002) Face recognition by independent component analysis. IEEE Transactions on Neural Networks 13(6): 1450–1464

[37] Draper B A, Baek K, Bartlett M S (2003) Recognizing faces with PCA and ICA. Computer Vision and Image Understanding 91(1–2): 115–137

[38] Ferreira A, Figueiredo M (2003) Class-adapted image compression using independent component analysis. In: IEEE International Conference on Image Processing -ICIP'2003, Barcelona, 2003

[39] Ferreira A, Figueiredo M (2003) Image compression using orthogonalised independent components bases. In: Proceedings of the IEEE Neural Networks for Signal Processing Workshop, Toulouse, 2003

[40] Puga A T, Alves A P (1998) An experiment on comparing PCA and ICA in classical transform image coding. In: Proceedings of Independent Component Analysis (ICA'98), Aussois, 1998, pp 105–108

[41] Wang Z (2003) Data compression in medical virtual reality. Dissertation, Shanghai Jiao Tong University (Chinese)

[42] Wong T T, Fu C W, Heng P A (2001) Interactive relighting of panoramas. IEEE Computer Graphics and Applications 21(1): 32–41

[43] Wong T T, Fu C W, Heng P A et al (2002) The plenoptic illumination function. IEEE Transactions. on Multimedia 4(3): 361–371

[44] Barlow H B (1961) Possible principles underlying the transformation of sensory messages. In: W.A. Rosenbluth (ed) Sensory Communication. The MIT Press, Cambridge, pp 217–234

[45] Attneave F (1954) Some informational aspects of visual perception. Psychol. Rev. 61:183–193

[46] van Hateren J H (1992) A theory of maximizing sensory information. Biol. Cybern. 68(1): 23–29

[47] Field D J (1994) What is the goal of sensory coding? Neural Computation

6: 559–601

[48] Field D J (1995) Visual coding, redundancy, and feature detection. In: Arbibm(ed) The Handbook of Brain Theory and Neural Networks. The MIT Press, Cambridge, MA

[49] Olshausen B A, Field D J (1996) Emergence of simple-cell receptive field properties by learning a sparse code for natural images. Nature 381: 607–609

[50] Wang Z, Leung C S, Zhu Y S (2004) Eigen-image based compression for the image-based relighting with cascade recursive least squared networks. Pattern Recognition 37(6): 1219 –1231

[51] Nimeroff J S, Simoncelli E, Doresy J (1994) Efficient re-rendering of naturally illuminated environments. In: Fifth Eurographics Workshop on Rendering, Darmstadt, 1994, pp 359–373

[52] Wang Z, Lee Y, Zhu Y S (2003) An improved optimal bit allocation method for sub-band coding. Pattern Recognition Letters 24(16): 3007–3013

[53] Frackowiak R S J, Friston K J, Frith C D et al (1997) Human brain function. Academic Press, San Diego

[54] Roy C S, Sherrington C S (1980) On the regulation of the blood-supply of the brain. The Journal of Physiology 11: 85–108

[55] Plum F, Posner J,Troy B (1968) Cerebral metabolic and circulatory responses to induced convulsions in animals. Arch Neurol 18(1): 1–13

[56] Posner J, Plum F, Troy B (1969) Cerebral metabolism. during electrically induced seizures in man. Arch Neurol 20: 388–395

[57] Fox P T, Raichle M E (1985) Stimulus rate determines regional brain blood flow in striate cortex. Annals of Neurology 17(3): 303–305

[58] Belliveau J, Rosen B, Kantor H (1990) Functional cerebral imaging by susceptibility-contrast NMR. Magnetic Resonance in Medicine 14: 538–546

[59] Ogawa S, Lee T M, Kay A (1990) Brain magnetic resonance imaging with contrast dependent on blood oxygenation. Proc Natl Acad Sci 87: 9868–9872

[60] Belliveau J, Kennedy D, Mckinstry R (1991) Functional mapping of the human visual cortex by magnetic resonance imaging. Science 254: 716–

719

[61] Ogawa S, Tank D, Menon R (1992) Intrinsic signal changes accompanying sensory stimulation: Functional brain mapping with magnetic resonance imaging. Natl. Acad. Sci. Proc. 89: 5951–5955

[62] Ogawa S, Menon R S, Tank D (1993) Functional brain mapping by blood oxygenation level-dependent contrast magnetic resonance imaging: A comparison of signal characteristics with a biophysical model. Biophys. J. 64: 803–812

[63] Frackowiak R S J, Friston K J, Frith C D et al (1997) Human Brain Function. Academic Press, San Diego

[64] McKeown M J, Sejnowski T J (1998) Independent component analysis of fMRI data: Examining the assumptions. Human Brain Mapping 6(5–6): 368–372

[65] Biswal B B, Ulmer J L (1999) Blind source separation of multiple signal sources of fMRI data sets using independent component analysis. Journal of Computer Assisted Tomography 23(2): 265–271

[66] Calhoun V D, Adali T, Pearlson G et al (2001) Spatial and temporal independent component analysis of functional mri data containing a pair of task-related waveforms. Hum Brain Map 13: 43–53

[67] Cichocki A, Kasprzak W, Skarbek W (1996) Adaptive learning algorithm for principal component analysis with partial data. Proc Cybernetics Syst 2:1014–1019

[68] Wang Z, Lee Y, Zhu Y S et al (2003) An improved sequential method for principal component analysis. Pattern Recognition Letters 24(9–10): 1409–1415

[69] Wang Z, Wang J J, Childress A (2005) CRLS-PCA based independent component analysis for fMRI study. In: 27th Annual International Conference of the IEEE Engineering in Medicine and Biology Society, Shanghai, 2005

[70] Wang Z, Lee T, Fiori S (2003) An improved sequential method for principal component analysis. Pattern Recognition Letters 24(9–10): 1409–1415

[71] Junqua J C, Haton J P (1995) Robustness in automatic speech recognition: Fundamentals and applications. Kluwer Academic, Boston

[72] Choi S, Hong H (2002) Multichannel signal separation for cocktail party

speech recognition: A dynamic recurrent Network. Neurocomput 49(4): 299–314

[73] Parra, Spence C (2000) Convolutive blind separation of nonstationary sources. IEEE Transactions Speech Audio Process 8(3): 320–327

[74] Kawamoto M, Matsuoka K, Ohnishi N (1998) A method of blind separation for convolved non-stationary signals. Neurocomputing 22(1–3): 157–171

[75] Buchner H, Aichner R, Kellermann W (2005) A generalization of blind source separation algorithms for convolutive mixtures based on second-order statistics. IEEE Transactions on Speech and Audio Processing 13(1): 120–134

[76] Saruwatari H, Kawamura T, Nishikawa T et al (2006) Blind source separation based on algorithm fast-convergence algorithm combining ICA and beamforming. IEEE Transactions on Speech and Audio Processing 14(2): 666–678

[77] Matsuoka K, Nakashima S (2001) Minimal distortion principle for blind source separation. In: Proc. ICA 2001, San Diego, 2001, pp 722–727

[78] Kolossa D, Orglmeister R (2004) Nonlinear postprocessing for blind speech separation. In: ICA 2004, Berlin, 2004, pp 832–839

[79] Kokkinakis K, Nandi A K (2006) Multichannel blind deconvolution for source separation in convolutive mixtures of speech. IEEE Transactions on Speech and Audio Processing 14(1):200–212

[80] James C J (2005) Independent component analysis for biomedical signals. Physiol. Me. 26: R15–39

[81] Kong W, Vanderburg C R, Gunshin H et al (2008) A review of independent component analysis application to microarray gene expression data. Biotechniques 45(5):501–520

[82] Lee S I, Batzoglou S (2003) Application of independent component analysis to microarrays. Genome Biology 4(11):R76.1–R76.21

[83] Schachtner R, Lutter D, Theis F J et al (2007) How to extract marker genes from microarray data sets. In: Proceedings of the 29th Annual, Lyon, 2007, pp 4215–4218

[84] Chinappetta P, Roubaud M C, Torrésani B (2004) Blind source separation and the analysis of microarray data. Journal of Computational Biology 11(6): 1090 –1109

[85] Himberg J, Hyvärinen A, Esposito F (2004) Validating the independent components of neuroimaging time-series via clustering and visualization. NeuroImage 22: 1214 – 1222

Index